1st International Conference on Food and Agriculture 2018

AA006241

IOP Conference Series: Earth and Environmental Science
Volume 207

Bali, Indonesia
20 – 21 October 2018

ISBN: 978-1-5108-7948-5
ISSN: 1755-1307

Printed from e-media with permission by:

Curran Associates, Inc.
57 Morehouse Lane
Red Hook, NY 12571

Some format issues inherent in the e-media version may also appear in this print version.

Printed by Curran Associates, Inc. (2019)

For permission requests, please contact the Institute of Physics at the address below.

Institute of Physics
Dirac House, Temple Back
Bristol BS1 6BE UK

Phone: 44 1 17 929 7481
Fax: 44 1 17 920 0979

techtracking@iop.org

Additional copies of this publication are available from:

Curran Associates, Inc.
57 Morehouse Lane
Red Hook, NY 12571 USA
Phone: 845-758-0400
Fax: 845-758-2633
Email: curran@proceedings.com
Web: www.proceedings.com

Table of contents

Volume 207

1st International Conference on Food and Agriculture 2018
20–21 October 2018, Bali, Indonesia

Accepted papers received: 12 November 2018
Published online: 7 December 2018

Preface

Papers

Agriculture Engineering and Biotechnology

Agroindustry and Agribusiness

Food Science and Technology

Organic Agriculture

IT for Agriculture

Renewable and Novel Energy Sources

The Other

IOP Conference Series: Earth and Environmental Science

The 1st International Conference on Food and Agriculture (ICoFA)

IOP Conf. Series: Earth and Environmental Science 207 (2018) 011001 doi:10.1088/1755-1315/207/1/011001

INTRODUCTION OF ICoFA 2018

Nowadays, the population all over the world is about 7.4 billion. Huge quantities of high quality food and agricultural products are needed every day. Therefore, forceful agricultural knowledge and innovation systems assure that the global food system contribute to decent supplies of high quality food and agricultural products, in a sustainable way. The International Conference on Food and Agriculture (ICoFA) will be a paramount and valuable event for scientists, researchers, students and practitioners to meet and discuss recent issues in food and agriculture sectors. The conference will cover a series of presentations and discussions in plenary, concurrent and poster sessions. ICoFA is dedicated to present high quality researches delivered by reputable Keynote Speakers and invited speakers from Asia region.

The 1st International Conference on Food and Agriculture was held on 20-21 October 2018 in Nusa Dua, Bali, Indonesia. Hosted by Politeknik Negeri Jember, the event was intended to provide technical forum and research discussion on food, agriculture, and how technology is effectively employed for sustainable development of food and agriculture. It is aimed to bring researchers, academicians, scientists, students, and practitioners together to participate and present the latest research findings, developments, and applications related to various aspects of agriculture engineering, organic agriculture, agribusiness, animal nutrition, animal production, veterinary Science, food science and technology, food safety, food security and sovereignty, IT for Agriculture, renewable energy and other researches and studies related to agriculture with the theme of "Current Innovation and Implementation of Modern Technology in Food and Sustainable Agriculture".

The conference was participated by 234 participants from whole part of Indonesia from Sabang to Merauke, from Malaysia, Thailand, India, Mozambique, and Saudia Arabia. There are 186 papers for oral presentations and the committee has selected 66 papers out of 186 papers to be published to IOP Conference series : Earth and Environmental Science.

Peer review statement

All papers published in this volume of *IOP Conference Series: Earth and Environmental Science* have been peer reviewed through processes administered by the proceedings Editors. Reviews were conducted by expert referees to the professional and scientific standards expected of a proceedings journal published by IOP Publishing.

Potential of *Beauveria bassiana* Lowland Isolates against *Spodoptera litura* in Tobacco Plant

D N Erawati[1]*, I Wardati[2] and S Humaida[3]

[1,2,3] Department of Agricultural Production, Politeknik Negeri Jember, Jl.Mastrip PO Box 164 Jember - Tel. +62-0331-333532, Fax. +62-0331-333531

*corresponding author: dyah_nuning_e@polije.ac.id

Abstract. The study of the ability of *Beauveria bassiana* lowland isolates as biological control of *Spodoptera litura* larvae needs to be done because of the specific characteristics of biological control that are site specific and host specific. The research objectives are: 1) analyze *B. bassiana* lowland isolates which were effective in suppressing *S. litura* attacks in the field and 2) analyze the ability of *B. bassiana* lowland isolates to increase the production of Kasturi tobacco. The study was conducted using Randomized Block Design (RAK) with 4 replications. Treatment consisted of application: (1) *B. bassiana* lowland isolate strain 715 0.4 gram/liter, (2) *B. bassiana* lowland isolate strain 725 0.4 gram/liter, (3) *B. bassiana* lowland isolate strain Jember spore density 10^9 spores/ml, (4) *B. bassiana* lowland isolate strain Jombang spore density 10^9 spores/ml, (5) *B. bassiana* highland isolate strain Temanggung spore density 10^9 spores/ml and (6) synthetic chemical insecticides 2 gram/liter. The result showed that *B. bassiana* lowland isolate strain Jombang succeeded in suppressing leaf damage due to *S. litura* attack 50% - 60% compared to other *B. bassiana* isolates and has the potential to be developed into biological control products against *S. litura* in lowland Kasturi tobacco plants.

1. Introduction

Spodoptera litura larvae can reduce the productivity of Kasturi tobacco plants by eating leaves marked by holes in the affected part of the plant. Entomopathogen fungi *Beauveria bassiana* had the high capability as biological controllers of tobacco leaf-eating larvae and is known to have so many hosts but still has specific host limitations and specific locations. Some research has shown that the application of *B. bassiana* had the high mortality of *S. litura* larvae between 50% - 90% [1], [2], [3].

The ability of *B. bassiana* to control pests from the order of Lepidoptera such as *S. litura* can still be increased through genetic activity [4]. The pathogenesis of entomopathogen fungi is supported by enzyme during the invasion period [5]. The hyphae of *B. bassiana* are able to penetrate into the host cuticle due to secreting the extracellular type of enzymes like lipase, protease and chitinase [6]. The study about *B. bassiana* pathogenicity has been widely investigated but studies on the ability of *B. bassiana* lowland isolates are still very limited. Therefore, the study to find *B. bassiana* lowland isolates that are most effective against *S. litura* larvae in Kasturi tobacco plants needs to be done to be able to increase the production of tobacco plants while still taking into account the quality of the environment.

The research objectives are (1) analyze *B. bassiana* lowland isolates which were effective in suppressing *S. litura* attacks in the field and (2) analyze the ability of *B. bassiana* lowland isolates to increase the production of Kasturi tobacco.

2. Material and Methods

The study has been carried out at the Plant Protection laboratory and tobacco field at the Jember State Polytechnic (*S 809'34,884 "E 113043'24.7944"*) from May to November 2018. The research used a single factor of Randomized Block Design and 4 replications. Treatments: (1) *B. bassiana* lowland isolate strain 715 0.4 gram/liter, (2) *B. bassiana* lowland isolate strain 725 0.4 gram/liter, (3) *B. bassiana* lowland isolate strain Jember spore density 10^9 spores/ml, (4) *B. bassiana* lowland isolate strain Jombang spore density 10^9 spores/ml, (5) *B. bassiana* highland isolate strain Temanggung spore density 10^9 spores/ml and (6) synthetic chemical insecticides 2 gram/liter.

2.1. Preparation of B. bassiana

B. bassiana isolates were collected from several lowlands: (1) strain 715 was isolated from Coleoptera and strain 725 was isolated from Lepidoptera in Kaliwining Jember with altitude 45 meters above sea level (the product form of freeze-dried spores) [7], (2) strain Jember was isolated from Coleoptera in Sumbersari Jember with altitude 89 meters above sea level (the product form of corn rice multiplication formulation), (3) strain Jombang was isolated from Lepidoptera in Mojoagung Jombang with altitude 44 meters above sea level (the product form of corn rice multiplication formulation) and (4) strain Temanggung was isolated from Coleoptera with altitude 675 meters above sea level (the product form of corn rice multiplication formulation).

2.2. Preparation of Kasturi Tobacco Plant

Preparation and implementation procedures in concert with the method of Tobacco Good Practice (GTP). Prepare the land for tobacco plants Kasturi to choose a flat land, open and a height of 89 meters above sea level. The Kasturi tobacco seed which is a type of chopped tobacco that is widely planted in the Jember area is planted and transplanted at the age of 45 days in a polybag containing planting media (topsoil: compost: sand) which is suitable for the growth. Placement of tobacco polybags with a single spacing (100 cm x 45 cm) and in line with the research layout. Maintenance tobacco plants are conducted in pursuance of Standard Operating Procedures for technical procedures include covering tobacco maintenance, watering, fertilizing, cultivating and controlling plant pests and diseases [8].

2.3. Application of Beauveria bassiana

The application procedure to prepare the *B. bassiana* from each strain lowlands and highlands follow with a predetermined concentration then mixing the solution with an adhesive compound until smooth and homogeneous. Applications treatments by spraying a solution in line with the determination of the tobacco crop research plots each week. Applications entomopathogen fungus carried out in the afternoon to avoid the effect of sunlight on *B. bassiana* and chemical application is done in accordance with standard operating procedures of the manufacturer.

2.4. Observation

The parameters observed in this study are (1) the intensity of pest attack on new leaves is open, which is calculating the percentage of damage to the leaves due to pest attack before and after the control application. The formula for calculating damage intensity:

$$P = \frac{\sum (n \times v)}{N \times Z} \times 100\%$$

P = leaf damage intensity (%)
n = number of leaves that have the same damage category
v = score value/category based on the area of the affected leaves
N = number of leaves observed per plant
Z = the highest damage score / category value

Value scale / damage scoring [9]:
0: no pest attack
1: damage area of leaves more than 0- 25%
3: leaf damage area of more than 25 - 50%
5: leaf damage area is more than 50 - 75%
7: leaf damage area is more than 75 - 100%

(2) Growth of tobacco plants on plant height and (3) Leaf growth include the number of leaves, leaf length, leaf width.

3. Result and Discussion

3.1. Leaf Damage Intensity

S. litura larvae were found since tobacco plants aged 9 days after planting and the population fluctuated during the vegetative phase of the plant which then decreased after tobacco plants entered the generative phase. The decline in population is thought to be due to the influence of treatment applications also related to the plant growth phase because *S. litura* attacks the leaves of plants that develop a lot in the vegetative phase of the plant. The direct impact of attacks is seen on the leaves. Pests attack by eating leaves that are marked by holes in the affected part of the plant. In severe attacks will cause harvest failure.

Table 1. The average intensity of Kasturi tobacco leaf damage due to *S. litura* attack

Treatment	14 dap	28 dap	42 dap
P1 (lowland isolate strain 715)	2.72 a	6.22 b	6.58 b
P2 (lowland isolate strain 725)	0.70 a	2.89 a	5.33 b
P3 (lowland isolate strain Jember)	6.93 b	7.48 b	4.61 ab
P4 (lowland isolate strain Jombang)	0.00 a	4.74 ab	2.43 a
P5 (highland isolate strain Temanggung)	0.93 a	6.31 b	4.73 ab
P6 (chemical insecticide)	1.10 a	1.05 a	1.64 a

The number accompanied by the same alphabet in the same track shows no significant difference based on the LSD (5%)

The level of damage Kasturi tobacco leaves aged 14 days after planting as a result of the attack of *S. litura* showed that the application *B. bassiana* lowland isolate strain Jombang was able to suppress the attack rate to 0% which is not significant with other treatments, except the highest level of leaf damage on the application *B. bassiana* lowland isolate strain Jember reach 6.9%. The level of damage increases with age leaf plants because the leaves are elongated growth and widening seen on plants 28 days and 42 days after planting (Table 1). The average damage of Kasturi tobacco leaves at 14, 28 and 42 days after planting respectively 2.06%, 4.78% and 4.22%. There was an increase in average leaf damage at 28 days after planting but the average leaf damage decreased at 42 days after planting. The application of *B. bassiana* lowland isolate strain Jombang succeeded in suppressing leaf damage due to the attack of *S. litura* reach 50% - 60% compared to the application of other *B. bassiana* isolates. The treatment of insecticides is not significantly different from the treatment of *B. bassiana* lowland isolate strain Jombang.

Chemical insecticide active ingredient used lambda-cyhalothrin 106 grams/liter and thiamethoxam 141 grams/liter with transmilar working mechanism into the network at the same plant are toxic contact and stomach poison to the target pest. Quickly absorbs into the plant tissue that is not easily washed away by rainwater [10]. Treatment *B. bassiana* lowland isolates strain Jombang was isolated from the order Lepidoptera pests at an altitude of 44 meters above sea level shows a lower level of leaf damage. This shows that the power of entomopathogen fungi in suppressing the level of pest attack is influenced by the high similarity of the place and host suitability. This experiment arranged at altitude 89 meters sea level including lowland area. One characteristic of *B. bassiana* as a biological controller is the specific nature of the host and the specific locations that have limited because only able to infect insects suitable host [11]. Entomopathogenic fungi have bio-trophic linkages with insect host and can act as a saprophyte [12]. Conidia entomopathogen fungi that enter the body of the host insect through feeding larvae will be digested and the infection sometimes does not occur before excretion due to incompatibility of intestinal specific virulence factors [4]. The fungus infects the host insect through the cuticle penetration into the host by way of adhesion [11]. Pathogenicity of entomopathogen fungi depend on the activity of lipase, protease and chitinase which will degrade the integument of insects [12]. Virulence ability is strongly influenced by extracellular enzyme activity. Chitinase is an enzyme that has the most role in decomposing the chitin layer from insect cuticles [6].

3.2. Plant Height

Plant height is one measure of plant growth is affected by photosynthesis rate. The rate of photosynthesis can be approached by counting the number of leaves. The process of photosynthesis will produce primary metabolites used for plant metabolism resulting from the growth and the development of plants. Ability to process light energy leaves accepted into carbohydrates by photosynthesis at temperatures referents and its value is affected by plant genetic and environmental conditions [13].

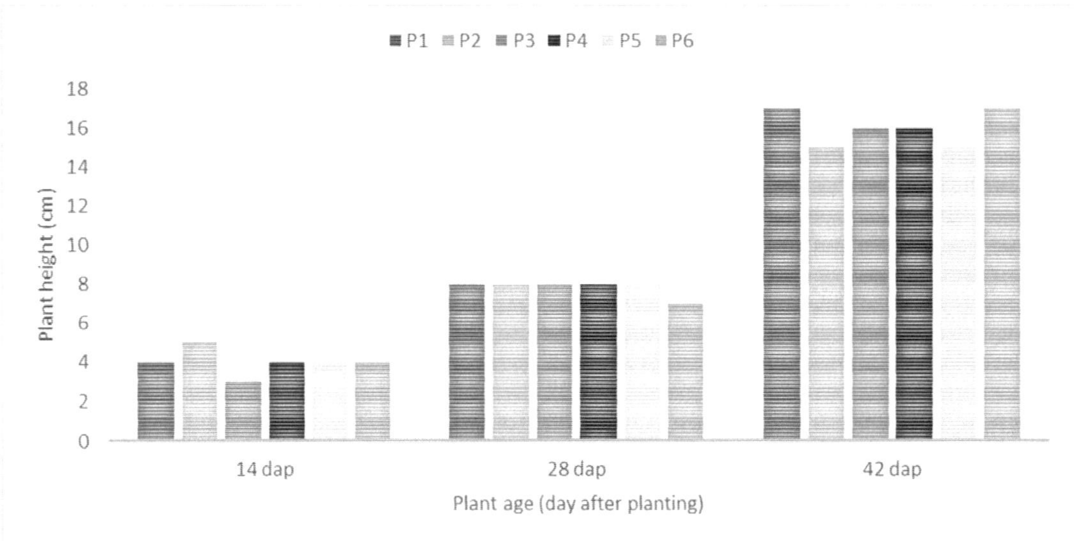

Figure 1. B. bassiana lowland application against the Kasturi tobacco plant height

Application of *B. bassiana* all isolates do not affect the height of tobacco plants as shown in Figure 1. This outcome indicates that the biological control application is still able to maintain plant growth. Biological controllers with *B. bassiana* will increase the natural stability of the ecosystem and support the existence of other natural enemies on the land and research plots. Several types of natural enemies other than biological control were applied to research plots. The existence of natural enemies in the ecosystem will support natural control running in balance and pest populations can be maintained at a non-detrimental boundary.

The application of biological control in an ecosystem will increase biodiversity in the ecosystem. Biodiversity causes a dynamic interaction between supporting components so that a balance condition is formed. In dynamic equilibrium conditions, there is no type of organism that becomes dominant because each type of organism is naturally controlled by other organisms and other natural control components [14]. Utilization entomopathogen fungus is one of alternative pest control. Biological control applications of fungi and selective insecticides will reduce dependence on chemical insecticides and reduce residual effects on agriculture [15].

3.3. Leaf Growth

Tobacco production is in the form of leaves that are widely used for industry. Tobacco Kasturi has low nicotine content of 1-3% in the leaves. Tobacco plant growth can be assessed through leaf growth, including the number, length and width of leaves.

Table 2. *B. bassiana* application for the average growth of the number, width and length of Kasturi tobacco leaves

Treatment	Leaf width (cm)			Leaf number (leaf blade)			Leaf length (cm)		
	14 dap	28 dap	42 dap	14 dap	28 dap	42 dap	14 dap	28 dap	42 dap
P1 (lowland isolate strain 715)	5.18	9.43 bc	13.03	3.71	5.00	8.83	9.49	16.95	23.96
P2 (lowland isolate strain 725)	5.44	10.7 abc	12.32	4.54	5.21	8.13	10.04	18.70	22.38
P3 (lowland isolate strain Jember)	5.22	9.54 bc	11.22	3.83	4.75	8.63	9.99	17.17	22.73
P4 (lowland isolate strain Jombang)	5.36	10.71 ab	12.44	4.17	5.38	8.63	10.11	18.66	24.00
P5 (highland isolate strain Temanggung)	4.79	8.71 c	12.32	3.96	4.83	8.63	9.16	15.67	22.86
P6 (chemical insecticide)	5.51	11.64 a	13.54	4.33	5.04	8.46	10.26	20.00	24.59

The number accompanied by the same alphabet in the same track shows no significant difference based on the LSD (5%)

Leaf number and leaf length at 14, 28 and 42 days after planting is not affected by the treatment of *B. bassiana*. It shows that the application of entomopathogen fungi does not affect the growth and development of Kasturi tobacco plants. Meanwhile, application *B. bassiana* affect leaf width at 28 days after planting. Application of chemical insecticide, *B. bassiana* lowland isolate strain Jombang and *B. bassiana* lowland strain 725 shown has a wider leaf width compared to other treatments.

Based on Table 2 shows the treatment only affects the width of the leaf at 28 days after planting. Treatment applications to give effect to the width of the leaves for larvae of *S. litura* attack by eating the leaves of tobacco plants. Perforated tobacco leaves due to the feeding activity of *S. litura* larvae. This will increase the risk of decreasing the production of tobacco plants considering the results of tobacco plants are leaves that become raw materials for cigarettes, drugs or cosmetics. *S. litura* known as the armyworm because in one egg groups contain about 350 eggs and egg reaches 2000-3000 the number of all the grains. The eggs will hatch after 3-5 days [16].

The width of the large leaves contained in the application showed that the two strains of *B. bassiana* lowland isolate strain Jombang and *B. bassiana* lowland isolate strain 725 minimize damage to tobacco leaves so that it has the potential to be developed as a biological control for *S. litura* larvae in Kasturi tobacco plants in the lowlands. The virulence ability of *B. bassiana* to resist the immune response of host insects is supported by several kinds of proteins and enzymes through molecular and physiological mechanisms [6]. Host insect integument is degraded through the ability of enzymes activity such as lipase, protease and chitinase [12].

Leaf number, leaf width and leaf length at 42 days after planting are not affected by the application. It shows the treatment was able to maintain the growth and development of leaf tobacco plant including B. bassiana highland isolate strain Temanggung as one strain originating from the highlands. The ability of *B. bassiana* highland isolates strain Temanggung to keep the average growth of the number, width, and length of tobacco leaves proved that *B. bassiana* have the power to adapt to the altitude. *B. bassiana* can adapt to different environments with the adjustment and activation of sets of genes that have been defined correctly [17]. The virulence of *B. bassiana* against *S. litura* larvae can be significantly improved through genetic transformation. Conidia result of the genetic transformation is spread widely by the vector so it can function as a controller *S. litura* [4].

Conidia viability of entomopathogen fungi can be maintained when formulated in the form of pure conidia for more easily applied and has a higher accuracy compared to other traditional methods [18]. Entomopathogen fungi more effectively applied to the pests that have gregarious eating patterns, such as *S. litura* as it will accelerate the spread of infection. It can be applied in the integrated pest management program [15].

4. Conclusion

B. bassiana lowland isolate strain Jombang succeeded in suppressing leaf damage due to *S. litura* attack 50% - 60% compared to other *B. bassiana* isolates and has the potential to be developed into biological control products against *S. litura* in lowland Kasturi tobacco plants.

Acknowledgment

The research was carried out with funds from Politeknik Negeri Jember for PNBP research activity 2018 with contract number SP DIPA-042.01.2.401005/2018.

References

[1] Erawati D N 2006 *Jurnal Ilmiah Inovasi* **6** (3):228-235
[2] Gupta S D and Kumar B 2014 *International Journal of Plant Science* **9** (1):97-100
[3] Agrawal S and Simon S 2017 *International Journal of Current Microbiology and Applied Science* **6** (8):1992-1996
[4] Qin Y, Ying S H, Chen Y, Shen Z C and Feng M G 2010 *Applied and Environmental Microbiology* **7** (14):4611-4618
[5] Moorthi V P, Balasubramanian C and Selvarani S R A 2015 *Springerplus* **4** (1):681
[6] Dhawan M and Joshi N 2017 *Journal of Microbiology* (48):522-529
[7] Sulistyowati E and Junianto Y D 2000 *Produksi dan Aplikasi Agents Pengendali Hayati Hama Utama Kopi dan Kakao* (Jember: Pusat Penelitian Kopi dan kakao Indonesia) p 26
[8] Direktorat Perlindungan Tanaman 2014 *Pedoman Uji Mutu dan Efikasi Lapangan Agens Pengendalian Hayati (APH)* (Jakarta : Direktorat Perlindungan Perkebunan, Kementrian Pertanian) p 115
[9] Sygenta 2017 Product brands ALIKA® (www.syngenta.com)
[10] Chalid N I 2005 *Pedoman Penerapan PHT Pada Agribisnis Tanaman Cabai* (Surabaya: Dinas Pertanian Jawa Timur) p 90
[11] Gabarty A, Salem H M and Ibrahim A A 2014 *Journal of Radiation Research and Applied Sciences* **7** (1): 95-100
[12] Mondal S, Baksi S, Koris A and Vatai G 2016 *Pasific Science Review A: Natural Science and Engineering* (18):85-99
[13] Djumali and Nurnasari E 2012 *Buletin Tanaman Tembakau, Serat dan Minyak Industri* **4** (1):10-20
[14] Erawati D N and Wardati I 2016 *Jurnal Perlindungan Tanaman Indonesia* **12** (2):1-5

IOP Conf. Series: Earth and Environmental Science **207** (2018) 012001 doi:10.1088/1755-1315/207/1/012001

[15] Koshravi R, Sendi J J, Zibaee A and Shokrgozar M A 2015 *Journal of King Saud University-Science* (27):49-53
[16] Kalshoven, L G E 1981 *The Pest of Crops in Indonesia* (Jakarta: Ichtiar Baru-van Hoeve) p 701
[17] Xiao G, Ying S H, Zeng P, Wang ZL, Zhang S, Xie X Q, Shang Y, St Leger R J, Zhao G P, Wang C and Feng M G 2012 *Scientific Reports* 2 (483)
[18] Oliviera D G P, Pauli G, Mascarin G M and Delalibera I 2015 *Journal of Micrbiological Methods* (119):44-52

IOP Conf. Series: Earth and Environmental Science **207** (2018) 012002 doi:10.1088/1755-1315/207/1/012002

The effects of different concentration of sucrose and various auxin on in vitro shoot and microtuber formation of red potato (*Solanum tuberosum*, L. var Desiree)

S L Asmono[1*], V K Sari[1] and Djenal[1]

[1]Department of Agricultural Production, Politeknik Negeri Jember, Jalan Mastrip PO BOX 164, Jember, 68101, Indonesia

* sepdian@polije.ac.id

Abstract. The experiment was aims to study the effect of different sucrose concentrations (30 , 90, 120 and 150 g.L^{-1}) and various types of auxin (IAA, IBA, NAA) on in vitro shoot and microtuber formation of Red Potato (*Solanum tuberosum*, L. var Desiree). Young stem cutting from planlet cultured on half strength MS medium. All types of auxin use the concentration of 0.25 mg.L^{-1}. The research conducted a completely randomized design (CRD) factorial. The data obtained were analyzed by analysis of variance (ANOVA) and DMRT test at 0.05. Parameters observed included growth of microtuber and shoot growth, such as number of shoots and shoot length. The observations result for 3 months, show that the higher the concentration of sucrose, can stimulate tuber formation but suppress shoot growth. Early tuber initiation (36,7 DAP) and optimum number of tuber was observed in 150 g.L^{-1} sucrose, whereas at 30g.L^{-1} had no effect to tuber induction. In addition, high number of microtuber was observed in NAA compare to other auxin. Observational data on shoot growth showed that the higher concentration of sucrose will inhibit bud formation. The 90 g.L^{-1} sucrose concentration showed the best number of shoots, which averaged 1.5 shoots/plantlet. But at a concentration of 30g/L sucrose, explants are only able to form callus. In the bud length parameter, IAA gave the best response that is an average of 3cm/shoot.

1. Introduction

The high productivity of potato is still constrained by availability of quality potato seeds, conventional cultivation techniques, as well as limited places with high temperatures and temperatures suitable for potato cultivation. In general, potato cultivars cultivated by farmers are highland potatoes (700 masl) such as Granola and Atlantic. But there are also potatoes that can grow well in medium plains such as Desiree cultivars [1][2]. That potatoes are also called "Red potatoes" because they have reddish bulbs [3]. These cultivars can potentially be developed to meet national potato needs.

The process of developing these cultivars must be supported by the provision of superior seeds through in vitro culture techniques. Through this technique, tubers (G0) and micro shoots can be produced in large quantities in a relatively short time. The composition of the media and growth regulating substances play an important role in spurring the growth of tubers and potato shoots in vitro. In addition, sucrose concentration is also one of the factors that need to be considered. Some researchers claim that sucrose is able to induce potato micro tubers [4]–[6]. Therefore, this study aims to determine the effect of some sucrose concentrations on in vitro shoots of Desiree potatoes in inducing micro bulbs.

2. Methods

This research was held at the Plant Tissue Culture Laboratory, Department of Agricultural Production, Jember State Polytechnic. Material used in this study were sucrose (30, 90, 120 and 150 g.L^{-1}), various types of auxin (0.25 mg.L^{-1} IAA, IBA, NAA), young stem cutting from planlet, half strength MS medium. The research conducted a completely randomized design (CRD) factorial. The data obtained were analyzed by analysis of variance (ANOVA) and DMRT test at ($p<0.05$). Parameters observed included percentage of callus, microtuber and shoot formation, and also growth of microtuber and shoot, such as number of shoots and shoot length.

3. Results and discussion

3.1 Emergence, number and fresh weight of microtubers

Table 1. Comparison of means and interaction of Auxin type and sucrose levels in the media for microtubers emergence (DAP).

Auxin (0,25 mg.L^{-1})	Sucrose Level 30 g.L^{-1}	90 g.L^{-1}	120 g.L^{-1}	150 g.L^{-1}	Mean of Auxin
IAA	0.0	45.6	46.8	40.6	55.8 [a]
IBA	0.0	45.0	45.4	40.2	55.2 [a]
NAA	0.0	35.6	34.6	29.2	47.4 [b]
Mean of Sucrose	0.0 [a]	42.1 [c]	42.3 [c]	36.7 [b]	

[a] at 30g.L^{-1} no tuber appears

[b] The number followed by the same letter are not significantly different at ($p<0.05$) level of Duncan's test.

Table 2. Comparison of means and interaction of Auxin type and sucrose levels in the media for microtubers number per planlet.

Auxin (0,25 mg.L^{-1})	Sucrose Level 30 g.L^{-1}	90 g.L^{-1}	120 g.L^{-1}	150 g.L^{-1}	Mean of Auxin
IAA	0.0	0.2	0.1	0.7	0.3 [a]
IBA	0.0	0.4	0.4	0.9	0.4 [a]
NAA	0.0	1.1	1.5	1.5	1.0 [b]
Mean of Sucrose	0.0 [a]	0.6 [ab]	0.7 [b]	1.0 [c]	

[a] The number followed by the same letter are not significantly different at ($p<0.05$) level of Duncan's test.

Table 3. Comparison of means and interaction of *Auxin type* and sucrose levels in the media for fresh wight of microtubers (g).

Auxin (0,25 mg.L^{-1})	Sucrose Level 30 g.L^{-1}	90 g.L^{-1}	120 g.L^{-1}	150 g.L^{-1}	Mean of Auxin
IAA	0.0	1.2	0.4	0.6	0.6
IBA	0.0	1.2	1.4	1.2	1.0
NAA	0.0	1.6	1.8	1.8	1.3
Mean of Sucrose	0.0 [a]	1.3 [b]	1.2 [b]	1.2 [b]	

[a] The number followed by the same letter are not significantly different at ($p<0.05$) level of Duncan's test.

Figure 1. Emergence of microtuber

Figure 2. 90 days old microtubers of red potato

Based on the ANOVA test, giving some sucrose concentrations combined with several types of auxin did not show a real effect. The emergence of microtubers (**Figure 1**) was first seen in the media with the addition of NAA and 150 $g.L^{-1}$ sucrose at 29.2 DAP (**Table 1**). In the treatment media also obtained the highest number of tubers, which is an average of 1.5 tubers per plantlet **(Tabel 2).** Both of these factors show a very real influence on a single basis. The formation of micro tubers is strongly influenced by sucrose as an energy source. Tubers on potato plants are also a place of energy storage for plants. In the process of micro tuber formation, sucrose as an external source of carbon is able to stimulate cell development to form tubers [7], [8]. The higher the concentration of sucrose affects the process of cell differentiation set the Sink tissue [9][10]. This is seen in the media with the addition of 30 $g.L^{-1}$ of sucrose is not able to stimulate tuber formation, only grows callus.

Type of auxin also affects tuber formation. NAA is able to accelerate tuber formation faster compare to other types of auxin tested. NAA act to increase the formation of nodes in potato stems which then become a place for tubers to grow[11]. The role of NAA is also seen in the process of cell division and enlargement at the stolon, then swelling occurs at the stolon tip[12].

In the parameters of fresh weight of micro tubers, the effect of interactions between auxin types and some sucrose concentrations also did not show any significant differences, nor did auxin factors singly. But the surcose factor alone affects the fresh weight of tubers (**Table 3**), although between 90-150 g.L-1 sucrose concentration was not significantly different based on the results of Duncan's test. The difference is seen only at a concentration of 30 g.L-1 which is no tuber that grows. Of the several concentrations tested, the fresh weight of tubers that were induced in a medium with a concentration of 90 g.L-1 showed a value greater than the level of other concentrations. The greater the concentration of sucrose is thought to be the cause of reduced tuber cell volume. Sucrose as a carbon and energy source for plants, also plays a role in the process of regulating osmotic pressure, therefore there is a change in osmotic potential so the absorption of water becomes reduced to increase the volume of tuber cells [10].

3.2 Number and length of shoots

Table 4. Comparison of means and interaction of *Auxin type* and sucrose levels in the media for number of shoots.

Auxin (0,25 $mg.L^{-1}$)	Sucrose Level				Mean of Auxin
	30 $g.L^{-1}$	90 $g.L^{-1}$	120 $g.L^{-1}$	150 $g.L^{-1}$	
IAA	0.0	2.0	2.2	1.4	1.4
IBA	0.0	2.2	1.6	1.0	1.2
NAA	0.0	3.0	1.6	2.0	1.7
Mean of Sucrose	0.0 [a]	2.4 [c]	1.8 [bc]	1.5 [b]	

[a] The number followed by the same letter are not significantly different at ($p<0.05$) level of Duncan's test.

Table 5. Comparison of means and interaction of *Auxin type* and sucrose levels in the media for shoot length (cm).

Auxin (0,25 mg.L^{-1})	Sucrose Level				Mean of Auxin
	30 g.L^{-1}	90 g.L^{-1}	120 g.L^{-1}	150 g.L^{-1}	
IAA	0.0 a	6.6 f	5.1 e	0.3 ab	3.0 ^{a}
IBA	0.0 a	0.5 b	0.5 b	0.2 ab	0.2 ^{b}
NAA	0.0 a	2.9 d	1.3 c	0.3ab	1.1 ^{b}
Mean of Sucrose	0.0 ^{a}	3.2 ^{b}	2.3 ^{b}	0.3 ^{a}	

[a] The number followed by the same letter are not significantly different at ($p<0.05$) level of Duncan's test.

Figure 3. Shoots formation in MS medium containing 90 g.L^{-1} sucrose

ANOVA analysis results showed that the interaction factor between sucrose concentration and some auxin types did not significantly affect the number of shoots (**Table 4**), but had a significant effect on shoot elongation. The interaction between IAA and 90 g.L-1 of sucrose produces the longest shoot, which is an average of 6.6 cm (**Table 5**). The results of the analysis also show that the level of sucrose concentration alone affects the number and length of shoots. Sucrose at a concentration of 90 g.L^{-1} is able to produce better shoots (**Figure 3**), which is an average of 2.4 shoots per explant as well as the length of shoots from that concentration on average 3.2 cm. The higher the concentration of sucrose added is thought to suppress shoot growth. Seen on sucrose with a concentration of 150 g.L^{-1} produces the least number of shoots and is shorter than other concentrations. High sucrose concentration can quickly accumulate on the shoots of explants so that it stimulates stolon formation, not bud formation[13] .

The role of auxin shows significant differences in shoot elongation. Whereas in the formation of shoots, some types of auxin show unreal differences. The role of IAA in bud elongation looks better because the resulting shoots are longer than other types of auxin, which is an average of 3 cm. IAA is a hormone that is also produced endogenously by plants [14]. It is suspected that the role of IAA added to the media is able to interact with IAA produced by the plant itself, so that it is more optimal to stimulate stem lengthening [15].

4. Conclusions

From this study it can be concluded that the combination of several types of auxin and sucrose concentration did not show any significant differences in the formation of micro tubers and shoots, but the interaction between the two was seen in the stem length parameters. The single auxin and sucrose factors greatly influence the tuber formation process. The use of NAA and 150 g.L^{-1} sucrose was able to bring up the first tuber, which was 36.7 DAP and produced more tubers than the other treatments, namely an average of 1 tuber per plantlet. But at 90 g.L^{-1} sucrose, it produces more bulbs than higher concentrations of sucrose. Visually, the media with the addition of 30 g.L^{-1} sucrose only occurs in the formation of callus which is unable to form tubers or shoots. The most shoots (2.4 shoots per explant) and the longest (3.2 cm per shoot) were seen in the media with an addition of 90 g.L$^{-1.}$ From the auxin factor, IAA is better able to stimulate stem elongation than other types of auxin.

5. Acknowledgements

The authors acknowledged the financial support by DRPM RISTEK DIKTI 2018.

6. References

[1] Prabaningrum L, Moekasan T K, Sulastrini , Handayani T, Sahat J P, Sofiari E, and Gunadi N 2014 *Monografi Balai Penelitian Tanaman Sayuran, Kementerian Pertanian*
[2] Wardiyati T, Dawam M, and Rofiq M 2016 *Cakrawala* 81–88
[3] Setiadi 2009 *Niaga Swadaya*
[4] Khuri S and Moorby J 1995 *Ann. Bot* .**75** 95
[5] Donnelly DJ, Coleman W K and Coleman S E 2003 *Am. J. potato Res.* **80** 103
[6] Rahman M H, Islam R, Hossain M, Islam M S 2010 *J. Agric. Technol.***6** 7
[7] El-Sawy A, Bekheet S, and Aly U I 2007 *J Agri Biol* **9** 675
[8] Struik P C and Wiersema S G 2009 *Wageningen Academic Pub*
[9] Pruski K, Astatkie T, Duplessis P, Lewis T, Nowak J, and Struik PC 2003 *Am. J. Potato Res.***80** 183
[10] Wang L and Ruan Y L 2013 *Front. Plant Sci.***4** 163
[11] Hossain M, Listanto E, Rasheed M M, Hermana M, and Hassans M K 2007 *J.Bangladesh SOc. Agri Sci Techno* **4** 219
[12] Stallknecht G F 1985 *Potato Physiol* 231
[13] Roumeliotis E, Kloosterman B, Oortwijn M, Kohlen W, Bouwmeester, Visser R G F, and Bachem C W B 2012 *J. Exp. Bot.***63** 4539
[14] Davies P J 2010 *Springer* 1–15
[15] Yang T, Davies P J, and Reid J B 1996 *Plant Physiol.***110** 1029

Design and performance test of a batch system rotary vacuum dryer with a 50-liter capacity to dry basidiomycota class mushrooms

S Djamila[1*], Iswahyono[1] and A Bahariawan[1]

1) Agriculture Engineering Department, Politeknik Negeri Jember, Mastrip Po Box 164 Jember , 68101, Indonesia

*Email: umi.djamilasy@gmail.com

Abstract. The community's most common drying process is drying under the sun for at least two to three days, depending on the weather. Research aimed to resolve the weather dependency of this drying process by drying mushroom in the oven at 40-degrees Celsius and drying process using tray dryer under a vacuum condition and obtained optimum cap and stalk drying results at 60-degrees Celsius under 20 kPa vacuum pressure. However, the shelf drying process resulted in differences in the moisture content. To resolve this problem, a dryer that can operate at a low temperature by vacuuming. The objectives of this research were to design and test the performance of a rotary vacuum dryer with a batch system for drying mushrooms from the Basidiomycota class. The results of functional test and performance tests using 1000 g of mushrooms were obtained: vacuum pressure 16 cmHg, temperature of drying chamber (60 ± 1) oC, drying time 16 hours, rpm drying cylinder 6-8 rpm, and final mushroom moisture content: 13% wb.

1. Introduction

The mushroom has easy damaged characteristics if it is stored in open air for 2-3 days because the high moisture content in it. Therefore, the growth and microbial activity and polyphenol oxidase enzyme activities continue (Agrocendawan Persada, 2011). The activity of these enzymes causes chemical changes, such as the appearance, taste, texture, and quality of the mushroom. To inhibit the activity of these enzymes can be done by reducing the moisture content through the drying process.

Drying process is one way to reduce the moisture content of the material. Therefore, it becomes durable because microbial growth and unwanted reactions become inhibited. Also, drying process makes the added value of mushrooms increases. This is shown by the selling price of dried oyster mushroom is Rp. 90.000/kg, yet the fresh oyster mushroom is only Rp. 18.000/kg (Robin, 2011)

The most used drying technique in the community is drying in the sun. The advantage is that, besides not requiring expensive fees and special expertise, also the drying capacity is unlimited. However, this method is less effective because it relies heavily on weather conditions and takes a 2 days long (Sulistyowati, 2004) and produces less hygienic products because the product is contaminated with dust or contaminants in the air (Raharjo, 2010).

Some researchers have tried to overcome the dependence of weather on drying natural ways such as Widyastuti (2015) has done drying oyster mushrooms using an oven at 40 oC for 48 hours resulted dried mushrooms with the best color that is pure white. Izan (2012) has conducted ear mushrooms studies on tray dryers but there is still a lack of mushrooms which are placed on tray 1 dried faster than those placed in tray 2, 3 and 4. As the result, it makes the final moisture content gradation occurs. Oyster mushrooms besides having high moisture content are also resistant to heat. For drying process Asgar (2013) used a rack dryer with a vacuum at 20 kPa of 60 oC produces the most optimal drying with drying rate of 0.11 gr / min.

Because there are some weaknesses in drying process that has been studied, it is necessary to design rotary vacuum dryer machine bacth system with the aim to produce dried mushrooms with uniform moisture content and browning. Also, the damage to nutritional content due to high temperature can be suppressed and drying time can be accelerated.

2. Materials and methods

2.1. Material and tool

The material that is used in the construction of the prototype Rotary Vacuum Dryer Batch System: 2 mm stainlees plate, the stainless steel plate has a hole size of 5 mm, 1 mm stainless plate, diameter copper pipe ¼ ", dried mushroom material.

The tools used include: welding machine welding, plate roll, hand drill, grinding, vacuum pressure gauge, temperature gauge and thermocouple

2.2. Design and build a rotary vacuum dryer with a batch system

Design analysis includes functional design and structural design that is equipped with engineering designs. Functional design is used to determine the components that is needed to make a Rotary Vacuum Dryer Batch System. While the structural design is used to determine the size of the components in accordance with the needs and strength of the material that is used.

Functional design analysis:

a. The vacuum dryer tube functions to produce the room conditions according to the set pressure
b. The drying cylinder that is equipped with flight functions to storage the material that will be dried and rotate during the drying process with certain interval time.
c. The condenser functions to condense the water vapor that is released during drying process
d. The water jet vacuum pump functions to suck air in the drying chamber so that the pressure becomes low and also to suck water vapor from the dried material
e. The operating control unit functions to activate the vacuum device and heating unit
f. The heating source of the bulb serves as a heating source
g. The electric motor serves to rotate the drying cylinder during the drying process
h. Reducer functions to change the rpm of the electric motor to the drying cylinder

Structural Design Analysis:

a. The drying chamber is designed in a cylindrical shape so that the drying cylinder can be rotated and the use of plate material is thinner. The dimensions of the drying chamber are 46 cm in diameter, 80 cm long and able to work at a vacuum pressure of 10 cmHg.
b. The drying chamber door is made of glass so that it can monitor the drying process.
c. The drying chamber heater uses 100 watts of 10 bulb to produce a low temperature <100 oC
d. The temperature that is used for drying is 60 oC
e. Drying chamber pressure during drying process is 60 Pa

2.3. Performance test

After the Rotary Vacuum Dryer Batch System Prototype is designed and constructed, then a performance test is performed to verify whether the criteria that is required in the dryer system can be qualified or not. The best criteria of drying system are determined by its ability to reach a drying temperature of 60 oC, vacuum pressure of 60 Pa and the moisture content of dried mushrooms are reaching 13% wb.

3. Results and discussion

3.1. Rotary vacuum dryer with a batch system

The Rotary Vacuum Dryer Batch system is made to work in a vacuum condition and there is a process of reversing the material that is dried automatically, so that the drying component consists of: dryer chamber, drying cylinder equipped with flight, heating system, water vapor cooling system, water jet vacuum pump, electric motor, temperature control panel and reduction system. The dryer arrangement can be seen in figure 1, resulting the design can be seen in figure 2.

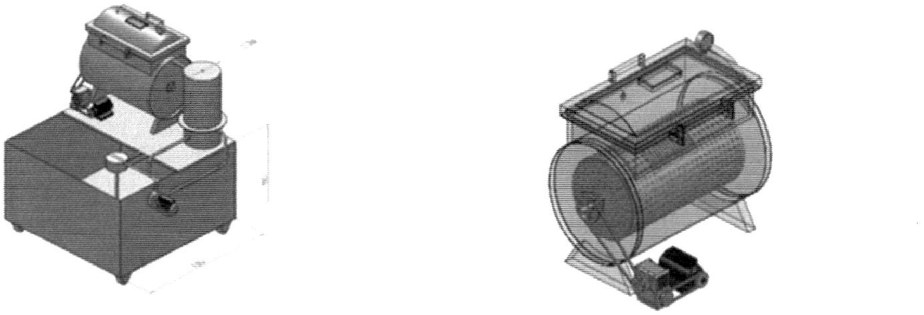

Figure 1. Design of prototype of a rotary vacuum dryer with a batch system

Figue 2. The rotary vacuum dryer with a batch system

3.1.1. Dryer chamber

The drying chamber is cylindrical so that the drying cylinder can rotate and the material that is used not too thick. Therefore it is necessary to be accounted the thickness of the stainless steel plate that is needed to be able to withstand the vacuum power of the pump. Pressure on the wall of chamber occurs because of the difference between the pressure inside the drying chamber and the pressure outside the drying chamber is P_{atm}. The pressure in the drying chamber is very small compared to the pressure outside the drying chamber. So that, there is an inward pressure loading. The drying chamber material is chosen from the stainless steel that is not easily corrosive, the tension stress of the stainless steel material σy is 370 MPa and the safety factor n is used 1.67 (Gere *et al*, 1987), with the strength data can be found of stress σ_i.

$$\sigma_i = \frac{\sigma_y}{n} = 221{,}56 MPa \qquad (1)$$

If the vacuum pressure is made p_v = 10 cmHg or 13.3 kPa and atmospheric pressure p_{atm} = 101.3 kPa, then the drying chamber wall pressure $\Delta p = p_{atm} + p_v$ = 114,3 kPa

The diameter of the drying chamber is 46 cm and the length is 80 cm, so the thickness of the wall can be searched by considering several types of loading.

Radial load:

$$\sigma_i = \frac{F}{A} = \frac{\Delta p \pi d l}{2 x_{dp} l} \qquad (2)$$

$$x_{dp} = \frac{\Delta p \pi d l}{2\sigma} = 0{,}22\ mm \qquad (3)$$

Axial Load:

$$\sigma_i = \frac{F}{A} = \frac{\Delta p 0{,}25 \pi d}{2 x_{dp} d} \qquad (4)$$

$$x_{dp} = \frac{0{,}25 \Delta p \pi d}{\sigma_i} = 0{,}035\ mm \qquad (5)$$

From the results of calculating the required radial plate thickness is 0.22 mm and the required thick axial loading is 0.035 mm, because the thickness of the plate that is used is greater (2 mm) than the thickness of the calculation, the plate of the drying chamber is qualified to withstand external loads.

3.1.2. Dryer chamber door

The drying chamber door is made of glass. The glass value σ_{ult} is known to be 10 x 102 Mpa. Garmo *et al*. (1984) said that if the stress used is σ_{ult} then the safety factor that used is 2.8 with these data then it can be searched σ_i (Gere *et al.*, 1987):

$$\sigma_i = \frac{\sigma_{ult}}{n} = 3{,}57x\ 10^2\ \text{Mpa} \qquad (6)$$

The thickness of the door is determined by using equation 4 by changing the variable x_{dp} to S_{pp} .

$$S_{pp} = \frac{0{,}25\ \Delta p d}{\sigma_i} = 0{,}321\ mm \qquad (7)$$

Because the calculation of the strength of the material result shows the required glass door thickness is 0.321 mm while the thickness of the glass used is 10 mm, it can be concluded that the drying door adequates to the material strength requirements.

3.1.3. Vacuum pressure

To make the vacuum condition of the drying chamber, water jet pump is used that works with the Bernoulli principle, where the water jets of pump that is through the pipe produces a venture effect or vacuum. By using 7 or 8 nozzles, a special suction pipe the air until the pressure in the drying chamber drops to 7.52 cmHg which causes the boiling point of the dried material to drop

3.1.4. Heating source

Considering the volume and temperature of the drying chamber 60 oC, that 100 watts of 10 bulbs are used, it is resulting in an energy of 1000 w.

3.1.5. Temperature control

The temperature of the drying chamber is maintained by 60 oC by using a closed loop control system by controlling the load on off control so that the accuracy of the control is more guaranteed and costs less.

3.2. Performance of rotary vacuum dryer with a batch system

The pressure drop in the early minutes can take place quickly because pumping air from the drying chamber to free air is easier because the pressure difference inside and outside is not too large. The vacuum pressure that able to be produced from a water jet pump is less than the maximum, which still shows 16 cmHg.

The temperature of the drying chamber is set at a temperature of 60 oC and it is controled automatically uses relay connector to maintain the set temperature. To reach the temperature according to the setpoint, it takes 40 minutes and changes in temperature after reaching set point ± 1 oC (figure 3).

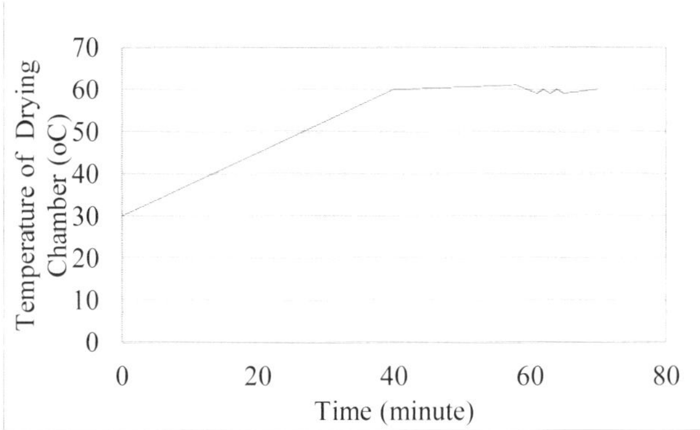

Figure 3. Reaching the temperature 60 oC at 40 minutes

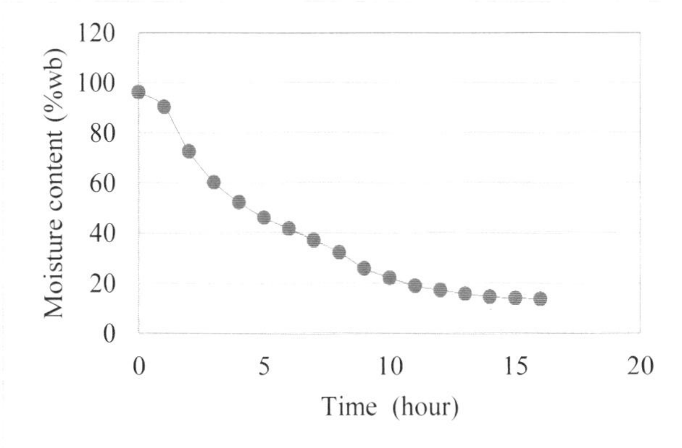

Figure 4. Drying rate

The magnitude of the decrease in the moisture content of the mushroom during drying process as shown in figure 4. It can be seen that the tool is able to reduce the moisture content from 96% to 13% wet basis (wb) within 16 hours, then the drying rate is 63.65 gH2O / hour (figure 5).

(a) (b)

Figure 5. (a) Fresh mushrooms (b) Dried mushrooms

4. Conclusion

Technical characteristics of functional test results and performance tests using 1,000 g of oyster mushroom obtained: Vacuum pressure (16 cmHg), temperature of drying chamber (60 ± 1) °C, drying Time Process (16 hours), rpm drying cylinder (6-8 rpm), final mushroom moisture content (13% wb).
It is necessary to test engine performance to see the actual capacity and characteristics of the Rotary Vacuum Drying with a Batch System.

Acknowledgment
The authors acknowledge the financial support from PNBP Politeknik Negeri Jember, Indonesia.

References

[1] Asgar, A., Zain, S., Widyasanti, A., dan Wulan, A., 2013, *Kajian Karakteristik Proses Pengeringan Jamur Tiram (Pleurotus sp.) Menggunakan Mesin Pengering Vakum (Characteristics Study of Drying Process of Oyster Mushrooms (Pleurotus sp.) Using Vacuum Dryer)*, J. Hort. Vol. 23 No. 4, 2013
[2] Agrocendawan Persada. 2011, *Pengaruh Kadar Air Pada Kualitas Jamur Tiram, diunduh tanggal* 17 Oktober 2011, <http:// Jamuraceda.Com/Tips-Memilih-Jamur-Kadar-Air.Html>.
[3] De Garmo, E.P. Black, J.T., Kohser, R.A., 1984, *Material and Prosesses in Manufacturing, six edition*, Mc. Millan Pub Co, New York, USA
[4] Gere, J. M., Timoshenko, S. 1987, *Mechanic of material, second edition*, Van Nostrand reinhold, Hongkong
[5] Kutovoy, V, Nikolaichuk, L & Slyesov, V., 2004, *The Theory of Vacuum Drying, International Drying Symposium*, vol. A, pp. 26627.
[6] Muhammadiah, M.S. Budi I.S., Erizal, Leopold O.N, Naresworo, N., 2012, *RekayasaMesin Pengering Vakum dengan Tekanan Terkendali,* Jurnal :Teknologi, Volume 15, No. 1, April 2012.
[7] Perumal, R., 2007, *Comparative Performance of Solar Cabinet, Vacuum Assisted Solar And Oven Drying Method*', Thesis, Natural Resources Technology Depostment, University Montreal, Kanada.
[8] Pinedo, A, Fernanda, E, Abraham, D & Zilda, D., 2004, *Vacuum Drying Carrot : Effect of Pretreatments and Parameters Process,* Int. Drying Symposium, vol. C, pp. 2012-26
[9] Raharjo, S., 2010, *Uji Kinerja Cabinet Dryer dengan Sistem Tray Dengan Pengurangan Kadar Air Pada Jamur Tiram*, Thesis, Teknik Kimia, Universitas Diponegoro, Semarang

IOP Conf. Series: Earth and Environmental Science **207** (2018) 012004 doi:10.1088/1755-1315/207/1/012004

Optimization of Design Heat Exchanger to Reduce Fouling Resistance in Milk Pasteurization

T Budiati[1] and T R Biyanto*[2]

[1] Food Engineering Department, Politeknik Negri Jember, Jember 68101 East Java, Indonesia

[2] Department of Engineering Physics, Institut Teknologi Sepuluh Nopember Surabaya, Surabaya 60111 East Java, Indonesia

*Email: trb@ep.its.ac.id

Abstract. Milk pasteurization process used Heat Exchanger (HE) to heat fresh milk at certain temperature and time period. Fouling in heat exchanger during milk pasteurization has been identified as a major problem in the efficiency of energy. Due to insufficient temperature during process, foodborne pathogens may not be killed properly and contaminate the food which make the human health adverse. Other problems may arise in fouling which is losing the efficiency during operation, increasing cost of pumping and downtime that affecting the efficiency of production cost. The fouling resistance depends on milk properties, heat exchanger design technologies and actual operating conditions. The objective of this research was to optimize the heat exchanger design of milk pasteurization by considering series-parallel connection to minimize fouling. The research method was using recent stochastic algorithms Duelist Algorithm, Killer-Whale Algorithm and Rain-Water Algorithm for fouling mitigation in milk pasteurization. The optimization results show the series-series connection reduce fouling up to 16% at the tube side (Rft) and reduce up to 5% at shell side.

1. Introduction

In the food industries, varieties of heat treatments such as milk pasteurization are performed. In milk in pasteurization process, milk is heated and changed the milk stabilities. A milk solids deposit is formed at both of heat-exchanger surface. Even the deposit does not affect to the milk decomposition significantly, it affected to the thermal and hydraulic impact which cause the increasing of operation cost and maintenance cost. Thus, fouling is considered as the worst problem in dairy processing plants. In the crude refineries, heat exchanger network required annual cleaning. In the dairy industries, the best practice of clean interval is every 5-10 hours. The operating cost in the U.S. pasteurized milk industries due to fouling has been approximation in $140 million annually [1]. However, these were not included the sterilized milk.

The root causes of milk fouling forming have been studied by several studies. The deposit composition has been analyzed. Furthermore, the chemical changes during milk heating have been clearly known which was caused by the presence of proteins and β-lactoglobulin. Lalande *et al.* [2] revealed that heat denaturation of the protein could promote the milk deposit forming. De Jong *et al.* [3] used β-lactoglobulin kinetics reaction to study the fouling deposition that was occurred in plate heat exchangers. The result of those study described that the deposit had close correlation with reaction rates of protein.

The study concerning on chemical reactions fouling was reported by Paterson and Fryer [4]. The study reported that the milk fouling was classified as chemical reaction fouling with high rate reaction rate. Therefore, the rate of fouling forming in milk pasteurization depended on the fouling reaction. Belmar-Beiny *et al.* investigated the fluid temperature and velocity effects to the fouling whey protein [5]. This study shows the increasing fouling rate was occurred when the milk was heating up in heat exchanger. Two possible fouling type mechanisms are depending on mass and energy transfers and reaction.

Fouling mitigation techniques commonly performed by designing higher efficiency of heat exchanger and applies cleaning scheduling of heat exchanger periodically. In cleaning scheduling of heat exchanger, the loss due to fouling can be avoided [6,7,8]. However, the frequency should be optimized to obtain highest saving. Alternatively, design of new heat exchanger or retrofit existing heat exchanger with larger heat transfer area and higher efficiency can also effective way to performed. Nevertheless, the capital cost of the HEN will be increase [6,9,10].

Design heat exchanger can be performed at internal geometries of heat exchanger and/or at parallel series of heat exchanger connection. The objective of this research was to optimize the design of milk pasteurization process by considering series-parallel connection to minimize fouling. The class of optimization problem was Mixed Integer Non Linear Programming (MINLP). The effective and common algorithm of optimization to solve class of MINLP is algorithm of multi-variable stochastic optimization [11]. Recently, this algorithm consists of the new algorithms, namely Duelist Algorithm [12], Killer-Whale Algorithm [13] and Rain-Water Algorithm [14] due to the effectively achievement of global optimum solution.

2. Literature Review

2.1 Series-parallel connections of heat exchangers

Connection of series-parallel HE affects to mass flowrate and inlet-outlet temperature different. The different of temperature and mass flowrate in the heat exchangers will affect to rate of fouling [2]. Series and parallel connection of heat exchanger can be identified based on the outlet and inlet of heat exchanger. The series connection will affect to temperature difference reducing and it may be decrease the fouling rate as show in figure 1. The parallel connection is usually used to increase production capacity as shown in figure 2.

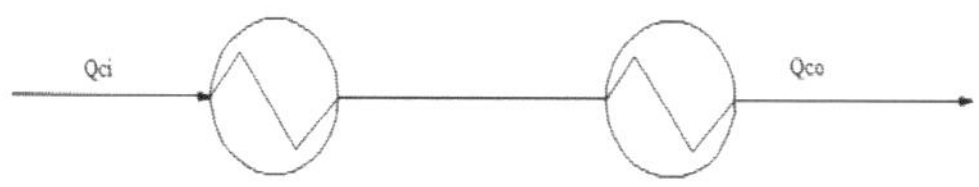

Figure 1. Series connection of heat exchangers

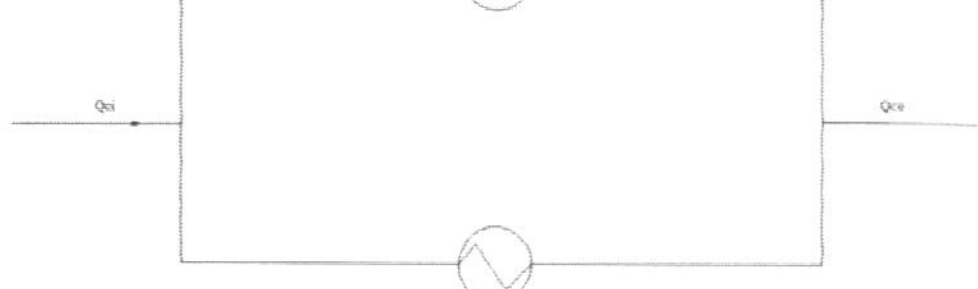

Figure 2. Parallel connection of heat exchangers

2.2 Model of Heat Exchanger
Heat exchanger modeling is used to determine tube and side heat transfer coefficients, overall heat transfer coefficient, heat duty and pressure drops.

2.2.1 Heat Transfer Calculation in Shell Side The method used to calculate heat transfer coefficient in shell side is Bell-Delaware's Method. Shell side crossflow area (Sm) is given by equation 1.

$$S_m = L_{bc}\left[L_{bb} + \frac{D_{ctl}}{L_{tp.eff}}\left(L_{tp} - do\right)\right] \quad (1)$$

Where : Sm = Shell side crossflow area (m^2); Lbc = Central baffle spacing (m); Lbb = Bundle to shell clearance (m); Dctl = Bundle diameter (m); Ltp.eff = Effective tube pitch (m); Ltp = Tube pitch (m); do = Tube outside diameter (m);

Shell side mass velocity (Gs) and Reynold number (Res) is given by equation 2 and equation 3.

$$G_s = \frac{m_s}{S_m} \quad (2)$$

$$Re_s = \frac{G_s\, d_o}{\mu_s} \quad (3)$$

Where : Gs = Shell side mass velocity (kg/ s m^2); ms = Shell side mass flow rate (kg/s); Re_s = Shell side Reynold number; μ_s = Shell side fluid viscosity (kg/m s).

The shell side Prandtl number (Prs) is given by equation 4.

$$Pr_s = \frac{\mu_s C_{ps}}{k_s} \quad (4)$$

Where : Prs = Prandtl number; Cps = Shell side specific heat (J/kg ^oC); Ks = Shell side thermal conductivity (W/ m ^oC); μ_s = Shell side fluid viscosity (kg/m s)

Ideal heat transfer coefficient (hi) is given by equation 5.

$$h_i = \frac{j_i C_{ps} G_s (\emptyset_s)^n}{Pr_s^{\frac{2}{3}}} \quad (5)$$

The term ji is the ideal Colburn as equation 6, 7 and 8.

$$j_i = 1{,}73 R_{es}^{-0{,}694} \text{ when } 1 \leq Res \leq 100 \quad (6)$$

$$j_i = 0{,}717\, Re_s^{-0{,}574} \text{ when } 100 \leq Res \leq 1000 \quad (7)$$

$$j_i = 0{,}236\, Re_s^{0{,}346} \text{ when } 1000 \leq Res \quad (8)$$

The term (ϕs) is viscosity correction factor, which accounts for the viscosity gradient at the tube wall (μw) versus the viscosity at the bulk mean temperature (μs) of the fluid as equation 9.

$$(\emptyset_s)^n = (\frac{\mu_s}{\mu_{sw}})^{0,14} \tag{9}$$

Where, hi = Ideal heat transfer coefficient (W/ m^2 ^OC); ji = Ideal Colburn; μ_{sw} = Viscosity at bulk mean temperature (kg/m s);

Shell side heat transfer coefficient is given by equation 10

$$h_s = h_i J_c J_l J_b J_s J_r \tag{10}$$

Where, hs = Shell side heat transfer coefficient (W/ m^2 ^OC); hi = Ideal heat transfer coefficient (W/m^2 ^oC); Jc = Segmental baffle window correction factor; Jl = Correction factors for baffle leakage effects for heat transfer; Jb = Correction factors for bundle bypass effects for heat transfer; Js = Correction for unequal baffle spacing at inlet and/or outlet ; Jr = Correction factor for adverse temperature gradient in laminar flow

2.2.2 Heat Transfer Calculation in Tube Side The method used to calculate heat transfer coefficient in tube side is Bell-Delaware's Method. Tube side flow area is given by equation 11.

$$A_t = \frac{\pi\, d_i^2\, N_t}{4} \tag{11}$$

Tube inside diameter is given by equation 12.

$$d_i = d_o - 2\, t_w \tag{12}$$

Number of tube is given by equation 13.

$$N_t = \frac{0{,}78\, d_{ctl}^2}{L_{tp}^2} \tag{13}$$

Where : At = Tube side flow area (m^2); di = Tube inside diameter (m); do = Tube outside diameter (m); Nt = Number of tube; tw = Tube thickness (m); Dctl = Bundle diameter (m); L_{tp} = Tube pitch (m)

Tube side mass velocity (Gt) is given by equation 14.

$$G_t = \frac{m_t\, N_p}{A_t} \tag{14}$$

Where, Gt = Tube side mass velocity (kg/ s m^2); mt = Tube side mass flow rate (kg/s); Np = Number of tube pass

Reynold number is given by equation 15.

$$Re_t = \frac{G_t\, d_i}{\mu_t} \tag{15}$$

Where, Ret = Tube side Reynold number; μt = Tube side fluid viscosity (kg/m s)

Prandtl number is given by equation 16.

$$Pr_t = \frac{Cp_t\, \mu_t}{K_t} \tag{16}$$

Where, Cpt = Tube side specific heat (J/kg ^{O}C); Kt = Tube side thermal conductivity (W/ m ^{O}C)

Tube side heat transfer coefficient is calculated by CollBurn equation as equation 17.

$$h_t = 0{,}023\ Re_t^{\,0,8} Pr_t^{\,0,4} \left(\frac{K_t}{d_i}\right)\left(\frac{\mu_t}{\mu_w}\right)^{0,14} \tag{17}$$

Where, ht = Tube side heat transfer coefficient (W/m^2 ^{o}C)

2.2.3 Calculation of Overall Heat Transfer Coefficient Heat transfer rate is affected by overall heat transfer coefficient, area of heat transfer, and logarithmic mean temperature difference as show in equation 18-21.

$$\text{cond} = \frac{d_o \left(\ln\left(\frac{d_o}{d_i}\right)\right)}{2\, K_{cs}} \tag{18}$$

$$\frac{1}{U_f} = \frac{d_o}{d_i\, h_t} + \frac{d_o\, R_{ft}}{d_i} + \text{cond} + R_{fs} + \frac{1}{h_s} \tag{19}$$

Where, Uf = Overall heat transfer coefficient under fouling conditions (W/m^2 ^{o}C); Cond= Conduction heat transfer (m^2 ^{o}C /W); Kcs = Thermal conductivity of tube material (W/ m ^{O}C); R_{fs}= Shell side fouling resistance (m^2 ^{o}C /W); R_{ft} = Tube side fouling resistance (m^2 ^{o}C /W)

Heat duty is given by equation 20.

$$Q = U_f\, LMTD_{corr.}\, A_o \tag{20}$$

Where, Uf = Overall heat transfer coefficient under fouling conditions (W/m^2 ^{o}C); $LMTD_{corr.}$ = Log Mean Temperature Difference corrected (^{o}C); A_o= Area of heat transfer (m^2)

Log Mean Temperature Difference corrected ($LMTD_{corr.}$) is given by equation 21.

$$LMTD_{corr.} = LMTD \,.\, F \tag{21}$$

Where, LMTD = Log Mean Temperature Difference (^{o}C); F = Correction factor

LMTD is the average temperature difference between hot and cold fluid. LMTD is calculated by the equation 22.

$$LMTD = \frac{(T_{h,i} - T_{c,o}) - (T_{h,o} - T_{c,i})}{\ln\left(\frac{T_{h,i} - T_{c,o}}{T_{h,o} - T_{c,i}}\right)} \tag{22}$$

Area of heat transfer is given by equation 23.

$$A_o = \frac{22}{7} d_o L_{ti} N_t N_s \tag{23}$$

2.2.4 Calculation of Pressure Drop Pressure drop in tube side utilize Pethukov and Popov's Methods as equation 24 and 25.

$$\Delta P_t = \left[2 \text{ x f } \frac{\text{L x Np}}{d_i} + 2 \text{ x Np}\right] \text{ x } \rho_t \text{ x } V^2 \tag{24}$$

$$f = (1{,}58 \ln Re_t - 3{,}28)^2 \tag{25}$$

Where : ΔP_t = Tube side pressure drop (kg/cm^2); f = Friction factor; L = Tube length (m); ρ_t = Tube side fluid density (kg/m^3); V = Fluid velocity (m/s)

Pressure drop in shell side utilizing Bell Delaware's Methods as equation 26 and 27.

$$\Delta P_s = \frac{2 f G_s^2 D_s (N_b+1)}{\rho_s d_o \left(\frac{\mu_s}{\mu_w}\right)^{0{,}14}} \tag{26}$$

$$f = e^{0{,}576 - 0{,}19 \ln Re_s} \tag{27}$$

Where : ΔP_s = Shell side pressure drop (kg/cm^2); f = Friction factor; ρ_s = Shell side fluid density (kg/m^3)

2.2.5 Calculation of Fouling Resistance Calculation of fouling resistance utilize Fryer and Slater model [2] as equation 28.

$$\frac{dRf}{dt} = \left[\frac{4.85}{\text{Re}} x10^{13} \exp\left(\frac{-87x10^5}{RT}\right)\right] x \left[\frac{1}{U_O}\right] \tag{28}$$

3. Recent Stochastic Algorithms

3.1 Duelist Algorithm
Biyanto [12] has developed Duelist Algorithm (DA) which was inspired by the capabilities being the winner, loser and champion in duelist. These included an evolutionary computation technique which was tested by a total of 11 different optimization problems and showed to be more robust than other recent stochastic algorithms.

3.2 Killer-Whale Algorithm (KWA)
Biyanto [13] has built KWA which was inspired by the life of Killer Whale during hunting the prey and has been compared with others algorithm , such as GA, CMA-ES, SA and ICA. This was tested by a total of 6 objective functions with . cauchy noise, gaussian noise, and uniform noise.

3.3 Rain-Water Algorithm (RWA)
Biyanto [14] has built RWA which was based on the pattern of free fall rain water that were from atmosphere to the lowest earth surfaces. RWA was found to be capable in providing the better performances compare to other four algorithms specifically on the speed of convergent [14].

4. Research Method

In general, the flow chart of research methodology of series-parallel heat exchanger design using recent stochastic algorithms for fouling mitigation in milk pasteurization was shown in figure 3.

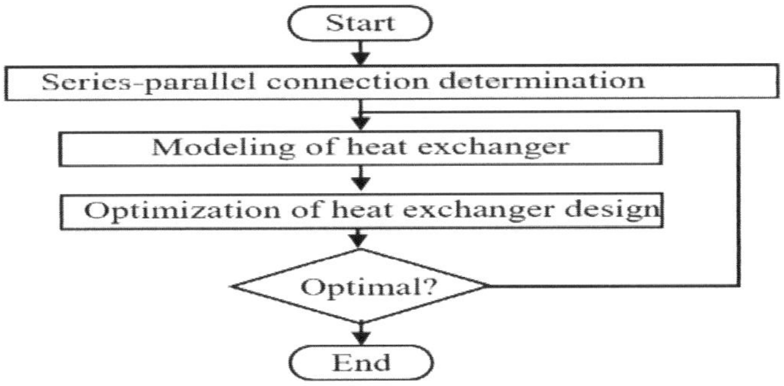

Figure 3. Optimization of heat exchanger design flow chart

5. Results and Discussions

Since series-series connection of heat exchanger provided lowest fouling resistance, the rest discussions only focus on series-series connection. In this research, modeling of series-parallel heat exchanger have been developed and optimization of heat of heat exchanger design for milk pasteurization have been performed using recent stochastic optimization algorithms.

In present study, KW, DA, and RWA were applied to optimize fouling resistance in milk pasteurization design. The optimized variables are shell inside diameter (Ds), tube outside diameter (do), and the number of baffles (Nb). The objective function during iterations can be seen in the results below. These algorithms exhibited similar optimization results. However, the time to reach optimum value are varies due to code and parameters for each algorithm as shown in figure 4.

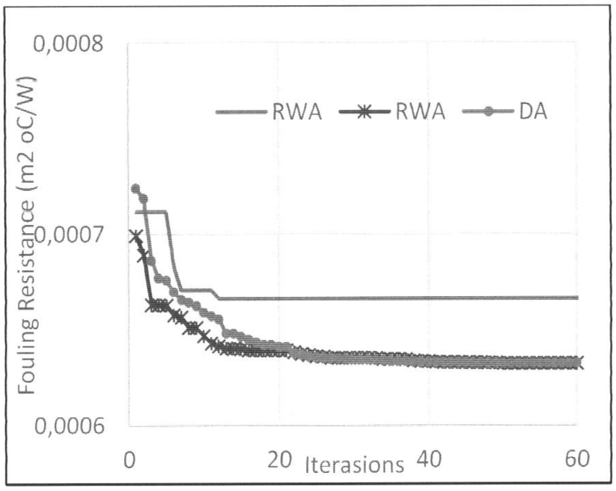

Figure 4. The minimized objective function of fouling resistance during iterations.

These results show that DA, KWA and RWA can optimize fouling resistance in milk pasteurization design due to their robustness. Biyanto [12, 13, 14] revealed that DA, KWA and RWA was more robust compared to other algorithms. The difference between them is speed to achieve global optimum.

Table 1. The Effect of Series-Series Heat Exchangers

Variables	Unit	Design	Optimization	Difference
Uf	w/m^{2o}C	342.55	364.84	7%
Ao	m^2	101.67	96.96	-5%
Ps	psi	04.59	04.28	-7%
Pt	psi	04.58	04.43	-3%
Rft	m^{2o}C / W	0.00048 72	0.0004622	-5%
Rfs	m^{2o}C / W	0.00008 64	0.0000722	-16%

Detail optimization results are tabulated in table 1. The table 1 shows fouling resistance at tube and shell side after optimization reduce 16% and 5%, respectively. It performances were obtained due to optimization in the proper selection of tube side outside diameter, shell side diameter and number of baffle. Moreover, the size of heat exchanger reduces 5% and differential pressure reduces up to 7%. The size of heat exchanger reducing will reduce capital cost. The reducing of differential pressure will reduce the price of pump and reduce the operational cost of pumping fluids.

6. Conclusions

By using DA, KWA and RWA for the optimization of fouling resistance in milk pasteurization, the series-series connection reduce fouling up to 16% at the tube side and up to 5% at shell side.

7. References

[1] Sandu C and R K Singh 1991 Energy Increase in Operation and Cleaning Due to Heat-Exchanger Fouling in Milk Pasteurization *Food Technol.* **23** 84

[2] Lalande M, Tissier J P, and Corrieu C 1985 Fouling of Heat Transfer Surfaces Related to b-Lactoglobulin Denaturation During Heat Processing of Milk *Biotechnol. Prog.* **1** 131

[3] De Jong P, Bouman S, and Van Der Linden H J 1992 Fouling of Heat Transfer Equipment in Relation to the Denaturation of plactoglobulin *J. SOC. of Dairy Technol.* **45** 3

[4] Paterson W R and Fryer P J 1988 A Reaction Engineering Approach to the Analysis of Fouling *Chem. Eng. Sci.* **43** 1714

[5] Belmar-Beiny M T, Gotharn S M, Paterson W R and Fryer P J 1993 The Effect of Reynolds Number and Fluid Temperature in Whey Protein Fouling *J. SOC. of Dairy Technol.* **19** 119

[6] Smaïli F, Vassiliadis V S, Wilson D I 2001 Mitigation of fouling in refinery heat exchanger networks by optimal management of cleaning *Energy Fuels* **15** 1038

[7] Sanaye S and Niroomand B 2007 Simulation of heat exchanger network (HEN) and planning the optimum cleaning schedule *Energy Convers. Manage.* **48** 1450

[8] Georgiadis M C, Papageorgiou L G, Macchietto S 1999 Optimal cyclic cleaning scheduling in heat exchanger networks under fouling *Comput. Chem. Eng.* **23** S203

[9] Yeap B L, Wilson D I, Polley G T and Pugh S J 2005 Retrofitting crude oil refinery heat exchanger networks to minimize fouling while maximizing heat recovery *Heat Transfer Eng.* **26** 23
[10] Ebert W and Panchal C B 1995 Analysis of Exxon crude-oil-slip stream coking data *Fouling mitigation of industrial heat exchangers* (San Luis Obispo, CA, United States)
[11] Biyanto T R 2016 Thermal and Hydraulic Impacts Consideration in Refinery Crude Preheat Train Cleaning Scheduling Using Recent Stochastic Optimization Methods *Applied Thermal Engineering* **108** 1436
[12] Biyanto T R, Fibrianto H Y, Nugroho G, Listijorini E, Budiati T and Huda H 2016 Duelist algorithm: an algorithm inspired by how duelist improve their capabilities in a duel, *Int. Conf. in Swarm Intelligence: Advances in Swarm Intelligence* **2** pp 39
[13] Biyanto T R 2017 Killer Whale Algorithm: An Algorithm Inspired by the Life of Killer Whale *Procedia Computer Science* **124** 151
[14] Biyanto T R 2017 Rain-Water Algorithm: Newton's Law of Rain Water Movements During Free Fall and Uniformly Accelerated Motion Utilization *International Conferencecon Advanced Mechatronics, Intelligent Manufacture, and Industrial Automation (ICAMIMIA 2017)* **1** 107

The impact of land use changes on carrying capacity of sampean watershed in Bondowoso Regency

S Sugiyarto[1,*], B Hariono[2], R Wijaya[2], P Destarianto[3], A Novawan[4]

[1]Department of Agricultural Production, Politeknik Negeri Jember, Indonesia
[2]Department of Agricultural Engineering, Politeknik Negeri Jember, Indonesia
[4]Department of Information Technology, Politeknik Negeri Jember, Indonesia
[5]Department of Language, Communication, and Tourism, Politeknik Negeri Jember, Indonesia

[*]corresponding author : sugiyarto@polije.ac.id

Abstract. Sampean watershed which covers Bondowoso and Situbondo areas is a watershed which has a role in the rice provision for Bondowoso and East Java. The development activities in Sampean watershed both upstream and downstream are classified into highly intensive, while the population growth is quite high. The changes in land use, the increases of residential areas in upstream, midstream and downstream areas, have contaminated Sampean watershed with pollutants which could affect the agricultural products. Related to that, the specific objectives of this study were: 1) To find out the changes in land use in Sampean watershed area; 2) To know the total land requirements and the area of rice equivalent land available in Sampean watershed area to support the people's life; and 3) To know the carrying capacity of Sampean watershed in Bondowoso region. The comparison method to compare between availability and land requirements was used as a tool to analyze the carrying capacity of Sampean watershed area so that it could provide an overview of the impact of land use change. The results of data analysis showed that the Land Carrying Capacity of Bondowoso regency reached surplus around 206,708 ha. Meanwhile, the areas with deficits in Carrying Capacity were Grujugan sub-district (-1,213 ha), Tamanan (-147 ha), Tlogosari (-3,181 ha) and Bondowoso (-15,422 ha).

1. Introduction

Watershed is a land which conjoins a river and its tributaries, in which its functions are to store and to drain water from rainfall and to channel it naturally to the lake or sea. The land boundary becomes the topographical separators while the boundary of the sea up to water area are still affected by land activities. Basic consideration of the environmental carrying capacity in spatial planning is from the Law of Republic of Indonesia No. 26 Year 2007 on the Spatial Planning, and No 32 Year 2009 on the Environmental Management. A step that must be taken in the preparation and determination of spatial planning is to determine the conceptual direction on the development that will be achieved through the environmental carrying capacity. Environmental carrying capacity is the environment ability to support the lives of human-beings and other living things. Preservation of environmental carrying capacity is a series of efforts to protect the environment ability against the pressure of change and/or negative impacts caused by an activity, so that it can still be able to support the lives of human-beings and other living things [1].

Regional economic development is a part of carrying capacity concepts for human populations that has been applied since 1960s. Various consumption habits make it difficult to predict the carrying capacity of the earth for human-beings. Therefore, the environmental carrying capacity benefits for both the population of human beings and for the difference of the consumption level influenced by the production and consumption of technology. Carrying capacity models are the primary vehicle for the estimation of a population's self-sufficiency. From a resource perspective, the most important parameters determining carrying capacity are basic human needs essential for a population's physical survival including food, water, shelter and energy [2].

Sampean watershed has both social and economic functions. In Bondowoso and Situbondo area, Sampean watershed is an urban watershed that has a strategic role in the national context, which needs to be managed specially. Construction activities in it, both upstream and downstream, are classified into highly intensive while the population growth is quite high. The changes in the land use and the increase in residential areas in the upstream midstream and downstream have contaminated Sampean watershed with pollutants. The pollutant sources came from domestic waste, industrial waste, agricultural waste, and livestock waste [2].

Due to its emergence, the environmental carrying capacity of Sampean watershed needs to be investigated to know to which extent it provides benefits or carrying capacity for the people's lives. The results of this study could be a recommendation for the policy makers on the land use along Sampean watershed. Carrying capacity ratio is a construction planning tool that provides an overview of the relationship between population, land use and the environment. Carrying capacity analysis can provide information needed in assessing the level of land capability in supporting all activities in that area [3].

According to the Minister Regulations on Environment No. 17 Year 2009, consisting of the Guidelines in Determining the Environmental Carrying Capacity on Regional Spatial Planning of the Minister of Environment, environmental carrying capacity is divided into 2 (two) components, namely supportive capacity and waste capacity (assimilative capacity). In the guidelines, the study of environmental carrying capacity is limited to the capacity natural resources supplies, especially in relation to the capacity, the availability, and the need for land and water in a space/area [4].

The Guidelines for Determining Environmental Carrying Capacity in Regional Spatial Planning was carried out based on 3 (three) approaches, they were: (1) the ability of the land in allocating space utilization, (2) the comparison between the availability and requirement of the land, and (3) the comparison between the availability and need of water. Determining land carrying capacity was conducted by comparing the availability and the need of land. Land availability was determined based on the data on the actual local total production of each commodity in an area, by adding up the products of all commodities in the region. Meanwhile, the need of the land was calculated based on the need of decent living [5].

The determination of water carrying capacity was done by comparing the availability and need of water. Water availability was determined by using runoff coefficient method based on the information of land use and annual rainfall data. Meanwhile, water need was calculated from the conversion results on the need of decent living. The surplus situation indicated that the availability of water in an area was sufficient, whereas the state of deficit showed that the area was not capable to meet the need for water. In order to meet the need of water, environmental functions related to the water system must be preserved. The calculation result of this method could be used as an input or consideration to the preparation of spatial planning and evaluation of spatial utilization for the sake of providing sustainable water resources [6].

2. Working Methodology

Environmental carrying capacity is the carrying capacity of the natural environment based on the biomass of plants and animals which can be collected and captured per unit of area and time around that area [4].

IOP Conf. Series: Earth and Environmental Science **207** (2018) 012005 doi:10.1088/1755-1315/207/1/012005

A. Research Area

The research areas took place around the Sampean River Basin Area in Situbondo and Bondowoso Regencies covering 6 observation stations started from February to October 2017.

B. Data Collection Method

The data used in this research were primary data and secondary data. Primary data was collected through direct observation and questionnaire in the field and secondary data was obtained from periodic reports of various agencies and the previous researches. The equipment used to process the obtained data were stationery and counting tool, a camera, and a set of computers. The data used in this study can be seen in table 1.

Table 1. The Used Data

No	The Type of Data	The Source of Data
1	Water quality, Industry type alongside of Sampean river	Field Observation
2	Demographic Data	Department of Demography and Civil Registration
3	Map (topography, land closure, administration)	Environment Agency
4	Land Use Changes	Regional Development Planning Agency

C. Analysis Method

The impact of the land-use change on land carrying capacity in Sampean watershed can be seen in the form of analysis as follows [4][7] :

- The Calculation of Land Availability
  The calculation was performed through the following stages:

  1) The Calculation of Land Supply

$$SL = (\sum(Pi \times Hi) / (Hb \times Ptvb) \quad (1)$$

Notes:
SL : Land Availability (ha).
Pi : Actual Production of Each Type of Commodity
Hi : Unit price for each type of commodity at producer level
Hb : Unit price of rice (Rp / kg) at producer level.
Ptvb : Rice Productivity (kg/ha).

In this calculation, the conversion factor used to equalize non-rice product with rice was the price.

  2) The Calculation of Land Demand

$$DL = N \times KHLL \quad (2)$$

Notes:
DL : The total demand of land is equivalent to rice (ha).
N : Total population (people)
KHLL : The area of land needed for decent living needs per population.

3. Results and Discussion

Administratively, Sampean watershed area consisted of Jember regency (1 district/ 4 villages) covering 4.115,36 Ha (3.05%); Bondowoso regency (23 districts/ 195 villages) covering about 120.397,63 Ha (89.6%); Situbondo regency (4 districts/ 33 villages) covering 10.520,02 Ha (7.73%). The total area of Sampean watershed was 135.033 Ha with 64 km main river length.

A. Land Carrying Capacity at Bondowoso Regency

The number of Sampean river observation points in this research were coded by *KSP*. There were 6 KSPs that can be seen in table 2.

Table 2. Sampean river research observation points

No	Location code	Village	Districts
1	KSP. 0	Penanggungan	Maesan
2	KSP. 1	Tegalmejin	Grujugan
3	KSP. 2	Tenggarang	Tenggarang
4	KSP. 3	Taman	Wonosari
5	KSP. 4	Besuk	Klabang
6	KSP. 5	Kotakan	Situbondo
7	KSP. 6	Widuri	Prajekan

Bondowoso regency has 1.560,10 km^2 wide with various topography conditions such as flatland, hills, and mountains. Bondowoso regency is located in three watershed areas, namely Sampean watershed, Deluwang watershed, and Banyuputih (Kalipahit) watershed.. There were 119 water springs spread all over the area and three hot springs in Sempol districts which was mostly functioned to fulfil the needs for daily water, irrigation, fishery, and tourism.

The land use with that composition was distributed in 23 districts areas in Bondowoso regency with the widest area located in Sempol districts (20.720 ha), followed by Seumberwringin districts, Cermee districts, Botolingo districts, and Tlogosari districts. The dominant land use in the widest districts was forest, dry land, and farming. Whereas, the narrowest districts were Bondowoso districts with 2.315,80 ha followed by Sukosari districts, Tenggarang districts, and Tamanan districts. These districts were very potential to have a high number of population density.

Bondowoso regency was divided into 23 districts, 209 villages, and 10 sub-districts. The land carrying capacity analysis of Bondowoso regency was shown in table 3. The main livelihood of Bondowoso regency inhabitants was in the farming sector. It is shown by the majority of the household earnings which was predominated by those from farming sector. Based on table 3, the land carrying capacity of Bondowoso regency had reached surplus of 54.276 ha.

The analysis of 23 districts at Bondowoso regency depicted that there were several districts which had surplus in land carrying capacity while some had deficit in it as shown in table 4.

Table 3. Land Carrying Capacity Analysis at Bondowoso Regency.

Land Needs		
Factor	**Unit**	**Value**
Total Population	Soul	756.989
Decent life land area	Ha	0,17
Land needs	**Ha**	**126.868**
Status		
Factor	**Unit**	**Value**
Land Availability (SL)	Ha	181.143
Land Needs (DL)	Ha	126.868
Carrying capacity Status	Surplus, If SL>DL Deficit, If SL<DL	**SURPLUS**

Table 4. Land Carrying capacity Condition of each district at Bondowoso regency.

No	Districts	Land Carrying capacity	
		Surplus (Ha)	Deficit (Ha)
1	Maesan	964	
2	Grujugan		**1,213**
3	Tamanan		**147**
4	Jambesari	1,001	
5	Pujer	68,261	
6	Tlogosari		**3,181**
7	Sukosari	34,655	
8	Sumber Wringin	1,769	
9	Tapen	27,031	
10	Wonosari	582	
11	Tenggarang	28,513	
12	Bondowoso		**15,422**
13	Curahdami	10,874	
14	Binakal	1,029	
15	Pakem	3,129	
16	Wringin	26,594	
17	Tegal Ampel	751	
18	Taman Krocok	6,665	
19	Klabang	5,246	
20	Botolinggo	1,571	
21	Sempol	0	
22	Prajekan	1,644	
23	Cerme	6,437	

B. Land Use

The land use in Bondowoso regency indicated there were some changes as shown in the table 5 below [6]:

Table 5. Land use changing at Bondowoso regency 2011 - 2016

No.	Land Use	2011 (Ha)	2014 (Ha)	2016 (Ha)
1	Fresh Water	220.97	220.27	219,70
2	Bush	25,948.75	25,944.35	25,941,74
3	Rocky Hill	25.27	25.27	25,27
4	Forest	22,573.57	22,573.57	22,570,70
5	Estate	14,387.67	14,353.58	14,301,91
6	Settlement Lands	10,694.88	11,166.83	12,468,42
7	Grass	2,837.84	2,828.38	2,818,46
8	Rice Field Irrigation	36,687.68	36,353.10	35,583,14
9	Rainfed Rice Field	11,675.51	11,591.76	11,390,05
10	Farmlands	3,0552.1	30,547.13	30,284,85
Total		**155,604,24**	**155,604.24**	**155,604.24**

From the table 5, it was known that the changes of land use in Bondowoso regency was decreased on fresh water, bush, forest, garden, grass, rice field irrigation, rainfed rice field, and farmlands.

The further analysis showed that the changing of land use that was quite significant happened in settlement sector, which increases as much as 1773,54 ha (1.1398% total wide) and rice field irrigation sector that was decreased by 1104,54 ha (0.7095% total wide) spreading in some districts including Sampean watershed area as shown in table 6.

Table 6. The Percentage of Land use changing at Bondowoso regency

No.	Land Use	The Change of Land Use (Ha)			
		2011-2014	%	2011-2016	%
1	Fresh Water	-0.7	-0.0004	-1.27	-0.0008
2	Bush	-4.4	-0.0028	-7.01	-0.0045
3	Rocky Hill	0	0.0000	0	0.0000
4	Forest	0	0.0000	-2.87	-0.0018
5	Estate	-34.09	-0.0219	-85.76	-0.0551
6	Settlement Lands	471.95	0.3033	1,773.54	1.1398
7	Grass	-9.46	-0.0061	-19.38	-0.0125
8	Rice Field Irrigation	-334.58	-0.2150	-1,104.54	-0.7098
9	Rainfed Rice Field	-83.75	-0.0538	-285.46	-0.1835
10	Farmlands	-4.97	-0.0032	-267.25	-0.1717

4. Conclusion

Land carrying capacity in Bondowoso regency gained surplus.The areas with deficits in Carrying Capacity were Grujugan sub-district (-1,213 ha), Tamanan (-147 ha), Tlogosari (-3,181 ha) and Bondowoso (-15,422 ha). The changing of land use in Bondowoso regency had decreased except on the rocky hill and settlement lands. The settlement sector has experienced the biggest land use change while the rice field irrigation experienced a decrease in land use change.

5. References

[1] Effendi H., 2015, Puslitbang Quality and Enviromental Laboratory, Ministry of Environment and Forestry, Jakarta, 16 October 2015.

[2] Hopfenberg, R. and Pimentel, D., 2001, Environment, Development and Sustainability, 3:1-15.

[3] Sari I.K., Lily Montarcih Limantara, Dwi Priyantoro, 2011, Irrigation Journal Vol. 2 No. 1: 29 - 41.

[4] Alvan Pahuluan, Tri Retnaningsih Soeprobowati, Hady Hadiyanto, 2017, Journal of Ecological Engineering,2017

[5] R M Subekti, D S A Suroso, 2018, IOP Conference Series: Earth and Environmental Science.

[6] Asmaranto R, Ery Suhartanto, Mike Yuanita, 2011, Irrigation Journal Vol. 2 No. 1: 79 - 85

[7] Saraswati S.P, Sunyoto, Bambang Agus Kironoto dan Suwarno Hadisusanto, 2014, Human and Environmental Journal Vol. 21, No.2 : 129-142.

[8] Bappeda Bondowoso, 2016.

IOP Conf. Series: Earth and Environmental Science **207** (2018) 012006 doi:10.1088/1755-1315/207/1/012006

Characterization of Morphology from Orchid Vanda sp. as a Genetic Information Source for Preservation and Agribusiness of Orchids in Indonesia

K Kasutjianingati[1] and R Firgiyanto[1]

[1] Departement of Agricultural Production, State Polytechnic of Jember, Jl. Mastrip, Jember, Indonesia

E-mail: kasutjiningati@yahoo.com

Abstract: The aim of the study was to characterize the collection of endangered Indonesian Vanda orchids based on morphological characters. The information obtained is useful for the conservation of germplasm and for the development of superior Vanda orchid breeding programs in the future. The research activity was conducted from May to November 2018 at the Biosciences Laboratory and Plant Laboratory - State Polytechnic Jember, East Java, Indonesia. The experimental design used in this study was a non factorial completely randomized design with three replications. The treatment in this experiment consisted of fourteen Vanda orchid species from various regions in Indonesia. Observation variables based on qualitative morphological characterization. The results of the experiment showed that the coefficient of similarity in morphological characters ranged from 0.33 to 1.00. The lowest value is found between Vanda Helfola and Arades. The results of the dendogram from grouping based on similarity coefficients are divided into two groups, namely the Vanda group consisting of all types of groupings at coefficients of 0.67 - 0.73 and Vanda groups which group at coefficients 0. 39

1. Introduction

Orchid (Orchidaceae) is an ornamental plant that is favored by many Indonesians because of the beauty of flowers [1] [2]. Orchid has 736 genus with 26,000 species. One genus of orchids that is cultivated by many people is the Vandaceous genus. Orchid Vanda has a wide distribution area in Indonesia [3]. The development of vanda orchids in Indonesia is still very limited because in general Vanda orchids are sold in the market directly from the forest [4]. This condition is feared to have an impact on the level of preservation as reported by Muharyati *et al.* [5], that the existence of Vanda orchids began to be threatened with extinction and even some of its species have been classified as CITES (Convention on International Trade in Endangered Species of Wild Fauna and Flora) in the appendix II category. The extinction was also supported by the existence of forest destruction due to natural disasters and human activities [6]. Therefore, efforts are needed to preserve conservation by conducting conservation in situ and ex situ, for example in the orchid nursery Experiment owned by Research Institutions and Universities.

Conservation activities begin with exploration and characterization of Vanda orchids species with the aim of knowing the level of diversity and superiority. This activity is an initial effort to collect native germplasm resources from Indonesian species that will be used as elders in conventional plant breeding activities as well as biotechnology by combining various levels of diversity including intra species, inter species, segregation and mutant plants [7], [8], [9]. One example is the assembly of new superior varieties from Vanda orchids by The Indonesian Ornamental Crop Research Institute using Vanda tricolor orchids as their parents to pass on the fragrance of flowers to the results of their crosses [10].

Characterization of Vanda orchid morphology according to Nurmalinda *et al* [11] includes the shape of the stem which is straight, slender, and not tubular. Flower stalks come out from the sides of the stem alternately. The length of the flower stalk is about 50 cm in size with the number of flowers ranging from 5 to 15 flowers. Based on the leaf shape Vanda orchids are divided into three groups, namely (a) Terer/pencil of Vanda which is terrestrial with direct sunlight requirements, (b) Strap leaves of Vanda that are epiphytic with little shade and (c) Semiteret of Vanda which is semi-intermediates Vanda from the combination of the two previous types of vanda with direct sunlight requirements.

Therefore, the basic knowledge of Vanda orchid morphology diversity is important to know because it is related to breeding activities for conservation or agribusiness purposes. This research activity aims to find out information about the decline of natural Vanda orchid characters that are still very little reviewed and known by many people. Based on the results of this characterization will be obtained information about various superior properties in these plants which can be used as propagation materials and parent broods in plant breeding so that the combination of these properties will appear on the breeding results.

2. Methods

This research activity was carried out in May to November 2018 at the Plant Laboratory and Bioscience Laboratory - State Polytechnic of Jember, Jember, East Java, Indonesia at an altitude of 90 m above sea level. The tools used in this study include rulers, shovel runners, scissors, colling boxes, stationery, sprayers and cameras. The material used is 14 types of Vanda sp. from various regions in Indonesia, media, fertilizers, pesticides and vitamins. The experimental design in this study used a completely randomized design with treatment of fourteen Vanda sp orchids from various regions with three replications.

Variable qualitative observations of Vanda orchid morphology refer to the orchid characterization guide from the Ornamental Crops Research Institute [12] including general crop performance (leaf cross section), leaf performance (leaf shape; leaf tip shape; leaf edge shape; surface texture of leaves and leaf symmetry) (Table1). Analysis data use NTSYS programme Version 2.0.

3. Result and Discussion

Characterization is an activity in the conservation of germplasm to determine morphological properties that can be used to distinguish between accessions, assess the magnitude of genetic diversity, identify varieties, assess the number of accessions and so on. Orchid stems were observed based on the type of growth. There are 2 types of orchid growth, namely monopodial and simpodial. The monopodial stem is a stem that extends upwards and there is only one stem, while the sympodial stem is the stem towards the side and in one plant there are more than one stem. The type of growth in all orchids from this study (Vanda species) is monopodial [7]. In this study, an analysis based on the morphological characters of 14 Vanda orchid assays was carried out from several regions (**Table 1**). The Fourteen types of Vanda orchids include Vanda Feotida (**Figure 1**), Vanda Scandens (**Figure 2**), Vanda Tricolor Sumbing (**Figure 3**), Vanda Helvola (**Figure 4**), Vanda Tri Color Dieng (**Figure 5**), Vanda Tri Color Cianjur (**Figure 6**), Vanda Purva (**Figure 7**), Rhynchostylis Retusa (**Figure 8**), Vanda Devoogtii (**Figure 9**), Vanda Arculata (**Figure 10**), Vanda Arades (**Figure 11**), Vanda Limbata (**Figure 12**), Vanda Celebica (**Figure 13**), and Vanda Merbabu (**Figure 14**). The character is scoring and the results of scoring are then analyzed using NTSYST to get a similarity coefficient. After obtaining the similarity coefficient, a dendogram was arranged to determine the classification of the

IOP Conf. Series: Earth and Environmental Science **207** (2018) 012006 doi:10.1088/1755-1315/207/1/012006

orchid types. The dendogram from the grouping results is presented in **Figure 15**. The character grouping occurred probably due to the influence of environmental factors in which the vanda orchid was obtained, as stated by [13], according to the results of his research on wild orchids in Thailand.

Figure 1. Vanda Feotida

Figure 2. Vanda Scandens

Figure 3. Vanda Tricolor Gunung Sumbing

Figure 4. Vanda Helvola

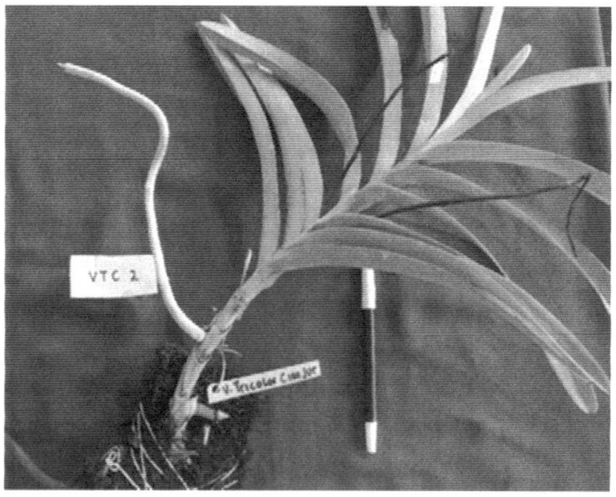

Figure 5. Vanda Tricolor Dieng, Central Java

Figure 6. Vanda Tricolor Cianjur, West Jawa

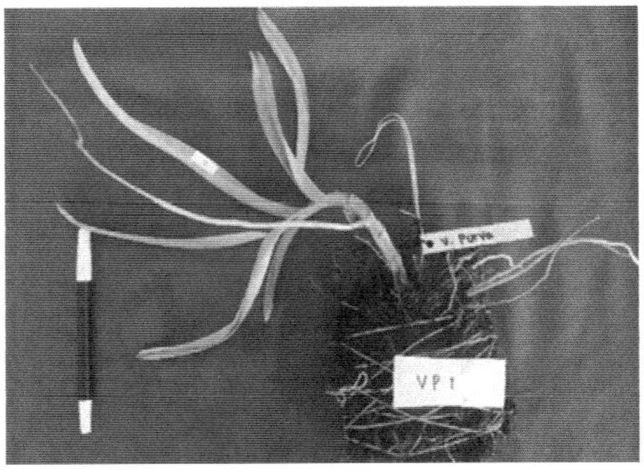

Figure 7. Vanda Purva

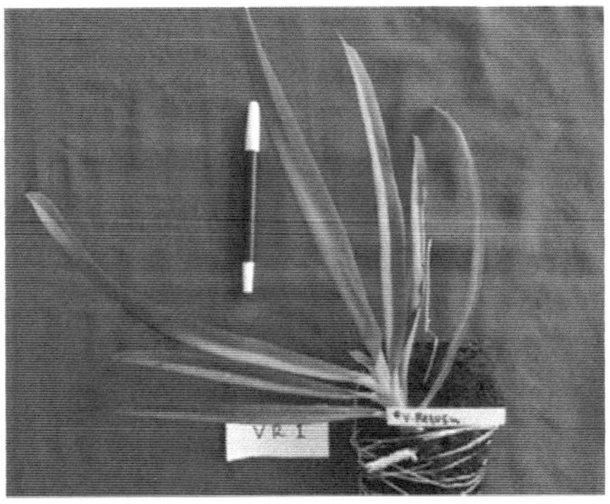

Figure 8. V. Rhynchostylis Retusa

Figure 9. Vanda Devogtii

Figure 10. Vanda Arculata

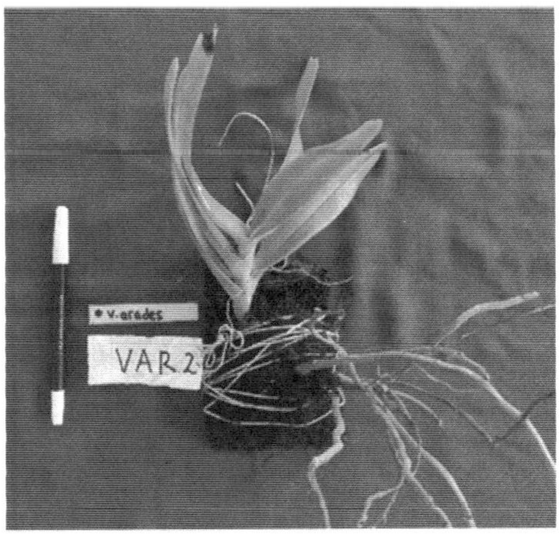

Figure 11. Vanda Arades

Figure 12. Vanda Limbata

IOP Conf. Series: Earth and Environmental Science **207** (2018) 012006 doi:10.1088/1755-1315/207/1/012006

Figure 13. Vanda Celebica

Figure 14. Vanda Tri Colour Merabu, Central Java

Table 1. Types and origin of collections from 14 accessions of Vanda sp.

No	Accession	Location of Accession Origin
1	Vanda Feotida	Pagar Alam, North Sumatra
2	Vanda Scandens	Kalimantan
3	Vanda Tricolor Gunung Sumbing	Gunung Sumbing, Central Java
4	Vanda Helvola	West Java
5	Vanda Tri Color Dieng	Dieng, Central Java
6	Vanda Tri Color Cianjur	Cianjur, West Jawa
7	Vanda Purva	East Java
8	Rhynchostylis Retusa	East Java
9	Vanda Devoogtii	East Java
10	Vanda Arculata	Sulawesi
11	Vanda Arades	East Java
12	Vanda Limbata	East Java
13	Vanda Celebica	Pasuruan, East Java
14	Vanda Gunung Merbabu	Gunung Merbabu, Central Java

In **Figure 15** shows that dendogram is divided into two main branches in the simple matching coefficient kooefesien between 0.33 - 1.00 that is group I at 0.67 and group II at 0.39 where the division of this group is based on 13 Vanda orchid morphology character identified (cross section of leaf, leaf shape, leaf composition, leaf edge, leaf surface texture, leaf symmetry, rooting type, leaf length, leaf width, leaf thickness and flower position). In detail the classification of Vanda orchids from the exploration of several areas can be seen in **Table 2.**

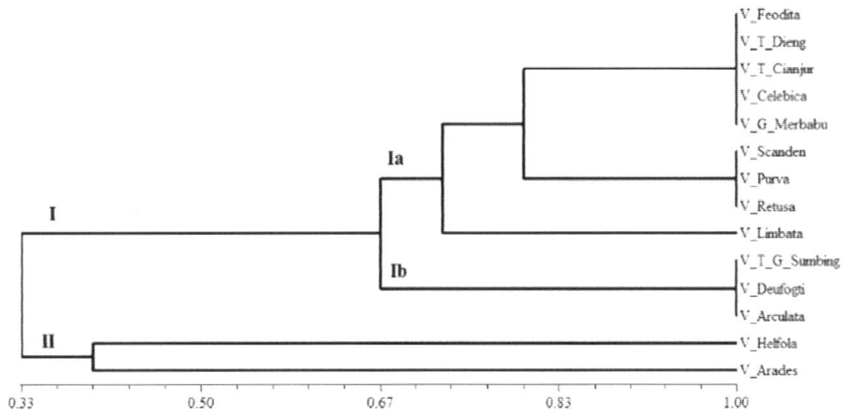

Figure 15. Dendrogram Similarity coefficients of 14 Vanda orchid accessions based on morphological character.
(information: V_Feodita = Vanda Feotida, V_Scanden = Vanda Scandens, V_T_G_Sumbing = Vanda Tricolor Gunung Sumbing, V_Helvola= Vanda Helvola, V_T_Dieng = Vanda Tri Color Dieng, V_T_Cianjur = Vanda Tri Color Cianjur, V_Purva = Vanda Purva, V_Retusa = Vanda Rhynchostylis Retusa, V_ Deufogti = Vanda Devoogtii, V_Arculata = Vanda Arculata, V_ Arades = Vanda Arades, V_ Limbata = Vanda Limbata, V_Celebica = Vanda Celebica, and V_Merbabu = Vanda Gunung Merbabu)

The separation of this group was due to very clear differences in character, Vanda helfola and Vanda arades were in group II. Explanation of differences in the cross-sectional characters of Vanda helfolia leaves with tridentate character, Vanda arades classified as bilaterally compressed while others (group I) are included in teret. In the character shape of Vanda helfola leaves classified as elliptic, Vanda arades is classified as oblong leaf shape, while Vanda group 1 is generally linear. In the Vanda helfola leaf tip character is tridentate character, Vanda arades has retuse character, while Vanda group 1 is generally emorginate. Furthermore, the length of Vanda helfola leaves and Vanda arades were relatively short (15-17 cm) compared to the length of Vanda orchid leaves belonging to group I (22 - 38 cm).

Table 2. Orchid accession groups based on dendrogram

Main group	Sub group	Accession
I	Ia	Vanda feotida, Vanda tricolor Dieng, Vanda tricolor Cianjur, Vanda celebica, Vanda Gunung Merbabu, Vanda scaden, Vanda purva, Rhynchostylis Retusa, Vanda limbata
	Ib	Vanda tricolor Gunung Sumbing, Vanda daufogti dan Vanda arculata
II		Vanda Helfola dan Vanda Arades

In group I it was divided into two sub-groups: sub-group 1a on the coefficient of diversity 0.73 and sub-group 1b on the coefficient of diversity 0. 67. Subgroup 1a consisted of three other types of groups, namely the first group (V. feodata, V. Dieng tricolor, V. tricolor Cianjur, V celebica, V Gunung Merbabu); Second group (V scanden, V purva, V retusa); and the third group is Vanda limbata. The character equation that unites groups is growth type, leaf cross section, flower position, leaf shape, leaf tip, leaf arrangement, leaf edge, leaf surface texture, leaf symmetry, rooting type, leaf length, leaf width, leaf thickness. Sub group I b consists of Vanda tricolor Sumbing mountain; Vanda daufogti and Vanda arculata. Similar to group 1a, the two types of Vanda from group I have unifying character similarities, namely growth type, leaf cross section, flower position, leaf shape, leaf tip, leaf arrangement, leaf edge, leaf surface texture, leaf symmetry, root type, leaf length, leaf width, leaf thickness.

4. Conclusion

Exploration results from 14 types of Vanda orchids have similarity values based on morphological characters that are quite high from 0.22 to 1.00. The lowest value is found in Vanda Helfola and Vanda Arades. The highest values were found in Vanda feotida, Vanda tricolor Dieng, Vanda tricolor Cianjur, Vanda celebica, Vanda Gunung Merbabu, Vanda scaden, Vanda purva, Rhynchostylis Retusa. Dendogram results of grouping based on similarity coefficients are divided into two main groups, the first group consists of Vanda groups which are in coefficients of 0.67 - 0.73 and the second group consists of Vanda groups at coefficients 0. 39. The first group is divided into two subgroups namely Vanda which grouped on diversity coefficients 0.73 and Vanda which grouped on the coefficient of diversity 0. 67.

The relationship between Vanda orchids based on the distribution area still needs to be expanded by using characters in addition to using qualitative characters, quantitative, anatomical, cytology to molecular level, so that more comprehensive data will be obtained.

Acknowledgment

The author would like to thank Ristekdikti for the BOPTN research fund in 2018 through the Research and Community Service Center of Jember State Polytechnic.

References

[1] Firgiyanto R, Aziz S A and Sukma D 2016 *J. Agron. Indones.*, **44** 204–210
[2] Krishardianto A and Sukma D 2017 *Bul. Agrohorti.* **5** 167–175
[3] Hartati S, Sumijati, Pardono and Cahyono O 2014 *J. Ilmu Ilmu Pertan.* **29** 31–34
[4] Hartati S, Cahyono O and Lestari N P 2015 *J. Sustain. Agric.* **32** 24
[5] Muharyati Y, Defiani M R, Putu N and Astiti A 2015 *J. Biol. Sci.* **71** 66–71
[6] Dwiyani I, Purwantoro A, Indrianto A and Semiarti E 2012 *J. Bumi Lestari*, **12** 93–98
[7] Purwantoro A, Ambarwati E and Setyaningsih F 2005 *J. Ilmu Pertan.* **12** 1–11
[8] Huang Y, Li F and Chen K 2010 *Biochem. Syst. Ecol.* **38** 93–102
[9] Romeida A, Sutjahjo S H, Purwito A and Sukma D 2012 *J. Agron. Indones.* **40** 218–224
[10] Winarto B, Dewanti M and Pramanik D 2013 *J. Hort.* **23** 114–128
[11] Nurmalinda S, Kartikaningrum N, Hayati Q and Dyah W 2011 *J. Hort.* **21** 372–384.
[12] Balai Penelitian Tanaman Hias 2007 *Panduan karakterisasi tanaman anggrek* (Jakarta Indonesia: Departemen Pertanian) pp 36
[13] Anuttato S, Boonruangrod R, Kongsamai B and Chanprame S 2017 *J. ISSAAS* **23** 20-32

Mathematical Model of the Water Quality in Kalibaru Watershed

B Hariono[1*], R Wijaya[1], M F Kurnianto[1], Sugiyarto[2], M J Wibowo[3], S Anwar[3]
[1] Department of Agricultural Technology, Politeknik Negeri Jember
[2] Departement of Agricutural Production, Politeknik Negeri Jember
[3] Engineering Departement, Politeknik Negeri Jember

*budihariono1966@gmail.com

Abstract. The Kalibaru Watershed is one of the watersheds which needs to be concerned because it is suspected to have decreased water quality, due to the high activity of the households, industry, livestock, and agriculture. This condition implies that the water quality as the embodiment of the environmental capacity of Kalibaru Watershed needs to be studied, so that it can be known how far it benefits the lives of the society.The aim of the research were to determine the water quality related to physical properties (pH), chemical properties (BOD, PO_4P, NO_2N) and microbiological properties (*Fecal coli*) from observation stations upstream, central and downstream of Kalibaru River watershed. The research approach framework was carried out by: (1) data collection to describe water quality problems in the Kalibaru watershed from the Sampean Baru River Area Water Resources Management Center in Bondowoso. The data such as the results of laboratory analysis of physical, chemical and microbiological properties in 3 (three) observation points of the Kalibaru watershed. The results of the analysis were processed using the regression method became a model for decrease in water quality as seen from the 3 parameters above and compared with the Republic of Indonesia Government Regulation No. 82 of 2001. Data analysis of the above parameters from January to December 2017 with the Polynomial equation of order 6. Research parameters include pH, BOD, PO4N, NO2N with MAXR method. The parameters of pH observation stasion in upstream, middle and downstream get result R Max respectively are 0,76; 0,60; 0,57. BOD parameter values respectively are 0,59: 0,42: 0,18. PO4P paramater values respectively are 0,52; 0,47; 0,48. NO2N paramater values respectively are 0,66; 0,71; 0,77. Fecal Coli parameter values respectively are 0,57; 0,85; 0,87.

1. Introduction

Some researches related to the mathematical model of changes in river water quality had been conducted, one of them was done [1] in the Cisadene watershed or Daerah Aliran Sungai (DAS) with Biochemical Oxygen Demand (BOD), Chemical Oxygen Demand (COD), and fecal coli parameters with the results in the form of changes in the quality level of BOD, COD, and Fecal coli that are caused by the population growth, so that it has implication for increasing industrial waste and domestic waste.

The condition of changes in water quality in the Cisadane watershed also occurs in the Kalibaru watershed, where the economic activities such as industrial, agricultural, plantation and fisheries along the Kalibaru watershed, from upstream, midstream and downstream are quite intensive with high population growth. This condition implies the entry of pollutants into the river body. Therefore, a mathematical model of changes in river water quality is very important to be carried out as an effort to predict the quality of water that can be used by the community along the watershed in accordance with its function.

The observation stations were done in upstream, midstream and downstream respectively in Kalibaru Manis Village of Kalibaru District, Karangharjo Village of Glenmore District and Karangdoro Village of Gambiran District, Banyuwangi Regency. The parameters observed were physical parameter (pH), chemical parameter (BOD, PO_4P, NO_2N) and microbiological parameter (Fecal coli), so that a mathematical model was obtained to predict a more comprehensive decrease in water quality.

2. Literature Review

Literature review that will be used for this research are

2.1. Water Pollution and Pollution Sources

Water pollution is the entry/inclusion of living things, energy and or other components into the river water from industrial activities, agriculture, plantations and fisheries which cause the quality of the river water decreases to a certain level so that the water cannot meet its intended function [2], where the pollution sources into the river water can be divided into domestic source and non-domestic source [3].

2.2. Criteria, Status, and Quality Standard of WaterStatus Air

Water quality is the desired concentration limit so that water has a quality that is worth to use in accordance with its function. Water quality standard is set by the government by including restrictions on the concentration of various water quality classifications according to their functions [4]. Criteria for water quality based on their class [5], are: (1) First class, intended as raw water for drinking water, and or other uses with water quality requirements that are the same as those uses; (2) Second class, is intended as a water recreation infrastructure/facility, freshwater fish cultivation, livestock, water for agriculture, and or other uses with the same water quality requirements as those uses; (3) Third class, intended as the cultivation of freshwater fish, livestock, agricultural water, and or other uses with the same water requirements as those uses; (4) Fourth class, intended as agricultural water, and or other uses with the same water quality requirements as those uses.

2.3. Regression Analysis

2.3.1. Definition of Regression

Regression is a correlation of two or more variables of observation values which can be interpreted in two forms [6] that are: (1) The position of the average population from the value of a variable, (2). If there is limited data available, regression is an adjustment of the function of the data.

2.3.2. Function of Regression

The function of the regression equation is as follows: (1) Description of the data, which is when the regression equation is still in the stage of data searching and benchmarking, (2) cause and effect relationship, (3) controlled experiment if there are factors that are difficult to control but can be expected to affect factor Y so that it functions as a regression analysis that is used as a comparative investigator and (4) Preparation of models and patterns of relationships of many variables X1, X2, X3, ..., Xk with variable Y, regression to find the most appropriate relationship or model with only involves a few of the variables X1, X2, X3, ..., Xk.

2.4. Regression Analysis Model

Regression analysis model for decrease in water quality in the Kalibaru watershed is explained by [7], including:

2.4.1. Linear

Is a straight line equation that can be written in the form: $Y = a + bX$.

2.4.2. Polynomial
The general form of a polynomial equation is: Y = anXn + an-1Xn-1 + ... + a1X + a0
Where: an, an-1, ..., a1, a0 are polynomial constants/coefficients; n is a non-negative integer; x is an independent variable that can be used to predict; and y is the dependent variable.

2.4.3. Exponential
Exponential function is formed as: f (X) = ax ...

Where: x is an independent variable that can be used to predict; a is a positive constant; and y is the dependent variable.

There are three types of exponential functions y = ax. If 0 <a <1, the exponential function goes down; if a = 1, the function is constant; and if a> 1, the function rises.

2.4.4. Logarithm
If a> 0 and a ≠ 1, the exponential function f (x) = ax is a function of decrease or increase, and therefore one-to-one. So it has an inverse function f-1, which is called a function of logarithms with principal number a and denoted by log, if we use the inverse function form:

log aX = y, ay = x

Where: x is an independent variable that can be used to predict; a is a positive constant; y is the dependent variable.

2.5. Adjusted R^2
Regression equation that has been obtained, can be used to estimate the equations generated by the data, and can be continued by assessing the good or bad suitability of the model with the data by evaluating the method of Relation Coefficient, Largest R^2. R^2 as the correlation coefficient or determinant coefficient (determination). The closer R^2 to number 1, the better the data match to the model [8].

3. Working Methodology
3.1. Research Location
This research was conducted in the Kalibaru Watershed with observation points in upstream, midstream and downstream respectively located in Kalibaru Manis Village of Kalibaru District, Karangharjo Village of Glenmore District and Karangdoro Village of Gambiran District, Banyuwangi Regency.

3.2. Tool and Material
The research tool used was Excel 2010 software. The materials used were: (1). physical parameter (pH), chemical parameter (BOD, PO4P, NO2N) and microbiological parameter (Fecal coli). Calculated Analytical Hierarchy Process (AHP).

3.3. Research Method
3.3.1. Research Approach Framework
The research approach framework was carried out by: (1) data collection to describe water quality problems in the Kalibaru watershed from the Sampean Baru River Area Water Resources Management Center in Bondowoso. The data such as the results of laboratory analysis of physical, chemical and microbiological properties in 3 (three) observation points of the Kalibaru watershed. The results of the analysis were processed using the regression method became a model for decrease in water quality as seen from the 3 parameters above and compared with the Republic of Indonesia Government Regulation No. 82 of 2001. The data from this research came from secondary water quality data in the Kalibaru Watershed from the Water Resources Management Unit of Bondowoso which measured the water quality [9].

3.3.2. Data Analysis Method

1. Water Quality Analysis

The water quality data of the Kalibaru watershed were analysis of physical, chemical and microbiological properties according to Government Regulation of Republic of Indonesia No. 82 of 2001 as shown in Table 1 [10].

2. Regression Analysis

The method used in analyzing the data of water quality in Kalibaru watershed was the 6th Polynomial Model regression analysis method; it was based on the MAXR concept of the 6th polynomial model that met the requirements. Regression analysis method was able to estimate the value of water quality in Kalibaru watershed at the observation station, although there was no measurement made at the station. Regression analysis by using Microsoft Office Excel 2010 and the produced regression analysis model could be analyzed in details [11].

Table 3.1 The Variable of Water Quality, Analysis and the Tools used.

No	Parameters	Unit	Quality I	Quality II	Quality III	Quality IV	Analytical Model
1	pH	-	6 s/d 9	5 s/d 9	5 s/d 9	5 s/d 9	potentiometer
2	BOD	mg/L	2	3	6	12	open reflux
5	PO4-P	mg/L	0.2	0.2	1	5	spectrometry
6	NO3-N	mg/L	10	10	20	20	spectrometry
Microbiology							
1	TotalColi	total/100ml	5000	5000	5000	5000	MPN method
2	Fecal Coliform		5000	5000	5000	5000	MPN method

4. Experiment and Result

4.1. Physical Characteristic

The result of monitoring the upstream, the midstream and the downstream respectively at observation stations was located in Kalibaru Manis Village of Kalibaru District, Karangharjo Village of Glenmore District and Karangdoro Village of Gambiran District, Banyuwangi Regency. It showed that the average pH data of 6.85; 6.95 and 6.85 were listed in Graph 1; 2 and 3. The graph changes in pH at 3 stations were drawn in Figure 4 and the average value of both observation and prediction was available on Figure 5.

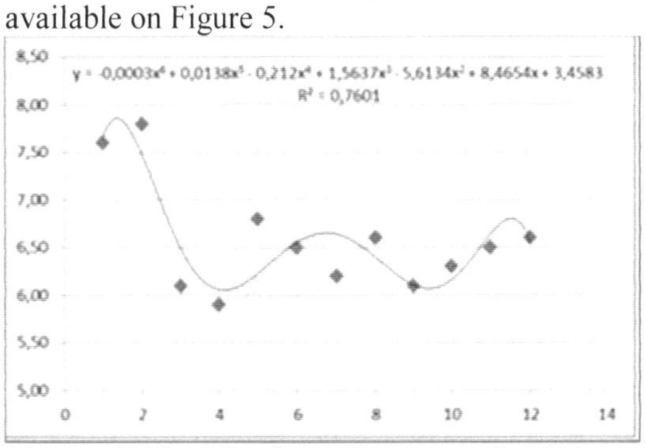

Figure 1. The Distribution Pattern of PH in the Period of January - December 2017 at the upstream observation station.

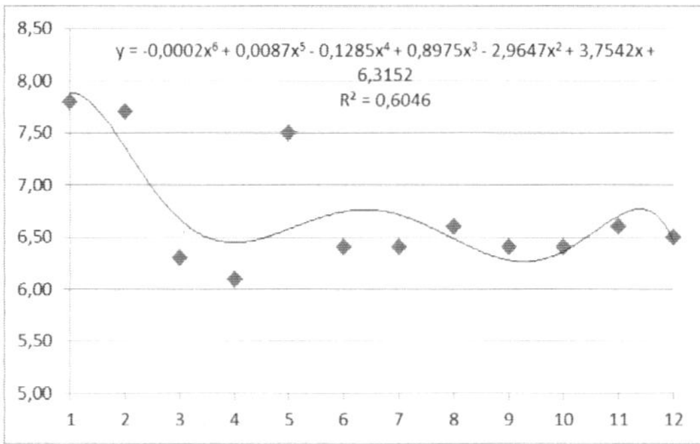

Figure 2. The Distribution Pattern of PH the Period of January - December 2017 at the midstream observation station.

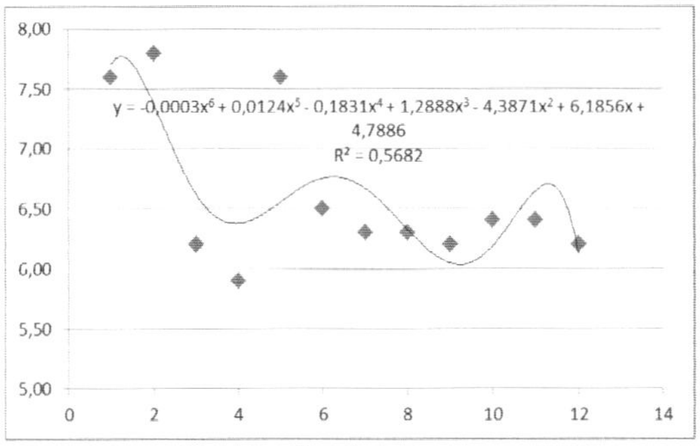

Figure 3. The Distribution Pattern of PH January - December 2017 Periodat the downstream observation.

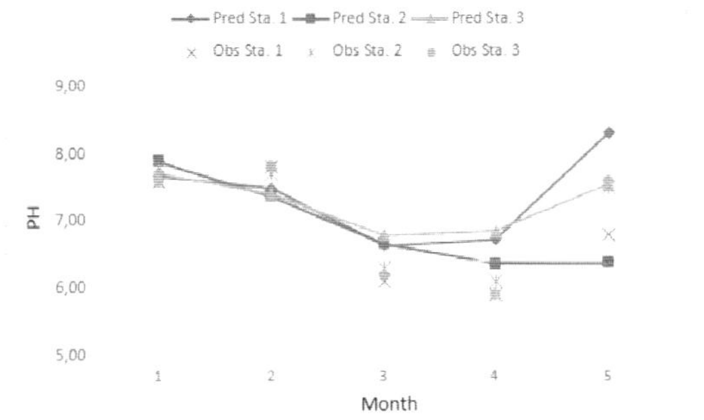

Figure 4. The Distribution Pattern of PH Observation and Prediction in the period of January - December 2017.

If it was compared to the data from Government Regulation of The Republic of Indonesia No. 82/2001, the pH value was still in the acceptable range, so that water quality I, II, III and IV reached 0 as their value. This means that the pH did not give any contributions to the contamination of Kalibaru watershed.

The mathematical model of pH value was based on the 6th polynomial equation, which were $y = -0{,}00030\ x^6 + 0{,}014\ x^5 - 0{,}21x^4 + 1{,}56\ x^3 - 5{,}61x^2 + 8{,}46\ x + 3{,}46$ with $R^2 = 0{,}76$; $y = -0{,}00020\ x^6 + 0{,}0087\ x^5 - 0{,}13\ x^4 + 0{,}90\ x^3 - 2{,}96\ x^2 + 3{,}75\ x + 6{,}32$ with $R^2 = 0{,}60$; $y = -0{,}00030\ x^6 + 0{,}012\ x^5 - 0{,}18\ x^4 + 1{,}29\ x^3 - 4{,}39\ x^2 + 6{,}19\ x + 4{,}79$ with $R^2 = 0{,}57$.

The changes of pH value in the upstream, the midstream and the downstream stations in accordance to MAXR method were 0.76; 0.60 and 0.57. This showed that Kalibaru watershed had no significant change in pH and was still in required limitation.

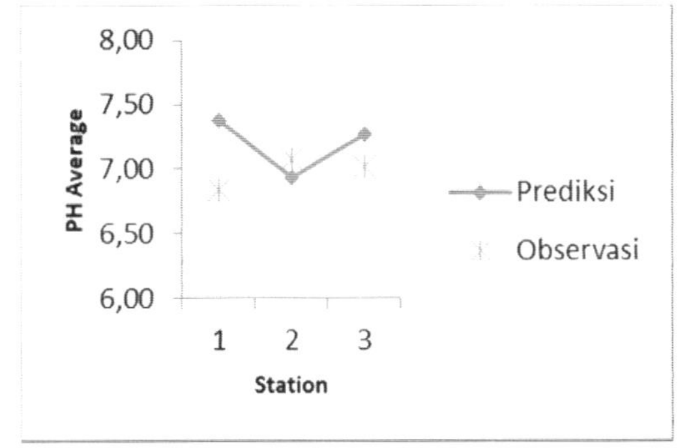

Figure 5. The Average Value of PH observation and prediction in the period of January - December 2017.

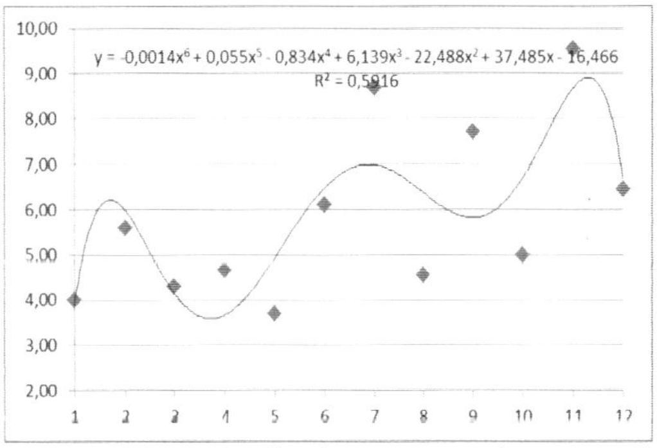

Figure 6. The Distribution Pattern of BOD mg/L in the Period of January - December 2017 at the upstream observation station.

The prediction result by using mathematical model in which PH value was obtained from the upstream (station 1), the midstream (station 2) and the downstream (station 3) were 7.37; 6.93 and 7.27. As they were compared to the average of PH value, the observation results which had consecutive values were 6.85; 6.95 and 6.85. Thus, it was clear that the average value of PH prediction was not as different as the average value of PH observation. The difference of the average value of PH prediction and observation obtained from the upstream, the midstream and the downstream were 0.53; 0.15 and 0.25. This mathematical model was capable to determine the average value of PH prediction since the difference between prediction and observation was less than 10% which was the limit/threshold value of mathematical model's feasibility to be used as a prediction.

4.2. Chemical Characteristic

4.2.1. BOD (mg/L)

The chemical characteristic that was used as the parameters covered: (1) BOD, (2) PO4P, and (3) NO2N. BOD contamination in the upstream, the midstream and the downstream in the observation stations showed the average value of 6.20 mg/L; 7.23 mg/L; 6,50 mg/L, as the requirements of the water quality III and IV were fulfilled according to the Government Regulation of The Republic of Indonesia No. 82/2001. BOD value in 3 observation stations was listed in the following Figure 6, 7 and 8. These data showed that BOD value from the upstream to midstream had an increase, whereas from the midstream to the downstream had a decrease. The increased BOD value was due to the existence of pollutant entering the river flow, meanwhile the decreased BOD value was caused by a decrease of pollutant in river flow.

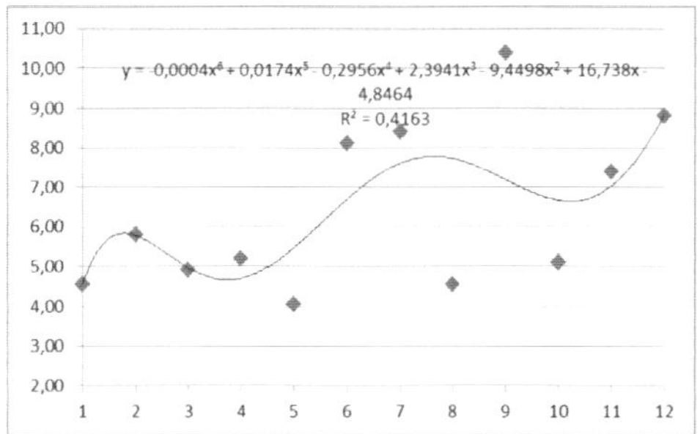

Figure 7. The Distribution Pattern of BOD mg/L in the Period of January - December 2017 at the midstream observation station.

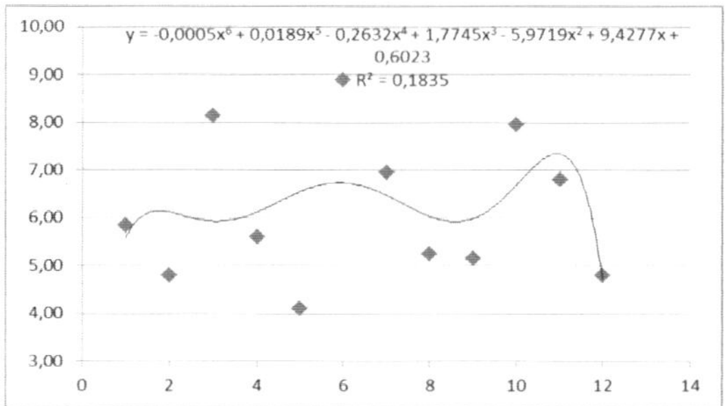

Figure 8. The Distribution Pattern of BOD mg/L in January - December 2017 Period at the downstream observation station.

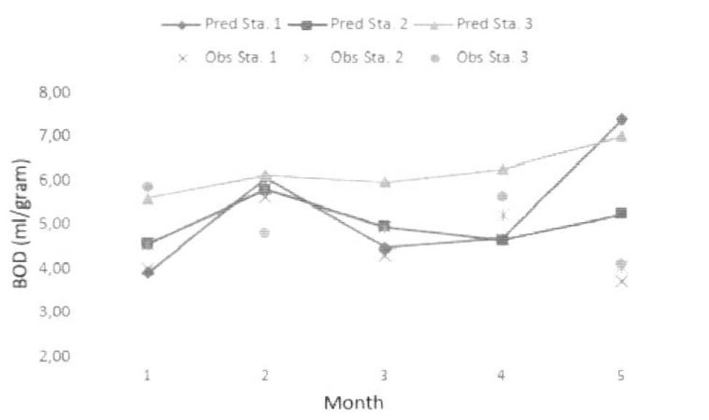

Figure 9. The Distribution Pattern of BOD Observation and Prediction from the upstream to the downstream.

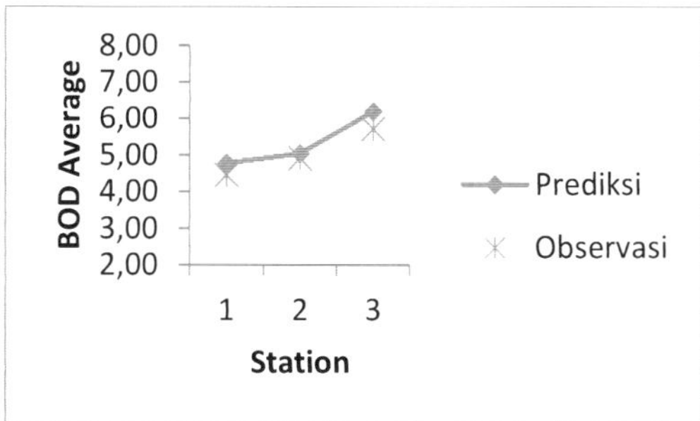

Figure 10. The Average Value of BOD Observation and Prediction from the Upstream to The downstream.

The mathematical model of BOD value adapted the 6th polynomial equation, $y = -0.0014 x^6 + 0.055 x^5 - 0.83 x^4 + 6.14 x^3 - 22.49 x^2 + 37.49 x - 16.47$ with $R^2 = 0.59$; $y = -0,00040 x^6 + 0,017 x^5 - 0,30 x^4 + 2,39 x^3 - 9,45 x^2 + 16,74 x - 4,85$ with $R^2 = 0,42$; $y = -0,00050 x^6 + 0,019 x^5 - 0,26 x^4 + 1,77 x^3 - 5,97 x^2 + 9,43 x + 0,60$ with $R^2 = 0,18$. The BOD value from the result of calculation by using mathematical model can be seen in Figure 9 and the average distribution value of BOD can be seen in Figure 10.

The changes in BOD value from the upstream, the midstream and the downstream stations based on MAXR method were 0.59; 0.42 and 0.18. The prediction result by using the obtained mathematical model, the average BOD value from the upstream (station 1), the midstream (station 2) and the downstream (station 3) were as much as 4.77; 5.03 and 6.19. As it was compared to the average BOD value of the observation result which reached the value of 6.20 mg/L; 7.23 mg/L; 6.50 mg/L, it could be seen that BOD average value of the prediction did not differ from the BOD average value of observation. The difference of BOD average value of both prediction and observation obtained in the upstream, the midstream and the downstream were 1.43; 2.2 and 0.31. This mathematical model was capable to be used as the determination of the average BOD value of prediction in the downstream station since the difference between the prediction and observation was less than 10% which was the limit/threshold value of mathematical model's feasibility to be used as a prediction.

4.2.2. PO_4P (mg/L)

The mean values of PO_4P (mg/L) in the upstream, the midstream and the downstream observation stations were 0.24 mg/L; 0,25 mg/L; 0.21 mg/L. The graph of PO_4P pollution reduction level was shown in Figure 11; 12 and 13. When compared to the water quality I along with the requirement of PO_4P content <0.2 mg/L, Kalibaru watershed could be used as the raw water quality I, II, III and IV. Phosphate values were constantly changing as it was because of the various pollution received by water and its coverage area. Phosphate contamination was caused by the pollution due to the anthropogenic activity, industry and livestock. The use of detergents, shampoo and soap from anthropogenic activities and industrial waste which were not neutralized caused foaming water and decreased the oxygen absorption. Based on the observations in the field, many people used the river for public bathing, washing facilities and latrines (MCK).

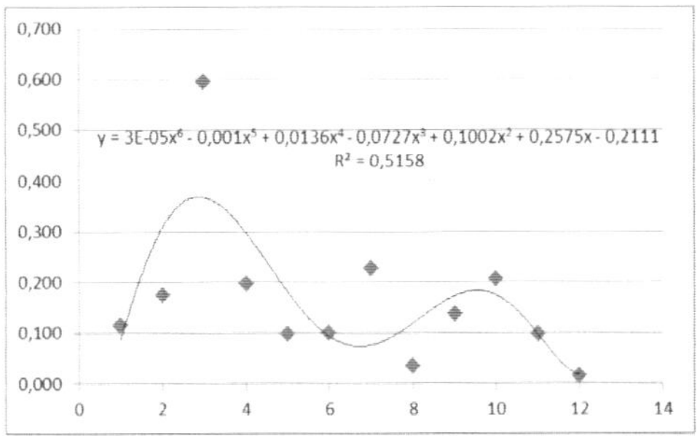

Figure 11. The Distribution Pattern of PO_4P mg /L in January - December 2017 period at the upstream observation station.

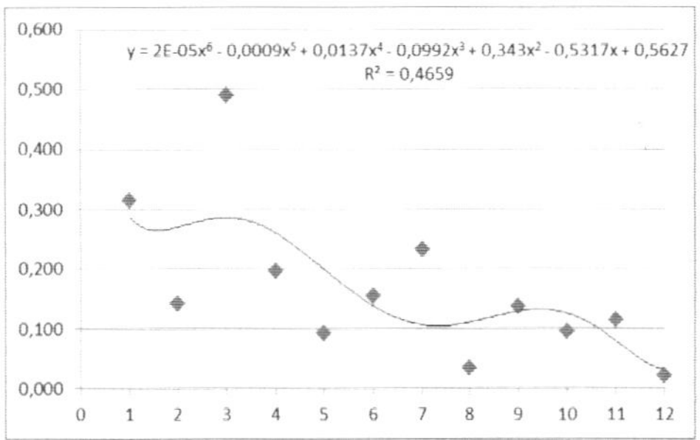

Figure 12. The Distribution Pattern of PO_4P mg/L in January - December 2017 period at the middle observation station.

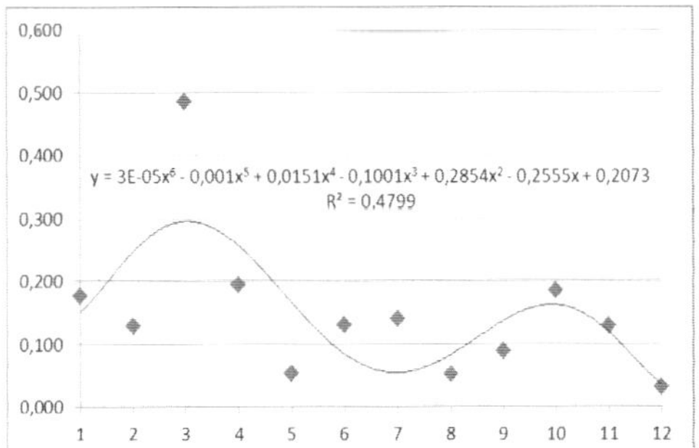

Figure 13. The Distribution Pattern of PO_4P mg/L in January - December 2017 Period at the downstream observation station.

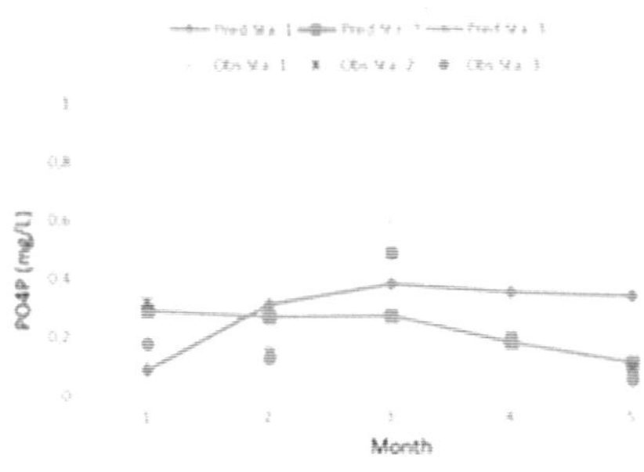

Figure 14. The Distribution Pattern of PO_4P Observation and Prediction from the upstream to the downstream.

Mathematical model of PO_4P pollution reduction in the upstream, the midstream and the downstream observation stations were $y = 3E\text{-}05\ x^6 - 0{,}0010\ x^5 + 0{,}014\ x^4 - 0{,}073\ x^3 + 0{,}10\ x^2 + 0{,}26\ x - 0{,}21$ with $R^2 = 0{,}52$; $y = 2E\text{-}05\ x^6 - 0{,}00090\ x^5 + 0{,}014\ x^4 - 0{,}099\ x^3 + 0{,}34\ x^2 - 0{,}53\ x + 0{,}56$ with $R^2 = 0{,}47$; $y = 3E\text{-}05\ x^6 - 0{,}0010\ x^5 + 0{,}015\ x^4 - 0{,}10\ x^3 + 0{,}29\ x^2 - 0{,}26\ x + 0{,}21$ with $R^2 = 0{,}48$. The value of PO_4P from the result of calculation by using mathematical model can be seen in Figure 14 and the average distribution value of PO_4P can be seen in Figure 15.

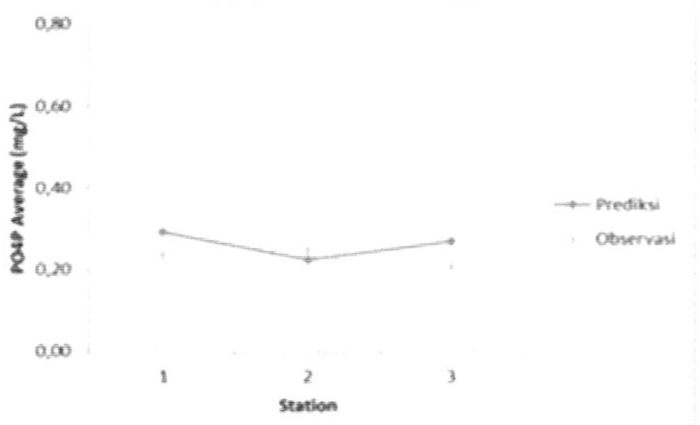

Figure 15. The Average value of PO_4P Observation and Prediction from the upstream to the downstream.

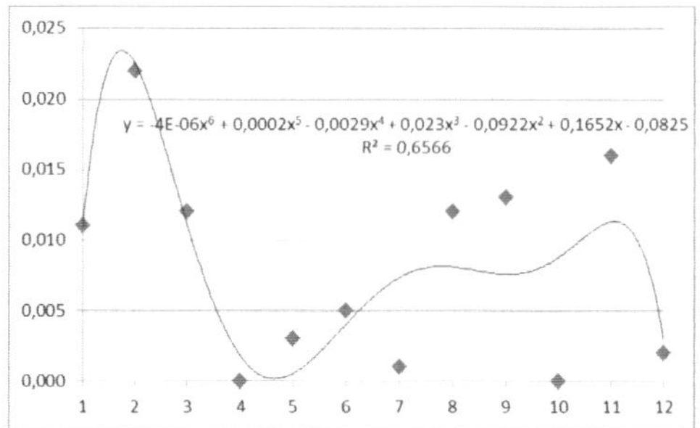

Figure 16. The pattern of distribution of NO_2N mg/L for the period of January - December 2017 at the upstream observation station.

The changes in PO_4P values from upstream, midstream and downstream stations based on the MAXR method were obtained 0.52; 0.47 and 0.48. The prediction results using mathematical model had been obtained the PO_4P average values from upstream (station 1), midstream (station 2) and downstream (station 3) respectively 0.29; 0.22 and 0.27. When compared to the PO_4P average value of the observation which had values of 0.24; 0.25 and 0.21, it can be seen that the average value of PO_4P prediction was not much different from the PO_4P average value of observation. The difference from the PO_4P average value of the prediction and observation obtained from upstream, midstream and downstream were 0.06; 0.02 and 0.06. This mathematical model could be seen in determining the prediction of average value of PO_4P because the difference between prediction and observation was less than 10% which was the threshold value of the mathematical model feasibility to be used as the determination of predictive values.

4.2.3. NO_2N (mg/L)

The average value of NO_2N (mg/L) pollution in upstream, midstream and downstream observation stations were 0.01 mg/L; 0.01 mg/L and 0.02 mg/L with a graph of NO_2N pollution reduction (mg/L) shown in Figure 16; 17 and 18. This value was still under the second quality water quality threshold required by the East Java Provincial Regulation No. 2 of 2008 of 0.06 mg/L. Sources of nitrite came from industrial and domestic waste. Natural water contained nitrite 0.001 mg/lt and should not exceed 0.06 mg/L.

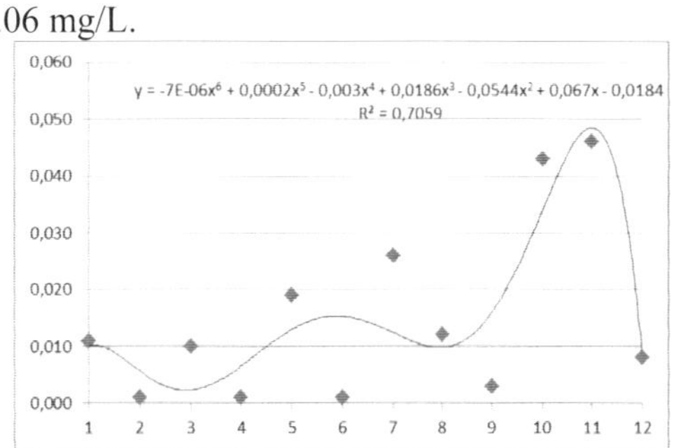

Figure 17. The distribution pattern of NO_2N mg/L for the period of January - December 2017 at the midstream observation station.

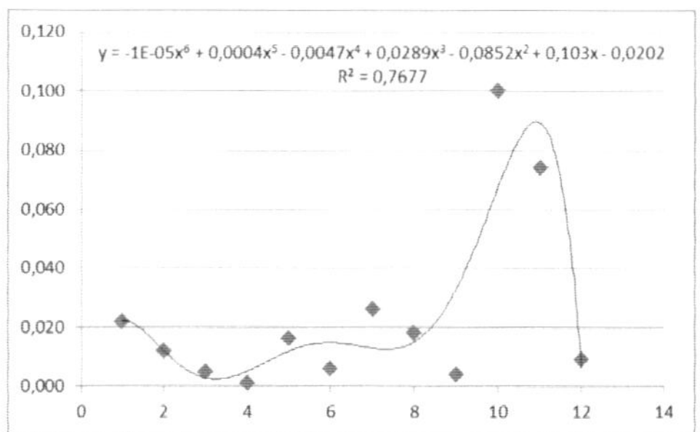

Figure 18. The distribution pattern of NO_2N mg/L for the period of January - December 2017 at the downstream observation station.

The mathematical models for NO_2N pollution reduction in upstream, midstream and downstream observation stations were $y = -4E\text{-}06\ x^6 + 0{,}00020\ x^5 - 0{,}0029\ x^4 + 0{,}023\ x^3 - 0{,}092\ x^2 + 0{,}17\ x - 0{,}083$ with $R^2 = 0{,}66$; $y = -7E\text{-}06\ x^6 + 0{,}00020\ x^5 - 0{,}0030\ x^4 + 0{,}019\ x^3 - 0{,}054\ x^2 + 0{,}067\ x - 0{,}018$ with $R^2 = 0{,}71$; $y = -1E\text{-}05\ x^6 + 0{,}00040\ x^5 - 0{,}0047\ x^4 + 0{,}029\ x^3 - 0{,}085\ x^2 + 0{,}10\ x\text{-}0{,}020$ with $R^2 = 0{,}77$. The NO_2N value from the calculation results using a mathematical model can be seen in Figure 19 and the average distribution value of NO_2N can be seen in Figure 20.

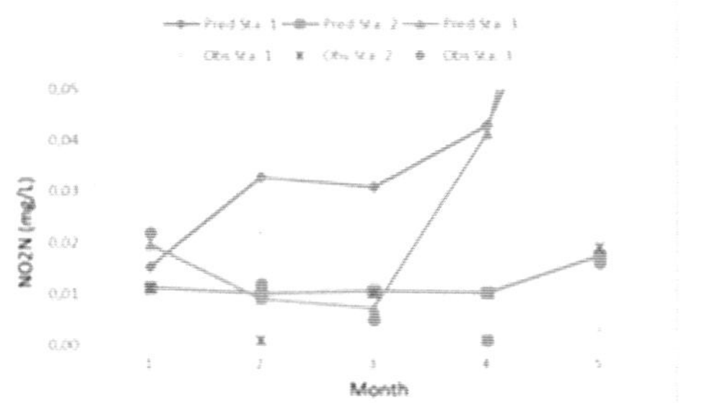

Figure 19. The pattern of NO_2N Observation and Prediction spread from upstream to downstream.

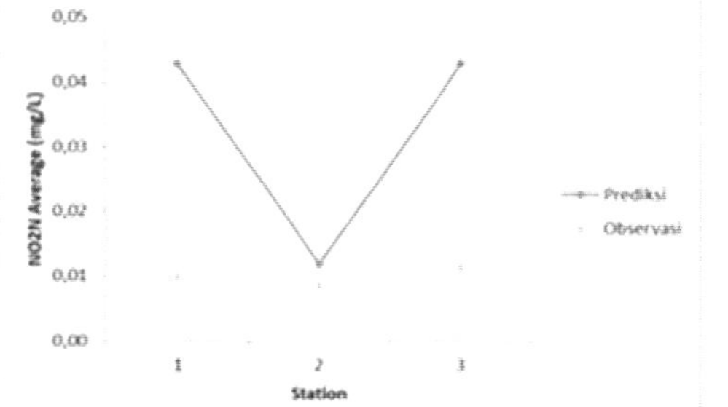

Figure 20. The average value of NO_2N Observation and Prediction from upstream to downstream.

Changes in NO_2N values from upstream, midstream and downstream stations based on the MAXR method were obtained 0.66; 0.71 and 0.77. The results of the prediction using mathematical models had been obtained the average NO_2N values from upstream (station 1), midstream (station 2) and downstream (station 3) respectively 0.04; 0.01 and 0.04. When compared to the average NO_2N value of the observation which had values of 0.01; 0.01 and 0.02 it could be seen if the average NO_2N value of the prediction was not much different from the NO_2N average value of observation. The difference between the NO_2N value of the average prediction and observation obtained from upstream, midstream and downstream were 0.03; 0 and 0.03. This mathematical model could be seen in determining the NO_2N average value of the prediction because the difference between prediction and observation was less than 10% which was the threshold value of the mathematical model feasibility to be used as a prediction value.

4.3. Microbiological Characteristic

4.3.1. Fecal coli (population/100 mL)

The average value of Fecal coli (population/100 mL) in upstream, midstream and downstream observation stations respectively were 29.8 population/100 mL; 22.4 population/100 mL and 18.8 population/100 mL. . The total amount of coliform was still under the criteria for class II river water quality of 5,000 mg/L. Fecal coli pollution graph is shown in Figure 21; 22 and 23.

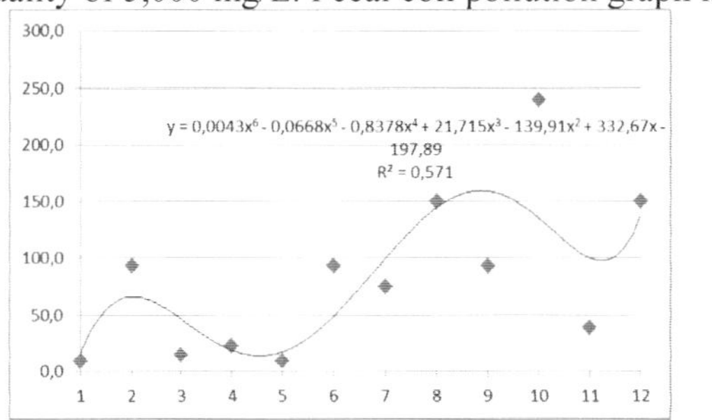

Figure 21. The pattern of the spread of Fecal coli for the period of January - December 2017 at the upstream observation station.

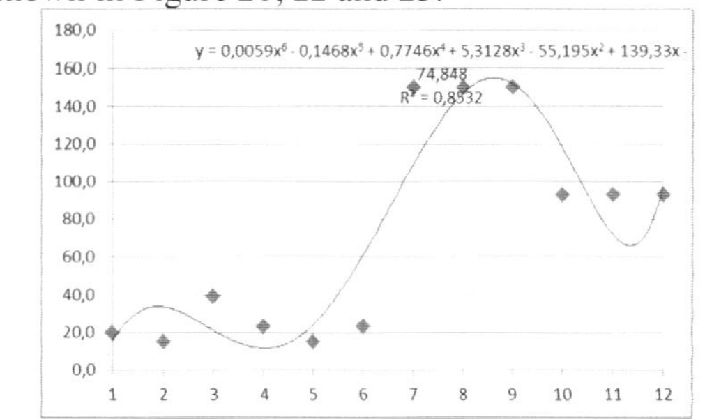

Figure 22. The pattern of distribution of Fecal coli for the period of January - December 2017 at the middle observation station.

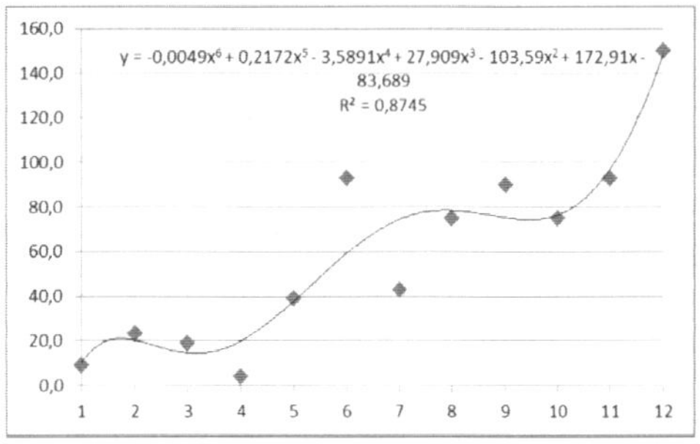

Figure 23. The pattern of distribution of Fecal coli for the period of January - December 2017 at the middle observation station.

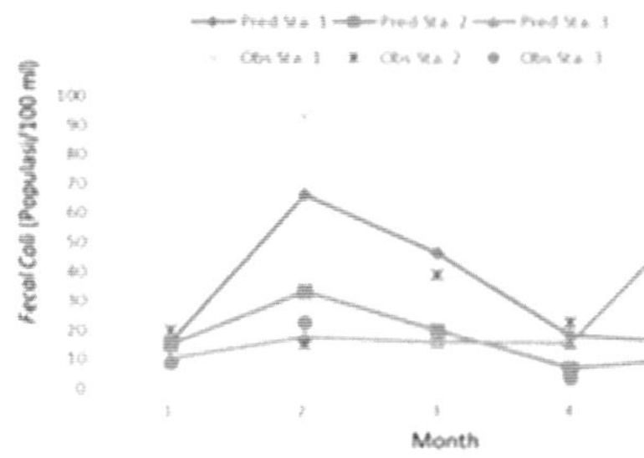

Figure 24. The pattern of spread of Fecal coli Observation and Prediction from upstream to downstream.

The mathematical model of Fecal coli decrease in upstream, midstream and downstream stations respectively were $y = 0.0043\ x^6 - 0.067\ x^5 - 0.84\ x^4 + 21.72\ x^3 - 139{,}91x^2 + 332{,}67\ x - 197{,}89$ with $R^2 = 0{,}57$; $y = 0.0059\ x^6 - 0.15\ x^5 + 0{,}77\ x^4 + 5{,}31\ x^3 - 55{,}19\ x^2 + 139{,}33\ x - 74{,}85$ with $R^2 = 0{,}85$; $y = -0.0049\ x^6 + 0.22\ x^5 - 3.59\ x^4 + 27.91\ x^3 - 103{,}59\ x^2 + 172{,}91\ x - 83{,}69$ with $R^2 = 0{,}87$. Fecal coli values from the calculation using mathematical model can be seen in Figure 24 and the average distribution of Fecal coli can be seen in Figure 25.

The changes in the Fecal coli value from upstream, midstream and downstream stations based on the MAXR method were 0.57; 0.85 and 0.87. The prediction results using the mathematical model had been obtained the Fecal coli average values from upstream (station 1), midstream (station 2) and downstream (station 3) respectively 32.29; 17,22 and 24,34. When compared to the average of Fecal coli, the results of observation which had a consecutive values of 29.8; 22.4 and 18.8, it could be seen that the Fecal coli average value of prediction was not much different from the Fecal coli average value of observation. The difference between the Fecal coli average value of the prediction and observation obtained from upstream, midstream and downstream were 2.59; 5.17 and 5.54. This mathematical model could be seen to be used in determining the Fecal coli average value of the predicted prediction at the upstream station that had difference in value between the smallest prediction and observation.

4.3.2 Total Coliform (population/100 mL)

Water quality can be determained from total coliform used the bacteria indicator. Coliform was founded a lot in warm-blooded animal's feces but it could also find in the water, soil, and vegetation environments. The measurement result showed that the coliform concentration in Kalibaru watershed was about 25-240 population /100ml. The distribution pattern of coliform value could be seen in Figure 17 and the coliform prediction and observation values could be seen in Figure 18. The value of total coliform was in water quality threshold.

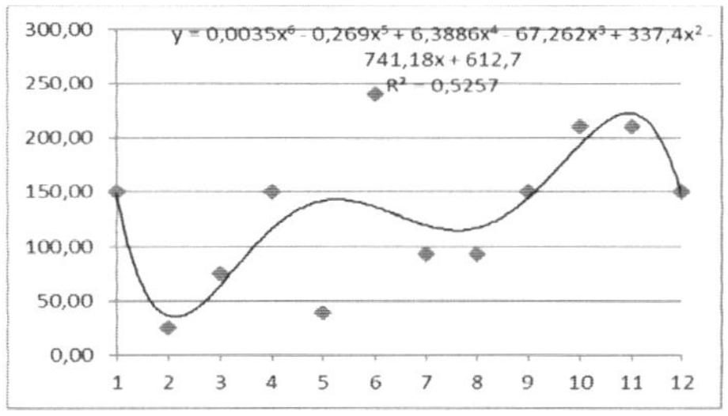

Figure 25. Coliform Distribution of Kalibaru watershed in January – December 2017 Period.

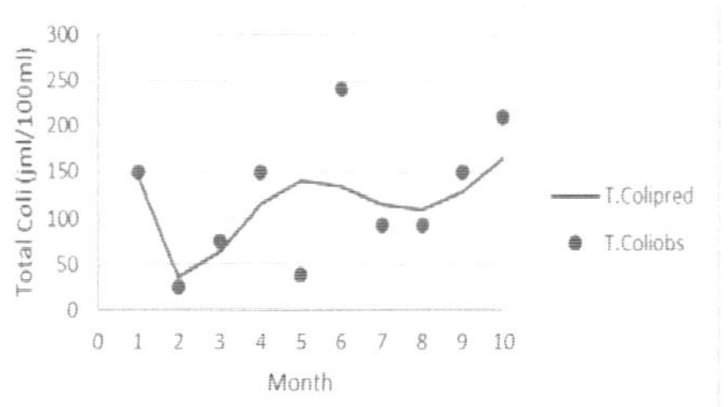

Figure 26. The Coliform Observation and Prediction in January December 2017 Period.

The change total coliform based on the MAXR method was 0.5257 and the gotten mathematical model was y = 0,0035x6 - 0,269x5 + 6,3886x4 - 67,262x3 + 337,4x2 - 741,18x + 612,7. The average total coliform prediction value was 116.3 population/100ml and the total coliform observation value was 113.17 population/100ml

5. Conclusion

From the research results on mathematical model to obtain the prediction value of the water quality of the Kalibaru watershed, it was concluded: The parameters of pH observation stasion in upstream, middle and downstream get result R Max respectively are 0,76; 0,60; 0,57. BOD parameter values respectively are 0,59: 0,42: 0,18. PO4P paramater values respectively are 0,52; 0,47; 0,48. NO2N paramater values respectively are 0,66; 0,71; 0,77. Fecal Coli parameter values respectively are 0,57; 0,85; 0,87.

Acknowledgments

Our thank goes to Department of Agricultural Engineering, Politeknik Negeri Jember, who has helped support for this research.

References

[1] Vollmer D, Costa D, Lin ES, Ninsalam Y, Shaad K, Prescott MF, et al 2015 JAWRA *Journal of the American Water Resources Association.*;**51(3)**:672-88.
[2] Regulation, G., 1990 concerning: Water Pollution Control. State Gazette of the Republic of Indonesia.
[3] Corominas L, Foley J, Guest J, Hospido A, Larsen H, Morera S, et al 2013 t. *Water research* ;**47(15):**5480-92.
[4] Permatasari PA, Setiawan Y, Khairiah RN, Effendi H, editors 2017 IOP Conference Series: Earth and Environmental Science: IOP Publishing.
[5] G. R. O. Q. And W. A. C. Control 2002 Water Quality And Water Pollution Control President Of The Republic Of Indonesia.
[6] Willer H, Lernoud J 2016: Research Institute of Organic Agriculture FiBL and IFOAM Organics International.
[7] Grover, C. and J. Stewart 2002 The work connection: The role of social security in British economic regulation. Palgrave Basingstoke.
[8] Sembiring, H., W. Raun, G. Johnson and R. Boman 1995 *Soil science* **159**, 81-89.
[9] Bagal MV, Gogate PR 2014 *Ultrasonics Sonochemistry* ;**21(1)**:1-14.
[10] Walukow A. F 2010 *Biological News*, vol. **10**, pp. 277 – 281.
[11] K. L. Taufik 2003 Faculty of Fisheries and Marine Science, IPB, Bogor.

Effect of ultraviolet and ultrasonic on potential antidiabetic activity of in vitro shoot cultures of *Orthosiphon aristatus*

D. Hunaefi[1,2*], N.D. Yuliana[1], I. Smetanska[3] and N. Gruda[4]

[1] Department of Food Science and Technology, Bogor Agricultural University (IPB), Indonesia
[2] South East Asian Food, Agricultural Science & Technology Center, Bogor Agricultural University (IPB), Indonesia
[3] University of Applied Science Weihenstephan- Triesdorf,Germany
[4] University of Bonn, Germany

*dashcbdk@apps.ipb.ac.id

Abstract. *Orthosiphon aristatus Boldingh*., a native tropical plant from Indonesia, is a medicinal herb that has been reported to possess an antidiabetic potential. However, study on this functional activity is limited especially on in vitro plant culture of *Orthosiphon aristatus*. The present study sought to investigate the effect of ultraviolet (UV) and ultrasonic (US) on this potential for *in vitro* shoot cultures of *Orthosiphon aristatus* (ISCOA). Methanol (70% v/v) extracts of ISCOA had the ability to inhibit alpha-amylase and alpha-glucosidase enzymes, dependent on the concentration of the extract. Among all treatment, highest inhibition of α-amylase and α-glucosidase activities were observed in combined treatment between UV for 60 minutes and US for 6 minutes. This was showed by the highest reduction by 0.4 point of IC50 inhibitory α-amylase and alpha-glucosidase activities in day 3 and in day 1 after exposure UV and US, respectively. The combination of UV and US has been shown to be effective in improving potential anti-diabetic properties of *in vitro* shoot cultures of Orthosiphon aristatus in which correlated with the increasing level of secondary metabolites.

1. Introduction

Orthosiphon aristatus is a native tropic-medicinal plant from Southeast Asia. It belongs to Lamiaceae family. This plant has been widely used as a traditional/natural alternative medicine in Indonesia, Malaysia and Thailand for the treatment of various diseases; for instance, gout, diabetes mellitus, hypertension, rheumatism, tonsillitis and menstrual disorder, and especially those affecting the urinary tract, that is for treating kidney ailments and bladder related diseases (18).

Many studies have indicated that *O. aristatus* exhibits several secondary metabolic active constituents. These compounds contribute to the plant's potential medicinal properties. However, the pharmacologically effects of *O. aristatus* have been recommended mainly due to its phenolics compounds (16) which are the most dominant constituent in the leaf. These phenolic compounds had been reported to be effective in reducing oxidative stress by inhibiting the formation of lipid peroxidation products in biological systems (17). One of the most important phenolic compounds in O.

aristatus is rosmarinic acid (RA) (1). This phenolic compound was thought to be the main contributor for the therapeutic effects and antioxidant activity of this plant (1).

RA is an ester of caffeic acid and 3, 4-dihydroxyphenyllacticacid (14) and it becomes one of the most targeted phytochemicals for development functional foods (44) due to its interesting biological activities, such as antiviral, antibacterial, anti-inflammatory and antioxidant (44). The development of functional foods containing RA from medicinal plants, herbs and spices that consist of naturally occurring antioxidants is preferable (38). Furthermore, this escalating demand of RA from medicinal plants has led us to do this research to increase the production of RA in Orthosiphon aristatus and to characterize the antioxidant properties of this plant extract which may provide useful information for better formulation of functional foods application.

Indeed, plant cell cultures that accumulate RA have been proposed for production of this phenolic compound (38, 50). And plant tissue cultures using plantlets that obtained from direct shoot organogenesis are generally preferred for mass production of the targeted compound as they are more genetically stable (12). Moreover, as a native tropic-medicinal plant, to grow *O. aristatus* in sub-tropic climate area was considered to be relatively difficult. Therefore, in vitro shoot cultures of Orthosiphon aristatus was used to be the media for this study.

Other reasons of using in vitro shoot cultures of *Orthosiphon aristatus* (abbreviated as ISCOA) are (1) this in vitro system may easily control biotic and abiotic stresses on the cultivar, (2) this system allows harvest of this plant at any condition regardless of climates condition and seasons, (3) and also through this system, rapid growth mass production can be achieved as well as be combined with sustainable secondary metabolites production (12, 46).

To improve secondary metabolites production of in vitro plant cultures, several strategies such as supplementation and physical stress elicitation has been used (5), for instance, ultraviolet (UV) as physical stress elicitation had been used to improve targeted secondary metabolite production in several plant cell cultures (4). Similarly, the utilization of ultrasound/ultrasonic (US) to enhance secondary metabolites in plant cell cultures had been reported by several researchers (54).

Ibdah et al. (19) showed that under ultraviolet stress with high light irradiance (1200–1500 μmol m−2 s−1) on *Mesembryanthemum crystallinum* displayed a rapid cell- in betacyanins and acylated flavonol glycosides production in the upper leaf epidermis. Lin et al. (31) reported that US had the ability to increase the biosynthesis of ginsenoside saponins in Panax gingseng cell cultures by 75% as compared to the control. Furthermore, Wu and Lin (54) also showed that US treatment for two minutes elevated volumetric taxol by 1.5- to 1.8-fold of control. However, there is no report of using both physical stresses on in vitro cultures of *Orthosiphon aritatus*.

Biosynthesis of RA is associated with the elevated of PAL activity through the phenyl-propanoid pathway (36, 47). In addition, proline-linked pentose phosphate pathway which includes the phenyl-propanoid pathway had been also proposed to regulate RA production (28, 56). That is why, for better understanding the effect of ultraviolet and ultrasonic treatment on biosynthesis of RA and phenolics in in vitro shoot cultures of O. aristatus, the model in Figure 1 has been intended based on previous model (43, 56).

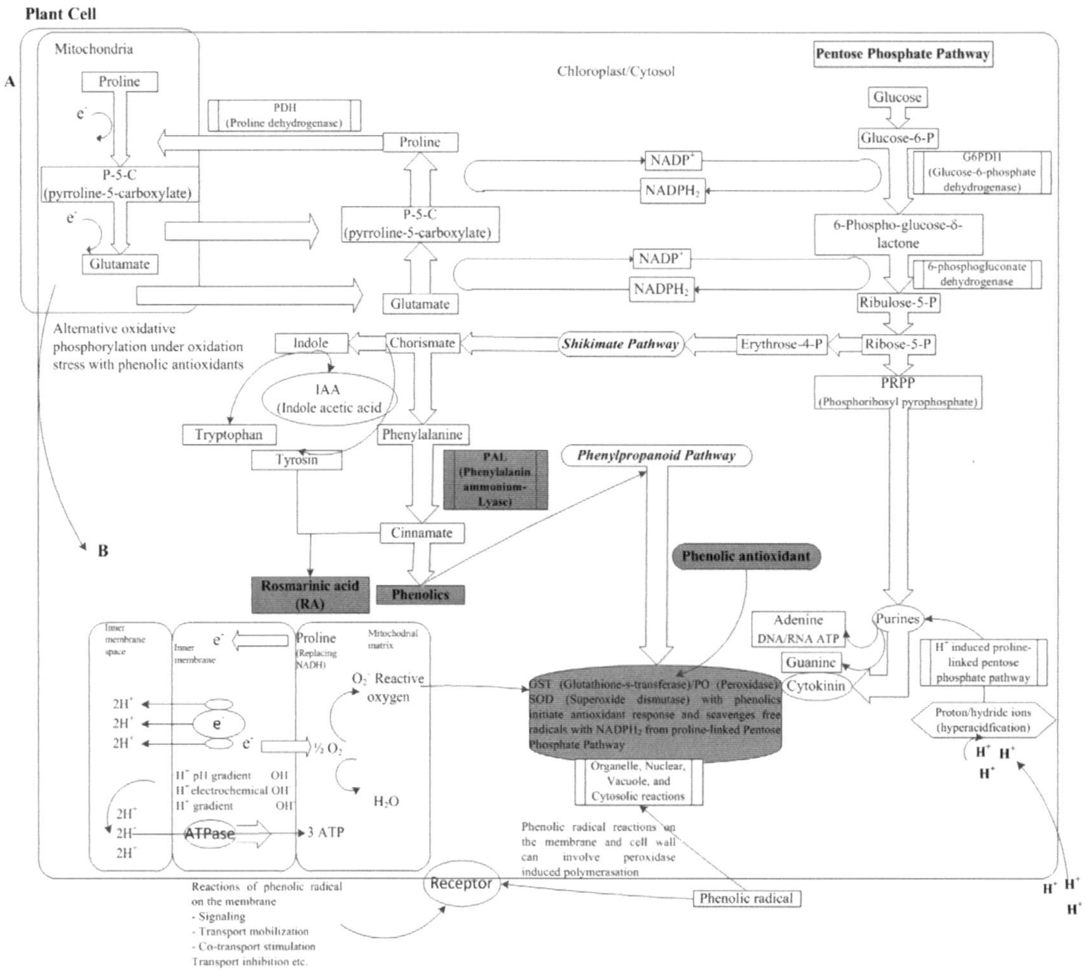

Figure 1. Proline-linked pentose phosphate pathway which includes the phenyl-propanoid pathway

Moreover, since antioxidant activity of plant extract are closely linked to their secondary metabolites as antioxidant component, therefore, the effect of the UV and US on antioxidant activity of ISCOA is also evaluated in our studies. One method of antioxidant activity analysis is not sufficient for the complete assessment of plant extract antioxidant potential due to the fact that antioxidant component in the plant extract may react differently to different antioxidant activity assay (8). For example, Wang and co-workers (52) found that some compounds, which have ABTS radical scavenging activity, may not show DPPH radical scavenging activity. Considering the limitation on using one antioxidant activity assay, thereby, in our experiment, determination of antioxidant activity was carried out using four different methods; DPPH free radical scavenging, Trolox Equivalent Antioxidant Capacity (TEAC), β-carotene/linoleic acid bleaching, and SOD-like antioxidant activity methods.

Lastly, the purposes of the study were (a) to investigate the effect of UV and US treatment on production of RA and other secondary metabolites with respect to inhibition activities of two enzymes: α-amylase and α-glucosidase activities and antioxidant properties of ISCOA, (b) to study the signals related to the pathway of secondary metabolites production (Figure 1) in in vitro shoot cultures such as glucose-6-phosphate dehydrogenase (G6PDH), phenylalanine ammonium-lyase (PAL), and peroxidase (PO) activity, and proline content were investigated, and (c) to further examine the profiles of phenolic acids of this plant after the treatment applied.

2. Material and methods

2.1. Plant Materials

Orthosiphon aristatus with article number of 600646 was purchased online from Pflanzenversand Hans-Günter Röpke (http://www.pflanzenkindergarten.de) for establishment of in vitro shoot culture.

2.2. Culture Media and Chemicals for analysis

MS Medium was obtained from E-Merck (Darmstadt, Germany). Pyrogallol, gallic acid, and ABTS (2,2'-azino-bis (3-ethylbenzthiazoline-6-sulphonic acid)) were obtained from Sigma-Aldrich Chemie GmbH (Steinheim, Germany); Quercetin Dihydrate ($C15H10O7 \cdot 2H2O$ □99% HPLC) from Fluka, Biochemika (Neu-Ulm, Germany); Folin-Ciocalteu's phenol reagent 2 M, and 1,1-diphenyl-2-picrylhydrazyl (DPPH) were purchased from Sigma-Aldrich Chemie GmbH (Buchs, Switzerland). The standards of phenolic acids; chlorogenic acid, p-hydroxybenzoic acid, vanillic acid, syringic acid, p-coumaric acid, sinapic acid, caffeic acid, ferulic acid, RA, cinnamic acid and gallic acid were obtained from Sigma Aldrich, Chemie GmbH (Steinheim, Germany). All used chemicals were analytical grade.

2.3. Establishment of ISCOA

Stems from the acclimatized mother plant were cut into 2.5 – 5 cm and sterilized in 70% ethanol for 20-30 seconds. They were then transferred into 10% sodium hypochlorite for 5 min, and washed three times in sterile water before being sown into 250 ml Erlenmeyer flasks on Murashige and Skoog medium (MS; Murashige and Skoog, 1962) for 4 weeks. Stems of sterile plants were cut into 1 – 1.5 cm pieces (±100 mg fresh weight.L-1) and placed in 250 ml Erlenmeyer flasks containing 50 ml liquid MS medium. ISCOA were transferred to a new medium every four weeks and maintained at 25 ± 2°C on a rotary shaker under light irradiation. The best line of ISCOA which was fourteen generation (sub-culturing) from the first establishment was used for our research.

2.4. Ultraviolet (UV) and Ultrasonic (US) treatment

As source of UV, we have used UV generator Paulmann L5/7/9/11.313 (Paulmann Licht Gmbh, Springe Völksen, Germany) with Phillips UVB lamp PL-S 9W/12/2P with emission range from 290-320 nm and peak at 302 nm (230 Volt □ 50 Herz and 0.17 A) (Phillips Electronics, Pila, Poland). For the US treatment, ultrasound bath, Bandelin Sonorex Digital 10 P (Bandelin electronic GmbH & Co. KG, Berlin, Germany) with double half-wave 35 kHz frequency automatic and ultrasonic peak output of 600 Watt with current consumption 3.4 A was used.

For the treatment, we have used ISCOA with the similar fresh weight (5.3 ± 0.04 g), length (4.0 ± 0.05 cm), number of leaves (6.0) and number of roots (4.0) of two weeks old and from 14th subculture. These ISCOA were subjected to the UV and US treatment in different combination (Table 1).

Table 1. Physical stresses treatment (UV and US) on ISCOA

Combination*	UV 0	UV 20	UV 40	UV 60
US 0	Control	UV 20	UV 40	UV 60
US 2	US 2	UV 20 – US 2	UV 40 – US 2	UV 60 – US 4
US 4	US 4	UV 20 – US 4	UV 40 – US 4	UV 60 – US 4
US 6	US 6	UV 20 – US 6	UV 40 – US 6	UV 60 – US 6

*0, 2, 4, 6, 20, 40, and 60 showed the time of the treatment in minutes and the name of the samples was named based on the time and combination physical treatment applied

The ISCOA in 250 ml flask were transferred into sterile Petri disc and subjected to UV irradiation for certain time (Table 1) under aseptic condition in clean bench. For US treatment, the ISCOA in 250 ml flask covered with sterilized plastic caps were directly subjected to US treatment in ultrasonic bath filled water (5 L) for a definite time mentioned in Table 1. Samplings were performed 1, 3, 7 and 14 days after the day of the treatment (day 0) and namely as day 1, day 3, day 7, and day 14.

2.5. Analysis of the fresh weight of ISCOA

2.5.1. Chloropyll a and b

Chlorophyll a and b in leaves were measured based on previous spectrophotometrically method by Lichtenthaler and Wellburn (30) using fresh weight basis. The chlorophyll content later expressed as mg per gram fresh weight (mg.g FW-1).

2.5.2. Glucose-6-phosphate dehydrogenase (G6PDH, EC 1.1.1.49), phenylalanine ammonium-lyase (PAL, EC 4.3.1.5), and peroxidase (PO, EC 1.11.1.7) activities

The extraction of G6PDH and its assay were determined spectrophotometrically at the wavelength 339 nm according to the previous method by Shetty et al. (45). The enzyme quantity was measured in nanomoles per minute per milligram protein (nmoles.min-1.mg protein-1). While the extraction of PAL and PO was assayed spectrophotometrically at OD 290 nm and 420 nm, respectively. Both assays were based on previous method described by Dörnenburg and Knorr (10). The activity of enzyme was expressed as micro-Units per gram of fresh weight (mU.g FW-1).

2.5.3. Proline content

Proline content of ISCOA leaves was measured on the spectrophotometer basis of a previously described assay used for plants Bates et al. (3). Proline content was expressed as micromoles per gram of fresh weight of ISCOA materials (µmoles.g FW-1).

2.5.4. Sample preparation and extraction

The plant was harvested and weighed and immersed in liquid nitrogen to avoid any possible enzyme degradation and to prevent phenolic compound volatilization then followed with freeze-drying. The lyophilized sample was grounded by flint mill (Retsch, Retsch GmbH, Haan, Germany) of 20000 rpm and 2 min to a fine powder. The dried powder of ISCOA was weighed of 100 mg and extracted with 5 ml of 70% methanol. And then it was placed in ultrasonic bath at 55°C for 2 hours. The mixture was centrifuged at 13000 rpm for 5 minutes and the supernatants were collected in different probe. The extraction was repeated twice until the final concentration of ISCOA extract was 10 mg.ml-1. The extract was stored at -20oC until further use. All samples and determination were prepared and measured in triplicate

2.5.5. α-amylase and α-glucosidase inhibitory activities

The α-amylase (EC 3.2.1.1) and α-glucosidase (EC 3.2.1.20) inhibitory activities of ISCOA extracts were analyzed according to the method as previous described by Ranilla et al. 2. The α-amylase inhibitory activity was measured spectrophotometrically at 540 nm, while α-glucosidase inhibitory activity was measured at 405 nm. Percentage of inhibition was calculated according to the equation:

$$\% \text{ Inhibition} = [(AC - AS)/AC] \times 100$$

where AC is the absorbance of the control, and AS is the absorbance of the samples. Appropriate dilution of the ISCOA extracts was made to calculate the IC50 of α-amylase and α-glucosidase inhibitory activities.

3. Results

We investigated the effect of UV and US treatment on α-amylase and α-glucosidase inhibitory activities in a model system, in vitro shoot cultures of *Orthosiphon aristatus*. The different between in vitro shoot cultures and in vitro tissue cultures had been explained by Schevchenko et al. (46).

Turning to the stresses applied, US treatment suppressed the growth of ISCOA as observed by significantly lower biomass and length appeared at day 3 post-exposure, whereas UV irradiation up to 60 min had no effect on the growth of ISCOA (Figure 2 a and Figure 2 b). Furthermore, although slightly lower in the content of chlorophyll a content of US 6 samples at day 1 after physical treatment, however, statistically, there was no significant different ($p > 0.05$) on the chlorophyll a and b of ISCOA leaves after the application of UV and US treatment (Figure 2 c and Figure 2 d). G6PDH, PAL, PO activities and proline content

In order to examine whether stress treatment involved in a signal transduction pathway leading to enhancement of RA synthesis. Therefore, G6PDH, PAL, PO activity and proline content were investigated. Figure 3 shows the intracellular G6PDH, PAL, PO activity; and proline content in ISCOA after the treatment was applied (1, 3, 7 and 14 days).

In term of G6PDH activity (Figure 3 a), this enzyme activity was not significant different after UV treatment as compared to control. However, the US 6 and the combined treatment of UV 60 and US 6 suppressed this enzyme activity at day 1 after treatment. The most prominent significant increase was observed on day 3 after the treatment (Figure 3 a). This enzyme activity dropped afterward and remained stable until day 14 of post-exposure. The intracellular PAL activity of ISCOA rose significantly after one day exposure to US 6 and UV 60-US 6 (Figure 3 b). It reached maximum at day 3 and declined slowly to day 14 of post-exposure. The UV treatment of ISCOA had no significant effect on PAL activity as compared to control. Whereas for peroxidase enzyme, only the combination (UV 60-US 6) treated samples showed a significant influence on PO enzyme activity. In contrast to PAL activity that reached maximum at day 3 after combined treatment UV 60 and US 6, PO activity in treated samples increased gradually and reached its maximum at day 7 (Figure 3 c).

Additionally, proline content increased (Figure 3 d) at day 1 after UV 60 treatment, while US 6 had no effect on proline content of ISCOA. The combined exposure of UV 60 and US 6 resulted in a rapid increase of proline content at day 1 and rose slightly at day 3. It dropped at day 7 and remained constant at day 14.

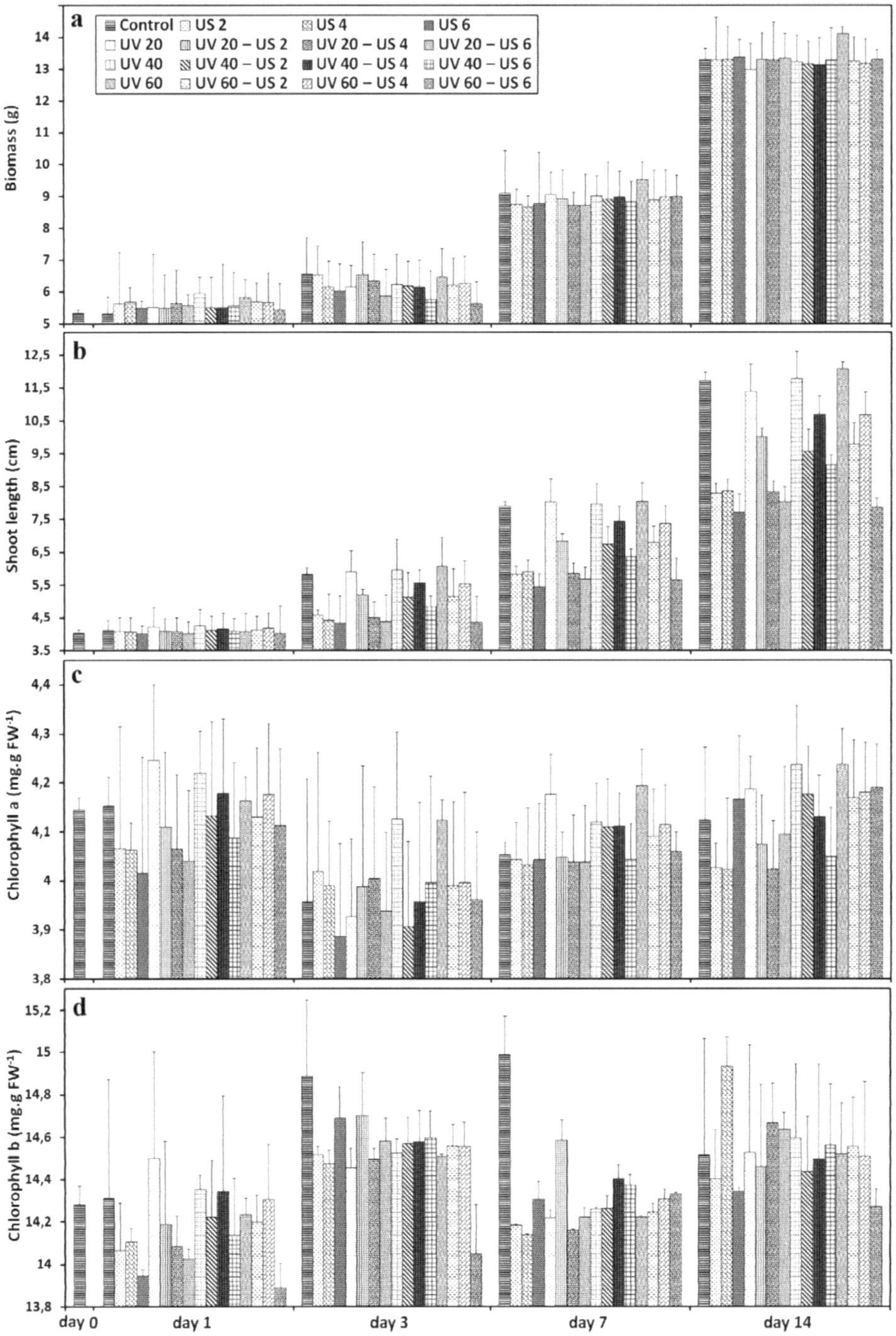

Figure 2. a. Biomass; b. shoot length; c. chlorophyll a; and d. chlorophyll b of ISCOA subjected to UV and US treatment.

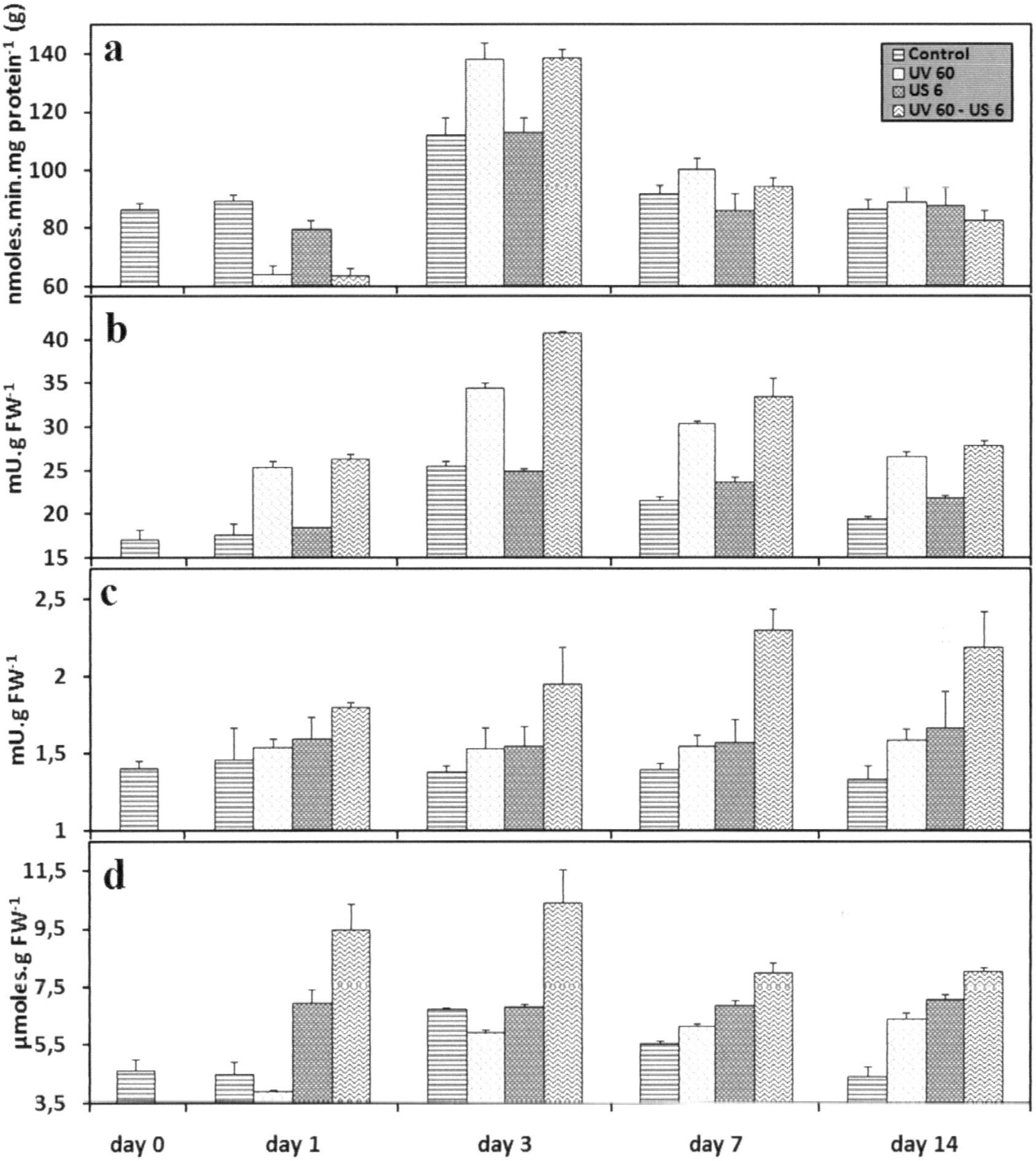

Figure 3. G6PDH, PAL, PO activity; and proline content in ISCOA after the treatment was applied (1, 3, 7 and 14 days)

α-amylase and α-glucosidase inhibitory activities

Results showed that methanol (70% v/v) extracts of ISCOA had activities to inhibit both enzymes dependent on the concentration of the extracts (Figure 4).

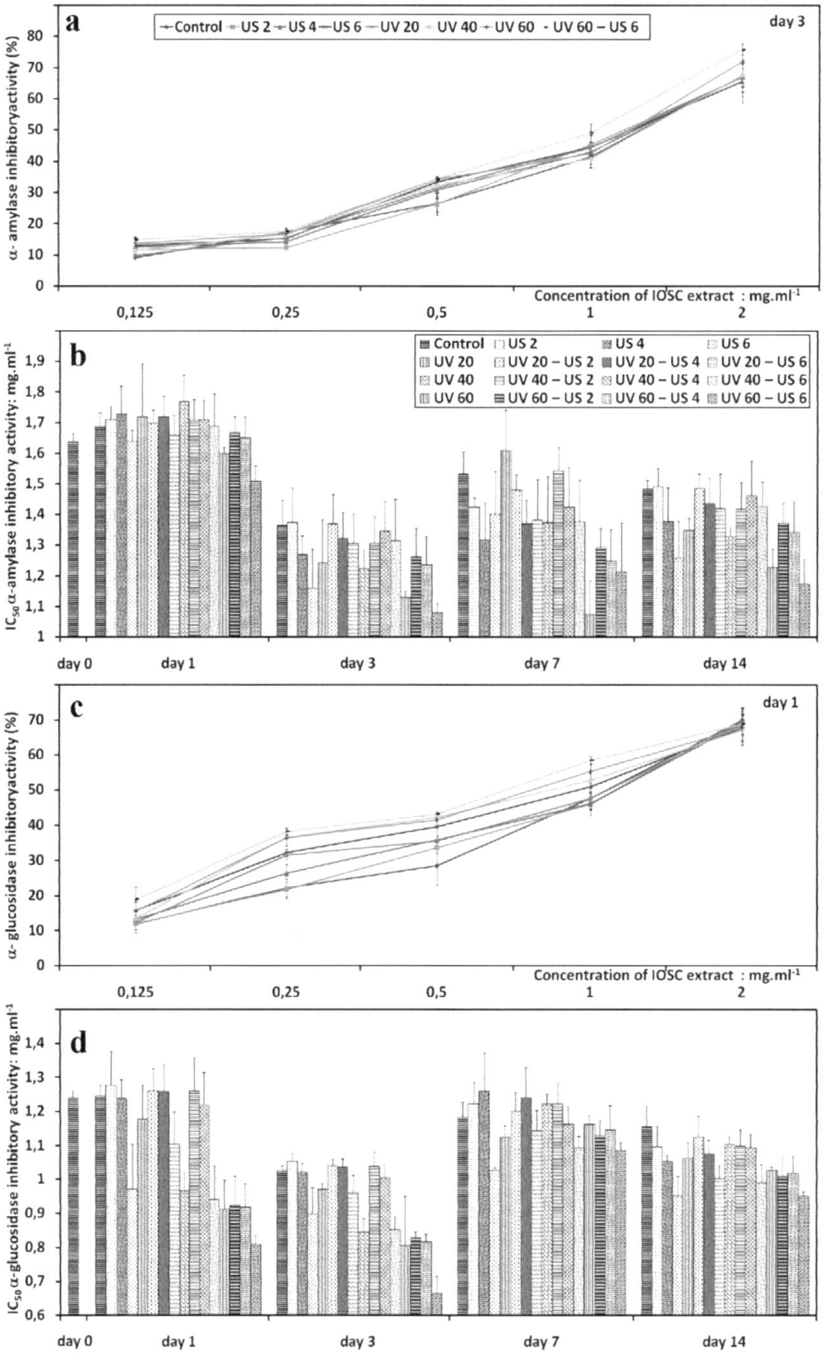

Figure 4. a. Concentration dependent ISCOA extracts α-amylase inhibitory activity to calculate IC50; b. IC50 α-amylase inhibitory activity of ISCOA extracts subjected to different physical stresses; c. Concentration dependent ISCOA extracts of α-glucosidase inhibitory activity to calculate IC50; and d. IC50 α-glucosidase inhibitory activity of ISCOA extracts subjected to different physical stresses.

Lower IC50 was achieved in α-glucosidase inhibitory activity as compared to α-amylase inhibitory activity. This means that ability of ISCOA extracts to inhibit the function of α-glucosidase is higher than α-amylase (Figure 4). The inhibitory activities of both enzymes attained the minimum level of IC50 at day 3 after treatment. Furthermore, results of these enzymes inhibition showed that inhibitory activity to α-glucosidase was more responsive to the physical stresses applied than α-amylase. Figure 7d showed that treatment with US 6 and UV 60 – US 6 resulted in significant lower of IC50 compared to control immediately after one day treatment while inhibitory of α-amylase appeared significantly different at day 3 post-exposure for UV 60-US 6 treatment.

4. Discussion

The inhibition of cell growth and reduction of cell viability after US treatment had been reported by several researchers (55). Similarly, several researchers reported that UV-B radiation exerts powerful stress effect on plants and it resulted in reduction of plant biomass (27). US and the combined treatment of US-6 and UV-60 treatment on biomass and shoot length was confirmed in our research. This phenomenon could be explained by morphological changes as reduced cell elongation as a result of plant adaptation to the treatment by lowering indole acetic acid (IAA) (26). Therefore, on the assumption that growth (biomass and shoot length) is an integral index of plant condition, a decrease in US and the combined of US 6 and UV 60-treated ISCOA growth due to the treatment can result from the changes in the physiological and biochemical characteristics (58).

In contrast to US effect, the biomass and shoot length was not hampered by the UV treatment alone. The similar effect of UV-B had been reported by Murali et al. (35) on total biomass and leaf area of soybean and by Bandurska et al. (2) on dry weight and water content of barley. Correspondingly, Lavola (29) reported that the exposure of UV-B for 3 hour per day on birch had no significant effect on shoot growth and did not result any leaf injuries. Interestingly, some enhancements on plant biomass had also been reported as a response to UV treatment (49). These phenomena were partially due to accumulation of secondary metabolites and leaf thickening (35) indicating that the UV induced formation of phenolic compounds had some antioxidant protective function (29).

The US and UV treatment resulted in slightly lower level (although no significant different) of chlorophyll a and b in ISCOA was noted. Similarly, Kováčik et al. (24) reported that chlorophylls a and b was found slightly lower after UV treatment. Recently, a decrease in total chlorophyll content after UV exposure was also reported by Anttila (2011). This change indicated as shift from primary metabolism to the secondary metabolism (11).

G6PDH (EC 1.1.1.49) is the key enzyme of pentose phosphate pathway. It plays important role to provide precursors in synthesis of phenolic secondary metabolite (45). And significantly higher G6PDH activity was observed in ISCOA after the combined treatment of US 6 and UV 60. The increase in G6PDH activity after UV-B radiation on turfgrasses plant had been reported by Sarkar et al. (42). However, no data was found about the changing in this enzyme activity after US exposure. Higher G6PDH activity of physical elicited samples can be explained by an activation of pentose phosphate pathway (42).

PAL (EC 4.3.1.5) had previously been shown to catalyze the incorporation of phenylalanine into phenolic compounds including RA, thus functioning as entry point enzyme in the initial step of the phenylpropanoid pathway that is responsible for the biosynthesis of these compounds (47) and the first stage in the formation of phenolics is deamination of phenylalanine catalyzed by PAL (2). In our experiments, the UV and US treatment significantly influenced the PAL activity as compared to the control. PAL activity in ISCOA leaves had the highest level at day 3 post-exposure. In Lithospermum erythrorhizon cell culture, PAL activity rose sharply after exposure to fix ultrasonic bath 38.5 kHz and reached the highest level at day 5 after exposure (31). The increase PAL activity of US-stimulated Taxus chinensis cells was also reported by Wu and Lin (54). Additional fact of induced PAL activity after the UV-B treatment was also observed on barley (2, 32) after the psychical stress treatment. PAL activity is a key factor responsible for increased accumulation of phenolics compound on barley leaves after UV-B radiation (32).

Meanwhile, peroxidase (EC 1.11.1.7) activity in plants had been reported to increase in response to the stress applied by Kwak et al. (27). This increase of PO activity was suggested being responsible for the formation of UV protection such as phenolic compounds (27). Similarly, a dramatic increase in peroxidase after 38.5 kHz US for 2 minutes had also been reported by Wu and Lin (53) as plant biochemical adaptation to US stress. The increment of this enzyme activity was confirmed in our study.

The increase in proline content of ISCOA leaves after UV and US treatment was pronounced. Proline is known as an osmolyte that accumulates in plant cells in response to stress (40). This amino acid synthesis is hypothesized to be able to regulate the pentose phosphate pathway in erythrocytes (28). Elevated proline levels in C. nivalis, turf grasses, and wheat seedlings after UV-B exposure had been previously reported by Duval et al. (11), Sarkar et al. (42) and Yao et al. (57), respectively. It has been hypothesized that stress induced proline accumulation in plants may reduce the damage by production of free radicals in thylakoid membrane of chloroplast (11). Various compensatory mechanisms that may be stimulated proline accumulation and plant survival are attributed to mitochondrial electron transport suppression (11). Although absence effect of UV treatment on barley had recently been reported by Bandurska (2), however, most of the report using UV as stress applied reported an increase in this amino acid (23, 57).

Application of physical elicitors such as US and UV may induce distinct changes in the plant secondary metabolism. Phenolics are one of the most widespread classes of plant secondary metabolites produced from the shikimate, acetate, and potentially the proline linked pentose phosphate pathways (11) for driving precursors (erythrose-4-phosphate) toward shikimate and phenylpropanoid pathway (Figure 1) (56). The significant increase in secondary metabolites, phenolic compounds in particular, after US and UV treatment as here reported, which is not surprising, are similar to finding by many investigators using UV (2) and US as abiotic elicitors. These facts demonstrate an important role of phenolic compounds in the protection cell against the effect of UV and US-physical elicitors.

Additionally, nowadays, research priorities on dietary management of type II diabetes are becoming more prevalent. Considering diet-linked challenge of type II diabetes, consumption of food and beverage rich in α-amylase and α-glucosidase inhibitors – so called hypoglycemic foods – are receiving more attention. α-Amylase and α-glucosidase are well-known key enzymes, playing a vital role in the management of hyperglycemia linked type II diabetes 1. Therefore, characterization of ISCOA extracts on α-Amylase and α-glucosidase inhibition activities were also analyzed.

Lower IC50 was achieved in α-glucosidase inhibitory activity as compared to □-amylase inhibitory activity. Interestingly, higher glucose-6-phosphate dehydrogenase, phenylalanine ammonium-lyase activity, peroxidase and proline content were also observed after the treatment applied which indicated an activation of secondary metabolism through the proline linked pentose phosphate pathway. The results indicated that the combination of UV and US has been shown to be effective in improving antioxidant properties of in vitro shoot cultures of Orthosiphon aristatus. The effectiveness was characterized by the highest increment in the concentration of secondary metabolites accumulation which resulted in the highest antioxidant activity.

With the results obtained in our study, it can be assumed that stimulation of secondary metabolites by US related to the metabolic activity and defense response of plant cell to the mechanical stress such as inducing energy-intense hydrodynamic and acoustic captivation and micro streaming which causes mechanical damage and shear stress to the cells under US treatment 34, 75 as well as various compensatory mechanisms under UV treatment that may involved in plant survival by activation of secondary metabolites production 48, 54.

The UV and US treatment can act as abiotic elicitors to induce plant defense responses 19 and provide a significant contribution to the antioxidant activity of the plant tissue 12, 78. The treatment resulted in a significant increase phenolic compounds which might be responsible for enhancing antioxidant activity: DPPH, TEAC, SOD-like activity and β-carotene/linoleic acid bleaching.

Stimulated antioxidant activity using β-carotene/linoleic acid bleaching method in C. nivalis extract after UV light exposure had been previously reported 48. Other works describing oxidative stress in Chlorella likewise report a markedly increase antioxidant activity after UV-B treatment 79 and in blueberries significantly higher antioxidant activity was detected in all range of UV radiation 77. Similarly, elevated activity of antioxidant enzymes: SOD, reactive oxygen species (ROS) and catalase (CAT), in Scenedesmus quadricauda 58 and turfgrasses 49 was affected by UV exposure and in Taxus chinensis was influenced after 38.5 kHz US for 2 minutes 36. Additionally, Kozak et al. 76 reported that antioxidant properties and flavonoids accumulation in soybean had increased as an adaptive response to UV radiation. These increasing antioxidant activity in US and UV stimulated ISCOA is a further proof that secondary metabolites (as antioxidant components) plays an important role in the cell survival under abiotic stress condition leads to increasing antioxidant activity and protection 37. Plants also synthesize low molecular antioxidants such as phenolics which may directly or indirectly scavenge ROS 58. These phenolics compounds posse protective function, for instance, they are potent UV-B absorbing compounds and they mitigate the effects of free radicals 64 as these were showed in our results by scavenging activity of DPPH free radicals.

The metabolism of phenolic substances has been intensively studied in plant cultures derived from different plant species (6). Here we report that the enhancement of phenolics accumulation in ISCOA as model system was directly proportional to the resistance of the plant to UV and US applied stresses and reflect the ability of Orthosiphon aristatus as a native medicine-tropical plant to occupy in open habitats (49). Longer stresses applied resulted in stronger response (51). It should be noted that the enhancement in phenolics production here was accompanied by an increase in PAL activity, a pivotal steps in the phenylpropanoid biosynthesis and this enzyme is one of the key factors for increased accumulation of these substances (20, 23). This increase of PAL activity was followed by an elevated concentration in proline content. And with respect to those marked increases, these were coupled with an increase in G6PDH enzyme activity. This enzyme is the first rate limiting enzyme to pentose phosphate pathway (Figure 1) (42) and Shetty (43) proposed that proline linked pentose phosphate pathway could stimulate both the shikimate and phenylpropanoid pathways. The modulation of this pathway (Figure 1) could lead to the stimulation of phenolic phytochemicals that can behave as antioxidants by trapping free radicals in direct interaction or scavenge them through a series coupled, antioxidant enzyme defense system reactions (42). UV light exposure had suggested modulating these compounds of C. nivalis by activation biochemical pathways related to proline metabolism (11). And in the our study, higher antioxidant activity in ISCOA due to UV and US treatment were observed in different antioxidant measurements and these phenomenon were associated with higher phenolics content accumulation. Moreover, it has been suggested that antioxidant response pathway in plants relies upon key NADPH-requiring enzymes similar to proline biosynthesis (43). These defensive mechanisms are believed to work through antioxidant response pathways involving peroxidases as well as phenolics biosynthesis (43). In the present study, higher peroxidase activity in ISCOA has been achieved after the physical stress treatment, therefore, these obtained results may reflect associated role with proline linked pentose phosphate pathway in the cytosol under stresses induced condition. These interesting findings provide stimulus for further research into deeply genetic responsible for this enhancement phenomenon by UV and US in ISCOA as a model system.

5. Conclusion

In conclusion, the present work has demonstrated the beneficial effects of UV and US treatment on secondary metabolites, enzyme activity, proline content and antioxidant properties of ISCOA as model study. The combination of UV and US has been shown to be effective in improving α-amylase and α-glucosidase inhibitory activities of ISCOA. The effectiveness was characterized by the lowering IC50 of α-amylase and α-glucosidase inhibitory activities. The results indicated that UV and US which can be considered as physical elicitation stressors may activate phenolic substances production in ISCOA through a proline linked pentose phosphate pathway. This mechanism, however, remained to be verified through further depth investigation focused on genetic expression that would be provide detail mechanism and understanding phenolics biosynthesis in ISCOA under the stresses applied.

References

[1]. Akowuah, Gabriel A., and Ismail Zhari. 2010. Effect of Extraction Temperature on Stability of Major Polyphenols and Antioxidant Activity of Orthosiphon stamineus Leaf. J. Herbs Spices Med Plants 16 (3-4):160-166.
[2]. Bandurska, Hanna, Małgorzata Pietrowska-Borek, and Małgorzata Cieślak. 2012. Response of barley seedlings to water deficit and enhanced UV-B irradiation acting alone and in combination. Acta Physiol. Plant. 34 (1):161-171.
[3]. Bates, L. S., R. P. Waldren, and I. D. Teare. 1973. Rapid determination of free proline for water-stress studies. Plant & Soil 39 (1):205-207.
[4]. Broeckling, Corey D., David V. Huhman, Mohamed A. Farag, Joel T. Smith, Gregory D. May, Pedro Mendes, Richard A. Dixon, and Lloyd W. Sumner. 2005. Metabolic profiling of Medicago truncatula cell cultures reveals the effects of biotic and abiotic elicitors on metabolism. J. Exp. Bot. 56 (410):323-336.
[5]. Cai, Zhenzhen. 2012. Effects elicitors on secondary metabolism of plant suspension cultures of Vitis vinifera and Malus domestica and their exudates, Department Method in Food Biotechnology, Berlin University of Technology, Berlin.
[6]. Cai, Zhenzhen, Anja Kastell, Inga Mewis, Dietrich Knorr, and Iryna Smetanska. 2011. Polysaccharide elicitors enhance anthocyanin and phenolic acid accumulation in cell suspension cultures of Vitis vinifera. Plant Cell Tiss. Org. Accepted manuscript (Online First):1-9.
[7]. Chumroenphat, Theeraphan, Intha Khanprom, and Luchai Butkhup. 2011. Stability of Phytochemicals and Antioxidant Properties in Ginger (Zingiber officinale Roscoe) Rhizome with Different Drying Methods. J. Herbs Spices Med Plants 17 (4):361-374.
[8]. Debnath, Trishna, Pyo-Jam Park, Narayan Chandra Deb Nath, Nadira Binte Samad, Hee Won Park, and Beong Ou Lim. 2011. Antioxidant activity of Gardenia jasminoides Ellis fruit extracts. Food Chem. 128 (3):697-703.
[9]. Devi, S. Asha, and Deepak Ganjewala. 2011. Antioxidant Activities of Methanolic Extracts of Sweet-Flag (Acorus calamus) Leaves and Rhizomes. J. Herbs Spices Med Plants 17 (1):1-11.
[10].Dörnenburg, Heike, and Dietrich Knorr. 1997. Evaluation of Elicitor- and High-Pressure-Induced Enzymatic Browning Utilizing Potato (Solanum tuberosum) Suspension Cultures as a Model System for Plant Tissues. J. Sci. Food Agric. 45 (10):4173-4177.
[11].Duval, Brian, Kalidas Shetty, and William Thomas. 1999. Phenolic compounds and antioxidant properties in the snow alga Chlamydomonas nivalis after exposure to UV light. J. Appl. Phycol. 11 (6):559-566.
[12].Fadhlina, Ani. 2008. The effects of temperature and fermentation time on selected chemical composition, antioxidant and quality of cat whiskers (Orthosiphon stamineus) tea, Chemistry, University Sains Malaysia.
[13].Germ, Mateja, Vekoslava Stibilj, Samo Kreft, Alenka Gaberščik, and Ivan Kreft. 2010. Flavonoid, tannin and hypericin concentrations in the leaves of St. John's wort (Hypericum perforatum L.) are affected by UV-B radiation levels. Food Chem. 122 (3):471-474.
[14].Gohari, Ahmad Reza, Soodabeh Saeidnia, Homa Hajimehdipoor, Maryam Shekarchi, and Abbass Hadjiakhoondi. 2011. Isolation and Quantification of Rosmarinic Acid from Hymenocrater calycinus. J. Herbs Spices Med Plants 17 (2):132-138.
[15].Harbaum-Piayda, B., B. Walter, G. B. Bengtsson, E. M. Hubbermann, W. Bilger, and K. Schwarz. 2010. Influence of pre-harvest UV-B irradiation and normal or controlled atmosphere storage on flavonoid and hydroxycinnamic acid contents of pak choi (Brassica campestris L. ssp chinensis var. communis). Postharvest Biol. Technol. 56 (3):202-208.
[16].Ho, Chun-Hoong, Ismail Noryati, Shaida-Fariza Sulaiman, and Ahmad Rosma. 2010. In vitro

antibacterial and antioxidant activities of Orthosiphon stamineus Benth. extracts against food-borne bacteria. Food Chem. 122 (4):1168-1172.

[17].Hollman, P. C. H., and M. B. Katan. 1999. Dietary Flavonoids: Intake, Health Effects and Bioavailability. Food Chem. Toxicol. 37 (9-10):937-942.

[18].Hsu, C. L., B. H. Hong, Y. S. Yu, and G. C. Yen. 2010. Antioxidant and Anti-Inflammatory Effects of Orthosiphon aristatus and Its Bioactive Compounds. J. Sci. Food Agric. 58 (4):2150-2156.

[19].Ibdah, M., A. Krins, H. K. Seidlitz, W. Heller, D. Strack, and T. Vogt. 2002. Spectral dependence of flavonol and betacyanin accumulation in Mesembryanthemum crystallinum under enhanced ultraviolet radiation. Plant, Cell Environ. 25 (9):1145-1154.

[20].Iwai, Mayuko, Mari Ohta, Hiroshi Tsuchiya, and Tetsuya Suzuki. 2010. Enhanced accumulation of caffeic acid, rosmarinic acid and luteolin-glucoside in red perilla cultivated under red diode laser and blue LED illumination followed by UV-A irradiation. J. Funct. Foods 2 (1):66-70.

[21].Kiong, Anna Ling Pick, Alvina Grace Lai, Sobri Hussein, and Abdul Rahim Harun. 2008. Physiological Responses of Orthosiphon stamineus Plantles to Gamma Irradiation. Am.-Eurasian J. Sustain. Agric. 2 (2):135-149.

[22].Komaraiah, P., P. B. K. Kishor, M. Carlsson, K. E. Magnusson, and C. F. Mandenius. 2005. Enhancement of anthraquinone accumulation in Morinda citrifolia suspension cultures. Plant Sci. 168 (5):1337-1344.

[23].Kováčik, Jozef, Bořivoj Klejdus, and Martin Bačkor. 2010. Physiological Responses of Scenedesmus quadricauda (Chlorophyceae) to UV-A and UV-C Light. Photochem. Photobiol. 86 (3):612-616.

[24].Kováčik, Jozef, Bořivoj Klejdus, František Štork, and Silvia Malčovská. 2011. Sensitivity of Xanthoria parietina to UV-A: Role of metabolic modulators. J. Photochem. Photobiol., B 103 (3):243-250.

[25].Kozak, R. G., R. A. Ricco, A. A. Gurni, A. D. Boveris, and S. Puntarulo. 1999. Antioxidant response of soybean cotyledons (Glycine max) to ultraviolet irradiation. Can. J. Plant. Sci. 79 (2):181-189.

[26].Kuhlmann, F., and C. Muller. 2009. Development-dependent effects of UV radiation exposure on broccoli plants and interactions with herbivorous insects. Environ. Exp. Bot. 66 (1):61-68.

[27].Kwak, Sang-Soo, Soo-Kyung Kim, Il-Hyun Park, and Jang R. Liu. 1996. Enhancement of peroxidase activity by stress-related chemicals in sweet potato. Phytochem. 43 (3):565-568.

[28].Kwok, Denise, and Kalidas Shetty. 1998. Efffect of proline and proline analogs on total phenolic and rosmarinic acid levels in shoot colones of Thyme (Thymus vulgaris L.). J. Food Biochem. 22 (1):37-51.

[29].Lavola, Anu. 1998. Accumulation of flavonoids and related compounds in birch induced by UV-B irradiance. Tree Physiol. 18 (1):53-58.

[30].Lichtenthaler, Harmut K., and Alan R. Wellburn. 1983. Determinations of total carotenoids and chlorophylls a and b of leaf extracts in different solvents. Biochem. Soc. Trans. 11:591-592.

[31].Lin, Lidong, and Jianyong Wu. 2002. Enhancement of shikonin production in single- and two-phase suspension cultures of Lithospermum erythrorhizon cells using low-energy ultrasound. Biotechnol. Bioeng. 78 (1):81-88.

[32].Liu, Lan, Dennis C. Gitz, and Jerry W. McClure. 1995. Effects of UV-B on flavonoids, ferulic acid, growth and photosynthesis in barley primary leaves. Physiol. Plant. 93 (4):725-733.

[33].Malanga, Gabriela, and Susana Puntarulo. 1995. Oxidative stress and antioxidant content in Chlorella vulgaris after exposure to ultraviolet-B radiation. Physiol. Plant. 94 (4):672-679.

[34].Mohdaly, Adel AA, Mohamed A Sarhan, Iryna Smetanska, and Awad Mahmoud. 2010. Antioxidant properties of various solvent extracts of potato peel, sugar beet pulp and sesame cake. J. Sci. Food Agric. 90 (2):218-226.

[35].Murali, N. S., and Alan H. Teramura. 1985. Effects of ultraviolet-B irradiance on soybean. VI. Influence of phosphorus nutrition on growth and flavonoid content. Physiol. Plant. 63 (4):413-

416.
[36].Ogata, A., A. Tsuruga, M. Matsuno, and H Mizukami. 2004. Elicitor-induced rosmarinic acid biosynthesis in Lithospermum erythrorhizon cell suspension cultures: Activities of rosmarinic acid synthase and the final two cytochrome P450-catalyzed hydroxylations. Plant Biotech. 21 (5):393–396.
[37].Olsson, L. C., M. Veit, G. Weissenböck, and J. F. Bornman. 1998. Differential flavonoid response to enhanced uv-b radiation in brassica napus. Phytochem. 49 (4):1021-1028.
[38].Petersen, Maike, and Monique S. J. Simmonds. 2003. Rosmarinic acid. Phytochem. 62 (2):121-125.
[39].Rajopadhye, Anagha A., and Anuradha S. Upadhye. 2012. In vitro Antioxidant and Hepatoprotective Effect of the Whole Plant of Glossocardia bosvallea (L. f.) D. C. against CCl4-Induced Oxidative Stress in Liver Slice Culture Model. J. Herbs Spices Med Plants 18 (3):274-286.
[40].Randhir, Reena, and Kalidas Shetty. 2007. Elicitation of the proline-linked pentose phosphate pathway metabolites and antioxidant enzyme response by ascorbic acid in dark germinated fava bean sprouts. J. Food Biochem. 31 (4):485-508.
[41].Rezaei, Ayatollah, Faezeh Ghanati, Mehrdad Behmanesh, and Manijhe Mokhtari-Dizaji. 2011. Ultrasound-Potentiated Salicylic Acid–Induced Physiological Effects and Production of Taxol in Hazelnut (Corylus Avellana L.) Cell Culture. Ultrasound in Med. Biol. 37 (11):1938-1947.
[42].Sarkar, Dipayan, Prasanta C. Bhowmik, Kwon Young In, and Kalidas Shetty. 2011. The role of proline-associated pentose phosphate pathway in cool-season turfgrasses after UV-B exposure. Environ. Exp. Bot. 70 (2–3):251-258.
[43].Shetty, K. 2004. Role of proline-linked pentose phosphate pathway in biosynthesis of plant phenolics for functional food and environmental applications: a review. Process Biochem. 39 (7):789-804.
[44].Shetty, Kalidas. 2001. Biosynthesis and Medical Applications of Rosmarinic Acid. J. Herbs Spices Med Plants 8 (2-3):161-181.
[45].Shetty, Preethi, Mokthar T. Atallah, and Kalidas Shetty. 2002. Effects of UV treatment on the proline-linked pentose phosphate pathway for phenolics and L-DOPA synthesis in dark germinated Vicia faba. Process Biochem. 37 (11):1285-1295.
[46].Shevchenko, Yaroslav, Iryna Smetanska, and Alexandra Wendt. 2010. Sprout culture of Stevia rebaudiana Bertoni. In: Stevia science, no fiction, Ed. J. Geuns: Euprint Heverlee.
[47].Sumaryono, W., P. Proksch, T. Hartmann, M. Nimtz, and V. Wray. 1991. Induction of rosmarinic acid accumulation in cell suspension cultures of Orthosiphon aristatus after treatment with yeast extract. Phytochem 30 (10):3267-3271.
[48].Susanti, Deny, Mohamad Osman, and Nurul Nadiah Draman. 2012. Natural Antioxidant Contents in the Seeds of Four Roselle Malaysian Varieties and a Kenaf: A Comparison. J. Herbs Spices Med Plants 18 (3):240-245.
[49].Tegelberg, R., R. Julkunen-Tiitto, and P. J. Aphalo. 2004. Red : far-red light ratio and UV-B radiation: their effects on leaf phenolics and growth of silver birch seedlings. Plant, Cell Environ. 27 (8):1005-1013.
[50].Thatoi, Hrudayanath, and Jayanta Kumar Patra. 2011. Biotechnology and Pharmacological Evaluation of Medicinal Plants: An Overview. J. Herbs Spices Med Plants 17 (3):214-248.
[51].Wang, Chien Y., Chi-Tsun Chen, and Shiow Y. Wang. 2009. Changes of flavonoid content and antioxidant capacity in blueberries after illumination with UV-C. Food Chem. 117 (3):426-431.
[52].Wang, Mingfu, Jiangang Li, Meera Rangarajan, Yu Shao, Edmond J. LaVoie, Tzou-Chi Huang, and Chi-Tang Ho. 1998. Antioxidative Phenolic Compounds from Sage (Salvia officinalis). J. Agric. Food. Chem. 46 (12):4869-4873.
[53].Wu, J., and L. Lin. 2002. Elicitor-like effects of low-energy ultrasound on plant (Panax ginseng) cells: induction of plant defense responses and secondary metabolite production. Appl. Microbiol. Biotechnol. 59 (1):51-57.

[54].Wu, J., and L. Lin. 2003. Enhancement of taxol production and release in Taxus chinensis cell cultures by ultrasound, methyl jasmonate and in situ solvent extraction. Appl. Microbiol. Biotechnol. 62 (2):151-155.

[55].Wu, J. Y., and X. C. Ge. 2004. Oxidative burst, jasmonic acid biosynthesis, and taxol production induced by low-energy ultrasound in Taxus chinensis cell suspension cultures. Biotechnol. Bioeng. 85 (7):714-721.

[56].Yang, Ronghui, and Kalidas Shetty. 1998. Stimulation of Rosmarinic Acid in Shoot Cultures of Oregano (Origanum vulgare) Clonal Line in Response to Proline, Proline Analogue, and Proline Precursors. J. Agric. Food. Chem. 46 (7):2888-2893.

[57].Yao, X., J. Chu, X. He, and C. Ba. 2011. Protective role of selenium in wheat seedlings subjected to enhanced UV-B radiation. Russian J. Plant Physiol. 58 (2):283-289.

[58].Zagoskina, N. V., G. A. Dubravina, A. K. Alyavina, and E. A. Goncharuk. 2003. Effect of Ultraviolet (UV-B) Radiation on the Formation and Localization of Phenolic Compounds in Tea Plant Callus Cultures. Russian J. Plant Physiol. 50 (2):270-275.

Determination of Water Quality Status of Stored Method in Tlogo Ampel Watershed

Saiful Anwar[1*], M Joko Wibowo[1], Sugiyarto[2], Budi Hariono[3], Rizza Wijaya[3]
[1] Department of Engineering, Politeknik Negeri Jember
[2] Departement of Agricutural Production, Politeknik Negeri Jember
[3] Department of Agriculture Engineering, Politeknik Negeri Jember

*sanwar2512@yahoo.com

Abstract. The research station is located in Wonokoyo Village, Kapongan Subdistrict, Situbondo Regency at 7°70'44 "South Latitude and 114°08'99" East Longitude. Common problems faced by watersheds in Indonesia are increased activities household, industrial, land conversion of which have a detrimental effect on water quality. This research is important to get a picture of water quality in Tlogo Ampel watershed so that it can be maintained for its sustainability. The evaluation of water quality in a Watershed is a complex thing which involves various influential parameters. This parameter study is divided into three parts that are physical property (Temperature and TSS), chemical property (pH, BOD, COD, DO, PO_4P, NO_3N, NO_2N), and microbiological property (total coliform and Fecal coli). Determining Water Quality Status, one of the methods used to determine the status of river water quality is by the Storet Method. With this Storet Method, we can know the parameters that meet or exceed the Water Quality Standard. The result of the research using Storet method shows the water quality of Tlogo Ampel watershed in the period of January - December 2017 for class I, II, III and IV quality respectively are bad, bad, medium and good. Water quality at Tlogo Ampel Watershed from the result of this research is good and recommended for livestock, fish cultivation and agriculture.

1. Introduction

Water is an important part of human life to support all the activities carried out. Monitoring of water quality is absolutely needed to prevent any deviation use of the water itself. One of the water sources which is widely utilized by Indonesian people is river. River is an open water space that can be utilized and accommodate the waste of households, industry, livestock, and agriculture activities around the river. Tlogo Ampel Watershed is located on Wonokoyo Village, Kapongan District, Bondowoso Regency. The water quality of Tlogo Ampel Watershed needs to be examined to find out its utilization for the surrounding society [1].

The evaluation of water quality in a Watershed is a complex thing which involves various influential parameters. This parameter study is divided into three parts that are physical property (Temperature and TSS), chemical property (pH, BOD, COD, DO, PO_4P, NO_3N, NO_2N), and microbiological property (total coliform and Fecal coli). The rapid development around Tlogo Ampel Watershed causes the changes of the cropping pattern and land use both for businesses and settlements which grow so fast, so that it causes the problem of water quality in the Watershed. The pollutant loads are extraneous to the nature or sourced from the nature itself that enter into an ecosystem and disrupt the function of ecosystem itself [2].

The sources of pollution material that enter into the water can come from the classified waste that are: (1) Point Source Discharges (point source) that is a point source or a pollutant source which can be known with certainty, can be in the form of a location such as industrial or domestic waste water and drainage channels. Waste water is the remainder of a business and/or an activity in the form of liquid (Government Regulation Number 82 of 2001); (2) Non-Point Source (distribution spread) come from the unknown source. Pollutants enter into the water through run off from the agricultural, residential, and urban areas [3].

This research aimed at finding out the water quality of Tlogo Ampel Watershed that became a reference for determining the water quality and the parameters that meet the water quality standard according to the Government Regulation Number 82 of 2001 concerning water quality management and pollution control.

2. Literature Review

Literature review that will be used for this research are

2.1 Watershed (DAS)

According to the experts, the notion of Watershed is very diverse, some call it drainage basin, watershed, or cacthment area. Based on those definitions, the Watershed can be defined as the unity of space which consists of abiotic elements (land, water, air), biotic elements (vegetation, animal, and other living organisms) and the human activities which interact and interdependence one to another so that it becomes an ecosystem unit, this means that if the links are already established, then the management of forest, land, water, society, and so on must concern on the role of the ecosystem components.

The land boundary is a topographic separator and the sea boundary until the water area that are still affected by the land activities (Law Number 7 of 2004). Therefore, the Watershed is a natural unit area which provides the production benefits as well as supplies the water through river, land water and/or spring to fulfill various life needs both for human and flora and fauna. The water utilization for life are for irrigation, agriculture, industry, household consumption, tourism, river transportation, and other needs [4].

2.2Water Resources

Watershed is one of water resources which is widely used by the community. To support this continuity, it is necessary to do a sustainable management. The management of water resources adheres to seven principles as follows [5]:

The Principle of Preservation, implies that the utilization of water resources is done by maintaining the preservation of water resources function sustainably. Principle of Balance, contains the understanding to always place social function, environmental function and economic function in harmony. The principle of Public Benefit, implies that the management of water resources is done to provide maximum benefit for public effectively and efficiently. The principle of integration and harmony, contains the understanding that the management of water resources is done in an integrated in realizing harmony for various interests by paying attention to the dynamic nature of water. The Principle of Justice, implies that the management of water resources is done equally to all levels of society in the territory of the country so that every citizen has the right to have equal opportunities to play a role and enjoy the results in real terms, by still providing protection to the layers of society whose economic level is lacking. The principle of independence, contains the understanding that the management of water resources is done by taking into account the ability and the superiority of the local norms and resources. The principle of Transparency and Accountability, implies that the management of water resources is done transparently and accountably.

2.3 Pollution Source
The pollution source in watershed can be divided into 2 categories, which are domestic and non-domestic.

2.3.1 Domestic
The domestic pollution source comes from village, market, city, terminal, hospital, and others. This pollution source later will produce domestic waste. Domestic waste is all waste that comes from bathroom, kitchen, laundry waste, hospital, pharmacy, and restaurant. This domestic waste contains organic materials in the form of liquid or solid, hazardous material, toxic material, dissolved salt, and bacteria.

2.3.2 Non - Domestic
Non-domestic pollution source is a pollution source comes from factory, industry, livestock, transportation, agriculture, fishery, and others. Agricultural waste consists of solid material of plant residual which are organic, pesticides, fertilizers that contain nitrogen, phosphorus, sulfur, minerals K and Ca. Meanwhile, the agricultural and livestock products waste are the by-products of fisheries and agriculture management.

2.4 Water Quality standard
Water quality standard is basically divided into four classes, namely 1) First class water, which is water that can be used for drinking water and/or other uses that require the same water quality, 2) Second class water, which is water that is designated to be used as water recreation infrastructure/facilities, freshwater fish cultivation, livestock, water to irrigate plants, and/or other uses that require the same water quality as these uses, 3) Third class water, which is water that can be used for freshwater fish cultivation, livestock, water to irrigate plants, and/or other uses that require the same water quality as these uses, 4) Fourth class water, which is water that is designated to be used to irrigate plants and/or other uses which require the same water quality as this uses [6].

2.5 Storet Method
Referring to the State Minister of Environment decree No. 115 of 2003 concerning Guidelines for Determining Water Quality Status, one of the methods used to determine the status of river water quality is by the Storet Method. With this Storet Method, we can know the parameters that meet or exceed the Water Quality Standard. In principle, Storet Method used to compare between water quality data and water quality standard that are adjusted to the designation to determine the Water Quality Status [7].

3. Research Methodology

3.1 Research Procedure
The research procedure was done through literature study to find any suggestions to support this research, the supporting data collection was the conventional data obtained from the institution which supported the objective and the target of the research, as well as field survey and report arrangement. In the survey preparation, there was a research area determination as well as tool and material preparation.

3.2 Research Location
This research was conducted at Tlogo Ampel Watershed with the observed station located at Wonokoyo Village, Kapongan District, Situbondo Regency.

3.3 Tool and Material
The used research tool was Excel 2010 software. The used materials were: (1) the physical properties data (Temperature and TSS), (2) the organic chemical properties data (pH, BOD, COD, DO, PO_4P and NO_3N, NO_2N) and (3) the microbiological properties (total coli and fecal coli).

3.4 Analysis Data
3.4.1 The Framework of the Research Approach
This research was conducted by collecting the needed data to know the water quality of Tlogo Ampel Watershed. Then, the water quality analysis was done by using storet method. Next, the trend determination from the storet water quality status of each monitoring station. The data needed in the study was water quality supporting data of the monitoring station at Tlogo Ampel Watershed from January-December 2017. The used parameter were temperature, TSS, BOD, COD, DO, PO_4P, NO_3N, NO_2N, Total coliform dan Fecal Coli.

3.4.2 Storet Method Data Analysis
The data obtained through Storet Method was analyzed and concluded by enclosing the literature as the support. The water quality class II based on the government regulation Number 82 year 2001 about Water Quality Management and Water Pollution Control was used as the reference for water quality properness. The water quality determination was based on the assessment system of US-EPA (Environmental Protection Agency). The procedures of water quality determination by using Storet Method as shown in Table 1 were as follow:

1. Collecting the data about water quality periodically.
2. Comparing the data from the measurement results of each water parameters to the threshold value which appropriate to the water class.
3. If the result fulfilled the water quality standard value (measurement result ≤ water quality standard), then it scored 0.
4. If the result did not fulfill the water quality standard value (measurement result > quality standard), then it scored 1.
5. Counting up all the negative scores from each counted parameter and deciding the quality status from the total.

Table 1. The Value System determination to decide the water quality status

Sample Quantity	Score	Parameter		
		Physic	Chemical	Biology
<10	Max	-1	-2	-3
	Minimum	-1	-2	-3
	Average	-3	-6	-9

After that, the classification was done based on the obtained value as follows:

1. Class A : Excellent, Score = 0 fulfilled quality standard
2. Class B : Good, Score = -1 s/d -10 light polluted
3. Class C : Moderate, Score = -11 s/d -30 moderately polluted
4. Class D : Bad, Score = ≥ -31 badly polluted

4. Experiment and Result

The water quality determination of Tlogo Ampel Watershed by using the decided parameters obtained from January- December 2017 was shown in Table 2. Table 2 showed that the physical parameter (Temperature and TSS) and microbiology (Total coliform and Fecal coli) were not the sources of the pollution at Tlogo Ampel Watershed. It could be seen that the data distribution obtained from January-December 2017 was still in suggested quality standard. The temperature parameter data tended to be constant every month, while the highest TSS parameter data was in December as much as 255.40

mg/L. The total coliform parameter reached the highest point in October-November as much as 210 population/100ml, while the Fecal coli parameter was high in October-November as much as 120 population/100ml.

The biggest pollution was caused by the chemical pollution such as BOD, COD, DO, and NO_2N. The data obtained from BOD parameter reached the highest point in May as much as 14.45 mg/L. The COD parameter tended to rise, yet it reached its highest point in November as much as 30.18 mg/L. The DO parameter tended to be constant every month but in April it decreased as much as 2.61 mg/L. The NO_2N parameter obtained from the research result was about 0-0.16 mg/L, therefore it can be said that the NO_2N content was not really different every month.

Table 2. The parameter observation data of water quality determination

PARAMETER	MONTH											
	JAN	FEBR	MARCH	APRIL	MAY	JUNE	JULY	AGUST	SEPT	OCT	NOV	DEC
Temperature	24,00	23,50	23,40	23,60	23,90	23,60	23,40	24,70	24,60	24,70	24,60	24,30
TSS	117,20	116,40	50,80	22,80	82,40	35,10	20,80	73,30	38,00	40,90	94,10	255,40
Ph	8,20	7,50	7,70	6,90	8,10	6,60	6,30	6,60	6,60	6,50	6,30	6,70
BOD	6,25	4,60	4,80	4,70	14,45	5,80	7,30	6,20	8,15	4,60	9,55	8,85
COD	16,87	18,28	22,55	26,14	34,21	14,75	19,84	18.94	24,38	11,76	30,18	27,19
DO	5.70	6,50	6,50	2,61	3,90	7,00	6,30	6,30	4,90	6,90	5,00	3,30
PO4-P	0,26	0,26	0,15	0,10	0,08	0,07	0,25	0,15	0,19	0,21	0,10	0,13
NO3-N	2,80	1,16	2,84	2,41	0,14	2,77	2,55	7,93	2,36	1,93	1,67	2,19
NH3-N	tt	0,44	tt**)	0,09	1,03	0,10	0,09	0,12	0,24	0,11	0,05	0,04
NO2-N	0.08	0,01	0,03	0,00	0,00	0,03	0,02	0,06	0,06	0,11	0,16	0,13
Total Coliform	150,00	25,00	75,00	150,00	39,00	240,00	93,00	93,00	150,00	210,00	210,00	150,00
Fecal Coliform	28,00	14,00	15,00	75,00	20,00	43,00	75,00	43,00	93,00	120,00	120,00	39,00

Storet method was very good to be used in deciding the water quality status of Tlogo Ampel Watershed. The measurement result of storet method in January-December 2017 could be seen in Table 3.

Table 3. The storet measurement result in January-December 2017 Period

No.	Parameter	Result			Score Class I	Score Class II	Score Class III	Score Class IV
		Max	Min	Average				
	Physic							
1	Temperature	24,700	23,400	24,050	-1	-1	-1	0
2	TSS	117,200	20,800	69,000	-4	-4	0	0
	Chemical							
1	pH	8,200	6,300	7,250	0	0	0	0
2	BOD	14,450	4,600	9,525	-10	-10	-8	-2
3	COD	34,210	14,750	24,480	-10	-2	0	0
4	DO	7,000	0,261	3,631	-8	-8	-2	0
5	PO4-P	0,264	0,069	0,167	-2	-2	0	0
6	NO3-N	7,932	0,140	4,036	0	0	0	0
7	NH3-N	1,026	0,088	0,557	-2	-2	0	0
8	NO2-N	0,075	0,002	0,039	-8	-8	-8	0

	Microbiology							
1	TotalColi	240,000	25,000	132,500	0	0	0	0
2	Fecal Coliform	120,000	14,000	67,000	-3	0	0	0
		Index Pollution			-48	-37	-19	-2

The water quality data of Tlogo Ampel Watershed was not classified into class I or class II as shown in table 3. This was shown by the pollution index that had been discovered. Therefore, the water in Tlogo Ampel Watershed was not suitable to be used as water consumption as well as freshwater fish cultivation. The total number of pollution index in class I was -48,if it was referenced to the classification value from US-EPA, then it was worse and badly polluted. The biggest contributor of the pollution in class I were BOD, COD, DO, and NO2N.

Table 4. Water quality class classification of Tlogo Ampel Watershed

DAS Tlogo Ampel	Water Quality Qualifications			
	class I	class II	class III	class IV
	-48	-37	-19	-2
	Bad	Bad	Medium	Good

Class II category obtained from the pollution index as much as -37, this could be categorized as a worse and badly polluted. The biggest contributor of the pollution in this class was dominated by BOD, DO, and NO_2N. In class III category which was allocated to be used for freshwater fish cultivation, livestock, and irrigate planting, and/or another usage that required the same water quality with the allocation got -19 on the total pollution index. This category was classified into moderate polluted. The contributed pollutant was dominated by BOD and NO_2N, while the other parameters were still in the determined threshold.

Class IV category which was allocated to be used for watering plants and another usage that required the same water quality with the allocation obtained -2 in its pollution index. This number was categorized as good and light polluted which meant that the water quality of Tlogo Ampel Watershed could be used for farming activity around the Watershed. The pollution in category IV originated only from BOD but still in the normal standard.

The result on this research can give recommendation about proper water use on Tlogo Ampel Watershed. Water quality on Tlogo Ampel Watershed is good for livestock and cultivation because it is included on category III. Water quality on Tlogo Ampel Watershed is very good for agriculture because it is included on category IV. The surrounding community can use this water to help their daily activities by building water streams using pipes and pumps for agriculture. In addition, it can also utilize riverbanks for freshwater fish cultivation

5. Conclusion

From the results of this study it can be concluded if water quality status at Tlogo Ampel Watershed is not recommended for water drink and water recreation. Water quality at Tlogo Ampel Watershed is good for livestock or fish cultivation and very good for agriculture.

Acknowledgments

Our thank goes to Department of Agricultural Engineering, Politeknik Negeri Jember, who has helped support for this research.

References

[1] Costa D, Burlando P, Priadi C 2016 *Sustainable Cities and Society* :**20**: 199-209.
[2] Sutriati, Armaita 2018 *Journal of Water Resources* **8**: 81-94.
[3] Sahabuddin, Hartina, Donny Harisuseno and Emma Yuliani 2014 *Water Resources Engineering Journal* **5**: 19-28.
[4] Bachrein, Saeful 2015 Bina Praja Journal: *Journal of Home Affairs Governance* **4**: 227-236.
[5] Pasandaran, Effendi 2017 *Agricultural Policy Analysis* **3**: 217-235.
[6] Macbub, Badruddin 2018 *Journal of Water Resources* **6**: 129-144.
[7] Novita Sari, Ratna, Titik Istirokhatun and Sudarno Sudarno 2017 *Journal of Environmental Engineering* 3: 1-9.

Determination of Agro-Industry Area Based on Cassava Commodity in Bondowoso Regency

D. Hermanuadi[1*], A. Brilliantina[2] and E.K. Novitasari[1]

[1]Department of Agricultural Engineering, State Polytechnic of Jember Mastrip Street PO BOX 164 Jember, Indonesia
[2]Department of Food Industry, State Polytechnic of Jember Mastrip Street PO BOX 164 Jember, Indonesia

*didiek_hermanuadi@polije.ac.id

Abstract. Bondowoso Regency has various potential commodities in agricultural sector, and one of them is cassava. It was seen from the high productivity and the amount of cassava production in Bondowoso Regency. The high productivity of cassava in Bondowoso, was supported by natural condition of Bondowoso regency with 27,67 percent suitable field for cultivation of cassava. The main purpose of this study is to determine the agroindustry area in Bondowoso Regency based on superior commodity of cassava. The first step was taken to determine the potential agroindustry of potential commodity is by using LQ method. Secondly, to determine the superior commodity based on cassava in agroindustry area is by using MCDM method. The result of the study showed that Wringin district is a potential area for the development of agroindustry based on cassava with superior commodity like Tape and it's processed products. **Keywords** : Agroindustry Area, Cassava, Bondowoso Regency, Tape, Superior Commodity

1. Introduction

Regional development is basically carried out by optimizing the natural resources through the development of local economic, based on basic economic activities in one area. Bondowoso Regency has the main potential in the agricultural sector which contributes quite large in the PDRB as well as for the labor absorption. Considered the number of livelihood-based population, the majority of the population works in the agriculture – based on home industry sector [2].

One sector that provided the largest contribution was the agricultural sector, which was 48.40% [2]. The data showed that the potential for agriculture was quite large which can be used as industrial raw materials. However, based on RPJMD of Bondowoso District 2013-2018, this potential had not been utilized optimally in industrial sector activities. This because the minimum optimization of small and medium industry in using the raw materials of best regional agricultural potential and the use of appropriate technology in industrial development. In addition, the development of agricultural sector in Bondowoso Regency had not been able yet to create more value for developing Bondowoso Regency.

Bondowoso Regency still has a relatively high level of unemployment,which is an issue in human resources development. This indicates that the development of agricultural sector as the basic sector has not been able to give more valuable contributions. Therefore, agricultural development must be followed by the development of complementary sectors and industriap sectors based on agricultural products (agro-industry), so that the additional sources aside from agricultural land could be obtained [8].

The government of Bondowoso Regency need to make policy priorities to keep the regional development run well. The determination of these policy priorities can be created by recognizing the superior commodity through the increased value of agricultural products. Each region had different superiorities to be optimally used. The prospect of increasing demand for value in agricultural products is an incentive to increase attention of agro-industries development in the context of economic growth and strategies to reduce poverty.

The purpose of this study was to discuss the application of the LQ method and the TOPSIS MCDM method, in identifying the potential of cassava – based in Bondowoso district. In addition, it is expected that the results of this study can be used as the basic for developing the potential of cassava in the Bondowoso area

2. Literature Review

Location Quotient (LQ) method to identify the best commodities was accommodated from Miller & Wright (1991), Isserman (1997), and Ron Hood [4]. According to Hood (1998), LQ is a simple and easy economic analysis tool with all its advantages and limitations. LQ method is one of base economic models, as the first step to understand the activity sector that supports the economic growth. LQ is used to measure the relative concentration or level of specialization of economic activities by using comparative approach through. This method presents a relative comparison of a sector ability in the area under investigation with the same ability in a wider area. The unit is used as a measurement to produce the LQ coefficient is the amount of labor, production, or other units that can be used as criteria.

Multiple Criteria Decision Making (MCDM) is a method of decision making to determine the best alternative of a number of alternatives based on certain criteria. Usually, the forms of the criteria, are measurements, rules or standards that are used in decision making. Based on its objectives, MCDM can be devided into 2 models i.e Multi Attribute Decision Making (MADM) and Multi Objective Decision Making (MODM). MADM is method that used to solve problems in a discrete space. Therefore, the assesment or selection of several alternatives in a limited number can be done by using MADM. While, MODM is a problem solving method in continuous space, such as problems in mathematical programming. Generally, it can be concluded that MADM is the right method to select the best alternative of a number alternatives, and MODM is used to designs the best alternatives [5]. The model used in this decision support system is Multi Attribute Decision Making - Technique for Order Preferred Similarity to Ideal Solution (MADM-TOPSIS). It is used for problem solving in rangking.

The TOPSIS method is a technique for ordering preferences based on similarity to an ideal solutions. An ideal solution (also called an ideal positive solution) is a solution to maximize the benefit and minimize the cost, while ideal negative solution (which is called anti-ideal solution) is used to maximizes the cost and minimizes the benefit. The best alternative is an alternatives which has the closest value to the ideal positive solution and the furthest from the ideal negative solution [9].

3. Research Method

3.1 Conceptual Framework

Regional development planning must ideally be supported by accurate data. In order to fulfill the accurate data, it is necessary to identify the potential of an area as an initial basis for further planning. The development of a strong agro-industry area must be in line with the vision and mission, and regional development policies. The strategy of developing best agroindustrycommodity is one of the efforts to accelerate the creation of comprehensive and simultaneous community welfare.

The initial step in formulating a best commodity development strategy was to identify the potential areato develop cassava-based agroidustry in Bondowoso Regency, then analyzing the best products to be developed. This approach was carried out by considering all aspects of supply (comparative advantage) and aspects of demand (competitive advantage) of regions that provided additional value and have a prospective for the development of best agroindustry of cassava-based. Identification of areas that have the potential to develop cassava-based agro-industry is done using the LQ method, while the MCDM method is used to determine the best commodities. The cluster approach was carried out to decide the policy of developing cassava-based superior agroindustry as an effort to increase additional value and regional competitiveness. The flowchart of the conceptual framework of this research is outlined in the form of a flow chart in Figure 3.1.

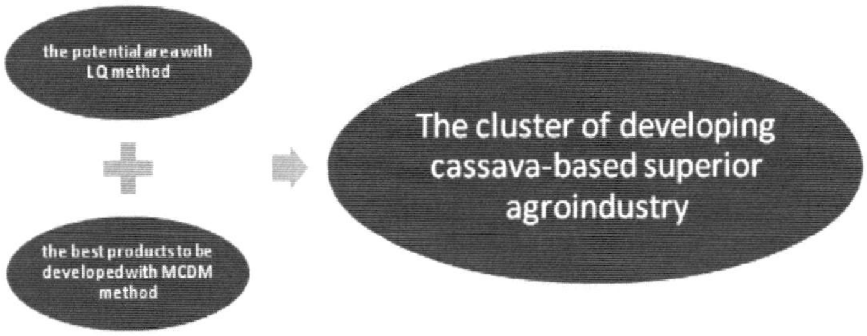

Figure 3.1 The flowchart of the conceptual framework

3.2 Place and Time of Research

This research activity was done in Bondowoso Regency, East Java. The research time was held from June to August 2018.

3.3 Data Collection Method

Data collected consist of primary and secondary data. Primary data were obtained by conducting field observations, interviews and distributing questionnaires to respondents. The technique used in taking respondents in this research was purposive sampling method by determining or deliberately choosing the example to be studied. Respondents in this research consisted of experts related to the agro-industry sector both from the internal government and from external parties as many as 15 respondents which included 3 people of Agriculture Departement, 4 people of Industry and Trade Departement, 5 practitioners, and 3 academicians. Secondary data was gained from library searching and department documents/ related institutions. Secondary data involved the statistical data from the Central Bureau of Statistics of Bondowoso Regency and other sources of literature relevant to the research topic.

3.4 Data Analysis Methods

The collected data was analyzed according to the research objectives to be achieved. The tools to help analyze and process the data were MS-Excel and Expert Choice 2000 programs.

To find out the development prospects of an area which was based on the comparative of the best potential,identifying the best commodityof basic and non-basic sectors using the LQ method was used. This approach was a comparison between the functions of relative production / range of commodity area at the regional level with the function of relative production / range of commodity area at a larger level. This can be formulated as follows:

$$LQij = \frac{Xij/Xi}{X.j/X..}$$

Where:

LQ*ij*: Quosien index of commodity location *i*in Bondowoso District
X*ij*: Total area size / commodity production *i* in the District to Bondowoso Regency
X*i*.: Total area size / total production of all commodities throughout Bondowoso Regency
X.*j*: Total area size / total commodity production*i* in East Java Province
X..: Total area size / total commodity production in East Java Province

Assessment criteria in determining the basis and non-basis degree sizes depended when the LQ value is greater than one (LQ> 1), the commodity belonged to basis commodity. Whereas if the value is equal to or smaller than one (LQ <1) means the commodity is included in non-basis commodity [10].

Multi criteria decision making from a problem that had n alternatives and m criteria. Each alternative was evaluated according to m criteria. All values / ranks assigned to alternatives with respect to the decision matrix denoted by the criteria weighting vector and fulfilling [1].

4. Results And Discussion

4.1 Analysis of Determination of Cassava-Based Agroindustry Potential Areas

In applying LQ method for the productivity of cassava plants, each unit was used for harvest area (ha) g and the amount of productivity. LQ results which based on harvest area aspect could meet the best criteria of the offer, since the harvest area was the result of the corresponding growth of plants with agroecological conditions which implicitly included elements (variables) of climate, physiography and soil type.

Table1 Static LQ districts in Bondowoso Regency

No.	Distric	Static LQ
1	**Wringin**	3,56
2	**Curahdami**	2,12
3	**Taman Krocok**	2,02
4	**Tegalampel**	1,44
5	**Prajekan**	1,42
6	**Botolinggo**	1,33
7	**Cermee**	1,26
8	**Binakal**	0,87
9	**Klabang**	0,81
10	**Jambesari DS**	0,78
11	**Sumber wringin**	0,40
12	**Tamanan**	0,32
13	**Grujugan**	0,20
14	**Pujer**	0,09
15	**Bondowoso**	0,08
16	**Sempol**	0,07
17	**Pakem**	0,04
18	**Sukosari**	0,04
19	**Maesan**	0,03
20	**Telogosari**	0,00
21	**Tapen**	0,00
22	**Wonosari**	0,00
23	**Tenggarang**	0,00

In the application of LQ towards the acquisition of areas that could be developed into cassava-based agroindustry areas based on aspects of harvested area, it was defined that LQ was the ratio between the relative share (share) of harvested area of cassava commodity at the regional level to the total area of harvested area of the sub-sector with a relatively broad share the harvest area of cassava at the national level against the total area of the national budget subsector. Through the above approach, it could be observed of whichdistricts could be developed to become a cassava-based agroindustry area in Bondowoso Regency (Table 1).

From Table 1 it can be seen that there are 7 districts in Bondowoso Regency could be developed into cassava-based agroindustry areas. These seven districts were Wringin district, Curahdami district, Krocok Park District, Tegal Ampel District, Pradjekan district, Botolinggo District, and Cermee District. Thosedistrict was chosen because the value of LQ was above 1, according to [3] LQ analysis is used to determine the commodities of the food crops sub-sector which has comparative advantages, with LQ criteria> 1. Furthermore, the kind of superior commodity analysis in order to be developed in the cassava-based agroindustry region was decided using the MCDM method.

4.2 Analysis of Determination of Cassava-based superiority Commodity in Bondowoso Regency

Determination the weight of decision's factors of agro-industry area was done by TOPSIS MCDM analyzing tool using Microsoft Excel. In this case there were several assessments that would be done by considering the values against the criteria. Alternatives in this case are cassava-based agroindustry in Bondowoso regency, namely: tapioca, MOCAF, chips, and tape and its processed.

There were several criteria used as design materials, namely market potential (C1), raw material (C2), additional value(C3), labor (C4), technology (C5), social culture (C6), and environmental impact (C7). Furthermore, these criteria would be used as factors to determine the superior products. The weight was a criterion that ought to exist in determining the decision to decide the superior product. The weighting values of each criterion are shown in table 3. Here was the decision making process with MCDM-TOPSIS:

a. Giving the value of each alternative (Ai) on each criterion (Cj) that had been determined, where the value was obtained based on the value of crisp, i = 1.2, ... m and j = 1.2, ... n.

Table 2. Value of Each Alternative

Alternative	C1	C2	C3	C4	C5	C6	C7
Tape	94	60	96	96	82	96	96
Keripik	102	54	70	102	88	104	104
Tapioka	94	58	70	98	78	102	82
Mocaf	86	56	62	78	78	102	78

b. Giving the weight value (W) on table 3.

Tabel 3. The Weight Value of Criteria

	Criteria	Weight
C1	market potential	0.17
C2	raw material	0.14
C3	add value	0.15
C4	labour	0.15
C5	technology	0.12
C6	socio-cultural	0.14
C7	environmental impact	0.14

c. Normalizing the matrix by calculating the normalized performance rating value (rij) of the alternative Ai on the Cj attribute. The normalization process of attributes to form a normalized matrix (R) and the multiplication between the weights with the value of each attribute to form a V matrix.

Table 4. VMatrix

Alternative	C1	C2	C3	C4	C5	C6	C7
Tape	1,353	0,688	1,162	1,162	0,825	1,113	1,113
Keripik	1,356	0,572	0,782	1,140	0,818	1,113	1,113
Tapioka	1,249	0,614	0,782	1,095	0,725	1,092	0,878
Mocaf	1,143	0,593	0,693	0,872	0,725	1,092	0,835

d. Determinating of positive ideal solutions and negative ideal solutions by using the value of matrix V. Positive ideal solutions were obtained by selecting the largest value from each criterion of all alternatives. On the contrary, a positive ideal solution was obtained by selecting the smallest value from each criterion of all alternatives.

Table 5. Ideal Solutions

Alternative	C1	C2	C3	C4	C5	C6	C7
A+	1,353	0,688	1,162	1,162	0,825	1,113	1,113
A-	1,143	0,593	0,693	0,872	0,725	1,092	0,835

e. Distance Determinatingconsidering the ideal solutions and preferences of each alternative. The value of the decision matrix and the ideal solution were used to determine the distance between the values of each alternative with the positive ideal solution and the negative ideal solution. Calculation of distance. The chosen alternative is an alternative that has the shortest distance from the positive ideal solution and the farthest distance from the negative ideal solution.

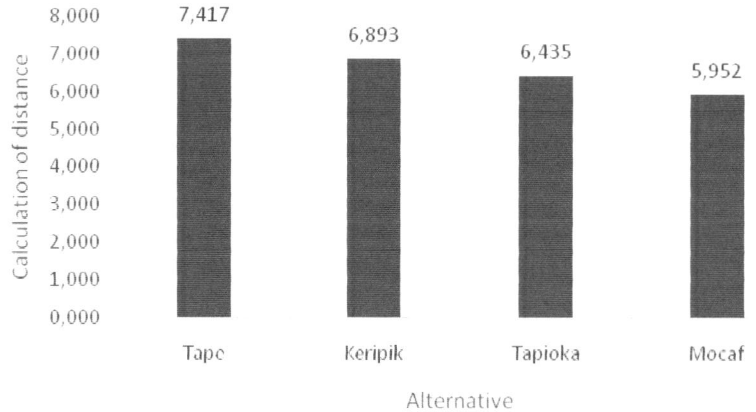

Picture 4.1 Alternative Ranking Results

Based on the value of the distance calculation for each alternative, the shortest distance from the positive ideal solution was cassava and processed tape, the farthest distance from the negative ideal solution was MOCAF product. So it can be concluded that the best product proposed to Bondowoso Regency government to be given the impetus to advance its agro-industry was tape and its processing.

5. CONCLUSION

Based on calculation results using TOPSIS LQ and MCDM method, it can be concluded that Bondowoso District could develop cassava-based agroindustry by forming a cluster in seven districts that are Wingin District, Curahdami District,Krocok District, TegalAmpel District, Pradjekan District,Botolinggo District, and Cerme District. While the best productswere in the form of tape and its processed.

ACKNOWLEDGEMENTS

This paper is dedicated to kemenristekdikti. This paper is part of the 2018 Higher Education grant applied research.

REFERENCES

[1] Ashtiani, B., Haghighirad, F., Makui, A., Montazer,G.A., 2008. Extension of Fuzzy TOPSIS MethodBased on Interval-valued Fuzzy Sets. *Applied Soft Computing*. Vol. 9, No.2, pp. **457-461**

[2] BPS Kabupaten Bondowoso. 2016. *Kabupaten Bondowoso dalam Angka Tahun 2016.* Kabupaten Bondowoso:Badan Pusat Statistik

[3] Hendayana, R. 2003. Aplikasi Metode Location Quotient (LQ) dalam Penentuan Komoditas Unggul Nasional, Balai Pengkajian dan Pengembangan Teknologi Pertanian, Bogor, *http://www.litbang.deptan.go.id/warta-ip/pdf-file/rahmadi-12.pdf*, 3 Maret 2013.

[4] Hood, R. 1998.*Economic Analysis : A Location Quotient, Primer,* Principal Sun Region Associates, Inc.

[5] Kusumadewi, S. Hartati, S. Harjoko, A. Wardoyo, R. 2006. Fuzzy Multi-Attribute Decision Making (Fuzzy-MADM). Yogyakarta: Graha Ilmu

[6] Miller, M. M., Wright, G.N. 1991.Location Quotient Basic Tool for EconomicDevelopmentAnalysis. *Economic Development Riview*, J. 9(2), 65.

[7] Morrissey, Karyn. 2014. Producing regional production multipliers for Irish marine sector policy: A location quotient approach, Ocean & Coastal Management, Volume 91, 2014, Pages 58-64, ISSN 0964-5691, https://doi.org/10.1016/j.ocecoaman.2014.02.006.

[8] Rustiadi, E. Dan S.Pranoto. 2007. *Agropolitan Membangun Ekonomi Perdesaan.* Bogor :Crespent Press.

[9] Wang, Y. M., & Elhag, T. M. S. 2006. Fuzzy TOPSIS method based on alpha level sets with an application to bridge risk assessment. *Expert Systems with Applications*, 31, pp. **309–319.**

[10] Widodo, Tri. 2006. *Perencanaan Pembangunan: aplikasi komputer (era otonomi daerah).* Yogyakarta:UPP STIM YKPN.

[11] Yang, J., 2008. Vendor Selection by Integrated Fuzzy MCDM Techniques with Independent and Interdependent Reletionship. *Information Sciences,* pp. **4166-4183**.

[12] Żak,J. Szymon Węgliński. 2014.The Selection of the Logistics Center Location Based on MCDM/A Methodology.*Transportation Research Procedia,*Volume 3,2014,Pages 555-564,ISSN 2352-1465,https://doi.org/10.1016/j.trpro.2014.10.034.

IOP Conf. Series: Earth and Environmental Science **207** (2018) 012011 doi:10.1088/1755-1315/207/1/012011

Quality Improvement of Vannamei Shrimp Production Process Using ISO 9001:2015

R Iskandar[1*] W Dhamayanthi[2] and I A A Pongoh[3]

[1]Agribusiness Management, Politeknik Negeri Jember
[2]Management Economics, Politeknik Negeri Jember
[3]Aquaculture, Politeknik Negeri Jember

*e-mail: ridwan.iskandar@polije.ac.id

Abstract. Every manufacturer who joins the trade blocks must always produce high quality products to compete in free market. This article presents the research results on implementation of quality management system in order to produce high quality products of fishery sector in Indonesia. Vannamei shrimp (Litopenaeus vannamei) is one of important high value of fish export commodities. Indonesia's vannamei shrimp production volume continues to increase but in reality it only supplies 12.1% of world's shrimp demand. This low supply has been linked to higher world competition in vannamei shrimp. In order to face the competition, it needs to improve the quality management system of production. Therefore, it needs a research to answer the competition problem. The study sample was chosen purposely from vannamei shrimp hatcher managers in East Java, Central Java, West Java, and Lampung and experts from universities. The implementation of study includes self assessment, readiness study, and implementation priority of ISO 9001:2015 clause. The research results show that some preparation elements of quality system are in compliance with international standards but there is still an important part of quality system not appropriate. Priority clauses in its application are Leadership, Enhancement, Performance Evaluation, Organizational Context, Operation, Planning, and Support.

1. Introduction

Indonesia has joined in trade blocs, such as APEC and MEA. This has the consequence of always producing high quality products. High quality products were obtained through the application of quality assurance concept, a quality system designed to ensure all stages of production activities work well. A definite guarantee to produce high quality products was a quality management system, such as ISO.

Indonesia was spurring various sectors to produce high quality products, one them was fishery sector. The GDP contribution of fishery sector to national GDP year 2014 was 2.34%, higher than 2013, while the contribution of agricultural sector was declining [1]. The vannamei shrimp (*Litopenaeus vannamei*) was high-value export commodities in fishery sector. Indonesian vannamei shrimp production increased 20.49% per year in period 2010-2014 [2]. According to [3], Indonesia has only supplied 30% of Japan's shrimp needs while world-level supply was only 12.1%. The not maximal supply was related to higher competition of world shrimp trade as a result of emergence of new producing countries.

The supply of Indonesian vannamei shrimp to global market was still likely to increase and compete with China and Thailand. With special efforts, Indonesia was optimistic to seize 10% -15% of needs of world markets. According to [4], to seize the above mentioned market, there were still many things that must be addressed, such as increasing the development of cultivators and seed producers should create products with best quality. Only quality commodity could win the market competition.

Production management should be improved to win the competition. Indonesia should improve the high quality production consistently. This could, among others, be pursued by upgrading the producers by implementing international quality standards. Vannamei shrimp production process was applied Indonesian National Standard (SNI). That was SNI 01-7252: 2006 standard about vannamei shrimp shed quality [5] and SNI 7311: 2009 about standard production of vannamei shrimp [6]. The previous research results indicated that application of this standard has a fair category. This means that some elements of SNI have been applied well, but there were still important elements that have not been consistent with these standards [7].

In relation to international quality commodity suppliers, it was necessary to change the application of high SNI quality standard toward ISO, the international quality standards recognized by various countries. The most up-to-date ISO series was a revised by 2015 series, therefore the ISO standard in question was ISO 9001: 2015.

Above description has encouraged this research with priority of upstream sector, the seed quality of vannamei shrimp for export quality. This research would analyze and develop the implementation of ISO 9001: 2015 on process of vannamei shrimp seeding to analyze the application level and at same time developed the correction efforts for vannamei shrimp farming agribusiness was to get ISO certified.

This research was conducted at vannamei shrimp production center in Indonesia. This area had many companies with "Good/Excellent" certificate and this area has been able to supply farmers in Indonesia.

2. Literature Review

2.1. Quality Control and Quality Assurance

To produce high quality products, producers must apply following Quality Management Principles: maintaining customer focus, optimizing leadership, using workers' participation, using process approach, making quality improvement commitments, using factual data in decision making, and fostering relationships with suppliers [8]. More specifically, [9] suggests that in order to obtain high quality products, companies need to apply the concept of quality control and quality assurance.

Quality Control means that every product must be produced in accordance with demand, where the approach was done through the application of Total Quality Management. It was an approach in running a business that tries to maximize employee participation and organizational competitiveness through continuous improvement. The effective of Quality Assurance includes three things: (1) an effective Quality Management System, (2) Periodic examination of system implementation to ensure system effectiveness, and (3) Periodic review of system to ensure the system continuously meets the changing demand.

2.2. ISO 9001:2015 Quality Management System

ISO 9001 documents explained argued that the adoption of a high quality management system is a strategic decision for any organization to improve overall performance and provide a foundation for sustainable development. In other words, ISO 9001 is a very definite guarantee and a standard guideline for design, manufacturing, sales and even service activities. ISO 9001 is a set of standards to leads the company to a high level of quality assurance in world-level business activities. The ISO 9001 documents also expressed the potential benefits to be gained by implementing ISO-9001 2015:

- Ability to deliver high quality products and services consistently.
- Increasing customer satisfaction.
- Knowing the risks and opportunities to achieve the goals.
- Proving that its quality managements system conforms to requirements of international standards.

[10] said that body certification give a certificate as a sign of a business activity applied sustainable quality system standards. The average economic profitability of a company with ISO 9001 was greater than non certified company. The products of certified company of a country would be easier to be accepted in other countries. It could be said that ISO 9001 was the international language produces

the goods. In addition, the study results indicated that companies performance with ISO 9001 certificate were better than without ISO 9001 certificate [11]. Similarly, performance point view in statistical analysis indicated that implementation of ISO 9000 significantly improves quality performance [12]. [13] concludes that ISO 9001 certification has a significant effect on quality outcome, business performance and customer satisfaction.

The advantages of ISO 9001 certification were greater access to foreign markets (especially those requiring ISO 9001) and compatibility with overseas suppliers. The process to achieve certification tends to improve the quality and uniformity of work and simultaneously also increase productivity which in turn could also increase competitiveness.

The most important thing that should be developed in application of ISO 9001 was quality documentation. All procedures, work instructions, formats, data, records, receipts, etc were written on paper. The quality documentation was described as a document pyramid, wider scope becomes at down in terms of details and volume.

The most important quality documentation was quality manual and quality system procedures. They were all reference to ISO 9001 activities. Quality manual were documents that contain company statements and commitments on application of ISO 9001. Quality manual were made by interpreting ISO 9001 clauses that were tailored to application conducted by company. The Quality System Procedure was a system to describe the methods used to implement the quality system specified in quality manual.

Sustainable production process is a controlled process to produce a high quality product. Each stage of the process will be executed with clear reference to quality standards, for example presented in Table 1 shows the nauplius and fries production process based on Indonesian National Standard.

Table 1. Seed Production Benchmark for Nauplius and Seed

No	Criteria	Unity	Size	Measurement time
1	Water temperature	^oC	29 - 32	Twice a day
2	Salinity : -Seed -Nauplius	0/00	29 – 34 31 – 34	Everyday
3	pH	-	7,5 – 8,5	Everyday
4	Dissolved oxygen, min.	g/l	5	Maximum every 3 days
5	Nitride, max.	g/l	0,1	Everyday
6	Pathogen bacterial, max.	Cfu/ml	10^3	Maximum every 3 days

Reference at Table 1 showed that the seed produced would have high quality and ISO 9000 ensures that standard reference was adhered properly.

2.3. Good Seeding Method

The vannamei shrimp seeding farming agribusiness was conducted based on Good Seeding Practices (CPIB) as a reference for fish seed business to produce high quality seeds. Good seeds were characterized by: rapid growth, uniform, high syntax, adaptive to enlargement environments, parasite-free and disease resistant, efficient in feed usage and contain no chemical residues and drugs which could harm humans and environment.

The study of quality seeds production was based on the compliance of production systems to CPIB. It was similar to implementation of ISO, difference lies in intended market scope. The problem was if the production system applied good CPIB, whether the production system was said to have implemented the good well ? This study does not compare the two but the two quality systems were mutually supportive. It could generally be said that application of CPIB could support the successful implementation of ISO, or vice versa. It depends on where we would analyze it. The CPIB generally contains following standards: seeding requirements, production processes, biosecurity implementation, and personnel management. In addition, it also contains the requirements of documentation system consisting of Standard Operational Procedure (SOP), form and record.

3. Research Method

This research was conducted at vannamei shrimp seed production center in East Java, Central Java, West Java and Lampung. Samples were taken purposively, consisting of vannamei shrimp farmers and experts from universities. The population is 26 hatcheries and the samples taken are 4 hatcheries. The study was conducted in April-August 2017.

3.1. Research Stages

a. Identifying the 9001:2005 clause and adapting to the vannamei shrimp seeding.
b. Improvement and refinement of the ISO 9001:2015 clauses that were still weakly applicable based on self-assessment results by employers.
c. Assessing the readiness to implement ISO 9001:2015 clauses.
d. Deciding the clauses implementation priority of ISO 9001:2005.

3.2. Flow Chart of Research

The research flowchart was structured as shown in Figure 1 below.

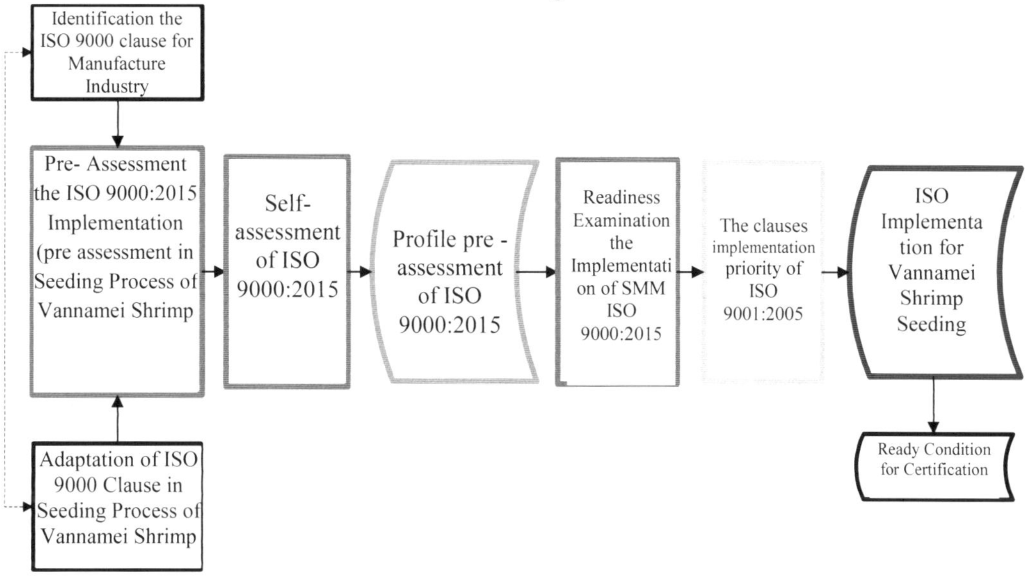

Figure 1. Research Flow Chart

3.3. Measurement and Development Method of Quality Management System

a) Identification of ISO 9001:2015 clause for manufacture and its adaptability into vannamei shrimp breeding process.
ISO clause identification was done by Arnold method [9], then the clauses were spelled out for vannamei shrimp seeding conditions.

b) Self-Assessment the Implementation of ISO 9001:2015.
This readiness assessment to apply the ISO uses checklist method. Choice the answers: (1) complete/exist/written/documented, score = 10, (2) incomplete/exist/ unwritten/undocumented, score = 5, and (3) none = 0. Assessment of 114 statements was made in 4 classes, namely: strong for 858-1140 score, fair for 572-857, weak for 286-571, and poor for 0-285 poor.

c) Readiness Study on Clauses Implementation.
Based on self assessment in point b) above, the performance calculation was done for each clause, as shown in Table 2. The category status of each clause was obtained. Clauses with low performance did not mean unnecessary of less important, they should be prioritized the improvement. The ISO 9001 standard was a unity, no clauses were categorized as more important than others.

d) Pairwise assessment in the Framework of Implementation Priority of ISO 9001:2015 Clauses. With respect to 7 clauses (Table 2), interests comparison between clauses is made within the Analytic Hierarchy Process (AHP) framework. AHP is one of multi criteria decision making method. In short, it is a method to derive ratio scales from paired comparisons. The expert choice system will get the priority implementation, but clause with high priority does not mean more important, although it still has to be prioritized the improvement.

Table 2. ISO 9001:2015 Clauses

Clauses	Sub-clauses
Organizational context	• Understand the Organization and its context • Requirement and expectations • Scope • Management system
Leadership	• Management Commitment • Policy • Roles, Responsibilities and Authorities
Planning	• Action to address risks and opportunities • Goals and plans to achieve them • Planning change
Supporter	• Resources • Competence • Consciousness • Communication • Documented information
Operation	• Operational control and planning • Product and service condition • Design and development of product and service • Control process for the product and service proved by external source • Production and service supply • Product and service release • Output non-conformance control
Performance evaluation	• Monitoring, measurement, analysis and evaluation • Internal audit • Management review
Improvement	• Non-conformance and Corrective Action • Continuous Improvement

Source: [8].

4. Results and Discussion

4.1. Self Assesment

Table 3 and figure 2 below showed the frequency distribution of 7 clauses.

Table 3. Score Based on Expert Assessment

No	Clause	Complete (x10)	Incomplete (x5)	None (0)	Total score	Average
1	Organization Context	13	7	0	165	8.25
2	Leadership	10	7	0	135	7.94
3	Planning	8	4	0	100	8.33
4	Supporter	11	12	0	170	7.39
5	Operation	8	11	4	135	5.87
6	Performance evaluation	0	10	3	50	3.85
7	Improvement	3	3	0	45	7.50
Total		53	54	7	800	7.02

Table 3 showed that Clause 5 and Clause 6 still contain "None" assessment. In other words, the application level of both clauses was low category even Clause 6 did not have "Complete" score.

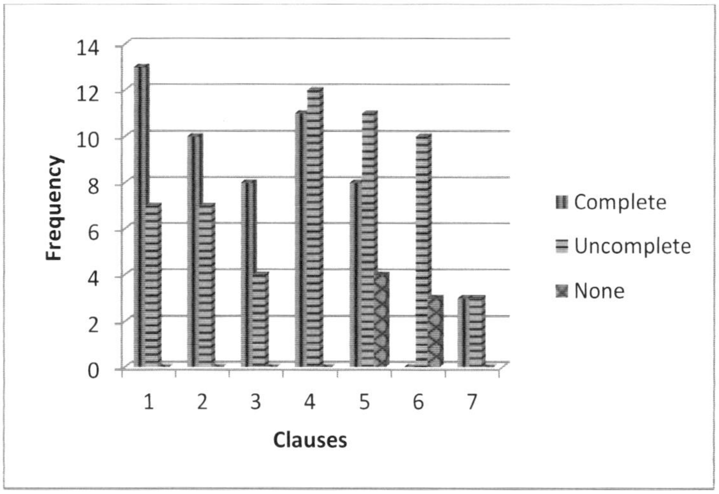

Figure 2. Self-Assessment Diagram for Implementation Readiness of ISO 9001:2015

The lowest average value was in Clause 6 (Performance Evaluation). It means that Clause 6 requires more improvement to improve the implementation of ISO 9001:2015 quality management system.

4.2. Examination the Readines of Clauses Implementation

Examination the readiness to implement clauses are based on score, category and self-assesment description, as shown in Table 4.

Final assessment result showed the total score for the readiness of vannamei shrimp agribusiness agribusiness company to implement ISO 9001: 2015 was 800, with following details:

- 53 components were considered complete x 10 = 530
- 54 components were considered incomplete x 5 = 270
- 7 components were loss x 0 = 0

Above total score shows that vannamei shrimp agribusiness company in Indonesia is included in fair category. The next analysis was to determine the performance value of each clause to determine the category of each clause in whole quality management system. The calculation result of performance value could be seen in Table 5.

Table 4. Overall Self Assessment Score, Category, and Overall Description

Score	Category	Description
858-1140	*Strong*	Most of company's system elements were consistent with ISO 9001 quality system standard. Areas with low score should apply improvements. It was also advisable to hold a pre-assessment of ISO 9001 registration agency.
572-857	*Fair*	Some elements of company's system were consistent with ISO 9001 quality management system standard. But there was still an important part of company's quality system that has not complied with standard or even none at all. Find exactly the field and apply the system in question; and get the services of ISO 9001 training specialists and consultants.
286-571	*Weak*	There were many company had quality management systems that inconsistent with ISO 9001 quality system standards. Companies have to do a lot of orientation and specific training on those standards. If the company was serious about ISO 9001 certification, arrange the implementing and controlling teams and get service from ISO 9001 training specialists and consultants.
0-285	*Poor*	Company quality system practically did not exist. Company should to carry out the review. Careful rules regarding training in Total Quality Management (TQM) tools and techniques were indispensable. It included ISO 9001 training and consulting, before the company invests time and effort in ISO 9001 certification.

Table 5. Performance Value of each Clause

Clause	Assessment Score	Maximum Value	Scale Value 100	Category
Organization Context	165	200	82.50	Strong
Leadership	135	170	79.41	Strong
Planning	100	120	83.33	Strong
Supporter	170	230	73.91	Medium
Operation	135	230	58.70	Medium
Performance evaluation	50	130	38.46	Weak
Improvement	45	60	75.00	Medium
TOTAL	800	1140	70.18	Medium

Clauses that within Strong category were Organizational Context, Leadership and Planning. This means that clauseslargely complied withISO 9001 quality system standard. The company had a quality policy that has been well described and implemented in production process. The organizational structure has been formed by showing the structure and work relationship management. The vannamei shrimp agribusiness manager and staff have conducted testing activities on effectiveness of production process.

Clauses within Fair category were Supporters, Operations, and Improvement. This indicates that some components of these clauses were consistent with ISO 9001 quality management system standards, but there were still important components of these clauses that were not in accordance with those standards. If the company would apply these clauses properly, it was necessary to initiate and undertake training under the guidance of ISO 9001 consultants for activities pertaining to this clause.

Clauses that within Weak category was application of Performance Evaluation clause. This means that there were many components of this clause that were inconsistent with ISO 9000 quality standard. All components of this clause were assessed by management but there was no written document. The Company did not set in writing the responsibility and authority to evaluate performance on all parts, especially in production department. The company did not have a written procedure to assess the performance of production process capability. Therefore, the company has to do a lot orientation and special training on the standard. In addition, company also needs to prepare the implementation and controlling teams assisted by ISO 9000 training specialists and consultants.

Performance value of each clause became benchmark to arrange clauses ISO 9001: 2015 grouping in according to level of readiness of its implementation. Table 6 showed more clearly the conditions of implementation.

Table 6. ISO 9001:2015 Clauses Grouping in According to Implementation Readiness Level

Category	Clauses	Score
Weak	Performance Evaluation	38.46
Fair	Operation	58.70
	Overall Performance	70.18
	Supporter	73.91
	Improvement	75.00
Strong	Leadership	79.41
	Organizational Context	82.50
	Planning	83.33

Table 6 showed that vannamei shrimp agribusiness in Indonesia has 3 clauses with strong category. It needs effort to strengthen the other 4 clauses to apply rightly all parts of ISO 9001: 2015 quality management system. The four clauses classified as moderate and weak were the clauses that need to be addressed. The overall revamping priority was done by prioritizing the clause with lowest category and value. It was merely to catch up the better clause of category and value without assuming one clause was more important than other.

4.3. Implementation Prioritization of ISO 9001:2015 Clauses

The criteria to determine the priority of clause implementation are the principles of integrated quality management, namely customer satisfaction, respect for everyone, sustainable improvement, and factual management [15]. The respondents' assessment results of integrated quality management principles can be seen in Figure 3.

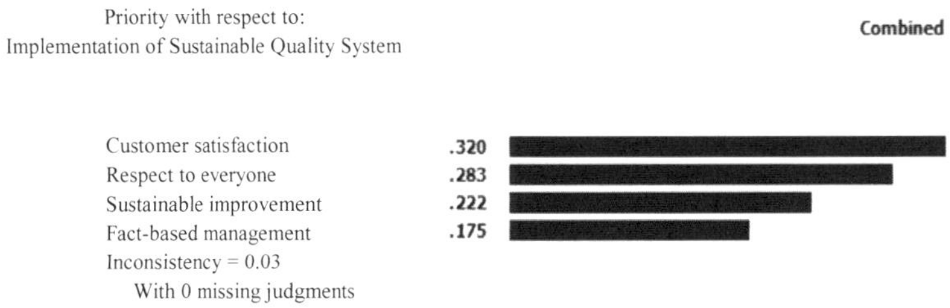

Figure 3. Priority Principles of Integrated Quality Management Implementation on Vannamei Shrimp Seeding Agribusiness

Analytical Hierarchy Process (AHP) results on the comparison of principles application of integrated quality management in selection of ISO 9001: 2015 implementation priority for shrimp vannamei shrimp agribusiness show that the principle that has the highest priority is the Customer Satisfaction with weight of 0.320. The principle of customer satisfaction becomes the main key for sustainable vannamei shrimp hatchery business because the demand determinant is only by customers. The higher customer satisfaction will increase the demand fries and ultimately the volume of sales can increase and affect the company's profits amidst intense market competition. It can be understood if customer satisfaction is considered a top priority that must be considered in implementation of ISO 9001: 2015. The Fact-based Management principle has the lowest priority principle because the agribusiness manager of vannamei shrimp farmers feels that steps to realize the Fact-based Management principle requires a complex understanding process and requires a systematic (plan-do-check-act) process and takes a long time to make it happen.

The implementation priority of ISO 9001:2015 clause with respect to Sustainable Quality System Implementation for Vannamei Shrimp Seeding Agribusiness in Indonesia is shown in figure 4.

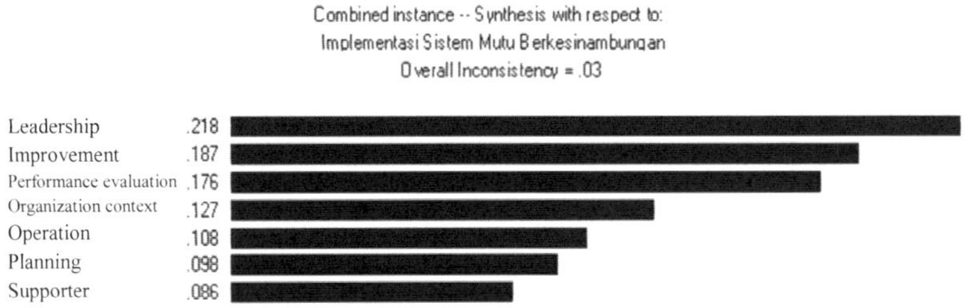

Figure 4. Implementation Priority of ISO 9001:2015 Clause Based on Sustainable Quality System Implementation

Figure 4 shows that priority weighting of ISO 9001:2015 clauses in implementation of sustainable quality management system for shrimp vannamei shrimp agribusiness is a Leadership clause with weight value of 0.218, next is Improvement (0.187), Performance Evaluation (0.176), Organizational Context (0.127), Operation (0.108), Planning (0,098) and Supporters (0.086). Leadership is the primary key in business operations for vannamei shrimp agribusiness companies because the company's leadership role will be able to determine the desired vannamei shrimp specification. It is necessary to open and expand market access amidst intense competition. Leadership in vannamei shrimp hatchery business can bring producer to market leadership, production, resources and price leader. In particular, leadership will lead the company to an improvement process related to process quality. Good vannamei shrimp agribusiness leadership has ensured ongoing performance evaluation process. Likewise good leadership has encouraged the improvement of relationships between communities and organizations, especially shrimp clubs related to disease control techniques and market expansion. This can improve the quality of production process. Furthermore, consistency of production process and planning also determined by organizational leadership in internal scope.

Likewise the Supporter clause as a clause with lowest rank of importance has been in good condition if under good leadership.

Overall, implementation of ISO 9001:2015 does not distinguish between clauses or all clauses. The priority is only to show that a clause must first be improved in its implementation to catch up with other clauses that have been implemented well.

5. Conclusion

a. The vannamei shrimp hatchery agribusiness in Indonesia had enough clauses with 3 Strong categories from 7 clauses. Therefore, vannamei shrimp hatchery agribusiness in Indonesia has applied almost half of ISO 9001:2015 quality management system. This good condition is supported by implementation of Indonesia National Standar quality management system as the adoption of ISO series. The other half clauses still need to be addressed by prioritizing the correction of lowest category clause. It is merely to catch up with better clauses of category without regard a clause more important than the other clause.

b. The vannamei shrimp seeding agribusiness companies in Indonesia were categorized in medium (fair) category to implement the ISO 9001:2015 quality management system. Some elements of company system were consistent with ISO 9001:2015 quality management system standards. However, there were important partsof agribusiness quality system that were inconsistent with standards or none. This could be overcome by applying improvements to these elements under the guidance and service of ISO 9001:2015 training and consultant specialists.

c. The ISO 9001:2015 clauses priority for the are Leadership, Enhancement, Performance Evaluation, Organization Context, Operations, Planning and Supporting clauses.

References

[1] Ministry of Maritime and Fisheries, "Marine and Fisheries in Figure 2014." The Center for Data, Statistics and Information, 2014.

[2] Ministry of Maritime and Fisheries, "Vaname Shrimp and Windu Shrimp are still the Mainstay of Indonesian Exports." Direktorat Jenderal Perikanan Budidaya, 2015.

[3] G. Maulana and D. Waraditya, "Export of Shrimp: Indonesia Still Leads Market in ASEAN," *Bisnis Indonesia*, Bandung, p. 3, 2014.

[4] Center for Coastal and Marine Resources Studies, "Towards the Number One Shrimp Manufacturer," *Bogor Agricultural University*, Bogor, 2015.

[5] "SNI 01-7252-2006 Benih Udang Vaname (Litopenaeus vannamei) Kelas Benih Sebar." Badan Standarisasi Nasional, 2006.

[6] "SNI 7311-2009 Produksi Benih Udang Vaname (Litopenaeus vannamei) Kelas Benih Sebar." Badan Standarisasi Nasional, 2009.

[7] H. Purnomo, "Implementation of SNI Quality Management System in Vannamei Shrimp Hatchery Efforts," Politeknik Negeri Jember, 2015.

[8] International Organization for Standardization, "Final Draft International Standard ISO 9001:2015." International Organization for Standardization, Geneve., 2015.

[9] S. Sundari, R. Iskandar, and E. T. Sule, "The Effect of Implementation Quality System on the Quality Culture of Farmers (Case Study on Edamame Production Industries)," *Mediterr. J. Soc. Sci.*, Oct. 2015.

[10] I. Heras, M. Casadesús, and C. Ochoa, "Effects of ISO 9000 Certification on Companies' Profitability: an Empirical Study," p. 8, 2016.

[11] E. K. Aba, M. A. Badar, and M. A. Hayden, "Impact of ISO 9001 certification on firms financial operating performance," *Int. J. Qual. Reliab. Manag.*, vol. 33, no. 1, pp. 78–89, Jan. 2016.

[12] A. Phan, M. Nguyen, H. Luong, and Y. Matsui, "ISO 9000 Implementation and Performance: Empirical Evidence from Vietnamese Companies," *Int. J. Product. Qual. Manag.*, vol. 18, no. 1, pp. 53–77, 2016.

[13] A. Al-Refaie, O. Ghnaimat, and M.-H. Li, "Effects of ISO 9001 Certification and KAAE on Performance of Jordanian Firms," vol. 6, no. 1, p. 9, 2012.

The main problems of na-oogst tobacco agribusiness in Jember

Muksin[1*], **Hari R**[1], **Tanti K**[1], **Titik I**[2]

[1]Agribusiness Department, Politeknik Negeri Jember, Indonesia
[2]Language, Communication and Tourism Department, Politeknik Negeri Jember, Indonesia

*Corresponding email: muksin.alyasini@gmail.com

Abstract. Tobacco business are business activities that have been entrenched in Jember district. Many farmers and large companies are depending their lives on tobacco farming. The Na-Oogst (NO) tobacco farming has been historically contributed to economy, social and cultural dynamics of local communities. Although Indonesia's tobacco business is worldwide, there are many problems in agribusiness activities. Based on this description, this study aims were: identifying problems found in the NO tobacco tobacco agribusiness chain, and analysing the actors related to the existing problems. This research is descriptive using qualitative research methods. Data collection was obtained from observations of actors and activities as well as in-depth interviews with informants in the Na-Oogst tobacco value chain. Data and information are analyzed by data reduction techniques, data presentation and verification retrieval through triangulation to draw conclusions. The results of the study show problems related to NO tobacco agribusiness were the ability of the actors, partnerships, and coordination linkages, and the main actors existing problems related to core actors are farmers, traders, and companies/ exporters.

1. Introduction

In East Java Province, 100% of Besuki Na-Oogst (NO) tobacco is for the export market . This commodity is not only for a business but also as a heritage that should be preserved [1]. According to the chairman of APTI (Indonesian Tobacco Farmers Association) said that in 2015 Jember exporters only agree to buy 50% of the farmers Na-Oogst tobacco harvest. According to The Head of the Disbunhut Jember District also said that planning to purchase tobacco in 2015 by holding companies in Jember district was 2,450 tons for NOTA tobacco (Na-Oogst early planting) and 650 tons for traditional Na-Oogst tobacco, then, the total will be 3,100 tons. Compared to 2014, the number of NOTA production was 5,891 tons and traditional Na-Oogst tobacco was 2,006 tons. It shows that the decrease of the purcase volume was 50 %. While, the plan absorbtion by exporters was 8,400 tons, but in the realization, traditional NOTA and Na-Oogst tobacco purchase was 3,602 tons. It means only 42,8% from the planned absrobtion [2].

The inability of absorbtion from most of tobacco farmers by exporters caused by a number of things includes the decline in the international tobacco market in Bremen, the shift in quality required by the international market, the number of Na-Ogst's tobacco stock in the company's warehouse is full and it can be occur because the company also planted its own tobacco.

These problems cannot be separated from supply chains and No tobacco agribusiness value chains and the actors involved. Therefore, this study aims were: describing the general chain of NO tobacco business activities, identifying the problems found in the NO tobacco agribusiness chain, and analysing the actors related to existing problems.

2. Method

This research approached Besuki NO tobacco agribusiness as a unified system including the relationship between elements and value chain actors and the core processes that take place in the value chain of tobacco agribusiness systems. Identification of the value chain and main obstacles found in NO tobacco agribusiness wascarried out by using a qualitative approach with descriptive research methods. This study was intended to uncover problems and obstacles in NO tobacco agribusiness. The confirmation was done to the description of the interview results with tobacco agribusiness key actors and review the relevant previous studies.

Data collection was obtained from observations of actors and activities as well as in-depth interviews (Depth Interview) with informants in the Na-Oogst tobacco value chain. Determination of informants was done through purposive sampling, with limits on the level of saturation or repetition of data and information obtained. It based on the criteria for involvement in NO tobacco agribusiness and the business. Data and information are analyzed by data reduction techniques, data presentation and verification retrieval through triangulation to draw conclusions. Research sites located on several sub-districts in Jember, including Kaliwates, Ajung, Sukowono, Ledokombo, Mayang, Wuluhan, Ambulu, and Balung.

3. Result and Discussion

3.1. A chain map and Tobacco Na-Oogst Agrubusiness Actors

The relationship between components in the tobacco agribusiness system must be a consideration in analyzing the effectiveness of a business in this Tobacco NO agribusiness case [3]. The general model of the Na-Oogst tobacco value chain system in Jember regency can be described as an order process from raw materials to the consumers, and various parties are involved in the whole activity.

Furthermore, from this model, derivative maps of different dimensions are created to dissect the value chain to know the various problems occured within through an analysis and assessment of the following maps.

The core process is the main process that must be passed by the raw material before it can be consumed. Based on the data and information obtained, the core processes consists of:

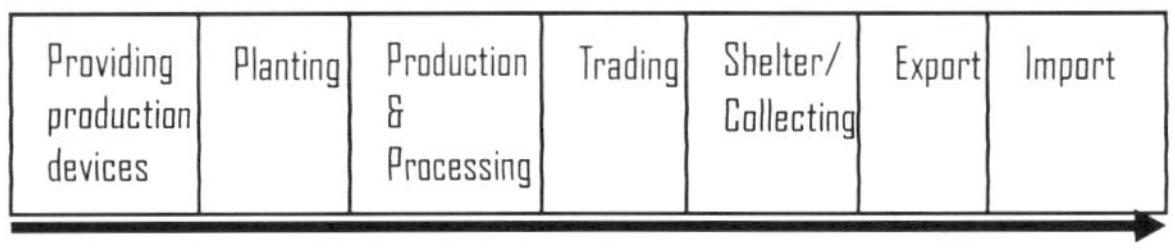

Figure 1: Core Process

Based on the figure above shows that NO tobacco agribusiness chain can be categorized into a simple value chain. The production process and processing into raw materials is one of the crucial activities to obtain the expected quality of tobacco. In general, the results of tobacco processing is cigarettes as omblad, degblad, and filler.

3.2 Agribusiness actors

The actors of agribusiness are subjects who did the main jobs in tobacco agribusiness and they have significant differences between actors. Based on the core process, the actors are: Farmers, Tobacco Traders (Belandang), Packaging Company, Exporter, Importer, and Consumers.

Based on the ability to influence, it shows that the main actors of trade are collectors called as "belandang", and face to face with farmers. However, there are other actors that can affect the overall agribusiness activities. The other actors are: Government, Employers' Association, Farmers' Associations or groups, Department of Industry and Trade, Agriculture and Plantation Services, Agency of Testing and Goods Quality Certification (UPT PSMB), Self-Help Companion, Financial Institutions, and KUTJ.

The roles of actors can be described in table below.

Table 1. Actors and Roles in Tobacco NO Agribusiness

No	Actors/Istitution	Roles
1	Regional Government / Department of Agriculture	• Making regulation of production process and post harvest • Implementating tobacco purchasing licenses • Facilitating and empowering farmer • Facilitaing partnership and supervision
2	Plantation Extension	• Facilitating and empowering farmer • Transferring technology • Giving information about resources
3	Industry and Trade Service	• Making regulation of the tobacco trade process • Facilitating and empowerment of entrepreneurs, traders / foreigners • Facilitating partnership and supervision
4	Agency of Testing and Goods Quality Certification (UPT PSMB)	• Testing tobacco quality and certification • Facilitating building capacity for extension management, and facilitators • Controlling quality of tobacco • Standardizing tobacco quality • Implementing sop for the quality entrepreneurs categories /
5	Entrepreneurs / Exporter	• Purchasing tobacco products • Increasing in added value of tobacco • Facilitating and empowering farmers & collectors / belandang • Giving sources of information for tobacco specific needs
6	Tobacco Entrepreneurs Association	• Giving information source for international tobacco dynamics demand • Providing cooperation to increase competitiveness • Doing collaboration to fulfill tobacco products
7	Traders / Collectors / Belandang	• Collecting product • Purchasing tobacco products • Providing company partners • giving sources of information for farmers
8	Farmers	• Producing tobacco • Increasing added value of tobacco • the fulfilling specified quality tobacco
9	Farmer Groups	• Giving sources of information from the outside environment for technology transfer • Bridging farmers and entrepreneurs • Facilitating farmer for the needs of production facilities

No	Actors/Istitution	Roles
11	Jember Tobacco Affairs Commission (KUTJ)	• Facilitating positive image of tobacco • Giving sources of information on the dynamics of the external environment for tobacco products • Providing mediation of tobacco actors problems
12	12 Financial Institutions	• Facilitating capital facilities for no tobacco agribusiness

3.3 Map of Problems and Potential Solutions

Tobacco agribusiness holds the potential welfare of farmers and rural communities [4]. In practice there are obstacles and difficulties as problems in tobacco industry. Some of these problems occur at the levels of planting, production and processing, and trading. These problems constraints the potential solutions can be described in table below.

Table 2. Map of Barriers and Possible Solutions

Dimension	Stages of activities				
	Input / Tobacco NO production Devices	Planting	Production and Processing	Trading	Export
Difficulties / Problems	• Input price instab-ility such as fertilizers and pestici-des • Availability of fertilizer • Powerless Farmers	• Weather and uncontrolled produc-tion environment • Poor cultiva-tion techni-ques • Low Facilita-tion of counseling	• Maximized Produc-tion techniques Uncontrolled Post-picking production techniques by farmer	• Difficulty of predicting Price/price instability • Low bargaining power of farmers • Difference/lack of clarity in quality criteria • Running of Implementation of standardizing	Difficulty of interact ing amongExporters importers, and Consumer
Solutions	• Giving Government regula-tion: Realization, supervision and evalu-ation • Empo-wering farmer groups	• Streng-thening counse-ling • Implementing Regula-tion planting: Realiza-tion, supervi-sion and evalua-tion	• Strengthening counseling • Providing The role of mediation (in the form of regula-tion) for the needs of planting area	• Providing Regulations for tobacco quality standards: realization, supervisio, evaluation • Giving an Advocacy for farmers and farmer groups in pricing	Providing Government facilitation for international trade advocacy

Based on this identification, main problems in the chain were found in the activities of cultivation, production, processing, trade and export activities. These problems were: instability input price such as fertilizers and pesticides; fertilizer availability; powerless farmers; uncontrolled weather and production environment; poor cultivation techniques; low facilitation of counselling; not maximized production; less controlled post-picking production techniques; difficulty on predicting price instability; low bargaining power of farmers; difference / lack of clarity in quality criteria; low running of implementation of standardization; and low interaction of exporters, importers, and consumer.

If further analyzed, there is a shared need in the value chain and no tobacco agribusiness actors that can be presented as follows: fulfillment of export needs; product quality and quantity; price certainty; upgradingof the weak chain; symmetrical information; no tobacco agribusiness sustainability; access to get capital from financial institutions; technology availability; continuity of meeting tobacco needs according to time; proportional margin acceptance; strengthening/empowering groups or associations of farmers; and quality of human resources.

Some efforts to overcome problems and fulfill common needs can be done through strengthening institutional coordination between actors by applying the principles of justice and equality.

4. Conclusion

The results of the study show: NO tobacco agribusiness chain can be categorized into simple one; Problems related to NO tobacco agribusiness are the ability of the actors, partnerships, and coordination linkages; also main actors are related problems to the core actors namely farmers, traders and companies/exporters.

Acknowledgements

Thank you for completing this research, which is a part of a multi-year study. Our thanks specially to the Ministry of Technology Research and Higher Education which has facilitated funding for activities. Our gratitude also goes to the Director of Politeknik Negeri Jember, the Jember District Agricultural Office, the Tobacco Entrepreneurs Association, the Tobacco Farmers Association and all parties who have helped the smooth running of the research activities. Hopefully this research will contribute to the development of Jember tobacco agribusiness.

References

[1] Bisnis Indonesia 2015 *Rencana Pembelian Tembakau Cerutu 2015 Menurun* Harian Bisnis Indonesia, dated on 31 Maret 2015

[2] PTPN X-MAGZ 2014 *Penelitian Tembakau Jember Tertantang Kejar Perbaikan Ukuran Dan Kualitas Tembakau.* Vol. 12 Edition Liputan April-Juni 2014

[3] Santoso, Kabul 2013 *Industri Hasil Tembakau Jatim Pasok 60 Persen Kebutuhan Nasional.* PTPN-X magazine, Vol. 008 Edition: April – Juni, 2013

[4] Santoso, Kabul 2014 *Tembakau dibutuhkan dan Dimusuhi.* Jember: Jember University.

Modeling of rainfall and fertilization factor of sugarcane productivity in Asembagus sugar factory Situbondo

I Harlianingtyas[1]*, D Hartatie[2] and A Salim[3]

[1,2,3]Department of Agriculture Production, Politeknik Negeri Jember, Mastrip street PO Box 164, Jember 68121, Indonesia

*Corresponding author: irma@polije.ac.id

Abstract. Sugarcane is one of the important plantation crops commodities in plantation sub-sector development in East Java. However, domestic sugar needs are increasing every year, even making the government decide on the policy to import sugar. This situation shows that sugar is a basic requirement in Indonesia. The profitability of sugar agro-industry is influenced by climate and price. Low price and production risks affect the company's sustainable resilience. Rainfall is an important climate element and determines the plant water balance. While the incidence of climate anomalies in Indonesia has been shown to predominantly affect agricultural production and food security, besides that the fertilization process is also a major factor influencing the production of sugar cane produced. Fertilizer can provide nutrients in the soil, so that sugar cane can grow well. Sugar production in Asembagus factory every year has fluctuations, some of the dominant factors are the influence of rainfall and fertilization. The results showed that rainfall(X_1) and fertilization(X_2) significantly affected sugarcane productivity(Y). The regression model is $Y = 33633 + 9{,}43X_1 - 2{,}10X_2$. The modeling results are expected to be used as a company recommendation to determine policies in increasing Asembagus sugar factory production and more productive sugarcane cultivation techniques.

1. Introduction

Sugarcane is one of the important annual crop plantation commodities in the plantation sub-sector development in East Java, which is to meet domestic sugar needs and export National Sugar Productivity since 2001 [1]. Domestic sugar needs are increasing every year, even making the government decide on the policy to import sugar. This situation shows that sugar is a staple in Indonesian. In 2002, the government was launched an accelerated increase in sugar productivity in order to achieve sugar self-sufficiency in 2007. Then it was revised by preparing a national sugar road map and targeting the achievement of direct sugar self-sufficiency (household consumption of 2.80 million tons) in 2009. However, the achievement of self-sufficiency is predicted to remain difficult to achieve if it is not accelerated with the construction of a new sugar factory, increasing the consumption of sugar directly above the sugar production capacity of sugar mills in Indonesia [2]. Sugar agro-industry is held hostage by the selling price of the product which is lower than the production cost so that it is powerless to face the latest developments. Increased productivity is relatively slow. Sugar agro-industry has also not been able to dismiss the notion that profitability is only obtained from a combination of agroclimatological support and prices because sugarcane must be influenced by agroclimatology even though technology can minimize risk while income is directly proportional to production and prices which mean low prices risk the sustainability of the company [3].

Rainfall which is an important climatic element and determines the plant water balance is very noticeable due to climate anomalies. While the incidence of climate anomalies in Indonesia has been shown to predominantly affect agricultural production and food security. Therefore the characteristics of climate anomaly variables need to be quantified (magnitude) so that the impact of climate anomalies can be anticipated earlier and minimized the risk [4]. During the critical phase of the plant, the amount of water given is greater. Each type of plant has a different critical phase. For the sugarcane plant, the critical phase is located at the time of shoot formation and vegetative growth at the age between 0 to 160 days [5]. Besides the rainfall factors that predominantly affect sugarcane production, the fertilization process is also a major factor affecting the production of sugar cane produced. Essential elements such as Nitrogen (N), Phosphate (P), and Potassium (K) are needed in sufficient quantities of sugar cane. Limited availability in the soil, causing these elements to be added through fertilization.

Asembagus factory is one of the sugar factories in Situbondo, which is included in the business unit of PT. Perkebunan Nusantara XI. Sugar production in Asembagus factory each year experiences fluctuations caused by several things, some of the dominant factors are the influence of rainfall and fertilization. This causes losses and decreases the efficiency of the company, because the amount of costs incurred is not balanced with the results received.

To see that the data can have a good impact on production, a statistical analysis is needed that can see the effect of rainfall and fertilizer on sugarcane production. One of the statistical analysis used is multiple linear regression analysis. Multiple linear regression analysis is a statistical method to predict the value of one response variable from more than one predictor variable. It is expected that regression analysis can provide a good model to see the influence of rainfall and fertilizer on sugar production. In line with the Master Plan for Research at the State Polytechnic of Jember Plantation Crops Production Study Program of 2018-2019, namely the technical improvement of the cultivation of plantation crops, this research is expected to provide a method of technical improvement of sugarcane cultivation and recommendations to Asembagus sugar factory Situbondo.

2. Method

2.1 Data Source

Data sources used in this study are secondary data at Asembagus sugar factory Situbondo, which includes rainfall data, fertilization, and sugar production data from 2007 to 2016. Data are taken in stages, because the data collection locations in the archive are different.

2.2 Operational Definition and Research Variables

The variables used in this study are predictor variables and response variables. Predictor variables consist of rainfall and fertilization in 2007 to 2016. While the response variable is sugar production from 2007 to 2016.

1. Rainfall (X1)
   Rainfall is the height of rainwater that collects in a flat place, does not evaporate, does not seep, and does not flow. Rainfall of 1 (one) millimeter means that an area of one meter of square in a flat place, there is one millimeter of water or one liter of water collected. The rainfall measured in this study is the rainfall that falls in Situbondo Regency, especially in Asembagus District.
2. Fertilizer (X2)
   Fertilization is the provision of ingredients intended to provide nutrients for plants. Generally fertilizers are given in solid or liquid form through the soil and absorbed by the roots of plants.

Fertilization carried out by Asembagus factory is 1 to 2 times in one year, the data is the total amount of fertilizer used per year that contain nitrogen (N), phosphate (P), and potassium (K).

3. Sugar Production (Y)
   Sugar production is the amount of sugar produced during harvest every year with unit of ton.

2.3 Step by step for analysis

The steps on multiple linear regression analysis is using the Ordinary Least Square (OLS) method. To get estimator of unbiased parameter until a multiple regression model is formed between rainfall and sugar cane production.

1. By using IBM SPSS 21 software, fertilization and rainfall variables are inputted as independent variables, and the variable of sugar production is entered as the dependent variable. Estimating the regression coefficient parameter with a significant 5%. This means that 95% of the estimated parameters describe the actual situation.
2. Create an appropriate regression model.
3. Make interpretations of the regression model.
4. Identifying the significance of the error, whether the normal distribution assumption is normal, identical and independent.
5. If you have fulfilled the assumptions, then the model obtained can be used.
6. Partial test.
   Hypothesis testing is partially carried out to determine the effect of each predictor variable on the response variable by comparing the T value with T of statistics table. Using significant level is $\alpha = 0,05$
   Hypothesis :
   $H_0: \beta_i = 0$
   $H_0: \beta_i \neq 0$
   Test Statistics :
   $$t = \frac{b_1}{\left\{\frac{\sum x_i^2}{n.\sum(x_i - \bar{x})^2}\right\}^{1/2} . s} = \frac{coef_x}{(SE\ coef)_x}$$
   Critical area :
   Reject H_0, if $|t| \geq t_{(n-k, \alpha/2)}$ or $P_{value} < 0,05$.
7. Perform simultaneous testing, by performing F test on the estimation results of the regression model. This test is used to determine the effect of the overall predictor variable on the response variable by comparing the F value with the F of statistics table.
   Hypothesis :
   $H_0: \beta_1 = \beta_2 = \cdots = \beta_m = 0$
   $H_0: minimal\ there\ is\ one\ of\ \beta_i \neq 0,\ where\ i = 1, 2, \ldots, m$
   Significant level : $\alpha = 0,05$
   Test statistic :
   $$F_{count} = \frac{R^2(n - k - 1)}{k(1 - R^2)}$$
   Critical area :
   Reject H_0, if $F_{count} \geq F_{\alpha;p;n-p-1}$ or $P_{value} < 0,05$.
8. Make interpretations of coefficient determination.
9. Conclusions.

3. Result and discussion

The data used to predict the regression model are rainfall, fertilization and production data from 2007 to 2016. By using the Ordinary Least Square (OLS) parameter estimation method, a regression model between rainfall factor (X_1) and fertilization (X_2) is obtained from the sugar production (Y) formed as follows :

$$Y = 33633 + 9{,}43X_1 - 2{,}10X_2$$

The regression model above can be interpreted as follows:

- ✓ The constants in regression equation explain that if rainfall variable (X_1) and fertilizer variable (X_2) are constant, then sugar production is 33633 tons of sugar.
- ✓ Regression coefficient of rainfall variable is 9,43 illustrates that rainfall (X_1) has a positive influence on the amount of sugar production in Asembagus sugar factory. This means that the higher the rainfall will increase sugar production in Asembagus sugar factory. Suppose the variable is constant fertilizer or 0, then each increase in 1 rainfall unit will increase sugar production by 9,43 tons.
- ✓ Regression coefficient of fertilizer variable is $-2{,}10$ illustrates that fertilizer (X_2) has a negative influence on the amount of sugar production in Asembagus factory. This means that the higher the fertilizer, the sugar production in Asembagus factory will decrease. Suppose the variable rainfall is constant or 0, then each increase in fertilizer dose is 1 unit then the sugar production of Asembagus factory will decrease by 2.10 tons.

To obtain the parameter estimation results with the OLS method, the Best Linear Unbiased Estimator (BLUE), the regression equation model must meet the classical assumptions. Deviations from classical assumptions cause deviations from parameter estimates both the magnitude of the constant and the magnitude of the regression coefficient.
Following are the results of the classic assumption of regression parameter estimation is:

1. Identical test (Glejser test)
   Assuming identical residual test can be done with Glejser test. The Glejser test is done by regressing between the value of the independent variable and absolute residual as dependent. Hypothesis for Glejser test is:
   $H_0 : \rho = 0$ (*identical residual*)
   $H_1 : \rho \neq 0$ (*residual not identical*)
   Significant level : α = 0,05
   Test statistics:

$$F_{count} = \frac{MSR}{MSE} = \frac{\left[\sum_{i=1}^{n}(|\hat{e}_i| - |\hat{e}|)^2\right]/p}{\left[\sum_{i=1}^{n}(|\hat{e}_i| - |\hat{e}|)^2\right]/(n-p-1)}$$

   Reject H_0, if $F_{count} \geq F_{\alpha;p;n-p-1}$ or $P_{value} < 0{,}05$.

Tabel 1. Result of Glejser test

Source	DF	SS	MS	F	P-value
Regression	2	8349230	4174615	2,42	0,159
Residual	7	12058551	1722650		
Total	9	20407781			

The results Glejser Test show that the *P-value* is 0,159 where this value is greater than 0.05. This means that failure to reject H_0 and assuming identical residuals are fulfilled.

2. Assuming normal distribution

Data normality tests can be done with the Kolmogorov – Smirnov test. The hypothesis for the Kolmogorov – Smirnov test is as follows:
H_0 : Residual is normal distribution
H_1 : Residual is not normal distribution
Significant level : α = 0,05

Test Statistics:

$$D = maks|F_0(x) - S_N(x)|$$

Where is :
$F_0(x)$ is a function of theoretical cumulative distribution while $S_N(x) = i/n$ is the cumulative probability function of observations from a random sample with i is observation and n is the number of observations.
Reject H_0, if $D > q(1-\alpha)$ and q is values based on the Kolmogorov-Smirnov table or *P-value*, where H_0 is rejected if the *P-value* < α.

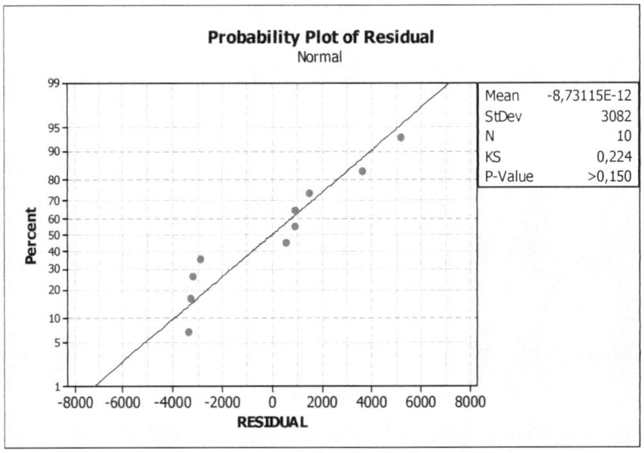

Figure 1. Normality test of Residual

Based on **Figure 1**, the $p - Value > 0{,}150$ is obtained, that is, the decision to accept H_0 is obtained, and concludes that the residual assumption of the regression with normal distribution is fulfilled. This is also seen from the plot image that spreads along the linear line.

3. Multicolinearity Test

Multicollinearity test is used to determine whether there is a strong correlation between the independent variables included in the formation of the model. To detect whether the linear regression model experiences multicollinearity can be examined using Variance Inflation Factor (VIF), if the VIF value > 10 means multicollinearity has occurred.

The results of the analysis show that the VIF value of each rainfall and fertilization variable is 1,002. This means that there is no multicollinearity between dependent variables.

4. Autocorrelation test

Autocorrelation test is used to test that linear regression model has a correlation between the interfering error in **t** period with error in the **t-1** period (before *t*). There are several ways to test autocorrelation. The first method is by looking at the ACF plot, and the second can be seen from the Durbin Waston (*DW*) value, if the *DW* value is located between *dU* and *(4 - dU)* or $dU \leq DW \leq (4 - dU)$ means that it is free from Autocorrelation, otherwise if the value of $DW < dL$ or $DW \leq (4 - dL)$ means that there is Autocorrelation. The *dL* and *dU* values can be seen in the Durbin Waston table, which is the *dL* value; *dU*, α, m,(k - 1). Description: m is the number of samples, k is the total of variables, and α is a significant level.

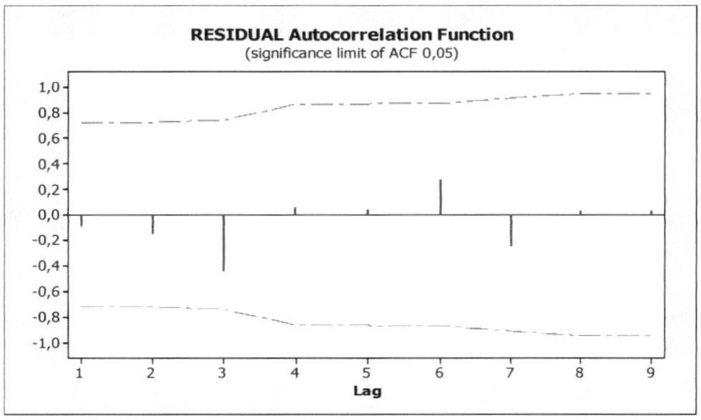

Figure 2. ACF Residual of Regression Model

Indications occur autocorrelation that is if *lag 1, lag 2, lag 3* and so on out of the boundary (red line). In **Figure 2** there is not a single lag that goes beyond the limit, meaning that by looking at the ACF plot there is no autocorrelation case. The second method is the Durbin Watson test.
Hypothesis :
$H_0 : \rho = 0$ (there is not auto correlation)
$H_1 : \rho \neq 0$ (there is auto correlation)
Critical area : failed to reject (accept) H_0, if $dU \leq DW \leq (4 - dU)$
Test Statistics :

$$d = \frac{\sum_{t-2}^{n}(e_t - e_{t-1})}{\sum_{t=1}^{n} e_t^2}$$

In the Durbin-Watson distribution table with *K* value (independent variable) = 7, the number of samples 103 and α is 5%, the value of dL = 0.467 and dU = 1.896, and the value of Durbin Watson is 1.885 and the value of *4 - DW* = 2.115. The value of 2.115 is greater than 1.896, meaning that there is no autocorrelation in the data.

5. Assumption of heteroscedasticity

Heteroscedasticity test is used to test whether in the confounding error liner regression model (e) has the same variance or not from one observation to another observation. To test Hiteroskedasticity, it can be seen from the significant value of Rank Spearman correlation between each independent variable and the residual. If the significant value is greater than α (5%) there is no heteroscedasticity, and vice versa if it is smaller than α (5%) then there is heteroscedasticity. The second way is to see visually from the residuals formed from the

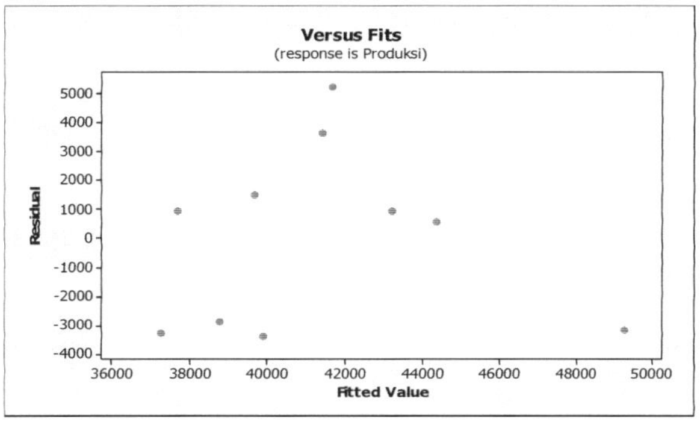

Figure 3. Scatter plot Residual and Production

Table 2. Correlation of Residual Rank Spearman with Independent Variable

		Rainfall	*Fertilizer*	*Residual*
Rainfall	*Correlation Coefficient*	1	-0,297	0,115
	P-value	.	0,405	0,751
Fertilizer	*Correlation Coefficient*	-0,297	1	0,079
	P-value	0,405	.	0,829

regression model whether to form a polat or not. Based on **Figure 3**, the points spread randomly and did not form certain patterns such as conical or linear. This means that there is no case of heteroscedasticity. In addition to seeing visually the heteroscedasticity test can be done by spearman rank correlation test between the independent variables and the residuals.

From **Table 2** Spearman rank correlation analysis results show that the correlation between rainfall and fertilizer variables has a low correlation coefficient of 0.115 and 0.079, respectively, with a P-value above α (0.05). In the rainfall variable P value is 0.751 and the fertilizer variable is 0.829. This means that the correlation between rainfall and fertilizer to the residual regression model is not significant. Thus, it can be concluded that there is no heteroscedasticity in the multiple linear regression model obtained.

3.1 Partial test (T test)

The results of data processing for partial tests shown in **Table 3** below.

Table 3. Partial test (T test) of rainfall and fertilizer variable

Variabel	*Koefisien*	*SE Koefisien*	*T*	*P-value*
Constant	33633	7496	4,49	0,003
Rainfall	9,430	3,086	3,06	0,018
Fertilizer	-2,102	5,588	-0,38	0,718

By comparing the T_{count} values in Table 3 with *T-table* $(T_{0,025;7})$ or by comparing the *P-value* with $\alpha = 0,05$. Based on the results of data processing obtained a decision that is on the variable rainfall value $T_{count} > T_{table}$ or $3,06 > 2,36$. In addition, it can be seen from the $P_{value} < \alpha$ or $0,018 < 0,05$, then H_0 is rejected, it means that rainfall has a significant effect on sugar production in Asembagus factory. In the second variable, fertilizer has a value of $T_{count} > T_{table}$ or $|-0,38| < 2,36$. In addition it can be seen from the value of $P_{value} < \alpha$ or $0,718 > 0,05$, then H_0 is accepted, meaning that fertilizer has an insignificant effect on sugar production in Asembagus factory.

3.2 Simultaneous test (F-test)

Results of data analysis of simultaneous test can be seen on Analyze of Variance (ANOVA) in **Table 4**.

Table 4. ANOVA of Rainfall and Fertilizer of Sugar Productivity

Source	df	SS	MS	F	Sig. value
Regress	2	117163678,911	58581839,456	4,798	0,049
Res. Error	7	85469959,253	12209994,179		
Total	9	202633638,164			

Based on the results in **Table 4** the analysis obtained the calculated F value is 4,798 while the F table or $F_{0,05;p;n-p-1} = 4,737$, then the calculated $F_{count} > F_{table}$ or 4,798 > 4,737, and can be obtained from $P_{value} < \alpha$ or $0,049 < 0,05$, it is decided to reject H0 means that the overall rainfall and fertilizer variables have a significant effect on sugar production in Asembagus factory.

The goodness of the regression model can be seen from the magnitude of the coefficient of determination or R square, based on the results of the analysis obtained the R square value of 0,578 or 57,8%. This value illustrates that the magnitude of the contribution of rainfall and fertilization variables to the ups and downs or variations in sugar production is 57,8% and the remaining 42,2% is the contribution of other variables not included in the model proposed in the study. While the R value of 0,760 means that the relationship between rainfall variables, fertilization and sugar production can be said to be strong or close because it is approaching to 1.

4. Conclusion

The regression model obtained is $Y = 33633 + 9{,}43X_1 - 2{,}10X_2$. Simultaneous testing shows that rainfall and fertilizer have a significant effect on sugar production in Asembagus factory, but partially only rainfall has a significant effect on sugar production at Asembagus factory. The model obtained also fulfills all the classical assumptions with a *R-square* value of 57.8%.

5. Acknowledgments

The authors acknowledged the financial support by PNBP State Polytechnic of Jember 2018.

6. References

[1] East Java Provincial Plantation Office 2011 Strategic Plan 2008-2013
[2] Directorate of Annual Crop Cultivation 2008 Sugar Self-Sufficiency Acceleration
[3] Suwandi A 2015 Sugar Industry Transformation Surabaya XI News.
[4] Estiningtyas W, Surmaini E, and Kharmila SH 2008 Journal of Meteorology and Geophysics **9** (2) : 65 – 77
[5] Hanum C 2012 *Plant Ecology* (Medan : USU Press)
[6] Indrawanto C, Purwono, Siswanto, Syakir M, and Widi R 2010 *Cultivation and Sugarcane Post Harvest* (Jakarta : ESKA Media)
[7] Soentoro N, Indiarto, and AMS Ali 1999 Scientific Research Journal of Bogor Agricultural University.
[8] Supriadi 1992 *Sugarcane Harvest: The twists and turns of the problem* (Yogyakarta : Kanisius)
[9] Draper N and Smith H 1981 *Applied Regression Analysis Second Edition* Translator : Sumantri B (Jakarta: Gramedia)
[10] Gujarati D 2003 *Basic Econometrics Fourth Ed* (New York: Mc Graw Hill)

IOP Conf. Series: Earth and Environmental Science **207** (2018) 012014 doi:10.1088/1755-1315/207/1/012014

Deficiency factors in rehabilitation and capacity building of local farmers national parks through interpretative structural modeling approaches

H Rujito[1], M M D Utami[2], Rizal[1] andH Y Riskiawan[3]

1 Department of Agribusiness Management, Politeknik Negeri Jember, Jl. Mastrip PO Box 164, Jember, Indonesia
2 Department of Animal Science, Politeknik Negeri Jember, Jl. Mastrip PO Box 164, Jember, Indonesia
3 Department ofInformation Technology, Politeknik Negeri Jember, Jl. Mastrip PO Box 164, Jember, Indonesia

e-mail: hrujito2@gmail.com

Abstract. The purpose of this study was made the structural model of the relationship between deficiency factors or constraints in Meru Betiri National Park (MBNP) rehabilitation and the empowerment of local farmers. Information from this research will be very useful for the consideration of policy development and its implementation in the rehabilitation and conservation of the area. The research method used a qualitative approach and the data and information are analyzed using Interpretative Structural Modeling (ISM). The procedure for obtaining information and data is carried out by depth-interviewing the head of the MBNP institution, the head of the agency of regional development, higher education experts, and the results of the Focus Group Discussion. The results showed that there were four factors of deficiencies that caused 12 other constraints. The four key deficiencies were the absence of a locally agreed organizational governance regulation, the lack of facilitation and assistance to local communities, not available technology for the development of regional potential yet, and the lack of synergy between related institutions. The implication of the results of this study is that the MBNP institution needs to formulate a policy to overcome these four key constraints so that other constraints can be overcome.

1. Introduction

In 1998-2000, the period of Indonesia experienced an economic crisis there was massive looting and illegal logging of teak forests in the conservation area of Meru Betiri National Park (MBNP) which was located in the province of East Java causing damage to 4023 ha. The identification of the causes of illegal logging by the MBNP institution shows that factors such as poverty, scarcity of employment, low education, ownership of narrow land are the cause of the looting of forests in the conservation area.

Considering these problems the MBNP Institute cooperates with communities around the area to rehabilitate damaged areas. Local communities around the area are educated to carry out rehabilitation and are expected to help maintain regional security and are given the right to plant intercrops among staple crops to increase their income. In addition, the rehabilitation partner farmers are empowered economically through the institutional group of rehabilitation partner farmers to build their capacity and independence

However, after the rehabilitation of the area was carried out for more than 12 years, the achievement of the target of rehabilitating the area was not yet according to the target set. Likewise, the capacity and independence of empowered local communities are still low. Partner farmer groups still depend a lot on government programs and also have not been able to function as effective institutions to grow social capital for mutual trust, mutual cooperation and places to seek a solution to common problems.

Actually, there are several models of national park management in several places in the world to be used as comparative studies in an effort to improve national park management [1]. Based on the experience of managing the national park shows that the involvement of the surrounding community to help manage the national park can be done through several alternative approaches such as Adaptive Management [2], Collaborative Management [3], and Adaptive Co-management[4] which is a general concept of national park management based on the framework of ecological and social systems in the region [5]. Managers of protected area authorities realize that there is a need to consider local ecological aspects and social dynamics in the national park area [6] and need to adopt an interdisciplinary approach. Nevertheless, national park managers acknowledge the uncertainty and difficulties in predicting the ecological and social dynamics that occur [7].

Good collaboration with local communities for the management of the National Park is needed in order to provide good results. As reported by [8]in Zimbabwe the collaboration between the managers of Hwange National Park and the people around the area, resulted in good management and reduced conflict with residents. Meanwhile, [9] reported that community empowerment through community-based projects is very effective in improving the lives of poor people in Nigerian villages through increasing productive assets, increasing community incomes and increasing the welfare of villagers in general in Nigeria. In order a community-based development approach to be effective in reducing poverty, it must be carefully planned and monitored.

The purpose of this research is to identify and model the shortcomings of the rehabilitation program and the capacity building of local communities in the MBNP area. so that the hierarchical relationship structure between the constraints causing the lack of achievement of the rehabilitation target and community capacity building can be explained. Information on the results of the research will be very useful for the consideration of improving policies in the rehabilitation and conservation of the MBNP area.

2. Materials and Methods

The participants of this study were the stakeholders of the MBNP which included the Head of the MBNP, the head of the agency of regional development, the Agriculture and Livestock Service Office, the Plantation and Forestry Service, the experts from universities, the heads of farmer groups, rehabilitation partners, village heads in the region, local community leaders, companion facilitators.

Measurement of data and information used in this study is a qualitative measure in accordance with the qualitative research approach followed by the conversion of data into quantitative data in the form of binary numbers to be analyzed and modeled using the ISM (Interpretative Structural Modeling) method approach. The procedure for obtaining data and information in this study was carried out in several ways such as in-depth interviews with the heads of MBNP institutions and related technical services, university experts, and obtained through Focus Group Discussion (FGD) with stakeholders in the field such as rehabilitation partner groups, facilitating facilitator of village heads and local community leaders.

The method of analysis and modeling used in this research is the method of Interpretative Structural Modeling (ISM) a method proposed by [10]which was originally used as a problem-solving method based on the identification and understanding of relationships between components within a complex system. Many researchers use this model to describe the relationship between the elements that cause problems in the system [11]. In the ISM approach, qualitative and interpretive data are used to solve complex problems based on mapping the structure of relationships between components and then transformed into a multi-level structure model [12]. The basic idea of identification of the components of a system is based on the opinions and experiences of competent practitioners and experts who understand how to decompose a complex system into several elements and sub-elements so that a multi-level structure model can be built[13].

The implementation of the ISM method begins by using the opinions of experts through several techniques such as brainstorming, FGD, in-depth interviews to develop relationships between attributes to identify interactions between system components. Thus interviews are conducted with experts in the field (experienced practitioners) and academic experts.Four symbols are used to indicate the type of relationship between components i and j as explained below:

V: if the relationship happens is that i influence j but not vice versa

A: if the relationship that occurs is that j influence I but not vice versa

X: if the relationships that occur are in j both influence each other

O; if the relationship that occurs i and j does not influence each other

These symbols (V, A, X,and O) are used to express opinions given by experts about the relationship between components of the system that are thought to cause problems. Then the information about the relationships between the components is arranged in a matrix of relationships between components called the Self Structural Interpretative Matrix (SSIM).

The next step of the symbols in the SSIM matrix is converted into a binary matrix form to compile the Reachability Matrix (RM) with the following rules:

If in SSIM the compound relation to component = V then ei = 1 and ej = 0

If in SSIM the compound relation to component ej = A then ei = 0 and ej = 1

If in SSIM the component ei relationship to component ej = X then eij = 1 and eij = 1

If in SSIM the ei component relationship to component ej = O then eij = 0 and eij = 0

Thus, an RM matrix is a matrix that explores the range of the influence of each individual attribute (component) both on itself and on other attributes that can be reached, and also its dependence on other attributes that can reach or influence it. While the range of influence from other attributes that precede it also consists of the range of influence on itself and its range on other attributes that can be achieved and its dependence on other attributes that reach it or influence it. In other words, the driving force of a particular attribute is the total number of attributes that can be reached or affected (including the range in itself) while the amount of dependence is the total number of attributes that can reach it. The highest ability to reach (influence) an attribute is set equal to the size of the quadrant (sector) identified as the top level in the ISM hierarchy. One of the important characteristics of the highest level criteria in the hierarchy of the ISM model is that it does not help to reach other criteria at the level above it (because it is already at the top level of hierarchy). Therefore, once the top level can be identified, it is immediately separated from the other criteria. The same process is repeated to explore the next level until the level of each criterion is found. Then, the level of attributes that have been identified is used to construct the diagram and the last model of the ISM [14].

In principle the steps for implementing the ISM method are as follows:

Step 1: Make a list of attributes that are being considered influential in dealing with performance; then the attributes listed are reviewed through a comprehensive literature review so that the validity is fulfilled.

Step 2: Gather the structural interaction matrix (SSIM) that has been obtained. Contextual relations between criteria are assessed by the opinions of experts. Every expert has hisown SSIM and interactions between experts will be avoided.

Step 3: Prepare a reachability matrix once the SIM collection process has been completed. This stage will convert the qualitative assessments of experts into binary code. Thus, these binary codes consist of an individual reachability matrix.

Step 4. Develop and partition (insulate) the levels of the reachability matrix. This means converting an individual reachability matrix which is the result of the assessment of the experts to then be

incorporated into the total reachability matrix. This aggregation process uses the average method to avoid extreme values in assessing relationships.
Step 5: build the ISM model. As outlined in the ISM methodology, the structural model of the criteria is built using the final multilevel partition and developed by eliminating transitivity.
Step 6: construct a diagram that states the driver power and the level of dependency after obtaining the total reachability matrix, where the horizontal axis states the size of the driver power of an element while the vertical axis states the level of dependency an element.

3. Results and Discussion

Based on the identification of some of the shortcomings of the rehabilitation system obtained through in-depth interviews with university experts and managers of MBNP and FGD that were followed by stakeholders and also by observing the socio-economic conditions of the community in the MBNP area there were 12 elements of the shortcomings in local community rehabilitation and development programs. in the MBNP area as follows: X1: low incomes of local communities, X2: lack achievement of rehabilitation target of MBNP area, X3: lack of participation and awareness of local communities, X4: lack of community learning process, X5: lack of social capital in local communities, X6: lack of capacity at the local community institution of MBNP area, X7: lack of adequate norms in managing local community institutions, X8: lack of facilitation or assistance to local communities, X9: lack of technology for community development of MBNP area, X10: lack of productivity of local community farming area, X11: lack of synergy between MBNP-related institutions, and X12: lack of self-reliance of local communities. Furthermore, relationship between the elements of deficiencies in the rehabilitation system and the capacity building of local communities based on the opinions of stakeholders and experts is expressed in the form of the SSIM (Structural Self Interaction Modeling) matrix as follows in Matrix SSIM (Structural Self Interaction Matrix) which states the relationship between elements of deficiency in the rehabilitation and capacity building of the local community of MBNP (Figure 1.)

No.	X2	X3	X4	X5	X6	X7	X8	X9	X10	X11	X12
X1	A	A	A	A	A	A	A	A	A	A	A
X2		A	A	A	A	A	A	A	A	A	V
X3			V	V	V	A	A	A	V	A	V
X4				X	X	A	A	A	X	A	V
X5					V	A	A	A	V	A	V
X6						A	A	A	V	O	V
X7							V	X	V	X	V
X8								V	V	A	V
X9									V	A	V
X10										A	V
X11											V
X12											

Figure 1 Matrix SSIM (Sutructural Self Interaction Matrix)

The letters (A, V, X,and O) in the SSIM Matrix are then converted to a binary code consisting of numbers 1 and 0 to become a matrix RM (Reachability-Matrix) (Figure 2.)

No.	X2	X3	X4	X5	X6	X7	X8	X9	X10	X11	X12
X1	0	0	0	0	0	0	0	0	0	0	0
X2	1	0	0	0	0	0	0	0	0	0	1
X3	1	1	1	1	1	0	0	0	1	0	1
X4	1	0	1	1	1	0	0	0	1	0	1
X5	1	0	1	1	1	0	0	0	1	0	1
X6	1	0	1	0	1	0	0	0	1	0	1
X7	1	1	1	1	1	1	1	1	1	1	1
X8	1	1	1	1	1	0	1	1	1	0	1
X9	1	1	1	1	1	1	0	1	1	0	1
X10	1	0	1	0	0	0	0	0	1	0	1
X11	1	1	1	1	0	1	1	1	1	1	1
X12	0	0	0	0	0	0	0	0	0	0	1

Figure 2 Reachability Matrix are conversions of elements SSIM matrix into binary code

Furthermore, the RM matrix which is the result of the transformation of the SSIM matrix is revised its consistency using transitivity rules so that the matrix RM has been revised its consistency. The RM Matrix can be seen as follows: Figure 3.

No.	X2	X3	X4	X5	X6	X7	X8	X9	X10	X11	X12	Drv
X1	0	0	0	0	0	0	0	0	0	0	0	1
X2	1	0	0	0	0	0	0	0	0	0	1	3
X3	1	1	1	1	1	0	0	0	1	0	1	8
X4	1	0	1	1	1	0	0	0	1	0	1	7
X5	1	0	1	1	1	0	0	0	1	0	1	7
X6	1	0	1	1	1	0	0	0	1	0	1	7
X7	1	1	1	1	1	1	1	1	1	1	1	12
X8	1	1	1	1	1	1	1	1	1	1	1	12
X9	1	1	1	1	1	1	1	1	1	1	1	12
X10	1	0	1	1	1	0	0	0	1	0	1	7
X11	1	1	1	1	1	1	1	1	1	1	1	12
X12	0	0	0	0	0	0	0	0	0	0	1	2
Dep	10	5	9	9	9	4	4	4	9	4	11	

Figure 3. The RM Matrix

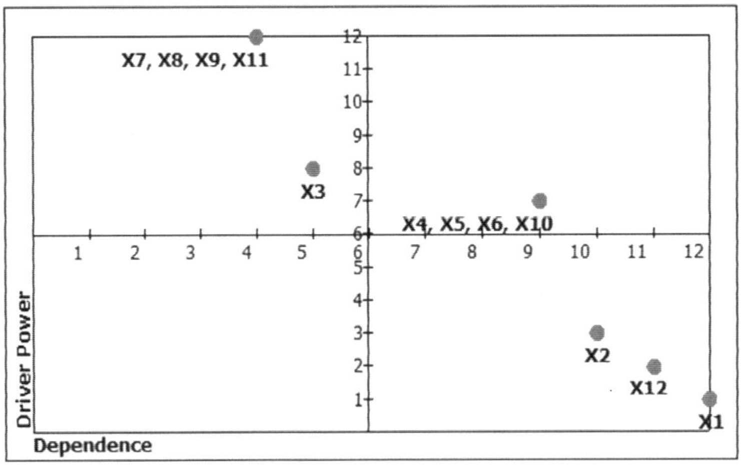

Figure 4. Structural model of all sub-elements Critical Factor "Disadvantages"

The next step is to map the deficiency elements based on the level of "driving force" and "dependency level" as in the following Figure 4. The four key deficiency elements are X7, X8, X9,and X11 whose position is in the same sector, namely in the IV sector or called Independent (not dependent on other elements) on interpretative structural models. The contextual meaning of each key deficiency interpretation is as follows: There is an element of X7 deficiency (= The absence of an instrument of governance rules for local community organizations that are mutually agreed upon by the TNMB), causing other problems to arise, one of which is when there is a distribution of assistance from the government (eg: capital, facilities, input resources, livestock assistance, etc.) to local community groups, so often the distribution of experiencing wrong targets means not falling to the poor who are entitled; but this assistance was more often accepted and enjoyed by the group administrators (who were relatively poor). As a result, poor group members who should be more entitled to receive aid become apathetic and are reluctant to participate in group activities. This condition causes no growth of social capital in the community. On the other hand, revolving capital assistance such as livestock, tools or machines or money and others are not clear in its development. This led to a waste of time, waste of energy and funds allocated by the government in the empowerment program of local communities.

The following is a diagram that is the result of conversion from the final output of the ISM analysis which illustrates the relationship between the elements of deficiencies of rehabilitation and empowerment of local farmers around the National Park as shown in Figure 5

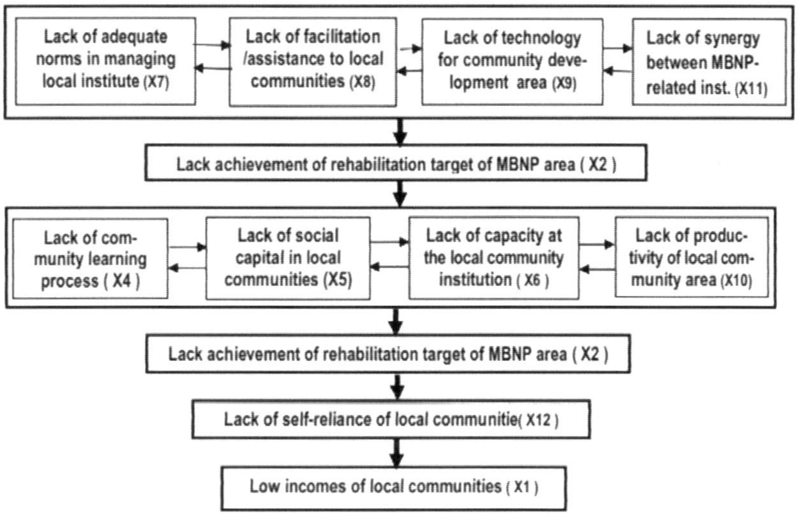

Figure 5. Structural diagram of elements critical factors "Weaknesses"

The reality in the field shows that the absence of a locally agreed upon governance regulation for local community organizations has made the distribution of benefits and accountability of the revolving capital assistance unclear. This situation will cause the institutional capacity of the local community to be difficult to develop and not the formation of social capital in the community concerned. The absence of social capital in a community will cause the independence and empowerment of the community to be low so that they cannot make a positive contribution to the region [15].

Actually this can be overcome by intensive assistance to the group to make a written rule that is agreed upon jointly by all group members by using several indicators as recommended by [16] whose indicators include clear boundary principles, distribution of benefits aimed at anyone, how the mechanism for regulating collective choice must be done, how is the activity to monitor it, what kind of sanctions are given for violations, how is the mechanism of conflict resolution, and how is the organization of joint ownership rights. By looking at the condition of the MBNP local community organization which is still lacking in power, it is necessary to increase facilitation and assistance compared to the previous one, including in making rules for local community institutional organizations.

Another key obstacle element is X8 (lack of facilitation and assistance by facilitators to local community groups of rehabilitation partners) which causes the local community group organization to not build its capacity. Tne indicator of the lack of capacity of local community groups is the lack of development of shared assets belonging to the community and also the absence of shared rules regarding the use of shared capital (both those that are self-supporting and those from government assistance) and the lack of social capital in the community. This situation naturally causes the participation of local communities into community groups as a place to work together and find common solutions to common problems to be low. The consequence is that there is no learning process for group members, no trust (mutual trust) between group members, no social capital in the group and conflict can easily occur. This situation can, of course, be overcome by intensifying group facilitation & intensive assistance.

Deficiency elements are the third key with the X3 notation (the absence of technology for the development of regional potential). The purpose of the lack of technology is the unavailability of appropriate technology that fits the needs of the community both its designation and capacity. For example for agricultural products that are quite abundant in conservation areas such as jackfruit, wood mushrooms, falcons, medicinal plants, bananas. But because the location of the market is far from the MBNP area and also the knowledge of the community for processing agricultural products is still lacking, and there are also no tools and machines for the handling and processing of these agricultural products, the agricultural products are often damaged or sold at low prices.

In addition to the need for technology for processing agricultural products, technology is also needed to open the isolation of local residents of the MBNP area in order to gain greater access to both access to marketing networks, technology, financial capital, agricultural inputs, market information, and others. So far, local residents of the MBNP area are relatively difficult to obtain market access, capital resources, technology because the location of the region is geographically rather remote and difficult to reach. To overcome perishable products and so far often suffer damage and can not be stored for a long time, it is necessary to design appropriate technology that is suitable for processing abundant agricultural products so that they can be stored longer and to increase added value. On the other hand, to increase the access of citizens in the MBNP area to other areas for access to capital, marketing networks, technology, input resources, etc., it is necessary to design and develop an information technology for the region and support the smart village model "Smart village" in the region.

The last key element about the lack of a system is X4 (lack of synergy between MBNP institutions and institutions in the regions related to MBNP). Based on the results of field research and observations in the field, there was a lack of coordination between the MBNP institutions and related institutions/agencies in Jember. This can happen because the MBNP institution is administratively under the direct control and supervision of the central government which in this case is the Ministry of Forestry, while the related agencies such as agency for regional development, Agriculture Agency, and Ministry of Trade. Because of the lack of adequate coordination between the relevant institutions, there was often an overlap of programs in the local community empowerment of MBNP because of the lack of equality of perception. This results in a lack of synergy between related institutions, especially in the MBNP regional development program so that its activities become inefficient and lack focus on the target.

This situation can be overcome if the MBNP institution as the regional management authority of the MBNP carries out an initiative to always coordinate with Bapekab and related agencies continuously and periodically. In this way, there will be a similarity in the perception and similarity of the focus on the target. Furthermore, from the four key constraints or 4 deficiency elements that are the key together cause X3 deficiency elements (lack of participation and awareness of local communities). This lack of participation and awareness of local communities causes other problems to arise such as the emergence of other deficiency elements such as X4 deficiency elements (lack of learning process in the community concerned), X5 deficiency elements (less formed Social Capital in the community concerned, X6 deficiency elements (lack of institutional capacity of the local community in the MBNP area and elements of X10 deficiencies (lack of productivity of local community farming).

Furthermore, the deficiency factors such as X4, X5, X6, and X10 together lead to other deficiency elements, namely X2 (less effective achievement of regional rehabilitation targets). This means that if you want to eliminate or overcome the deficiency of X2 elements, you must eliminate the deficiency elements X4, X5, X6, and X10 first. It can be explained that the deficiency factors such as X4, X5, X6, and X10 together cause other deficiency elements to occur, namely X2 (less effective achievement of regional rehabilitation targets). This means that if you want to eliminate or overcome the deficiency of X2 elements, you must eliminate the deficiency elements X4, X5, X6, and X10 first. Furthermore, with the existence of the X2 element, it will cause the emergence of element X12 (lack of independence of the local community of TNMB) which then in a chain raises element X1 (low income of local communities in the rehabilitation area).

Finally, to overcome the shortcomings in the TNMB area rehabilitation system and also in strengthening the capacity of local communities, what must be done is to eliminate key deficiency elements such as (X7, X8, X9, and X11) before removing other deficiencies.

Conclusion

The conclusions of the research results that have been described as follows: there were 12 deficiency factors in the rehabilitation, conservation and capacity building system of the MBNP local community which together have a negative effect on the effectiveness of achieving rehabilitation and capacity building of local communities in the MBNP area and encourage the emergence of a final element of lack of independence and income of local communities. This condition is certainly not in accordance with the objective of rehabilitating the MBNP area and developing the capacity of local communities around the rehabilitation area. The results of the analysis using the Interpretative Structural Modeling method which is based on the level of driver power and dependency on other elements, shows that there are four key deficiency factors that cause other deficiencies in the TNMB area rehabilitation system; The 4 key elements were X7: there is no instrument for governance arrangements for community institutional organizations to be agreed upon and jointly monitored, X8: lack of facilitation and assistance to local communities by facilitators, X9: there is no appropriate technology available according to the needs of developing the potential of the region, X11: lack of synergy between institutions related to MBNP the elements of deficiency with the lowest driver power and the highest dependence (meaning their existence is largely determined by the presence or absence of other deficiency elements) are X1 (low income of local people), and X12: lack of independence of local communities

Acknowledgment

The authors really highly appreciate to Politeknik Negeri Jember for a grant to research and present this paper on the 1st International Conference on Food and Agriculture 2018.

References

[1] Braasch M, García-Barrios L, Cortina-Villar S, Huber-Sannwald E, and Ramírez-Marcial N 2008*Environ. Model. Softw.***105** 153–170.
[2] BirgéH E, Allen C R, Garmestani A S and Pope KL 2018 *J. Environ. Manage.***183** 343–352
[3] Ansell C and Gash A 2008*J. public Adm. Res. theory***18** 4 543–571
[4] Fabricius C and Currie B 2015 Springer 147–179.

[5] McGinnis M D and Ostrom E 2014*Ecol. Soc.***19** 2
[6] Mathevet R, Thompson J D, Folke CandChapin F S 2016*Ecol. Appl.***26** 15–16
[7] Leys A J and Vanclay J K 2011*Land use policy***28** 3574–584
[8] Perrotton A, de Garine-Wichatitsky M, Valls-Fox H and Le Page C 2017*Ecol. Soc.*221
[9] Jayne T S 2003*Food Policy***28** 3 253–275
[10] Kumar S, Luthra S and Haleem A 2013*J. Ind. Eng. Int.***9** 1 6
[11] Attri R, Dev Nand Sharma V 2013 *Res. J. Manag. Sci.***2319** 1171
[12] Wu K-J, Tseng M-L, Chiu A S FandLim M K 2017*Int. J. Prod. Econ.* **190** 96–107
[13] Govindan K, Kannan D, Mathiyazhagan K, de S. Jabbour A B L and Jabbour C J C 2013*Int. J. Environ. Stud.***70** 4 477–493
[14] Lim M K, Tseng M-L, Tan K H and Bui T D 2017*J. Clean. Prod.***162** 806–816
[15] Gibbs D and Jonas A E G 2000*Geoforum* **31**3 299–313
[16] Fauziyah E and Sanudin S 2017*J. Wasian***4** 2 79–88

Development of Agroindustry Based on Region Superiority in The Efforts to Accelerate Economic Growth in Arjasa District

D E Putra[1] and A M Ismail[2]

[12] Agribusiness Management Department, Politeknik Negeri Jember, Mastrip PO Box 164 Jember, 68101, East Java, Indonesia

Email : dhanangeka@polije.ac.id

Abstract. This research attempts to analyze the condition of agroindustry in Arjasa District. Agroindustry index (I_A) is used to determine the superior agroindustry based on its raw material. Thus, the purpose of this study is the application of a strategic management model with the concept of integration between agroindustry and the potential resources. The Agroindustry Index Model (I_A) uses quantitative data on variable land area and production. Data were obtained from Department of Industry and Commerce, Central Bureau of Statistics, Food Crop Agriculture Service, Horticulture and Plantation in Jember Regency. Whereas, the exponential comparison method (MPE) to determine the superiority of the product is based on the assessment of three agro-industry experts, namely from Department of Industry and Commerce, Office of Cooperatives and Micro Enterprises and a lecturer of the agroindustry management study program, Politeknik Negeri Jember. Based on the results that agroindustry made from corn became the superior agroindustry in Arjasa District. Analysis of superior products using the exponential comparison method (MPE) shows that products corn flour is the top seed followed by corn rice, corn cake, emping, and animal feed.

1. Introduction

The added value of agricultural products is closely related to agro-industry, agro-industry is a cornerstone of hope for success in the agricultural sector. As an industry that processes agricultural products, the sustainability of agribusiness ultimately depends on agro-industry. Resources owned by a region can be optimized thanks to business touches from agro-industry, which can make less valuable products economically very expensive so that this is what determines that agro-industry is a pillar in the success of revitalization of agriculture which was launched in the dream of the President of the Republic of Indonesia.

Agricultural problems that are general in nature, namely products that are seasonal and quickly damaged; less or no innovation; very long marketing chain; the welfare of farmers who are concerned, need an appropriate strategy to get a solution to the existing problems. How to maximize the potential of existing resources, to later be used as an agro-industry that can develop economically well.

Based on data from the Central Bureau of Statistics in 2017 [1], in Jember Regency there was one village that was very lagging, 46 villages were underdeveloped, 114 villages were developing, and 60 villages were developed. The number of poor people in Jember Regency in 2017 has increased compared to the previous year. If in 2016 the number of poor people in Jember was 265,100, 2017 rose to 266,900. There was an increase of 0.03 percent. The number of underdeveloped villages and underprivileged families in Jember is an indication that economic development so far, has not touched the people of the lower layers so that with the existence of the crisis causes remote rural areas to become vulnerable so that they fall into poor areas.

The sub-district where many of the poor people are in Arjasa Subdistrict, the Central Bureau of Statistics data in 2017 states that the total poor population in 2011 was 17,387 and this number increased substantially with time and population growth. This poverty can also be seen from the indicator of the level of education of the community, from 39,463 inhabitants in the Subdistrict of Arjasa, as many as 18,213 people are residents who have never attended school and did not complete elementary school. This can also be seen from the number of villages left behind, as many as 6 out of 6 villages in Arjasa Sub-district are included in 46 underdeveloped villages in Jember Regency.

Although classified as underdeveloped areas, in general, the regions in Arjasa Subdistrict have economic potential that is feasible to be developed, this can be seen from economically to be developed as an agricultural and agribusiness area. Often these potentials do not develop optimally due to limited human resource capabilities, for example in processing, marketing, and capital ownership. Agroindustry is expected to be the economic generator of the community in Arjasa Subdistrict.

Rural development must be carried out with an approach that is appropriate to nature and characteristics. Rural development should follow the four great effort, another one interrelated and constitute the principal strategy of rural development, namely: First, the economic empowerment of rural communities. In this effort capital input and guidance on the use of technology and marketing are needed to enable and make the village community independent; Second, improving the quality of rural human resources in order to have an adequate basis for improving and strengthening productivity and competitiveness; Third, infrastructure development in rural areas. For rural areas, transportation infrastructure is an absolute necessity, because the transportation infrastructure will spur the backwardness of rural communities; and fourth, building rural institutions both formal and informal. Institutions needed by the countryside are the creation of good services especially to spur the rural economy such as financial institutions [2].

This research is very important because it is closely related to the establishment of agro-industry development strategies that optimize the potential of disadvantaged areas based on local resources so that the agro-industry can play a role in contributing to the value of the economy in Jember Regency. One of the hopes of regional autonomy is to increase and equal the welfare of its people, thus the regional government must formulate appropriate and directed development policies [3]. Referring to the things mentioned above, the purpose of this study are (1) designing an agroindustry index model to determine the superiority agroindustry (2) analysis of superiority products based on superiority agorindustry.

2. Method

Research on "Development of Agroindustry Based on Region Superiority in The Efforts to Accelerate Economic Growth in District Arjasa", is a type of descriptive research, using survey methods, data that has been collected is then compiled, analyzed, and explained so as to provide an overview of the phenomena that occur, explain the relationship, test the hypotheses and draw conclusions from the results of the analysis obtained.

Determination of regional superior agroindustry using Agroindustrial Index Method (I_A), which is a quantitative method designed to obtain a comparative value between the variables assumed as determinants of the agroindustry development system in a region. The index is defined as a numerical scale used to compare variables with other variables or with a number of reference numbers. The index is also defined as the number obtained and a formula, which is used to classify a data set [4]. The variables used as input models are (1) Land area, as an indicator of land availability in the preparation of raw material development strategies; (2) Total Production, as an indicator of raw material availability which is a consideration in industrial installed capacity planning;

The Exponential Comparison Method (MPE) is one method for determining the priority sequence of decision alternatives with multiple criteria. Exponential In using the Exponential Comparison Method there are several steps that must be carried out, namely: preparing decision alternatives to be chosen, determining criteria or comparisons of important decisions to be evaluated, determining the importance of each decision criterion, evaluating all alternatives on each criterion, calculating the total score or value of each alternative, and determining the priority order of the decision based on the score or the total value of each alternative [5].

Determining the importance of criteria is done by interviewing experts or through brainstorming agreements. Whereas the determination of alternative scores on certain criteria is done by giving the value of each alternative based on the value of the criteria. The greater the alternative value, the greater the alternative score. The total score for each decision alternative will be relatively different significantly because of the exponential function. The exponential function is one of the most important functions in mathematics, the function functions as a comparison material to make a decision.

3. Result and Discussion

To see the potential of industrial raw materials based on food crops, in table 1 through table 5 is presented the area of food crops and the production of each type of plantation commodity in Arjasa District and its development for several years (2014-2016), then detailed in each District Arjasa.

Table 1. Area of various food crop commodities

No	Type of Commodity	Area (Ha)		
		2014	2015	2016
1	Paddy	3339	3149	3270
2	Corn	265	174	541
3	Peanuts	50	33	59
4	Cassava	133	168	154

Source. BPS, 2015-2017

In table 1, it can be seen that the largest area of food crops is paddy commodity followed by corn, it is clear that there is a reduction in the area of food crops in paddy, corn and other crops. This is most likely due to a shift in land use, from plantation land to residential land, transportation and so on. follow the pace of population growth and the development of a variety of community activities or also due to policy changes. Next in table 2 is a description of the production of each type of food crop commodity.

Table 2. Production of various types of food crop commodities

No	Type of Commodity	Production (quintal)		
		2014	2015	2016
1	Paddy	180000	176520	184830
2	Corn	17260	11460	33220
3	Peanuts	650	430	910
4	Cassava	27050	29980	20840

Source. BPS, 2015-2017

All types of food crop commodities are spread throughout the village area although in varying amounts as shown in the next table.

Table 3. Area and production of food crops in each village in Arjasa District

Plant type: Paddy

No	Village	Plant Area Produce (Ha)	Production (ton)
1	Kemuning Lor	621	3150
2	Darsono	682	3470
3	Arjasa	476	2190
4	Biting	358	1780
5	Candijati	469	2460
6	Kamal	446	2050

Sumber. BPS Arjasa, 2017

Table 4. Area and production of food crops in each village in Arjasa District

Plant type: Corn

No	Village	Plant Area Produce (Ha)	Production (ton)
1	Kemuning Lor	44	278
2	Darsono	42	256
3	Arjasa	31	187
4	Biting	24	132
5	Candijati	22	118
6	Kamal	24	124

Source. BPS Arjasa, 2017

Table 5. Area and production of food crops in each village in Arjasa District

Plant type: Cassava

No	Village	Plant Area Produce (Ha)	Production (ton)
1	Kemuning Lor	5	86
2	Darsono	2	41
3	Arjasa	2	38
4	Biting	-	-
5	Candijati	-	-
6	Kamal	6	103

Source. BPS Arjasa, 2017

Paddy commodity is the most dominant of all, but because of the demand and consumption aspects that do not require added value processing, which is directly processed into rice. hence, rice is not included in food commodities that are analyzed in the determination of superior agro-industries. Determination of Agroindustry and products superiority

This study uses the type of raw material for food commodities as a superior commodity in Arjasa District. The types of agro-industries recorded were agro-industries made from corn, peanuts, and cassava. This analysis is carried out using the agroindustrial index (I_A) method. The variables used are land area and production, data obtained from the Central Bureau of Statistics of the Arjasa Subdistrict [6]. The superior agroindustry ranking for each variable is the rating potential mapping on the logarithm of the variable, the rating potential value used is 1 for the lowest up to 9 for the highest.

The land area variable is shown in table 6. Land area index (I_{LL}) index based on Pr value mapping on the logarithm of the land area value range, so that the leading rank is obtained based on land area index.

Table 6. Land area index of food commodities for agro-industrial raw materials

No	Material	Area (Ha)	**Log**	I_{LL} **(Pr/** land area**)**	**Ranking**
1	corn	541	2,73	9	1
2	peanuts	59	1,77	6	3
3	cassava	154	2,19	7	2

Source. Data processed, 2018

In table 6 it can be seen that the highest I_{LL} is in agro-industries based corn, followed by agro-industries based on raw materials for cassava and peanuts. The next analysis is the production variable as shown in table 7. The raw material production index (I_{PR}) is based on the calculation of the Pr value on the logarithm of the range of production quantities. Superior rankings based on the production index are agro-industries based corn.

Table 7. Index of land area for food commodities for agro-industrial raw materials

No	Raw Material	**Area (Ha)**	**Log**	I_{LL} **(Pr/land area)**	**Ranking**
1	corn	33220	4,52	9	1
2	peanuts	910	2,96	6	3
3	cassava	20840	4,32	8	2

Source. Data processed, 2018

Prioritized Ranking of agro-industry superiority

Based on the analysis that has been done, it can be seen that the value of the agroindustry index for each of the superior agroindustry in Arjasa Subdistrict is presented in table 8. It is known that agroindustry based corn, occupies the highest priority although the difference with cassava-based agroindustry is relatively close, which means that agroindustry based cassava can be used as an alternative of agro-industry.

Table 8. Sequence of superior ranking of agro-industry in Kecamatan Arjasa

Raw Material	Total index ($I_{LL}+I_{PR}$)	Ranking
corn	18	1
peanuts	12	3
cassava	15	2

Source. Data processed, 2018

Determination of superiority Products

The steps in determining the superior agro-industry products are selected from several existing products, namely products that are made from corn as superior. The determination technique used uses the exponential comparison method (MPE), which is one method for determining the priority of decision alternatives with multiple criteria. Previously field observations and expert surveys were conducted to determine product alternatives and their weighting. In table 9 is the result of in-depth interviews with experts[7].

Table 9. Alternative superior agro-industry products

Code	Product Alternative
A	Corn rice
B	Corn cake
C	Corn Emping
D	Cornflour
E	Marning
F	Animal feed

Table 10. Alternative assessment of superior products

No	criteria	Weight	Alternative product value					
			A	B	C	D	E	F
1	Market potential	9	7	6	6	8	6	6
2	Availability of raw materials	7	8	6	7	8	6	7
3	Value-added	6	9	8	6	9	4	5
4	Labor absorption	5	7	5	5	6	5	4
5	Existing technology	8	7	8	8	7	8	8
6	Profit	7	8	8	7	9	7	5
7	Environmental impact	6	7	7	7	7	8	9

Priority ranking of superior products

Based on the calculation of the alternative value of the product with the MPE technique, a superior product sequence is based on the highest value as shown in table 11.

Table 11. Calculation results with exponential comparison method (MPE)

Priority	Selected alternative	MPE value
Superior Product 1	Corn rice	147519516
Superior Product 2	Corn cake	50978609
Superior Product 3	Corn Emping	29614918
Superior Product 4	Cornflour	28669428
Superior Product 5	Marning	28304670

Source. Data processed, 2018

Calculations using the exponential comparison method (MPE) produce five superior products, namely, respectively; 1) corn flour; 2) corn rice; 3) corn cake; 4) corn chips and 5) animal feed.

4. Conclusion

The agroindustrial index method (IA) which utilizes quantitative data types can be used to determine the superior rank of agroindustry although the results are strongly influenced by the number of variables identified as model input. Based on the variables of land area and production from the entire identified agro-industry, placing agro-industries based on corn as the top seed. The superior product analysis using the Exponential Comparison Method (MPE) showed that corn flour products occupy the top seed followed by corn rice, corn cake, emping, and animal feed.

5. Acknowledges

The author would like to thank to Politeknik Negeri Jember in this case the Center for Research and Community Service (P3M) which has provided financial support to this research.

6. References

[1] Central Bureau of Statistics, "Provinsi Jawa Timur Dalam Angka." BPS Jawa Timur, 2017.
[2] A. Agarwal, R. Shankar, and M. K. Tiwari, "Modeling agility of supply chain," *Ind. Mark. Manag.*, vol. 36, no. 4, pp. 443–457, May 2007.
[3] G. Kannan, S. Pokharel, and P. Sasi Kumar, "A hybrid approach using ISM and fuzzy TOPSIS for the selection of reverse logistics provider," *Resour. Conserv. Recycl.*, vol. 54, no. 1, pp. 28–36, Nov. 2009.
[4] S.-P. Chen and W.-Y. Wu, "A systematic procedure to evaluate an automobile manufacturer–distributor partnership," *Eur. J. Oper. Res.*, vol. 205, no. 3, pp. 687–698, Sep. 2010.
[5] K. Kusrini, *Konsep dan Aplikasi Sistem Pendukung Keputusan*. Yogyakarta: Andi, 2009.
[6] M. Kuncoro, *Ekonomi Pembangunan*. Jakarta: Salemba Empat, 2006.
[7] G. B. C. Backus, G. T. Timmer, A. A. Dijkhuizen, V. R. Eidman, and R. P. King, "The impact of a decision support system for strategic pig farm planning on the advice of extension officers," *Comput. Electron. Agric.*, vol. 12, no. 1, pp. 51–64, Jan. 1995.

IOP Conf. Series: Earth and Environmental Science **207** (2018) 012016 doi:10.1088/1755-1315/207/1/012016

Design of Agroindustry Development Strategy Based on Fisheries Cacthe of Merauke Regency

R A Djamali[2], P Betaubun[2]

[1] Agribusiness Management Department, Politeknik Negeri Jember, Jember 68101 East Java, Indonesia
[2] Civil Engineering Department, Universitas Musamus Merauke 99611, Merauke, Papua, Indonesia

Email: abdoel_djamali@polije.ac.id

Abstract. The potential fisheries catche in Merauke Regency is very large and diverse. The contribution of fisheries catches to Gross Regional Domestic Product (GRDP) Merauke Regency in 2016 amounted to Rp. 1,729 trillion or 56.6% of the total GRDP of the agricultural sector. During this time, fishery catche originating from the sea, rivers and swamps are only sold to meet the needs of consumers in fresh form to the local market and only a small part for other regions. As an effort to increase the added value of fisheries catche, namely by the development of fishery products-based agro-industries. The research objective is to design a fisheries catche based agroindustry development strategy for Merauke Regency. The research method used is purposive sampling design research, expert system research methods, and analysis techniques with exponential comparison method (MPE). The conclusions in this study, namely: (a) Agroindustry priorities that need to be developed in succession: freezing of shrimp, snapper fillets, freezing tilapia fish, and fish crackers, (b) Prioritized business scale of Small and Medium Enterprises, (c) Support from the Government, namely regulatory policies for determining potential locations, facilitating ease of use attract investment in fisheries agro-industry, and the ease of obtaining credit

1. Introduction

Papua Island, namely the Province of West Papua, Papua and Papua New Quinea as an area that has a high biodiversity of marine fisheries, but freshwater fish is only about 400 species [1]. Merauke Regency is one of the areas with the highest fishery production potential in Papua Province. Fishing catches come from marine, river and swamp fisheries [1,2,3,4]. his area mostly has a flat and swampy topography because it is located in the Trans-Fly Coastal Lowlands with extensive wetland areas in the form of flood swamps from large rivers flowing inside [5]. Some of the major rivers that flow in Merauke Regency such as the Bian River, Digul River, Kumbe River, Maro River and Buraka River are areas that are potential as freshwater fishing areas.

Fish is one of the people's food ingredients to meet the needs of animal protein. Diverse types and kinds of fish, but only partially commonly consumed and have economic value or market value. Some types of fish catch levels of marine fisheries and inland waters that have high economic value in Merauke namely baramundi (*Lates calcarifer*), red snapper (*Lutjanus altifrontalis*), white prawn

(*Penaeus indicus*), barred spanish macherel (*Scamberomorrus commersoni*), mangrove crab (*Scylla serrata*), snakehead fish (*Ophiocephalus striatus*), and tilapia fish (*Oreochromis mossambicus*) [6,7].

The potential fisheries catche in Merauke Regency is very large and diverse. The contribution of fisheries catche to Gross Regional Domestic Product (GRDP) Merauke Regency in 2016 amounted to Rp. 1,729 trillion or 56.6% of the total GRDP of the agricultural sector [6].

During this time, fishery catche originating from the sea, rivers and swamps are only sold to meet the needs of consumers in fresh form to the local market and only a small part for other regions. As an effort to increase the added value of fisheries catche, namely by the development of fishery products-based agro-industries. The research objective is to design a fisheries catche based agroindustry development strategy for Merauke Regency

2. Methodology

2.1. Research methods

Merauke Regency was chosen as a research location which was conducted purposively. Where this area is a coastal area that has fishery yield potential both in the catch of the sea, rivers and swamps. This study uses an expert system approach that is a research system that uses several experts in the field of fisheries who are expected to be able to give consideration, assessment, and thinking in the topics studied. The expert system is a system of artificial intelligence, which is built on the basis of deep expertise of a certain subject area (obtained from experts and specialists in the area). Expert systems are one of the few types of artificial intelligence systems that are widely used and have found practical application. Expert systems are a progressive trend in the field of artificial intelligence [8]. The increased interest is determined by the possibility of their application to problems of different fields of human. Data collection was obtained through the Indept Interview process in a Focus Group Discussion (FGD) [9]. Who became an expert in this research were 3 researchers, 1 bureaucrat, and 1 fishery entrepreneur.

2.2. Analysis Technique

The analytical technique used in this study is the Exponential Comparative Method (ECM) one of the methods to determine the priority sequence of decision alternatives with multiple criteria. This technique is used as a helper for individuals making decisions to use a well-defined model design at the process stage.

The steps that must be taken in ECM analysis are: (a) determine alternative decisions, (b) establish criteria that will be evaluated, (c) determine the degree of importance of each decision criterion used, (d) assess all alternatives on each criterion, (e) calculate the total score of each alternative, and (f) determine the order of priority of the decision based on the total value of each alternative [9]. The score calculation formulation for each alternative in the exponential comparison method is

$$\text{Value Total } (TN_i) = \sum_{j=1}^{m} (RK_{ij})^{TKK_j} \quad (1)$$

Where,:

TN_i = value total alternative to -i

RK_{ij} = the relative importance of the criteria to-j for decision choices to-i

TKK_j = the relative importance of the criteria decision to-j; $TKK_j > 0$;

n = number of decision choices

m = number of decision criteria

The score range used for weighting and assessing each of the criteria used in this study is between 1-9

Determining the importance of criteria is done by interviewing experts or through brainstorming agreements. Determination of alternative scores on certain criteria is done by giving the value of each alternative based on the value of the criteria. Advantages of the ECM Method:

a. Reducing the bias that might occur in the analysis
b. Score scores that illustrate the order of priorities into large (exponential functions) result in a more real priority sequence of decision alternatives [9]

3. Results and Discussions

In order to make it easier to understand this research subject, it is necessary to begin by determining the definitions of each of the criteria used in this study, as follows:

Table 1. Definition of each criterion in decision making

Num	Criteria	Definitions
1	Market Potential	The potential market is how much the level of consumer interest in buying a product resulting from a company
2	Availability of Raw Materials	That is the ability of an area to meet the main standard needs in producing a product
3	Value Added product	How much the value of a product increases after going through the product processing process
4	Workforce Absorption	As a measure of how much work opportunities can be created if the company is running
5	Availability of skilled workers	As a measure of the ability of a region to prepare skilled manpower needs in accordance with company requirements
6	Availability of processing technology	The level of availability of processing fisheries processing technology
7	Socio-Cultural Condition Support	That is assessing the level of support of socio-cultural norms of the local community towards the establishment of the fishery product processing company
8	Impact on the Environment	How much impact does the establishment of fisheries products have on the surrounding environment
9	Availability of Supporting Infrastructure	Namely how much infrastructure is available that can support the growth of fisheries agro-industry
10	Government Support	namely the level of government partiality in spurring growth and attracting interest in investment in the fisheries agro-industry sector

Furthermore, the experts were asked to determine the weight of each of the criteria used in this study. The value given by each expert is averaged by using the Geomean method to minimize the bias between expert judgment. Furthermore, the average weight that has been averaged, can only be used for ECM analysis. The weights of each of the following criteria in table 2 below:

Table 2. Determination of the weight of each criteria

Num	Criteria	Expert					Weight
		I	II	III	IV	V	
1	Market Potential	9	9	9	8	9	8,79
2	Availability of Raw Materials	9	7	9	8	9	8,36
3	Value Added product	6	5	6	7	6	5,97
4	Workforce Absorption	7	8	7	7	7	7,19
5	Availability of skilled workers	6	6	7	6	6	6,19
6	Availability of processing technology	5	4	5	6	5	4,96
7	Socio-Cultural Condition Support	7	7	7	8	8	7,38
8	Impact on the Environment	5	6	5	6	5	5,38
9	Availability of Supporting Infrastructure	7	6	6	8	8	6,94
10	Government Support	5	6	4	4	4	4,54

Diversity of marine fisheries catch levels and public waters in Merauke and based on technical and economic considerations of experts then determined 7 agro-industry fisheries potential alternatives to be developed. For this reason, the agro-fisheries need to set priorities that need to be developed. Furthermore, the experts provide an assessment of each alternative agro-industries will be developed by each of 10 criteria to offer. The range of values between 1-9, where the value is given each expert is the result of an integrated assessment according to the background respective expertise. The results of the assessment, as follows:

Table 3. Alternative product values for each criterion

Num	Criteria	Weight	Alternative agroindustry fisheries						
			FS	FC	FT	DS	TE	DG	SF
1	Market potential	8,79	9	8	8	7	6	5	8
2	Availability of raw materials	8,36	8	7	8	8	6	8	8
3	Value added product	5,97	8	6	8	5	5	5	8
4	Workforce absorption	7,19	8	7	8	8	6	8	8
5	Availability of skilled workers	6,19	6	6	6	8	6	8	6
6	Availability of processing technology	4,96	7	7	7	6	6	6	7
7	Socio-cultural condition support	7,38	8	8	7	8	6	8	8
8	Impact on the environment	5,38	6	6	6	6	6	6	6
9	Availability of supporting infrastructure	6,94	6	6	6	6	6	6	6
10	Government support	4,54	7	7	7	7	7	7	7

Notes: FS=freezing of shrimph, FC=, FT= freezing of tilapia, DS=dried shrimp, TE=terasi, DG=dried gabus fish, and SF= snapper fillets

Then the results of the assessment in table 3, ECM analysis performed to determine the priority ranking of agro-fishery to be developed as in table 4, as follows:

Table 4. The total value of ECM

Num	Alternative agroindustry fsheries	Total value of ECM	Rank
1	Freezing of shrimph	244.477.215,87	1
2	Fish crackers	104.668.713,02	4
3	Freezing of tilapia	127.696.071,71	3
4	Dried shrimp	70.731.064,46	5
5	Terasi	11.434.360,21	7
6	Dried gabus fish	45.283.560,39	6
7	Snapper fillets	130.618.118,25	2

The ECM calculation results in table 4 show that the 4 main priorities of fisheries agro-industry that need to be developed in Merauke Regency are sequential, as follows: freezing of shrimph, snapper fillets, freezing of tilapia, fish crackers. Shrimp, snapper, tilapia fish, and other types of fish caught so far have only been sold in the form of fresh fish in the local market which must be sold in a very short time. This is because the fresh fish has decreased in quality over time in open space as long as it is not treated. Of course, fresh fish that are not sold for a long time will quickly reduce the selling price even in the next time it starts to rot and not sell well. Usually the freshness of the fish lasts between 5-6 hours, then the fisherman/trader will experience a loss.

The level of fish consumption in Merauke Regency, which is 34.77 per capita per year, is still far below the national level of fish consumption of 41.1 kg per capita per year [7]. However, actually to meet the needs of animal protein, indigenous Papuans are usually obtained from catches in the forest such as deer, pigs and kangaroos. This is a challenge consisting of continuing to suppress movements like eating fish. The impact of the habit of consuming prey animals from the forest, actually greatly disturbs the sustainability of the animal's life which is classified as protected and extinct animals.

Until now, in Merauke Regency there is still a very small number of small medium enterprise (UKM) is moving to process the fishery products. For this reason, the development of agro-industries that process fishery products is an effective solution which is certainly very useful to: (a) increase the added value of fishery products, (b) improve the competitiveness of fishery products, (c) open new employment opportunities, improve the welfare of fishermen and their families , (d) encourage the availability of fisheries products, provide support for increasing GRDP, (e) minimize the occurrence of social insecurity and poverty in rural areas, and (f) encouraging the growth of Industrial linkage from upstream to downstream.

Based on brainstorming with experts, it was agreed that the most appropriate business scale capable of running a fishery agro-industry is small medium enterprise (UKM). The advantage in the business scale is that it requires small investment, simple processing processing technology, and can be developed in the distric / village centers producing raw materials. Areas that have the potential to develop fisheries agroindustry are Merauke District and Naukenjerai District. Both districts in terms of distance are relatively easy to reach, the availability of infrastructure is relatively more adequate, and the center of fish catches. It is expected that the future will spur local economic growth and equitable distribution of public welfare.

The important thing that needs to be needed is support from the central and regional governments. For the sustainability and sustainability of fisheries agro-industry development, the government support is one of the driving forces to stimulate the growth of small medium enterprises in fisheries. The role of the government includes: regulatory policies to determine potential locations, facilitating the ease of use of attracting investment in the fisheries agro-industry, and the ease of obtaining credit.

It is expected that the results of this study for further research include multidisciplinary integrated business plan, spatial planning for fisheries agro-industry, and market mapping and marketing design

4. Conclusions

(a) Agroindustry priorities that need to be developed in succession: freezing of shrimp, snapper fillets, freezing tilapia fish, and fish crackers, (b) Prioritized business scale of small and medium enterprises, and (c) Support from the government, namely regulatory policies for determining potential locations, facilitating ease of use attract investment in fisheries agro-industry, and the ease of obtaining credit

5. References

[1] Allen G R 1991 *Field Guide to the Freshwater Fishes of New Guinea* Christensen Research Institute Madang - Papua New Guinea 268 p.

[2] Allen G R 1996 Freshwater Fishes of Irian Jaya. *In:* Kitchner, D.J. & A. Suyanto (eds). *Proceeding of the First International Conference on Eastern Indonesia-Australian Vertebrate Fauna*. Manado, Indonesia. p.15 – 21.

[3] Allen G R and Renyaan S Y 2000 Fishes of the Wapoga River System, northwestern Irian Jaya, Indonesia. *In:* Mack, A.L. & L.E. Alonso (eds). A Biological Assessment of the Wapoga River Area of Northwestern Irian Jaya, Indonesia. RAP Bulletin of Biological Assessment 9, Conservation International, Washington D.C. p.50-58

[4] Allen G R, Hortle K G and Renyaan S Y 2000 *Freshwater Fishes of the Timika Region New Guinea* PT Freeport Indonesian Company Timika 175 p

[5] Polhemus D A and G Allen G R 2007 Freshwater Biogeography of Papua. *In* Marshall, A.J. & B.M. Beehler (eds.). The Ecology of Papua Part I. Periplus Edition, Singapore. p.207-245

[6] Biro Pusat Statistik Kabupaten Merauke 2017 *Merauke dalam Angka* Merauke

[7] Biro Pusat Statistik Propinsi Papua 2018 *Papua dalam Angka* Jayapura

[8] Mansiya K, Alma Z, Torgyn M, Marzhan M and Kanat N 2014 The Methodology of Expert Systems *IJCSNS International Journal of Computer Science and Network Security.* 14 2

[9] Marimin 2005 *Teknik dan Aplikasi Pengambilan keputusan dengan Kriteria Majemuk* volume 2, Jakarta: Grasindo

Effects of zeolite in aflatoxin B1 contaminated diet on aflatoxin residues and liver histopathology of laying duck

I Sumantri[1], H Herliani[1], M Yuliani[2] and N Nuryono[3]

[1]Department of Animal Science, Faculty of Agriculture, University of Lambung Mangkurat, Banjarbaru-70714, Indonesia
[2]Banjarbaru Veterinary Office, Ambulung No. 24, Banjarbaru-70712, Indonesia
[3]Faculty of Mathematics and Natural Science, Universitas Gadjah Mada, Yogyakarta-55281, Indonesia

E-mail: isumantri@ulm.ac.id

Abstract. This research was conducted to study the effects of zeolite in reducing aflatoxin residues and liver histopathology of laying duck. Sixty-four of Indonesian local laying duck (*Anas platyrinchos* Borneo) were randomly allocated to 2 levels of aflatoxin B1 (AFB1) (low: 30 ppb; and high: 70 ppb) and 2 levels of zeolite inclusions (0 and 2%). The trial was conducted for 28 days and at the end of treatment, the ducks were sacrificed. Meat, liver, and egg samples were collected for AFB1 and aflatoxin M1 (AFM1) determinations. AFB1 and AFM1 concentrations were determined using ELISA analysis. Data were analyzed by analysis of variance using the general linear model of SPSS software. Liver samples were also analyzed for the histopathological study. Results showed that levels of AFB1 significantly ($P<0.05$) increase AFB1 concentration in the liver and egg. Zeolite inclusion did not significantly ($P>0.05$) reduce AFB1 and AFM1 concentrations in meat, liver, and egg. Examination of liver samples indicated moderate and severe liver pathology in the diet without zeolite. Therefore, it was concluded that zeolite inclusion in the high AFB1 contaminated diet does not reduce aflatoxins residue in the tissues but could prevent liver alteration of laying duck.

1. Introduction

Aflatoxin B1 (AFB1) is a toxic and carcinogenic substance produced by mainly toxigenic strains of *Aspergillus flavus* and *A. parasiticus* [1]. Factors of climate, the composition of the commodity, agronomic practices, harvesting, handling, and storage contribute on fungi to grow and produce mycotoxin [2]. Previous studies indicated a high occurrence of AFB1 in feed and feedstuffs collected in Indonesia [3-5].

Ingestion of AFB1-contaminated diet by animals will not only impact on their health and production but will also result in the excretion of AFB1 residues in the tissues, milk, and egg [6]. Duck is one of the sensitive animals to aflatoxin exposure that related to its liver biotransformation capacity [7]. Therefore, consumption of AFB1 contaminated diet will not only lead to a decrease in duck performance but potentially present aflatoxin residues in duck's tissues and egg.

Nowadays, the use of adsorbents, such as zeolite, in the feed is widely recommended to inhibit aflatoxin absorption in the gastrointestinal tract. This approach is more effective and applicable due to some reasons, such as relatively inexpensive, generally recognized as safe, and can be easily formulated in ration [8]. This research was objected to studying the efficacy of zeolite inclusion in AFB1 contaminated diet in reducing the adverse effects of AFB1 on laying duck.

2. Material and Method

2.1. AFB1 Contaminated Diet Production

AFB1-contaminated feed (CF) was produced by inoculation of *Aspergillus flavus* FNCC in a ground maize [9]. CF was estimated containing AFB1 at the level of 500 ppb. Then, CF was mixed into commercial feed to result in 2 levels of AFB1 in diet, namely low (30 ppb) and high (70 ppb).

2.2. Experimental Method

A total of 64 Alabio laying ducks (*Anas platyrinchos* Borneo) were used in the experiment. Ducks were weighed and randomly allocated to experimental units that consisted of 4 dietary treatments with 4 replications and 4 birds in each replication. The diets treatments were: AFB1 30 ppb+0% zeolite; AFB1 70 ppb+0% zeolite; AFB1 30 ppb+ 2% zeolite; and AFB1 70 ppb+ 2% zeolite. Experimental diet was provided restricted (150 g/d/bird) to ensure the amount AFB1 intake, whereas drinking water was provided *ad libitum*. The experiment was started when the ducks were aged 7 months with hen day average more than 60%. The experimental diet was conducted for 28 days. At the end of the experiment, ducks were sacrificed. Liver, thigh meat, and egg were collected for AFB1 and AFM1 content analyses. Liver samples were also examined for liver histopathology observation.

2.3. Histopathology Study

Representative liver samples were fixed in 10% buffered neutral formalin for histopathological study. Sections were cut at 5-micron thickness and stained by the hematoxylin and eosin method of Harris according to Manual Standard of Patologi Diagnose of Veterinary Laboratory.

2.4. Aflatoxin and Data Analysis

Aflatoxins contents of samples were determined using ELISA methods. ELISA kits used in the analysis were ELISA kit AgraQuant® Aflatoxin B1 (Romer Labs. Singapore) for AFB1 analysis and ELISA kit AgraQuant® Aflatoxin M1 Sensitive 25/50 (Romer Labs. Singapore) for AFM1 analysis. Data were analyzed using analysis of variance according to a completely randomized design. All statistical analysis was performed using software package SPSS version 18.0.

3. Results and Discussion

3.1. Aflatoxin B1 Residue

Results showed levels of AFB1 or zeolite have no significant effects on AFB1 residue in the liver and egg. However, zeolite 2% in the diet significantly reduced AFB1 levels in meat (Table 1.).

Table 1. The concentration of AFB1 residue in the liver, meat, and egg of laying duck (ppb).

Treatment	Liver^{NS}	Meat	Egg^{NS}
AFB1 30 ppb+0% Zeolite	1.48	0.57^a	1.37
AFB1 70 ppb+0% Zeolite	2.37	1.29^b	1.91
AFB1 30 ppb+ 2% Zeolite	1.72	0.69^a	1.49
AFB1 70 ppb+ 2% Zeolite	2.18	0.45^a	1.35

NS: Not Significant ($P > 0.05$)
[a, b] Means in each column with different superscripts are significantly different ($P < 0.05$)

AFB1 residues in meat, liver, and egg were detected in very low levels or almost in not detected levels for ELISA analysis. This study offered a low level of AFB1 contamination in experimental diet compare to many aflatoxin studies in poultry, such as 660-3000 ppb [10] or 1000-2000 ppb in broiler [11]. AFB1 residue in the animal product is a dose-dependent, thus low aflatoxin intake from feed will result in low AFB1 residue in the tissue, milk or egg [6].

Several studies showed the carry-over ratio of AFB1-feed into aflatoxin residues in the egg is very low [6]. In young laying hens, it was calculated to be 0.02% or almost not detected. However, AFB1 carry-over ratio is very high in the liver of poultry [12]. Study of [13] showed that after three weeks of trial, AFB1 levels in the liver of broiler fed different levels of AFB1 are similar. There is no difference of AFB1 levels between treatment in this study supports findings that hepatic metabolism of aflatoxin B1 become more efficient after three weeks of exposure and residue no longer accumulated in the liver.

Many studies have been conducted to study the use of adsorbents in the feed that will bind the aflatoxins and prevent its absorption in the gastrointestinal tract [14]. Several adsorbents, such as zeolite, bentonite, and synthetic aluminosilicates have the capability to bind aflatoxin by chemisorption mechanism that reduces aflatoxin bioavailability in the gastrointestinal tract. However, the efficacy of aflatoxin adsorbent has not always been demonstrated in an *in vivo* experiment.

3.2. Aflatoxin M1 Residue

Studies on AFM1 residue in duck's tissues and egg are very limited. AFM1 was not detected (< 0.01 ppb) in the egg of laying hens fed with a diet containing 2,500 ppb AFB1 for four weeks [15]. Similarly, negative detection of AFM1 in the liver also resulted in that experiment, confirmed that only small quantities of aflatoxins are likely to be stored in the hen tissues.

This study showed that AFB1 and zeolite levels have no significant effects on the levels of AFM1 in the liver, meat, and egg of laying duck. Our result confirmed that the highest level of AFM1 was found in the liver and the lowest was in the egg. And it was interesting, that AFM1 levels found in the liver of this study reach more than 100 ppt. The liver is considered the target organ for AFB_1 because it is the organ where most aflatoxins are bioactivated to the reactive 8,9-epoxide form, which is known to bind DNA and proteins, damaging the liver structures and increasing liver weight [14].

Table 2. The concentration of AFM1 residue in the liver, meat, and egg of laying duck (ppt).

Treatment	Liver^{NS}	Meat^{NS}	Egg^{NS}
AFB1 30 ppb+0% Zeolite	125.20	88.07	51.33
AFB1 70 ppb+0% Zeolite	129.15	90.77	47.81
AFB1 30 ppb+ 2% Zeolite	101.86	78.21	67.49
AFB1 70 ppb+ 2% Zeolite	140.85	95.79	41.06

NS: Not Significant ($P > 0.05$)

AFM1 is a hydroxylated metabolite of AFB1 in the liver or tissue cells which can be excreted by the animal through urine, faeces, milk and egg [6, 16]. This study indicated zeolite inclusion in the diet has no significant effect in reducing AFM1 levels in laying duck tissues and egg. However, this study confirmed that AFM1 levels will be found at the highest level in the liver and the lowest level is in the egg.

3.3. Liver Histopathology

Observation of liver histopathology indicated mild acute degeneration of vacuoles in the liver of ducks received low-level of AFB1 but this degeneration was severe in high-level of AFB1. In zeolite groups, mild vacuoles degeneration was found in low-level AFB1 and medium degeneration was in high-level of AFB1 (Figure 1). Hepatic lesions correlated with aflatoxicosis is described as a vacuolation of hepatic cells due to fatty metamorphosis. This metamorphosis is classified as degenerative changes of the liver [17].

Adsorbent inclusion in the diet has a protective effect against aflatoxin exposure. This experiment showed zeolite inclusion seems to reduce the adverse effects of AFB1 exposure as indicated in the result of liver histopathology study of the zeolite group.

Study of [18] found that in low levels of AFB1 (50 to 100 ppb), all livers samples showed histopathological alterations, with an accumulation of fat vacuoles, except the normal appearance of livers from broiler received bentonite in the diet. Study of [19] suggested that ducks are a very sensitive species for aflatoxin injury and it would appear that they are also prone to develop hepatic tumours. The time taken for the tumour induction was about 90 days after oral exposure of AFB1 and histopathologically they were categorized as hepatocellular carcinoma, cholangiocellular carcinoma, and chronic hepatitis.

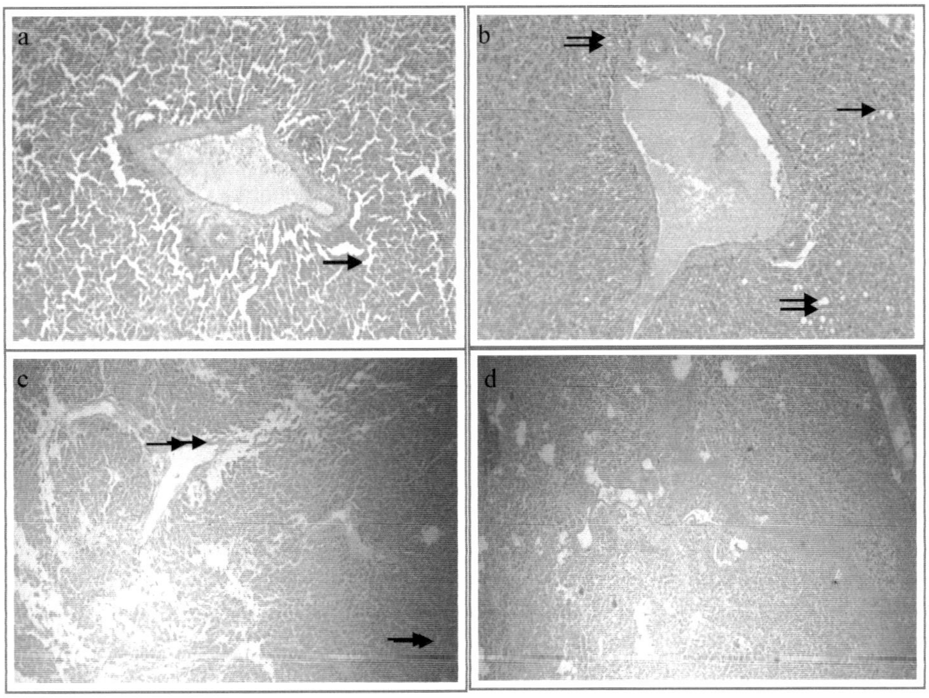

Figure 1. Acute degenerative hepatocyte in liver samples: a. Mild (AFB1 30 ppb+0% zeolite); b. Severe (AFB1 70 ppb+0% zeolite); c. Mild (AFB1 30 ppb+2% zeolite); d. Medium (AFB1 70 ppb+2% zeolite).

4. Conclusion

This study concluded that zeolite inclusion in AFB1 contaminated could not reduce aflatoxin residues in the tissues and egg but could prevent liver alteration of laying duck.

5. Acknowledgement

Authors thank the Indonesian Ministry of Research and Higher Education who supports this research through PTUPT Research Grant Project No. 119/UN8.2/PL/2017.

6. References

[1] International Agency for Research on Cancer (IARC) 2002 Monographs on the evaluation of carcinogenic risks to humans Number 82 *Some traditional herbal medicines, some mycotoxins, naphthalene and styrene* (Lyon: IARC) pp 171–300

[2] Bryden W L 2012 Mycotoxin contamination of the feed supply chain: Implications for animal productivity and feed security *Animal Feed Science and Technology* **173** 134 – 158

[3] Sumantri I, Agus A, Irawan B, Habibah, Faizah N and Wulandari K J 2017 Aflatoxins contamination in feed and products of alabio duck (*Anas platyrinchos* borneo) collected from South Kalimantan, Indonesia *Bulletin of Animal Science* **41(2)** 163-168

[4] Agus A, Sumantri I, Murti T W and Boehm J 2013 Survey on the Occurrence of Aflatoxin B1 Contamination in Dairy Ration and Its Carry Over into the Milk in Yogyakarta and Central Java Provinces of Indonesia *Book of Abstract ISM-MycoRed International Conference* (Apulia: International Society of Mycotoxin)

[5] Rodrigues I and Naehrer K 2012 A three-year survey on the worldwide occurrence of mycotoxins in feedstuffs and feed *Toxins* **4** 663-675

[6] Voelkel I, Schroer-Merker E and Czerny C P 2011 The carry-over of mycotoxins in products of animal origin with special regards to its implications for the European Food Safety Legislation *Food and Nutrition Science* **2** 852-867

[7] Diaz G J and Murcia H W 2011 Biotransformation of aflatoxin B1and its relationship with the differential toxicological response to aflatoxin in commercial poultry species *Aflatoxin – Biochemistry and Molecular*, ed R G Guevara-Gonzalez (Rijeka: Intech Pub) p 3

[8] Kutz R E, Sampson J D, Pompeu L D, Ledoux D R, Spain J N, Vazquez-Anon M and Rottinghaus G E 2009 Efficacy of Solis, NovasilPlus, and MTB-100 to reduce aflatoxin M1 levels in milk of early to mid lactation cows fed aflatoxin B1 *Journal of Dairy Science* **92** 3959 – 3963

[9] Agus A, Maryudhani Y B, Yunianta, Wedhastri S and Nuryono 2010 Production of crude Aflatoksin B1 using different isolates and substrates *Book of Abstracts ISM International Conference* (Lyngby: International Society of Mycotoxin) p 130

[10] Galarza-Seeber R , Latorre J D, Bielke L R, Kuttappan V A, Wolfenden A D, Hernandez-Velasco X, Merino-Guzman R, Vicente J L, Donoghue A, Cross D, Hargis B M and Tellez G 2016 Leaky gut in mycotoxins: Aflatoxin B1 does not increase gut permeability in broiler chickens *Frontiers in Veterinary Science* **3** 1-10

[11] Diaz G J and Sugahara M 1995 Individual and combined effects of aflatoxin and gizzerosine in broiler chickens *British Poultry Science* **36 (5)** 729-736

[12] Bintvihok A, Thiengnin S, Doi K and Kumagai S 2002 Residues of aflatoxins in the liver, muscle and eggs of domestic fowls *Journal of Veterinary Medicine Science* **64 (11)** 1037-1039

[13] Fowler J, Li W and Bailey C 2015 Effects of a calcium bentonite clay in diets containing aflatoxin when measuring liver residues of aflatoxin B1 in starter broiler chicks *Toxins* **7** 3455-3464

[14] Denli M, Blandon J C, Guynot M E, Salado S and Perez J F 2009 Effects of dietary AflaDetox on performance, serum biochemistry, histopathological changes, and aflatoxin residues in broilers exposed to aflatoxin B1 *Poultry Science* **88** 1444–1451

[15] Zaghini A, Martelli G, Roncada P, Simioli M and Rizzi L 2005 Mannanoligosaccharides and aflatoxin B1 in feed for laying hens: effects on egg quality, aflatoxins B1 and M1 residues in eggs, and aflatoxin B1levels in liver *Poultry Science* **84** 825-832

[16] Iqbal S Z, Jinap S, Pirouz A A and Faizal A R A 2015 Aflatoxin M_1 in milk and dairy products, occurrence and recent challenges: A review *Trends in Food Science and Technology* **46** 110-119

[17] Espada Y, Domingo M, Gomez J and Calvo M A 1992 Pathological lesions following an experimental intoxication with aflatoxin B1 in broiler chickens *Research in Veterinary Science* **53** 275-279

[18] Magnoli A P, Monge M P, Miazzo R D, Cavaglieri L R, Magnoli C E, Merkis C I, Cristofolini A L, Dalcero A M and Chiacchiera S M 2011 Effect of low levels of aflatoxin B1 on performance, biochemical parameters, and aflatoxin B1 in broiler liver tissues in the presence of monensin and sodium bentonite *Poultry Science* **90** 48–58

[19] Leenadevi T, Valsala K V and Rajan A 1995 Aflatoxin induced hepatocarcinogenesis in ducks*Mycotoxin Research* **11** 2-8

IOP Conf. Series: Earth and Environmental Science **207** (2018) 012018 doi:10.1088/1755-1315/207/1/012018

In Vitro Studies: Potential Of Natural Source Herbal As Anti Cholesterol Agent

R Widyani[1], **MH Hermawan**[2] **and K Tirtodihardjo** [3]

[1]Animal Husbandry, University of Muhammadiyah Cirebon
[2]Technic Industry, University of Muhammadiyah Cirebon
[3]Pharmacy Faculty, Gadjah Mada University, Yogyakarta

E-mail : retno.widyani@umc.ac.id

Abstract. Globalization era, must human life with practices. Human consumption Fast food that rich of fat availability everywhere daily. The change of lifestyle include eating habit. Fruit, vegetable and sport activity very limited in human life style the now era. This is caused obstruction blood vessel by fat very high, that cause stroke and heart coronary, the vitality of disease. This study aims to evaluate potential of Indonesian as a rich country of any fruit, leaf, (herbal) and fat emulsion activity of herbal. Perhaps can be a problem solving human now. May be so can formulated broiler chicken nutrition to produce meat healthy. The methods was in vitro, cholesterol content test with MU/INST/4 (GC) method at Centre of Agro Based Industry, Bogor. Material used are formulation of stroke care, heart care and green coconut water as control. The result of this study herbal have potentials as an anti-cholesterol agent. Average cholesterol content from heart care formula 23.2 mg/100 gram, stroke care formula 86.9 mg/100 gram and control green coconut water 130.7 mg/100 gram.

1. Introduction

Indonesia is the rich country with herbal plant such as fruit, leaves, stems, roots, rhizomes and tubers. Herbal plant are very beneficial for human health and beauty also animal health and production.

First step research was analysis the content nutrition of Persimmon fruit such as proximate analysis, mineral content, vitamin content and microbiology content [1]. Second step was biological research by giving formula to beef cattle traditionally maintained. The result this experiment indicated that Persimmon fruit and other herbal formula can reduced the number of worm eggs [2]. The third experiment was identification of microbial content at Persimmon fruit. This result indicated that Persimmon fruit auto fermentation content of *Bacillus sp* that have potential as an antibiotic, antifungal and anthelmintic [3]. The fourth step was in vitro studies potential formula Persimmon fruit with other herbal as lipid emulsion [4] and as an anti-cholesterol agent [5]

This experiment design to know the potential of herbal composition on heart care [R] and stroke care [R] with control green coconut water.

2. Material and Methods

2.1. Material

Material used in experiment are human plasma liquid that contain 200 mg/dl cholesterol, stroke care formulation, heart care formulation and green coconut water.

2.1.1. Preparation sample for human plasma liquid. 5 g of egg yolk dissolved in 100 ml of isotonic solution, and for control 5 g of egg yolk dissolved in green coconut water.

2.1.2. Preparation sample of Stroke Care formulation. Persimmon fruit (*Dyospiros kaki)* 50%, celery (*Apium graveolens*) 20%, dutch teak (*Guazuma umnifolia*) 20%, roselle *(Hibiscus sabdarifa)* 10%.

2.1.3. Preparation sample of Heart Care formulation.. Bitter plant (*Andropogus panniculata*) 35%, betel plant (*Piper bettle*) 25%, ginger rhizome (*Zingiber officinale*) 25%, nutmeg (*Myristica fragrans*) 15%.

2.2. Methods

2.2.1. Invitro Methods

2.2.1.1. Control
2.2.1.1.1. Take 5 g of egg yolk and dissolved in 500 ml of green coconut water, mixed.
2.2.1.1.2. Take 5 g of egg yolk and dissolved in 1000 ml of green coconut water, mixed
2.2.1.1.3. Take 5 g of egg yolk and dissolved in 1500 ml of green coconut water, mixed

2.2.1.2. Heart care ^{R}

2.2.1.2.1. Take 1 g heart care ^{R} for 5 g of egg yolks, dissolved in 100 ml of isotonic water
2.2.1.2.2. Take 2 g heart care ^{R} for 5 g of egg yolks, dissolved in 100 ml of isotonic water
2.2.1.2.3. Take 3 g heart care ^{R} for 5 g of egg yolks, dissolved in 100 ml of isotonic water

2.2.1.3. Stroke care ^{R}
2.2.1.1. Take 1 g stroke care ^{R} for 5 g of egg yolks, dissolved in 100 ml of isotonic water
2.2.1.2. Take 2 g stroke care ^{R} for 5 g of egg yolks, dissolved in 100 ml of isotonic water
2.2.1.3. Take 3 g stroke care ^{R} for 5 g of egg yolks, dissolved in 100 ml of isotonic water

2.2.2. Cholesterol content. [6]
2.2.2.1. Prepare GC according to sample standard
2.2.2.2. Inject the solvent 2 μl into the Colom
2.2.2.3. Solvent peak will be visible
2.2.2.4. Inject the sample 5 μl into the Colom
2.2.2.5. Read cholesterol levels in the sample

2.2.3. Statistic calculation: This experiment used average of dupplo laboratory test.

3. Result and discussion

Cholesterol is a metabolite containing sterol fat found in cell membranes and circulated in blood plasma. The normal human cholesterol content 200 mg/dl. Way of life, food consumption high fat, fried, fast food, overweight, minimalized sport, genetic can cause high blood cholesterol and atherosclerosis. Atherosclerosis in brain can cause stroke and in heart can cause heart coronary disease. Cholesterol can decrease with herbal consumption. Indonesia rich with many kind of herbal plant. Empirical case show that "jamu" can make human healthy and smart. "Jamu: is natural cultural heritage. Complex mechanism on jamu to make our health and beauty hereditary.

IOP Conf. Series: Earth and Environmental Science **207** (2018) 012018 doi:10.1088/1755-1315/207/1/012018

The development of herbal medicine from boiled with water, powder, simplifier, extraction and Nano technology. We mixed many herbal plant and processes to decrease blood cholesterol such as heart care and stroke care.
Composition herbal used in this experiment:

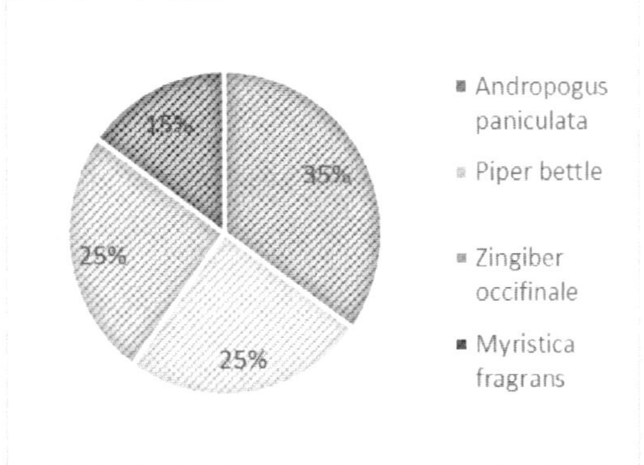

Figure 1. Heart care composition

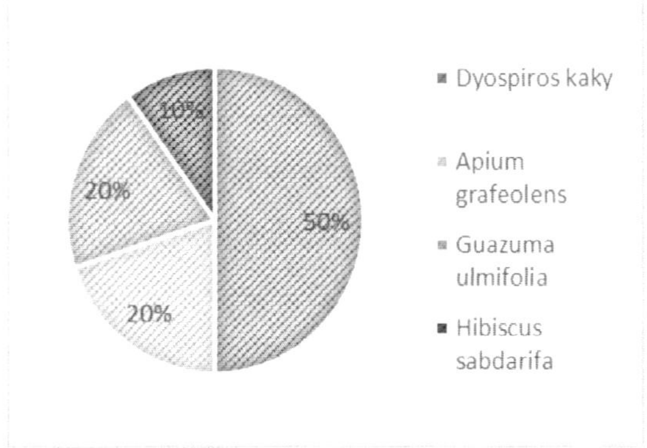

Figure 2. Stroke care composition

Herbal plant used in this experiment:

Heart Care			

Andropogus panniculata	*Piper betlle*	*Zingiber officinale*	*Myristica fragrans*
Sroke Care			

Diospiros kaky	*Apium grafeolens*	*Guazuma ulmifolia*	*Hibiscus sabdarifa*

Figure 3. Herbal Plant

Persimmon fruit (*Diospiros kaky*) content polyphenol can reduce cholesterol, prevent heart coroner disease, prevent hardening of the arteries, maintain blood pressure, maintenance of the blood vessel flexibility, stability of blood pressure the key of heart healthy [10]. Flavonoid in Sledri (*Apium grafeolens*) can blood circulation well and reduce blood pressure [11]. Jati belanda (*Guazuma ulmifola*) content alkaloid, substances and resin also flavonoid. These compound can inhibits fat and cholesterol absorption [12].
Cholesterol content in this experiment see in table 1.

Table 1. Cholesterol content in vitro study in human plasma liquid

Dose (mg/dl)	Cholesterol content (mg/100 g)		
	Heart Care	*Stroke Care*	*Control Green coconut water*
20	8.81	102	214
40	8.79	64.5	114
60	52.0	94.2	64.2
Average	23.2	86.9	130.7

The control this experiment used green coconut water, because from empirical study green coconut water also can eliminate vascular obstruction. Normal plasma human cholesterol content was 200 mg/dl and in this control with dose 20 mg/dl was 214 mg/100 g. Increasing the dose can decrease cholesterol content and the average 130.7 mg/100 g.

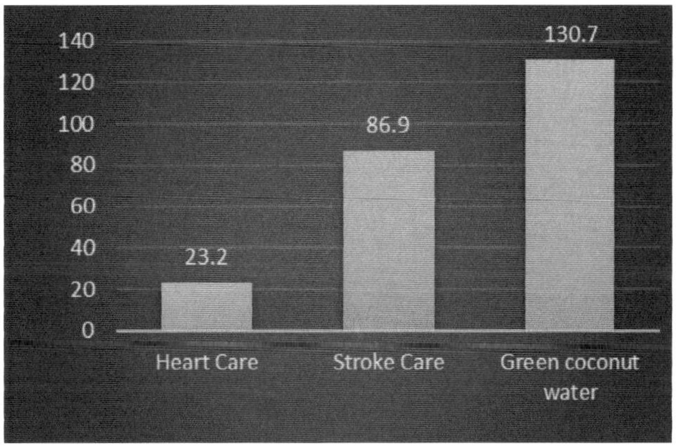

Figure 4. Cholesterol content in different herbal formulation

The average cholesterol heart care 23.2 mg/100 g and stroke care 86.9 mg/100 g. This indicated that heart care more efficient to decrease cholesterol content in the plasma blood human by study in vitro

4. Conclusion

The result of this study herbal combination have potentials as an anti-cholesterol agent. Average cholesterol content from heart care formula 23.2 mg/100 gram, stroke care formula 86.9 mg/100 gram and control green coconut water 130.7 mg/100 gram

5. Acknowledgments

One of us (Widyani, Hermawan and Tirtodiharjo) thanks to Deepublish Publisher Yogyakarta and Autoimun Care Indonesia Cirebon for fund this conference

IOP Conf. Series: Earth and Environmental Science **207** (2018) 012018 doi:10.1088/1755-1315/207/1/012018

References

[1] R Widyani and M Hermawan 2016 *Nutritional value of Persimmon yoghurt (Dyospiros kaki) as healthy soft drink to make healthy and fitness : An Analysis*. The 1st Conference Technology on Biosciences and Social Sciences. Andalas University. Padang
[2] R Widyani, M H Hermawan, F D Perwitasari, I Herawati. 2016 *Jurnal Ilmu Ternak*. Desember. *Vol* **16** *No* **2** 71-77.
[3] R Widyani, M H Hermawan, S M Noor, AETH Wahyuni, K Tirtodihardjo. 2017 *Potency of Persimmon Fruit (Dyospiros kaki) As An Organic Antibiotic, Antifungal and Anthelmintic on The Livestock : An Analysis* 7th Proceeding International Seminar Of Tropical Animal Production.. Gadjah Mada University. Yogyakarta. p 319-323
[4] Ratnani 2012 https://ulyadays.com/sambiloto/
[5] Anonimous 2015 *Lima kandungan daun sirih yang bermanfaat untuk kesehatan* https://khasiatdaun.com/kandungan-daun-sirih/.
[6] Rubiyanto, D 2018 *Teknik Dasar Kromatografi*. Deepublish Publisher, Yogyakarta.
[7] Anonimus 2006 *Kandungan dan Manfaat Tanaman Jahe* http://agroteknologi.web.id
[8] Sutomo 2006 http://repository.usu.ac.id/ bitstream/handle/ 123456789/ 29258/ Chapter%20II.pdf?sequence=4&isAllowed=y
[9] I W Sumardika and I M Jawi 2012 *J. Medicina, vol* **43** *no* **2** 67-71 Mei 2012.
[10] Anonimous 2018 *Kandungan dan manfaat buah kesemek* http://tanamanbuatobat. blogspot.com/2013/02.
[11] Anonimous 2018 *Seledri (Apium grafeolens L)* http://ccrc.farmasi. ugm.ac.id/ ?p 225
[12] Anonimous 2016 *Kandungan dan manfaat daun jati belanda untuk kesehatan* https://manfaatjahemerah.com/kandungan-dan-manfaat-daun-jati-belanda/.

Probiotics based on Local Microorganism as a subtitute of Antibiotic Growth Promotor (AGP) on Broiler productivity

U Suryadi[1*] and A F Prasetyo[1]

[1] Poultry Business Management Study Program, Department of Animal Husbandry, Politeknik Negeri Jember , Jl. Mastrip PO BOX 164, Jember, East Java, Indonesia

*E-mail: ujang_suryadi@polije.ac.id

Abstract. This research aimed to determine the effectiveness of the addition of probiotics based on local microorganisms (MOL) of cattle's rumen and chicken intestines in feed as a substitute for Antibiotic Growth Promotor (AGP). 200 day old chicks were used in this research. The method used the Completely Randomized Design with four tratments and five replications, such as P0 (without probiotics), P1 ((Feed + probiotics with a dose of 5 ml/Kg), P2 (Feed + probiotics at a dose of 10 ml/Kg), P3 (Feed + probiotics at a dose of 15 ml/Kg). The parameters of this research are performance and carcass quality. The results of microbiological analysis showed that probiotics based on MOL contained a concentration of *Bifidobacterium* 3.5 x 10^7 cfu/ ml, *Bacillus* 0.9 x 10^6 cfu/ml, *Streptococcus* 1.06 x 10^6 cfu/ml, and *Lactobacillus* 12.5 x 10^7 cfu/ml. The research showed that the adding of Probiotic was not significant ($P>0.05$) on feed intake, body weight gain, and carcass percentage but significantly ($P>0.05$) on mortality and Feed Conversion Ratio. Treatment without probiotic has a higher mortality (8%) than with probiotics (0,67%). The Feed Conversion Ratio by given probiotics 5 ml and 10 ml/kg feed was lower than treatment without probiotics.

1. Introduction

The given of antibiotics to broiler aimed at suppressing disease progression was also used as an effort to increase livestock productivity which was called the Antibiotic Growth Promoter (AGP). Using of AGP causes *Salmonella* resistance and leaves harmful residues in people who consume broiler.

Using of AGP in poultry feed Indonesia, start from January 1, 2018 has been banned giving to chickens. The ban on the use of AGP has caused farmers in Blitar to suffer losses, because the weight of broiler bodies has decreased to 40% from their original weight and decreased egg production capacity from 90% to 40% [1]. Cases that occur in chicken farmers in Blitar require countermeasures so that farmers do not experience loss.

AGP can be replaced by organic acids, enzymes, probiotics, prebiotics, synbiotics, and herbs [2]. Probiotics are microorganisms that are beneficial and can give effect to performance and nutrient digestibility [3], pathogen inhibition [4], and to manage of intestinal microflora [5]. The most commonly used bacterial probiotics are the strains of *Bifidobacterium, Enterococcus, Lactobacillus, Bacillus, Pediococcus* and *Streptococcus*. Some products contain viable yeast and other fungi in addition to bacteria [6], the bacteria are found in the rumen of cattle and are also found in the crop, small intestine and chicken ceca, especially *Lactobacilli* producing lactic acid and acetic acid. The presence of microbes from the rumen of cattle and the digestion of native chickens can be used as an opportunity to use as probiotics based on MOL (Local Microorganisms). So that it is expected to be used as a substitute for AGP to increasing broiler productivity, however it needs to be studied how far the MOL-based probiotic concentration affects the performance of the broiler. This research needed to be done because it is expected to help increase broiler productivity after experiencing a decrease due to the cessation of AGP use.

2. Materials and Methods

This research carried out at the Politeknik Negeri Jember from July to September 2018. This research used COBB 500 broiler strains which were not differentiated sexes by 200 (unsexed), harvest to 35 days. The starter period in brooding was maintained on 20 plots with a litter floor measuring 52 x 38 x 37 cm, each plot was given a 15 watt incandescent lamp which functions as a heater and lighting, and contains 10 DOCs. Finisher period, the plot was widened to 1 M^2 for 10 broilers. The cage plot used for placement of each treatment was randomized.

This research has used equipment; 1. O'haus scales with a capacity of 1300 g with an accuracy of 0.05 g, 2. Digital weighing capacity of 5 kg, 3. The function of higrotermometer and humidity, 4. Place was used to food and drink 5. Plastic was used to store feed samples. 6. Cleaning equipment includes: brooms, wipes, buckets, disinfectant sprayers. Starter period until the finisher used BRI Patriot Feed with 20% crude protein content, crude fiber maximum 5%, maximum fat 6%, calcium 0.9-1.1%, and phosphorus 0.7 - 0.9%. Treatment was given was probiotics made by using rumen bacteria and chicken intestines as local micro organisms (MOL). The manufacturing method is listed in (**Figure 1**).

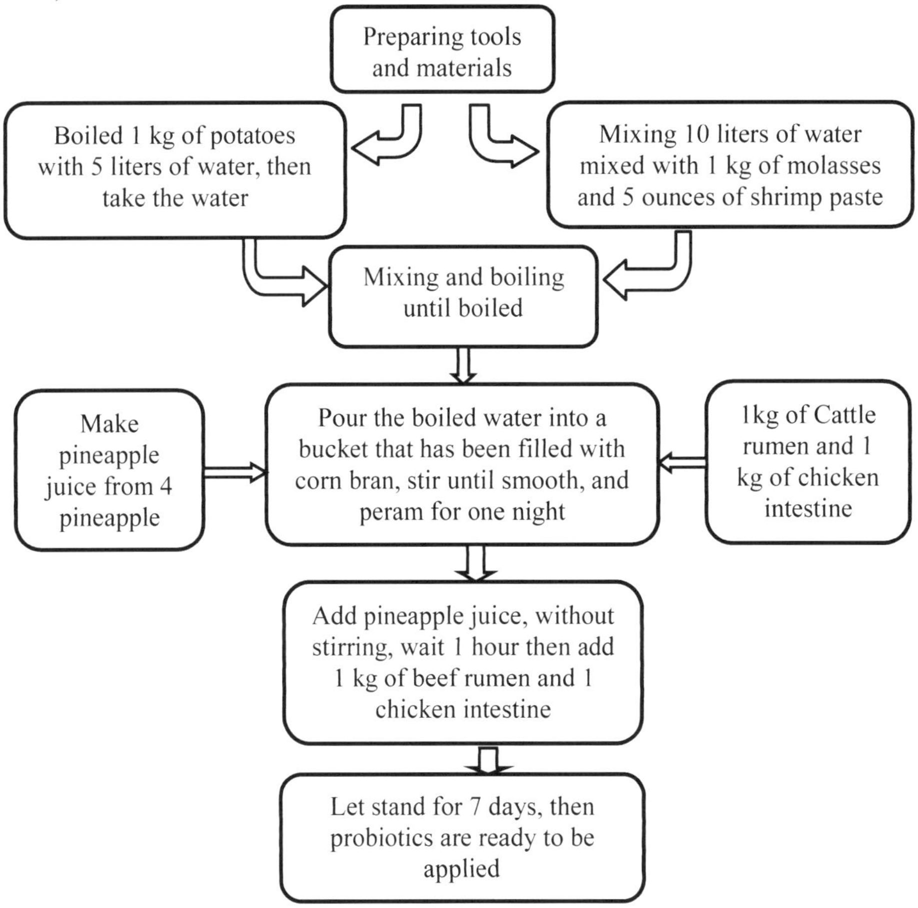

. **Figure 1.** Probiotic making scheme

Probiotic bacteria in MOL are identified by the method morphological and biochemical tests. The method morphological covering shape, colony color, and gram staining. Gram staining test was used to determine the bacterial class. Biochemical tests are was used to identify and detect a pure culture of bacterial isolation through its physiological properties with the method TSIA, MR-VP, citrate, motility and indole, catalase and urease.

Probiotic application, probiotics was given by 1th day until harvesting through feed with several stages. This research uses a completely randomized design method with four treatments and five replications. Each test contained ten broilers, the number of broilers used in this research were 200 heads.

Retrieval of data, the data taken for the first time was the concentration of probiotic micro organisms which include *Bifidobacterium, Bacillus, Streptococcus* and *Lactobacillus*. The second stage was observed the research parameters that include; Increased body weight, Feed consumption, Feed Conversion, and Percentage of carcass. Probiotic application through feeding according to treatment; P0 = Control Treatment (without probiotics), P1 = probiotics with a dose of 5 ml / Kg of feed, P2 = probiotics with a dose of 10 ml / Kg of feed, and P3 = probiotics with a dose of 15 ml / Kg of feed

3. Results and discussion

The probiotic microorganisms in the MOL of rumen cattle and chicken intestine given to broilers contained *Bifidobacterium* 3.5 x 10^7 cfu/ml concentrations, *Bacillus* concentrations 0.9 x 10^6 cfu/ml, *Streptococcus* concentrations 1.06 x 10^6 cfu/ml, and *Lactobacillus* concentrations 12.5 x 10^7 cfu/ml. Showed MOL made from cattle rumen and chicken intestines contain probiotic microorganisms so that they can be used as probiotics. The results showed parameters measured to determine the effect of adding probiotics based on local microorganisms (MOL) of rumen cattle and chicken intestine in feed as a substitute for Antibiotic Growth Promotor (AGP) in broilers listed in Table 1.

Table 1. Performance of broilers by given probiotics

Treatment	Parameter				
	Mortality (%)	Body Weight Gain (gr)	Consumption (gr)	FCR	Carcass (%)
P0	8,00^{a}	1808.86	3283.55	1.82^{a}	74.05
P1	0,00^{b}	1858.40	3237.69	1.74^{b}	72.02
P2	2,00^{ab}	1817.80	3157.69	1.74^{b}	73.63
P3	0,00^{b}	1769.02	3139.57	1.77^{ab}	73.48

The difference this letters superscripts showing a real differenced (P<0.05)

3.1. Effect of MOL addition to broiler mortality

In the **Table 1**. shows that the MOL based on the contents of cattle rumen and chicken intestine had a significant effect ($P < 0.05$) on broiler mortality. Broilers given MOL were significantly lower in mortality than broiler mortality rates that were not given MOL, it is suspected that the MOL probiotic microorganisms given to broilers could live in the broiler digestive tract because the pH conditions were in accordance with the needs of the microorganisms. The measurement results showed pH in proventiculus 5, ventricular 3.6, intestine 5.5. Isolates of Lactic Acid Bacteria 9A had good growth in the range of pH 2 - pH 4 after incubated for 3 hours, according to the time needed for food to pass through the stomach [7]. The average absorbance value of resistance to pH 2.0, 3.0, 4.0 and 5.0, *L. acidophillus* and *L. fermentum* is almost the same at each hour of incubation, meaning that the bacteria *L.acidophillus* and *L. fermentum* can survive on the digestive tract pH of poultry in proventriculus and ventriculus which has a very low pH (pH 2.0–3.0). B. fibrosolvens is a rumen bacterium that can be tolerates pH < 6 but possesses lower cellulolytic activity [8]. Rumen cellulolytic bacteria to live and develop require optimum between 6 and 9 is best, while a pH less than 5.5 affects fiber digestibility [9].

The low rate of broiler mortality in chickens given MOL was thought to be probiotic bacteria inside MOL can live in the digestive tract of broilers, especially *Lactobacillus* bacteria which can produce lactic acid by fermenting the carbohydrate feed into organic acids. Lactic acid fermentation appears to be effective in reducing the number of bacterial pathogens [10]. *Bifdobacteria* and other probiotic cultures that contribute to animal health through mechanisms such as competition with pathogenic bacteria, stimulate the immune system, increase production of short chain fatty acids, control intestinal function, prevent cancer and improve digestion and absorption of nutrients [11].

3.2. Effect of MOL addition to body weight increase

In this research probiotics on MOL did not has a significant effect (P> 0.05) on broiler body weight gain, this was presumably because the number of probiotic microbes contained in MOL was less than normal. Probiotic microorganisms given to livestock must be attached to the intestinal mucosa, and at least containing 3 x 10^{10} cfu/g [12], while microorganisms in MOL given to broilers contained *Bifidobacterium* 3.5 x 10^{7} cfu/ml concentrations, *Bacillus* concentrations 0.9 x 10^{6} cfu/ml, *Streptococcus* concentrations 1.06 x 10^{6} cfu/ml, and *Lactobacillus* concentrations 12.5 x 10^{7} cfu/ml. Probiotics do not give a significant effect on broiler body weight gain because the substrate in the feed given has a low crude fiber content that was a maximum of 5% so that the nutrients for probiotic microorganisms derived from the cattle rumen are less supportive for development. Body weight gain of broiler that given by feed with crude fiber 4.5% produced a weight gain of 1,737 grams significantly (P> 0.05) lower than the body weight gain that given by feed with crude fiber 6,0% is 1,814 grams. The adding of 0.25% starbio in broiler rations with 6% crude fiber produced the highest body weight gain compared to other treatments [13].

Another thing that causes the addition of probiotic MOL to body weight gain was not significantly, it was suspected that probiotic microorganisms were in a different environment from the previous environment such as pH conditions in the broiler digestive tract which are not in accordance with the pH in the rumen, namely pH in proventiculus 5, ventricular 3, 6, intestine 5.5. Food substrate was needed so that probiotic microbes can reproduce properly. Food substrate that supports the development of probiotic microbes in the digestive tract consists of ingredients that generally contain lots of fiber [14]. The impact of various probiotics in various locations or maintenance systems was possible because probiotics were a single factor, but many factors influence their performance. Factors that influence the performance of probiotics include (1) the composition of the host microbiota, (2) how to administer probiotics, (3) age and type of host, and (4) the quality and type of probiotics used [15].

3.3. Effect of MOL Addition to feed consumption

In this research probiotics on MOL did not has a significant effect (P> 0.05) on feed comsumption because probiotic can not develop well in broiler digestive tract because the substrate of feed contained low coarse fiber, so the micro organisms can not develop optimally so that the cellulase enzyme produced as a breaker of low crude fiber, this results in less digestion and absorption of feed in the small intestine.

3.4. Effect of MOL addition to feed conversion

Feed conversion is influenced by the ability to consume, metabolize nutrients in the ration and absorb these feed nutrients to increase body weight. The results of the research as listed in Table 2, showed that broiler feed conversion by given probiotics significantly (P <0.05) was lower than those not given probiotics, this can happen because the numbers in broilers given probiotics 5 ml/kg of feed show low consumption while the weight gain is high among treatments. Broilers by given MOL shows that the function of probiotics microorganisms had an effect on increasing feed digestibility. In microorganisms produce special enzymes that have the ability to break bonds. Breaking the bonds of complex molecules into simple molecules makes it easier to absorb feed in the digestive tract of living things.

3.5. Effect of MOL addition on carcass percentage

The addition of probiotics in the ration did not significant affect the percentage of broiler carcasses, this was due to the fact that broiler body weight at the end of maintenance showed unreal differences ($p > 0.05$). The addition of probiotics, prebiotics and a combination of both in the ration did not affect broiler meat formation, so it has not been able to significant increase the percentage of broiler carcasses between those added probiotics 68.04% and those without probiotics 66.56% [16].

4. Conclusions

MOL synthesized from cattle rumen and chicken intestine contains *Bifidobacterium* probiotic microorganisms with a concentration of 3.5×10^7 cfu/ml, *Bacillus* concentration 0.9×10^6 cfu/ml, *Streptococcus* concentration 1.06×10^6 cfu/ml, and concentration *Lactobacillus* 12.5×10^7 cfu/ml.

Probiotics based on local microorganisms from cattle rumen and chicken intestine affected gain significantly to the mortality and feed conversion ratio, but not significantly to the feed intake, body weight and percentage of broiler carcass.

5. Acknowledgements

Researchers would like to thank for P3M Politeknik Negeri Jember who have provided research funding sourced from PNPB so that this research can be done.

6. References

[1] Riady dan Erliana 2018 *Pakan Ternak AGP Dilarang, Ayam di Blitar Diserang Penyakit* https://news.detik.com.

[2] Sumanto 2016 *Awareness of the Growth Promotors Antibiotics (AGP) and Its Application in Chicken Feed* Indonesian Research Institute for Animal Production. Bogor P 462-468

[3] Apata D F 2008 Growth performance, nutrient digestibility and immune response of broiler chicks fed diets supplemented with a culture of *Lactobacillus bulgaricus* *J. Sci. Food Agric* **88** 1253–1258

[4] Higgins S E, Higgins J P, Wolfenden A D, Henderson S N, Torres Rodriguez A, Tellez G and Hargis B 2008 Evaluation of a *Lactobacillus*-based probiotic culture for the reduction of *Salmonella*Enteritidis in neonatal broiler chicks *Poult. Sci.* **87** 27–31

[5] Yu B, Liu J R, Hsiao F S and Chiou P W S 2008 Evaluation of *Lactobacillus reuteri* Pg4 strain expressing heterologous β-glucanase as a probiotic in poultry diets based on barley. *Anim. Feed Sci. Technol* **141** 82–91

[6] Yirga H 2015 The Use of Probiotics in Animal Nutrition. *J. Prob. Health* **3** 132

[7] Oozer R, Leplingard A, Mater D D G, Mogenet A, Michelin R, Seksek I, Marteau P, Dore J, Bresson JL and Corthier G 2006 Survival of *Lactobacillus* casei in the Human Digestive Tract After Consumption of Fermented Milk *Appl and Enviro. Microbiol* p 5615-5617

[8] Russell, J B and Wilson D B 1996 Why are ruminal cellulolytic bacteria unable to digest cellulose at low pH *J. Dairy Sci.* **79** 1503–1509

[9] Weimer P J 1996 Why Don't Ruminal Bacteria Digest Cellulose Faster *J. Dairy Sci.* **79** 1496-1502

[10] Russell S M, Fletcher D L, Pancorbo O C and Merka W C 1993 Effect of lactic acid fermentation on bacterial pathogens and indicator organisms in broiler processing waste. *Poult. Sci.* **72** 1573

[11] Ziggers D 2000 Tos, a new prebiotic drived from whey *Anim. Feed Sci. and Tech.* **5**

[12] Pal A, Ray L and Chattophadhyay P 2006 Purification and immobilization of an *Aspergillus* terreusxylanase: Use of continuous fluidized column reactor. *Int. J. Biotechnol* **5** 163 - 168

[13] Gunawan and Undari M M S 2003 Pengaruh Penggunaan Probiotik dalam Ransum terhadap Produktivitas Ayam *Wartazoa* **13** 94

IOP Conf. Series: Earth and Environmental Science **207** (2018) 012019 doi:10.1088/1755-1315/207/1/012019

[14] Patterson J A and Kristin M B 2003 Application of prebiotics and probiotics in poultry production Literature Review Pub. Med.
[15] Kompiang I P 2009 Pemanfaatan Mikroorganisme sebagai Probiotik untuk Meningkatkan Produksi Ternak Unggas di Indonesia. *Jurnal Pengembangan Inovasi Pertanian* **2** 2009
[16] Daud M 2006 Persentase dan Kualitas Karkas Ayam Pedaging yang Diberi Probiotik dan Prebiotik dalam Ransum *Jurnal Ilmu Ternak* **6** 126-131.

IOP Conf. Series: Earth and Environmental Science **207** (2018) 012020 doi:10.1088/1755-1315/207/1/012020

Personality assessment of different horse breeds trained for military purposes

T S M Widi[1*], G A Rakasiwi[1], T Nugroho[1] and N Widyas[2]

[1] Faculty of Animal Science, Universitas Gadjah Mada, Yogyakarta, Indonesia
[2] Department of Animal Science, Faculty of Agriculture, Universitas Sebelas Maret, Surakarta, Indonesia

[*]Email: widi.tsm@ugm.ac.id

Abstract. Behaviour traits play important role in horse used for military purposes. The aim of this study was to assess the personality of different horse breeds which trained for military purposes in Cavalry Detachment, Parongpong, Bandung, West Java. This study was performed on 40 trained horses which comprised of 20 Thoroughbred (Tbh), 10 Andalusian and 10 Unknown and on 25 non trained horse as control, which comprised of 5 Tbh, 15 Andalusian and 5 Unknown. The Horse Personality Questionnaire (HPQ), a 25 item rating method, was used to assess the horse personality of 3 horse breeds in two groups, by adjusting into 15 characters due to housing management circumstances. Principal component analysis (PCA) on HPQ data has identified five underlying personality components in observed horses. These were Dominance, Anxiousness, Stereotype, Liveliness, and Inquisitiveness. Further, Kruskal Wallis procedure was employed to observe the personality differences among breeds and training status. The PCA result showed that there are five principal components which explained most of the variance in the datasets (73.4% and 74.86% for trained and non-trained horses, respectively). Kruskal Wallis results showed that breed and training status affecting seven of the applied 15 HPQ characters: Apprehensive, curious, eccentric, equable, fearful, protective and suspicious. Equable, protective and suspicious traits were only affected by training status regardless the horse breeds. We conclude that the variation of horses' behaviour among breeds and training status existed.

1. Introduction

Horse (*Equus caballus*) has become part of human life for over 6000 years and serving various functions such as human's food, working, herding, sport, entertainment and even warfare (military/cavalry) [1]. Domestication of horses has many different purposes, resulting domestic horses has various morphological types, physical characteristics and breeds [2]. Selection process not only focus on physical characteristic such as speed, size and strength [2] but also on its behaviours [3,4].

Horse personality is important part in selection process, because personality is an indicator that makes individuals unique. It can be defined as the characteristics of individuals who have responsibility for a consistent patterns of feeling thinking and behaviour [5]. The phenomenon of how and why individuals have different personalities is an interesting thing to explore further. Personality assessment on horses can be done by using subjective questionnaires to observe individual behaviour in various situations [6–11] or conducting individual classifications and assessments [6,7,12–17].

Personality traits, especially in domesticated horses, play an important role because it influences their comfortable as riding partner [18]. In addition, there has no much research explaining the acceptance of the temperament and personality of the horse to its quality as a riding horse. Therefore, in the future, this sector needs to be scrutinized especially in terms of conducting temperament testing.

In general, horse personalities can be classified into six, Dominance, Anxiousness, Stereotype, Liveliness, Inquisitiveness and Protection [2]. Every horse breeds has a distinctive behaviour and different personality [19]. Lloyd's [2], identified that Anxiousness and Excitability are the most variation between breeds, while Protection and Dominance show small variations.

As a role in warfare, Cavalry is soldiers fighting on horseback on the battlefield. Although not widely used, nowadays, some countries still maintain small units of cavalry for ceremonial, patrol and educational purposes. Military horses are trained to have various special skills, so they are will be have unique personalities. Therefore, this study was aimed to assess the personality of different type of horse breed which trained for military purposes in Cavalry Detachment, Bandung, Indonesia.

2. Materials and Methods

A total of forty military trained horses was used in this study, which consisted of twenty Thoroughbred (Tbh), 10 Andalusian (A) and 10 Unknown (Un); and twenty-five non-trained horses which consisted of 5 Thoroughbred, 10 Andalusian and 5 Unknown. The horses belong to Cavalry Detachment, Parongpong, Bandung, West Java, Indonesia.

Questionnaire design. Horse Personality Questionnaire (HPQ) was used to assess horse personality in this study [19], a 25 rating method, adjusted into 15 characters (personalities) (**Table 1**) due to housing management circumstances. There was no different questionnaire format was used to assess for all breeds.

Table 1. Behavioural definitions of 15 personality components

No.	Personality component	Behavioural definition
1	Active	Constantly moving, does not like being still for long
2	Aggressive	Cause harm or potential harm to other individuals, display signs of aggression, both horse and human
3	Apprehensive	Seems to be anxious about everything, fears or avoids any kind of risk
4	Curious	Readily explores new situations
5	Eccentric	Shows stereotypes, unusual mannerisms and exaggerated behaviour
6	Equable	Easy to handle, calm way; not easily disturbed
7	Excitable	Over reacts to any change, easily excited, highly strung
8	Fearful	Retreats readily from others or from outside disturbances
9	Insecure	Hesitates to act alone; seeks reassurance from others
10	Irritable	Reacts negatively with little provocation
11	Protective	Prevents harm or possible harm to others
12	Slow	Moves and rests in a relaxed manner, moves slowly and deliberately, not easy hurried
13	Sociable	Seeks companionship of others
14	Suspicious	Does not trust others readily (human and horse), trusts few individuals
15	Stubborn	Does not give in easily, not very cooperative

Direct assessment using HPQ. Horses were scored on each personality using a 5 points with a score one represented no expression and a score five represented total expressions, thus three represented and 'average' or intermediate score. Horses were observed from 11.00 a.m. until 03.00 p.m., in three weeks. Each horse was assessed three times.

Calculation of component scores and statistical analysis. Calculation of the component scores for individual horse are using the horse personality model that developed by Lloyd et al. [19]. Eigen value criterion (i.e. eigenvalue should more than one), scree plots and variability each component were used to determine the extracted components. Data was subjected to PCA analysis by using SPSS program. Further, Kruskal Wallis procedure was employed to observe the personality differences among breeds and training status, using R programing language.

3. Results and Discussion

3.1. Personality assessment

Result showed that Principal Component Analysis (PCA) on Horse Personality data has identified five underlying personality components in observed horses. These were *Dominance, Anxiousness*, *Stereotype*, *Liveliness*, and *Inquisitiveness*. **Table 2** and **Table 3** show the traits for each personality components for trained horses.

Table 2. Personality traits of trained horses

Personality component				
Dominance	Anxiousness	Stereotype	Liveliness	Inquisitiveness
Equable	Aggressive	Eccentric	Active	Curious
Excitable	Irritable		Slow	
Stubborn	Protective			
Apprehensive	Sociable			
Insecure	Suspicious			
Fearful				
4.67[a]	2.85[a]	1.28[a]	1.15[a]	1.06[a]
31.14[b]	50.16[b]	58.70[b]	66.39[b]	73.44[b]

[a]Eigenvalue
[b]Cumulative (%)

Traits of *Equable*, *Excitable*, *Stubborn*, *Apprehensive*, *Insecure* and *Fearful* contributed the *Dominance* behaviour component in trained horses. While for non-trained horses, *Apprehensive*, *Curious*, *Stubborn*, *Protective*, *Suspicious* and *Insecure* contributed the Dominance behaviour component. Anxiousness behaviour component in trained horses were contributed by *Aggressive, Irritable*, *Protective*, *Sociable* and *Suspicious*; while for non-trained horse, were contributed by *Sociable*, *Equable* and *Eccentric* ones. Stereotype behaviour of trained horses was indicated by *Eccentric* trait, and non-trained horse was indicated by *Irritable* trait.

Table 3. Personality traits of non-trained horses

Personality component				
Dominance	Anxiousness	Stereotype	Liveliness	Inquisitiveness
Fearful	Sociable	Irritable	Active	Aggressive
Apprehensive	Equable		Excitable	
Curious	Eccentric		Slow	
Stubborn				
Protective				
Suspicious				
Insecure				
4.76[a]	2.15[a]	1.64[a]	1.44[a]	1.22[a]
31.83[b]	46.17[b]	57.13[b]	66.71[b]	74.86[b]

[a]Eigenvalue
[b]Cumulative (%)

Liveliness behaviour component of trained horses were contributed by *Active* and *Slow* traits; while for non-trained horses were contributed by *Active*, *Excitable* and *Slow* traits. *Curious* and *Aggressive* indicated the Inquisitiveness behaviour component of trained horses and non trained horses, respectively.

3.2. Differences among breeds

Table 4 shows Kruskal Wallis results on personality traits among breeds and between training statuses.

Table 4. Mean of personality traits among breeds and between training statuses of horses

Personality trait	Breed and training status	Mean	p value
Active	Andalusian trained	30.45	0.74
	Andalusian non-trained	33.43	
	Tbh trained	33.48	
	Tbh non-trained	39.50	
	Unknown trained	33.50	
	Unknown non-trained	30.45	
Aggressive	Andalusian trained	25.45	0.06
	Andalusian non-trained	38.07	
	Tbh trained	27.85	
	Tbh non-trained	38.70	
	Unknown trained	31.50	
	Unknown non-trained	50.80	
Apprehensive	Andalusian trained	27.40[bc]	<0.01
	Andalusian non-trained	43.27[ab]	
	Tbh trained	27.60[bc]	
	Tbh non-trained	45.40[a]	
	Unknown trained	22.60[c]	
	Unknown non-trained	43.40[ab]	
Curious	Andalusian trained	24.87[bc]	0.03
	Andalusian non-trained	42.60[a]	
	Tbh trained	31.10[abc]	
	Tbh non-trained	36.48[ab]	
	Unknown trained	38.15[ab]	
	Unknown non-trained	15.90[c]	
Eccentric	Andalusian trained	21.40[c]	0.03
	Andalusian non-trained	26.83[bc]	
	Tbh trained	35.83[ab]	
	Tbh non-trained	46.90[a]	
	Unknown trained	36.05[ab]	
	Unknown non-trained	43.40[ab]	
Equable	Andalusian trained	42.40[a]	<0.01
	Andalusian non-trained	22.60[b]	
	Thoroughbred trained	40.85[a]	

	Thoroughbred non-trained	19.90^{b}	
	Unknown trained	35.50^{a}	
	Unknown non-trained	16.10^{b}	
Excitable	Andalusian trained	32.20	0.44
	Andalusian non-trained	31.40	
	Thoroughbred trained	34.60	
	Thoroughbred non- trained	31.40	
	Unknown trained	28.20	
	Unknown non-trained	44.20	
Fearful	Andalusian trained	29.55^{bc}	<0.05
	Andalusian non-trained	40.40^{ab}	
	Thoroughbred trained	24.83^{c}	
	Thoroughbred non- trained	41.30^{ab}	
	Unknown trained	27.35^{bc}	
	Unknown non-trained	53.40^{a}	
Insecure	Andalusian trained	32.85	0.34
	Andalusian non-trained	39.33	
	Thoroughbred trained	28.83	
	Thoroughbred non-trained	40.80	
	Unknown trained	26.40	
	Unknown non-trained	36.40	
Irritable	Andalusian trained	25.60	0.06
	Andalusian non-trained	41.00	
	Thoroughbred trained	27.20	
	Thoroughbred non-trained	42.60	
	Unknown trained	30.40	
	Unknown non-trained	27.20	
Protective	Andalusian trained	26.50^{b}	0.03
	Andalusian non-trained	41.83^{a}	
	Thoroughbred trained	28.90^{b}	
	Thoroughbred non-trained	34.50^{ab}	
	Unknown trained	27.45^{b}	
	Unknown non-trained	45.50^{a}	
Slow	Andalusian trained	33.00	0.58
	Andalusian non-trained	30.87	
	Thoroughbred trained	34.60	
	Thoroughbred non-trained	33.00	
	Unknown trained	33.00	
	Unknown non-trained	33.00	
Suspicious	Andalusian trained	26.05^{b}	<0.01
	Andalusian non-trained	45.40^{a}	
	Thoroughbred trained	26.13^{b}	
	Thoroughbred non-trained	34.20^{ab}	

	Unknown trained	28.20^{b}	
	Unknown non-trained	45.60^{a}	
Stubborn	Andalusian trained	31.47	0.08
	Andalusian non-trained	41.37	
	Thoroughbred trained	29.18	
	Thoroughbred non-trained	35.10	
	Unknown trained	22.85	
	Unknown non-trained	44.50	
Sociable	Andalusian trained	38.90	0.15
	Andalusian non-trained	23.07	
	Thoroughbred trained	37.00	
	Thoroughbred non-trained	37.20	
	Unknown trained	35.10	
	Unknown non-trained	26.60	

^{a,b,c} different superscripts indicate significant differences between groups in each trait

The results showed that breed and training status affecting seven of the applied 15 HPQ characters: *Apprehensive*, *Curious, Eccentric*, *Equable*, *Fearful*, *Protective* and *Suspicious*. In all groups of breed and training status, the horses have about similar level of aggressive, however Andalusian are significantly more apprehensive regardless the training status. *Equable*, *Protective* and *Suspicious* traits were only affected by training status regardless the horse breeds.

All trained horses have a low level of *Apprehensive* and *Fearful*. We suspect this is related that trained horses more often interact with humans so that fear and anxiety are reduced. Military horses are trained to still calm in harsh conditions such as fireworks, war or tumultuous crowds [20]. Then, trained horses have a quick response to stimuli, result that they are not easily shocked and fear [21]. The trained thoroughbred has the lowest fearful level which shows that thoroughbred is one of the breed that has a calm characteristic [22]. Furthermore, *Equable*, *Protective* and *Suspicious* traits were only affected by training status regardless by the horse breeds. The trained horse has better calm way, didn't easy disturbed, lower protective and suspicious properties than non-trained horse. This traits are affected by various stimuli along the horse's live, for example, how many rider have been a riding its horse [23].

Andalusian horses have low *Eccentric* level compared to other horses in both trained and untrained conditions. *Eccentric* trait is an undesirable behaviour of military horse. We suspect that Andalusian horses do not have high eccentric traits genetically. This is because each breed carries the various characteristics [24]. On the other hand, training status on Andalusian horses can reduce the *Curious* trait of the horse but actually increases the unknown horse. It is showed that the response of each breed to the stimulus is different, resulting in different personalities [2].

4. Conclusion

Principal component analysis (PCA) on Horse Personality Questionnaire data has identified five underlying personality components in observed horses: *Dominance, Anxiousness*, *Stereotype*, *Liveliness*, and *Inquisitiveness* which were contributed by 15 different structure of traits for trained and non-trained horses. The variation of horses' behaviour among breeds and training status existed.

References

[1] Villalon A E 1999 *The circle of war in the middle ages: essay on medieval military and naval history* (Woodbridge, UK: The Boydell Press)

[2] Lloyd A S, Martin J E, Bornett-Gaucci H L I and Wilkinson R G 2008 *Appl. Anim. Behav. Sci* **112** 369–83

[3] Hislop J 1992 *Breeding for racing* (London, UK: The Kingswood Press)
[4] Houpt K A and Kusunose R 2000 Genetics of behavior *The Genetics of The Horse* ed A T Bowling and A Ruvinsky (Oxon, UK: CABI) pp 281–306
[5] Pervin L A and Cervone D 2015 *Personality: Theory and Research* (New York: John Wiley and Sons)
[6] Le Scolan N, Hausberger M and Wolff A 1997 *Behav. Process* **41** 257–66
[7] Anderson M K, Friend T H, Evans J W and Bushong D M 1999 *Appl. Anim. Behav. Sci.* **63** 11–24
[8] Morris P H, Gale A and Duffy K 2002 *Pers. Indiv. Differ.* **33** 67–81
[9] Momozawa Y, Ono T, Sato F, Kikusui T, Takeuchi Y, Mori Y and Kusunose R 2003 *Appl. Anim. Behav. Sci.* **84** 127–38
[10] Momozawa Y, Kusunose R, Kikusui T, Takeuchi Y and Mori Y 2005 *Appl. Anim. Behav. Sci.* **92** 77–84
[11] French J M 1993 *Appl. Anim. Behav. Sci.* **36** 249–57
[12] Wolff A, Hausberger M and Le Scolan N 1997 *Behav. Process* **40** 209–21
[13] Visser E K, van Reenen C G, Engel B, Schilder M B H, Barneveld A and Blokhuis H J 2003 *Appl. Anim. Behav. Sci.* **82** 279–95
[14] Visser E K, van Reenen C G, van der Werf J T N, Schilder M B H, Knaap J H, Barneveld A and Blokhuis H J 2002 *Physiol. Behav.* **76** 289–96
[15] Visser E K, van Reenen C G, Hopster H, Schilder M B H, Knaap J H, Barneveld A and Blokhuis H J 2001 *Appl. Ani. Behav. Sci.* **74** 241–58
[16] Seaman S C, Davidson H P B and Waran N K 2002 *Appl. Anim. Behav. Sci.* **78** 175–91
[17] McCall C A, Hall S, McElhenney W H and Cummins K A 2006 *Appl. Anim. Behav. Sci.* **96** 115–27
[18] Graf P, König von Borstel U and Gauly M 2013 *Züchtungskunde* **85** 129–42
[19] Lloyd A S, Martin J E, Bornett-Gauci H L I and Wilkinson R G 2007 *Appl. Anim. Behav. Sci.* **105** 205–22
[20] Munsters C C B M, Visser E K, Van Den Broek J and Sloet Van Oldruitenborgh-Oosterbaan M M 2013 *Animal* **7** 822–7
[21] von Borstel U U K, Duncan I J H, Lundin M C and Keeling L J 2010 *Appl. Anim. Behav. Sci. Anim. Behav. Sci.* **125** 124–31
[22] McGreevy P D and Thomson P C 2006 *Appl. Anim. Behav. Sci.* **99** 183–90
[23] Schork I G, de Azevedo C S and Young R J 2018 *PLoS One* **13** 1–18
[24] König v B 2013 *CAB Rev. Perspect. Agric. Vet. Sci. Nutr. Nat. Resour.* **8** 1–21

Effect of supplement Maggot Black Soldier Fly live on the percentage of carcass and weight of carcass of male Alabio ducks

A. Gunawan[1], S. Erlina[2], R. Samudera[1], M. Syarif D[1], M.Y. Noor[1] and A.X. Lantu[1]

[1]Livestock Study Program, Faculty of Agriculture, Islamic University of Kalimantan, Banjarmasin Indonesia
[2]Agribusiness Study Program, Faculty of Agriculture, Islamic University of Kalimantan, Banjarmasin Indonesia

E-mail: aamgunawan@yahoo.com

Abstract. Maggot black soldier fly (Hermetia illucens) contains complete nutrients and in a living condition is very favored by ducks. Therefore this study aims to study the giving of live maggots tried in ducks. The study used 60 male alabio ducks placed randomly into 20 cages. The design used is a completely randomized 2x2 factorial design with five replications. The treatment consisted of P1: 10% live maggot supplementation on 16% ration protein, P2: 10% live maggot supplementation on 21% protein ration, P3: 0% live maggot supplementation on 16% ration protein, and P4: 0% live maggot on protein ration 21%. The data obtained were analyzed of variance and Duncan's multiple range test. The results showed that the giving of live BSF maggot had a highly significant effect on the percentage of carcass and weight of carcass cuts, but did not affect the percentage of abdominal fat and the weight of the viscera parts. Ration protein levels of 16% and 21% did not affect the percentage of carcass, weight of abdominal fat, the weight of carcass cuts, and the weight percentage of viscera part of male alabio ducks (Anas platyrhynchos Borneo).

1. Introduction

Many Indonesians living in Kalimantan are working in the fields of animal husbandry and fisheries. The problem faced by these two business fields is the problem of feed, many people are classified as small and medium-sized businesses that go bankrupt due to soaring feed prices. Feed costs are very large in livestock or fisheries business, which is about 75% of the total production cost. Therefore, there is a great need for the availability of cheap feed, so that the business carried out by the community can continue continuously.

Maggot can be used to feed various types of fish and livestock [1] [2] [3] [4] [5] [6] [7] [8] [9]. Maggot can be given alive or meal, of course, this can be used as a new business for people in Kalimantan, where the population has a lot of fish and livestock, especially ducks. Especially at this time, the price of feed continues to increase.

Maggot *Hermetia illucens* or better known as maggot black soldier fly (BSF) is very well used as a source of protein feed ingredients. Maggot BSF contains about 45% protein with 35% fat content [10]. According to [11] ensure livestock grow well, the starter phase duck ration is prepared 21% protein, this protein content is enough as a growth promoter, which is supported by only 4% crude fiber and 2,900 kcal/kg of metabolizable energy. Unlike the finisher phase where the protein content is relatively low at 19% and the metabolizable energy of 3,000 kcal/kg can guarantee growth. Because finisher ducks need higher energy for meat growth and formation. Low protein ration (16%) certainly has a much cheaper price compared to high protein content (21%). Thus, live maggot supplementation on low protein content rations is expected to have the same quality as 21% protein ration. Protein supplements can come from animals or plants and can be given as a protein source in rations if the protein content is low. Great protein deficiency, cattle will experience an average growth loss of 6-7%, in laying hens can cause severe molting and egg production stops. On the basis of these considerations, the preparation of the ration requires additional protein that is rich in the content of essential amino acids.

Carcasses with a larger weight will also affect the profit at the time of sale. The pieces of the carcass are the body parts of poultry after being reduced by feathers, blood, head, feet and internal organs. Generally, restaurants cut duck carcasses into four parts, namely thighs, breast, back, and wings. Ducks that are cut are young ducks because of the soft texture of the meat. The addition of live BSF maggots is expected to affect the growth of Alabio ducks. Carcass cuts that have a high percentage of weight will also produce high profits and reduce feed costs by BSF live maggots. Therefore, the authors are interested in conducting a study on the effect of giving maggot live for the growth phase ducks on the percentage of carcass, abdominal fat, the weight of carcass cuts, and percentage of viscera parts of Alabio ducks.

2. Materials and methods

Maggot BSF lives, prepupa phase. Live BSF Maggot is obtained from cultivation using palm kernel meal media. This study used as many as 60 Alabio ducks, 3 weeks old and randomly divided into 20 cages. Each cage is filled with three ducks. The composition of the experimental ration is presented in Table 1, the nutrient content of the experimental ration are presented in Table 2.

Table 1. Nutrients Composition of Ingredient (As-fed)

Ingredient	Crude Protein (%)	M.E Kcal/kg
Concentrated duck[1]	38.00	3,300
Rice bran[2]	13.20	2,878
Corn[2]	9.42	3,182
Fish meal[2]	36.00	3,468

[1]PT. Wonokoyo Jaya Corporindo Surabaya
[2]Yield of laboratory analyze

Table 2. Nutrient Content of Experimental Rations

Feed Ingredient	Treatment ration	
	Low Protein	High Protein
Concentrated duck	19.72	25
Rice Bran	25	25
Corn	55.28	36.85
Fish meal	0	13.15
Total	100	100
Nutrient content		
Crude protein (%)	16	21
Metabolizable energy (kcal/kg)	3,129.27	3,173.11

The cage is made of materials such as boards, bamboo, zinc roofs with an area of 3 x 6 m. Inside the cage, there are 20 cages with a size of 100 x 90 x 50 cm. The base of the cage is made of bamboo with a slat system. The drinking place used is made of PVC longitudinal gutters and the place of feed used is a tub made of 20 wooden boards.

The equipment of this research such as basin, with a diameter of 50 cm and a height of 30 cm as many as 50 pieces are used to place palm kernel cake, machine for making pellet feed, digital scales of Oxone brand 5 kg with accuracy of 1 g, used to weigh feed, ducks, and equipment for cutting ducks and cutting carcass parts

The design used is a completely randomized design (CRD) 2x2 factorial pattern. The first factor was the give of live BSF Maggot with two levels, namely 0%, and 10%. The second factor is protein ration with two levels, namely high protein ration (21%) and low protein ration (16%). Each treatment combination uses five replications. The combination of treatment is:
P1 = Low Protein Ration (16%) + Live Maggot BSF (0%)
P2 = Low Protein Ration (16%) + Live Maggot BSF (10%)
P3 = High Protein Ration (21%) + Live Maggot BSF (0%)
P4 = High Protein Ration (21%) + Live Maggot BSF (10%)
If based on the results of the analysis of variance there is a significant effect, then the test was continued using Duncan's multiple range test [12].

The research procedure following: the basin is 20 pieces each filled with three kg of palm kernel meal and six liters of water, covered with dried banana leaves. Left in the open for three weeks. Maggots found in the basin were harvested, then weighed and given to ducks. Newly arrived ducklings are inserted in each cage before weighing a uniform initial weight. Ducklings are placed in a cage compartment each filled with three ducks. After that, ducklings are given research ration in accordance with the four types of treatment. Before being given, each treatment ration material was mixed with each other with the following procedure: each raw material was weighed according to requirements, each treatment was milled with a pellet machine until it was completely homogeneous.

The ration is given three times a day or ad libitum, ie in the morning, afternoon and evening. The ration is given again if inside the food is completely gone while drinking water is given ad libitum. All cages are equipped with feeder and drinker water to suit your needs, for lighting used 15-watt incandescent lamps. Male alabio ducks aged eight weeks fasted from feed about 6-12 hours then cut, just before being cut, the ducks were weighed, then the ducks were allowed to stand for about two minutes.

Cutting is done at the border of the neck and head, by cutting the jugular vein, carotid artery, trachea, and esophagus (Kosher method). After that, leave it for 1-3 minutes until the blood stops dripping. Furthermore, alabio ducks without blood are weighed, then dipped in hot water at a temperature of approximately 80 °C until the hair is easily revoked. Fur is pulled out manually, and ducks without fur are weighed. The carcass is obtained by separating the paws, neck, head, and viscera. The carcass is weighed, then cut the breast, wings, rear, back, and thighs. The viscera and giblets are cleaned and separate the parts of the viscera before being weighed.
Response Variable Measurement

1. Percentage of the carcass, calculated using the formula:

$$\frac{Carcass\ wei}{Slaughter\ weight}\ x\ 100\%$$

2. Percentage of abdominal fat, calculated using the formula: $\frac{\boldsymbol{abdominal\ fat\ weight}}{\boldsymbol{carcass\ weight}}$ x 100%

3. Percentage of breast weight, calculated using the formula: $\frac{Breast\ wei}{carcass\ wei}$ x 100%

4. Percentage of wing weight, calculated using the formula: $\frac{Wing\ weigh}{carcass\ weight}$ x 100%

5. Percentage of back weight, calculated using the formula: $\frac{Back\ weig}{carcass\ weigh}$ x 100%

6. Percentage of rear weight, calculated using the formula: $\frac{Rear\ weight}{carcass\ weigh}$ x 100%

7. Percentage of thigh weight, calculated using the formula: $\frac{Thigh\ weight}{carcass\ weight}$ x 100%

8. Percentage of viscera parts weight, calculated using the formula: $\frac{viscera\ parts\ weight}{viscera\ weight}$ x 100%

Observation data during the study was collected, then an analysis of variance was conducted to determine differences in the effect of treatment on the observed variables. Variance homogeneity test (Bartlett test) was carried out. If the results of the analysis of variance showed a significant or highly significant effect, the test was continued with the Duncan Multiple Range Test [12].

3. Results and Discussion

3.1. Percentage of Carcass.

Data from observation of carcass and abdominal fat percentage of each treatment are presented in Table 3.

Table 3. The average percentage of the carcass and abdominal fat of Alabio ducks aged 8 weeks

Variable	Protein (%)	Live maggot BSF (%)		Average (%) $\pm$ SEM
		0	10	
Carcass percentage	16 (Low)	56.00	68.96	**62.48 $\pm$ 2,29**
	21 (High)	56.75	66.83	**61.79 $\pm$ 2.07**
	Average (%) $\pm$ SEM	**56.37^a $\pm$ 1.17**	**67.89^b $\pm$ 0.90**	
% abdominal fat	16 (Low)	1.07	1.31	**1.19 $\pm$ 0.11**
	21 (High)	1.12	1.15	**1.13 $\pm$ 0.12**
	Average (%) $\pm$ SEM	**1.09 $\pm$ 0.09**	**1.23 $\pm$ 0.13**	

Description: The numbers followed by different superscript letters in the direction of the rows indicate a significantly different

Based on the results of variance analysis it can be seen that the administration of live BSF maggots in rations had a highly significant effect (P <0.01) on the percentage of male Alabio duck carcasses aged eight weeks, but no effect (P> 0.05) on the treatment of protein ration levels. The administration of maggot is very active in increasing the percentage of carcasses of Alabio ducks. Giving 10% of live maggot from feed given every day can increase the percentage of the carcass from 56.37% to 67.83%. According to [13] further reported that the percentage of the carcass was sent to the weight of the cut. In the opinion of [14], the percentage of the carcass is influenced by the rate of growth and quality of feed. The growth rate shown by the existence of life weight gain will affect the cut weight produced and will affect the percentage of carcass produced.

Reported of [2] replacing fishmeal with a maggot or maggot flour living as much as 10% of the feed given each day will result in better weight gain with the same carcass production. This is also supported by [15] usage rate of more than 10% results in low consumption and performance which is likely due to the low palatability due to the black color of the maggot. Furthermore, it was stated that rations containing unbalanced energy and protein could affect the percentage of carcasses.

The percentage of the carcass is related to gender, age, and body weight. Carcass weight will increase with increasing age and increasing body weight of ducks that are affected by nutrient intake during maintenance. It is suspected that the addition of live BSF maggots in feeding Alabio ducks will improve the quality of the ration so that the ration nutrients can be digested properly in the body to produce meat. This is consistent with the opinion of [16] that supplementary feed can increase final body weight and a better percentage of the carcass. According to [17], the percentage of the carcass is influenced by body weight and body fat when it reaches market conditions. The level of protein in the ration that does not affect the percentage of carcass caused by ducks used in this study has entered the grower period 3-8 weeks. According to [11], the duck period of grower aged 2-7 weeks is sufficiently given 16% protein ration, while metabolic energy is 3,000 kcal/kg.

3.2. Percentage of Abdominal Fat

Data from the observation of the percentage of abdominal fat of each treatment are presented in Table 3. Based on the results of variance analysis it can be seen that the administration of black soldier fly 0% and 10% had no significant effect on the percentage of abdominal fat in male Alabio ducks aged eight weeks, also with the treatment of protein levels in rations 16% and 21%, both of which give relatively the same results. The average percentage of abdominal fat from Alabio ducks from each treatment ranged between 1.07 - 1.31%. This small percentage of abdominal fat is thought to be closely related to the age of young cattle, which is eight weeks, so that fat accumulation during growth is still small. Reported of [18] states that abdominal fat is clearly visible when the ducks reach 9 weeks. It is suspected that the presence of abdominal fat is more influenced by slaughter age.

The results of this study were relatively low compared to the results of the study by [19] reported that the average abdominal fat of local ducks ranged from 3.08 to 4.12%. This is due to the balance between protein and energy from all treatments given. So that it allows consuming the same amount of protein and energy. The same protein and energy balance will produce relatively the same abdominal fat.

Based on the results of variance analysis it can be seen that the treatment of protein levels in this study did not significantly affect the percentage of abdominal fat. This is because the 16% low protein ration has fulfilled the needs of ducks in accordance with the standards provided by the Indonesian National Standard (SNI) which is 15%, while the 21% high protein ration is too excessive resulting in smelly and watery excreta. As for which affects the low consumption of additional feed in this study allegedly due to the low palatability due to the black color of the living maggot. The results of this study are in line with [20] body fat accumulation is influenced by the quality of the ration, the method of feeding and maintenance management. Abdominal fat results from accumulation during life and ration consumption.

3.3. Percentage of Carcass

Data from the observation of the percentage of carcass cuts of each treatment are presented in Table 4. The average weight percentage of carcass pieces consisting of breast, wings, back, rear, and thighs are presented in Table 4. The results of variance analysis showed that the administration of live BSF maggots had a highly significant effect ($P<0.01$) on weight percentage carcass parts. The level of protein in ration 16% and 21% does not affect the weight percentage of carcass cuts.

The average percentage of breast, wing, back, rear, and thigh in a row at the treatment rate of 10% live maggot administration was 19.92%; 14.18%; 15.41%; 14.33%; 21.84%, and without maggot 23.64%; 16.48%; 18.30%; 16.29%; 25.31%. The results of the analysis of variance showed that the treatment of live maggot supplementation showed a highly significant effect on the percentage of the weight of the breast, wings, back, rear, and thighs. Provision of live maggots results in a lower percentage of carcass pieces than without the administration of live maggot. This is easy to understand because giving live maggots results in a higher percentage of carcass, the carcass weight obtained is also greater, so that the weight percentage of the carcass pieces becomes smaller. Based on the observations in this study, it turned out that ducks preferred to consume maggot live first, after being exhausted, then consuming rations, resulting in better absorption of nutrients, especially protein and fat, which resulted in the formation of muscle tissue and fat. In addition, the fat-containing maggots are quite high at around 35% [10] which is able to be absorbed by the body even though it can reduce the level of feed consumption.

The average percentage of breast, wing, back, rear, and thighs respectively in the treatment rate of 16% protein ration giving was 21.36%; 15.22%; 16.75%; 15.60%; 23.92%, whereas in the treatment of protein level 21% was 22.20%, 15.44%; 16.96%; 14.87%; 23.23%. The level of protein ration that has no effect, this is because the 16% low protein ration has fulfilled the needs of ducks in accordance with the standards provided by the Indonesian National Standard (SNI) which is 15%, while the 21% high protein ration is too excessive resulting in smelly and watery excreta. Breast weight percentage obtained from the results of this study is greater than the study of [21] which obtained a percentage of the mojosari duck's breast section of 11.13%. This is easy to understand because of differences in measurement methods. This is more due to differences in rations given. The weight percentage of carcass pieces is illustrated in Figure 1.

Table 4. Average Weight Percentage of Pieces of Alabio Duck Age 8 Weeks

Variable	Protein (%)	Live maggot BSF (%)		Average (%) ± SEM
		0	10	
Breast percentage	16 (Low)	23.08	19.63	**21.36 ± 0.85**
	21 (High)	24.19	20.20	**22.20 ± 0.83**
	Average (%) ± SEM	**23.64^b ± 0,53**	**19.92^a ± 0,63**	
Wing persentase	16 (Low)	16.28	14.15	**15.21 ± 0.44**
	21 (High)	16.68	14.20	**15.44 ± 0.47**
	Average (%) ± SEM	**16.48^a ± 0.24**	**14.18^b ± 0.26**	
Back percentage	16 (Low)	18.22	15.28	**16.75± 0.63**
	21 (High)	18.37	15.54	**16.95 ± 0.57**
	Average (%) ± SEM	**18.30^b ± 0.34**	**15.41^a ± 0.38**	
Rear percentage	16 (Low)	16.85	14.34	**15.60 ± 0.48**
	21 (High)	15.42	14.32	**14.87 ± 0.27**
	Average (%) ± SEM	**16.14^b ± 0.33**	**14.33^a ± 0.21**	
Thigh percentage	16 (Low)	25.57	22.26	**23.91 ± 0.72**
	21 (High)	25.04	21.41	**23.22± 0.67**
	Average (%) ± SEM	**25.30^b ± 0.34**	**21.83^a ± 0.47**	

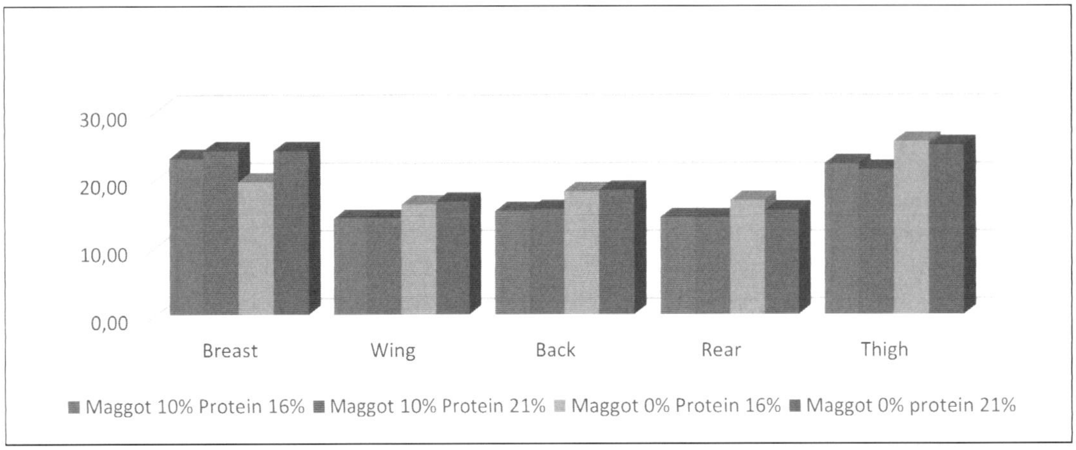

Figure 1. Weight Percentage of Pieces of Male Alabio Ducks Age 8 Weeks

3.4. Weight Percentage of Viscera Parts

The results of observations of percentages of viscera parts are presented in Table 5. The average percentage of liver weight, gizzard, heart, and intestine at 10% of live maggots was 28.475%; 36.775%; 6,435%; and 28.32% respectively. The results of variance analysis showed that the treatment of live maggot administration, as well as the provision of protein ration levels both, expressed no significant effect (P> 0.05) on the percentage of viscera.

Table 5. Average Weight Percentage of Parts Viscera of Alabio Ducks Age 8 Weeks

Variabel	Protein (%)	Live Maggot BSF (%)		Average (%) ± SEM
		0	10	
Liver percentage	16 (Low)	28.69	28.69	**28.69 ± 1.16**
	21 (High)	25.52	28.26	**26.89 ± 0.82**
	Average (%)± SEM	**27.10 ± 1.14**	**28.475 ± 0.89**	
Gizzard percentage	16 (Low)	35.28	35.58	**35.43 ± 1.34**
	21 (High)	37.63	37.97	**37.80 ± 1.10**
	Average (%)± SEM	**36.45 ± 1.44**	**36.77 ± 1.11**	
Heart percentage	16 (Low)	6.53	6.56	**6.54 ± 0.35**
	21 (High)	6.47	6.31	**6.39 ± 0.17**
	Average (%)± SEM	**6.50 ± 0.32**	**6.43 ± 0.23**	
Intestine percentage	16 (Low)	29.51	29.17	**29.34 ± 1.12**
	21 (High)	30.39	27.47	**28.93 ± 1.06**
	Average (%)± SEM	**29.95 ± 1.34**	**28.32 ± 0.66**	

The weight percentage of innards from the results of this study is higher than that of [21] where for Mojosari ducks the percentage of intestines was 3.15% and Tegal ducks were 2.81%, and according to [13] research results on male local ducks aged 8 weeks the percentage of liver 2.21%, heart 0.72%, and bile 4.74%. The difference in the results of this study is caused more by differences in measurement methods. In the study of [21] and [13], all noncarcass weight is divided by cutting weight, while in this study the divider is the weight of viscera.

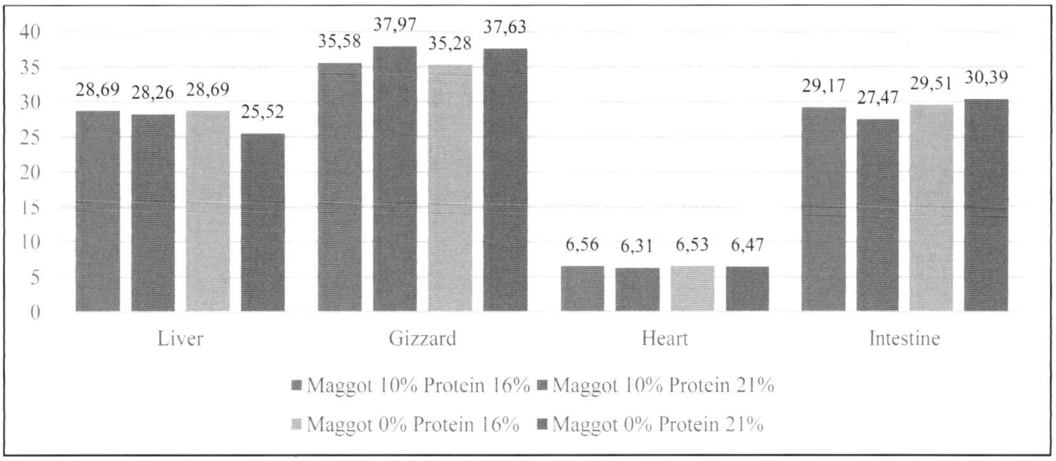

Figure 2. Percentage of Percentage of Male Viscera of Alabio Ducks Age 8 Weeks

Maggot (*Hermetia ilucens)* life BSF used in this study is thought to have calcium (Ca) which has a better biological value than calcium (Ca) in fish meal, conditions of dietary calcium intake meet the achievement of the most active calcium absorption level in the mechanism transport in the small intestine (upper duodenum and jejunum), which was expressed in an increase in dietary calcium digestibility using live Maggot BSF, [22] reported that the calcium content in Maggot was quite high, had a digestibility rate of 88%. The increasing digestibility of calcium (Ca) with the increasing addition of maggot of BSF life in the ration indicated that calcium (Ca) contained in the living maggot BSF had better biological value [23].

4. Conclusion

The provision of live BSF maggots at 10% of the amount of ration given can result in a higher percentage of carcass but produce a lower percentage of carcass pieces than 0% BSF maggot (without maggot). The level of protein ration of 16-21% resulted in the percentage of carcass, abdominal fat, carcass pieces and the percentage of parts of the innards that were relatively the same in male alabio ducks aged 8 weeks.

Acknowledgments

The author would like to thank the Director of Research and Community Service, Directorate General of Kemristekdikti Research and Development Strengthening and General Chairperson of the Kalimantan Islamic University Foundation Management Board Muhammad Arsyad Al Banjari who has provided research budget assistance to the authors.

References

[1] Agunbiade J A 2007 *J. Poult. Sci.* **44** 3 278–282
[2] Téguia A, Mpoame M and Mba J A O 2002 *Tropicultura*, vol. 20, no. 4, pp. 187–192, 2002.
[3] Awoniyi T A M, Aletor V A and Aina J M 2003 *Int. J. Poult. Sci.* **2** 4 271–274
[4] Bodri M S and Cole E R 2007 *Georg. J. Sci.* **65** 2 82
[5] Bondari K and Sheppard D C 1981 *Aquaculture* **24** 103–109
[6] Hem S, Toure S, Sagbla C and Legendre M 2008 *African J. Biotechnol.* **7** 8
[7] Haryati E S and Pranata A 2011 *Univ. Hassanudin, Makassar* 1–14
[8] Kardana D, Haetami K and Suherman H 2012 *J. Perikan. Kelaut.* **3** 4
[9] Torang I 2013 *J. Ilmu Hewani Trop. (Journal Trop. Anim. Sci.* **2** 1 12–16

[10] Olivier P A 2000 *Wide Integration of Specialized Crop and Livestock Production* **18**
[11] Council N R 1994 *National Academy Press,* Washington, DC.
[12] Steel R G D and Torrie J h 1993 Prinsip dan prosedur statistik suatu pendekatan biometrik *PT. Gramedia, Jakarta*, 1993.
[13] D. R. T. P. Karkas And D. A. N. L. Non-Karkas Percentage Of Carcass, Non-Carcass And Abdominal Fat Of Eight Week Old Local Male Duck
[14] Soeparno E 2005 Ilmu dan Teknologi Daging *Gadjah Mada University Press,* Yogyakarta
[15] Bamgbose A M 1999 *Indian J. Anim. Sci.* **69** 12 1056–1058
[16] Supriyati K I P 2002 *JITV* **7** 150–154
[17] Rizal Y 2000 *J. Peternak. dan Lingkung.* **6** 02
[18] Elfeil W, Soliman E and Sobeih M 2011 *Epidemiological studies on environmental pollution in poultry fa*rms
[19] Tugiyanti E and Yuwanta I T 2000 *Studi penggunaan sumber energi pakan terhadap karkas dan foie gras unggas air lokal Indonesia* Universitas Gadjah Mada
[20] Ramadhan K, Huda N and Ahmad R 2014 *J. Food Sci. Technol.* **51** 2 256–266
[21] Armissaputri N K, Ismoyowati I and Mugiyono S 2014 *J. Ilm. Peternak.* **1** 3
[22] Finke M D 2013 *Zoo Biol.* **32** 1 27–36
[23] Adedokun S A and Adeola O 2013 *J. Appl. Poult. Res.* 22 3 600–608

Precaution in Introducing Double-Muscled Exotic Breeds into Indonesian Cattle Population

N Widyas[1], S Prastowo[1], T S M Widi[2] and I G S Budisatria[2*]

[1]Department of Animal Science, Sebelas Maret University, Surakarta INDONESIA
[2]Faculty of Animal Sciences, University of Gadjah Mada, Yogyakarta INDONESIA

*E-mail: budisatria@ugm.ac.id

Abstract. Indonesian has been striving to obtain meat self-sufficiency since the later decade and the most successful program so far is crossbreeding which was able to increase the national meat production. Recent attempt is to introduce double muscling (DM) sires as the male genetic resources in crossbreeding program. Before the program is implied, this paper will present a simulation study underlying the theoretical insight concerning the utilization of DM sires. This is important because DM was an expression of muscular hypertrophy (*mh*), a specific mutation in bovine MSTN gene which down-regulate the muscle fiber disposition; and thus, has specific inheritance and expression pattern.

Mating simulation with 10 DM sires and 1000 dams showed that the first-generation offspring F1_Batch1) will all carry one copy of *mh*-gene. Assumption that 90% of F1_Batch1 were survived to adulthood with 50:50 sex ratio is made. The frequency of normal (+/+) individuals decreased whilst the frequency of heterozygous individuals carrying *mh*-gene mutations increased overtime for discrete and overlapping generations scenarios. There were new DM individuals emerging in overlapping-generation scenario. Calves carrying *mh*-gene were larger, gestation period was longer and prone to calving difficulties. DM dams also developed narrower birth canal due to the excess of muscle mass. Introduction of DM cattle is tempting to increase meat production. If DM cattle introduction is unavoidable, precautions are to be make regarding the reproduction of carrier cows. All F1 have to be slaughtered and must not be allowed to reproduce. Otherwise the *mh*-gene mutation would spread in the population and disturb the calf production. DM introduction will pose a major threat in the Indonesian cattle population especially from reproductive aspect.

1. Introduction

Indonesian has been striving to obtain meat self-sufficiency since the later decade and the most successful program so far is crossbreeding [1,2]. It has long been offered as a faster way to achieve genetic improvement in livestock, especially in large populations such as country level. Crossbreeding practice is conducted by introducing superior exotic breeds into the currently existing genepool of a species. Our local genetic resources are more robust to tropical environment [3,4] as a mechanism of adaptation, but less superior in producing meat and/or milk; therefore, crossbreeding is seen as a promising solution for the lack of meat production problem. The implementation of such program, however, need careful assessments and considerations since it bore the potential to disturb the local cattle population.

Indonesia have been importing male meat-type *Bos taurus* for artificial insemination (AI) semen source since early 1980's. Although far from being sufficient, crossbreeding program through AI technology managed to increase the population of meat-type cattle in Indonesia and thus boosting the national meat production [5]. Inspired by this success, Indonesian government now plan to introduce Double Muscling (DM) cattle through its semen and embryo transfer (ET) [6,7]. The conventional taurine is naturally of a bigger size compared to Indonesian locals which are mostly *Bos indicus* and *Bos javanicus* [3,4]. On the other hand, the DM cattle, aside from being naturally of taurian breed, it possesses the double muscling phenotype which is an expression of muscular hypertrophy (*mh*), a specific mutation in bovine myostatin (MSTN) gene [8–10]. The presence of this mutation is indeed causing DM cattle to develop more muscle both in mass and fiber size compared to conventional cattle, but its inheritance pattern is nowhere near genetically additive. Further, there are side effects which need to be considered when a decision to include DM cattle for national commodity is made especially from the female reproduction aspect. Introduction of DM cattle is tempting to increase meat production. However, thorough studies regarding this decision are to be made to avoid the unwanted effects. We conduct a literature and simulation study on the spreading of MSTN mutation in a hypothetical cattle population along with its inheritance pattern. We try to write this article in a practical way that it is readable and understandable to all of the stake holders.

2. The genetics of double-muscling trait

The *mh* allele is the causal mutation responsible for muscular hypertrophy; which phenotype manifestation is excess of muscles in both mass and numbers [9,11,12] resulting in cattle with extraordinary body conformation. This mutation occurred naturally in the form of 11-bp deletion in the MSTN gene and can be found in the most of the taurian breed populations such as Belgian Blue, Galician blond, Charolais, Angus and Piedmontese [8,9,11–14]. The frequency of the mutant allele, however, vary with regards to the historical aspects among populations [15]. The meat type DM cattle such as Belgian Blue (BB) were originally dual purpose breed in the beginning [16,17]. Selection program in favour of DM phenotype which was accumulated for generations had created a beef type Belgian Blue homozygous for the *mh*-allele (*mh/mh*) [8,14,17].

Preceding researches reported that there were two alleles: wild type (+) and mutated (*mh*) MSTN gene involved in the expression of DM [18]. It is therefore in the populations of dual-purpose BB and Piedmontese, there were three genotypes segregated at this particular locus: +/+ (conventional cattle), +/*mh* and *mh/mh* [16,17,19]. Owing to the fact that the major cause of DM phenotype is a mutation in MSTN gene, the inheritance pattern of this trait does not strictly follow infinitesimal model as in complex traits. Complex traits behave in an additive manner following normal distribution because their expressions were governed by very many genes [20,21]. The DM phenotype, however, is affected by a single gene, thus its expression only depends on the number of *mh* allele present in an individual.

Straight and crossbreeding experiment had been conducted using DM individuals. In Piedmontese straight-breeding, homozygous *mh/mh* and heterozygous +/*mh* were significantly heavier and produce more carcase compared to their conventional counterpart [19]. In crossbreeding experiments, however, DM-sired cattle were not always superior in term of growth traits compared to the conventional-sired ones [22,23]. Scientific articles regarding the proportion of offspring exhibiting DM traits in the breeding experiment also existed. In DM × DM mating, the proportion of calves with DM phenotype were around 77 – 86%; whereas in DM × N (normal conventional cattle) mating were around 42 – 45% [24,25]. It is also important to notify that the phenotype of the heterozygote cattle could be anywhere between normal to heavily DM [11]. MSTN is not only functioned as a major gene controlling muscle growth which mutation can cause DM phenotype. MSTN also interacted with other genes; thus, mutation in MSTN could affect the product of the epistasis and causing large variation in the cattle's phenotype [11,12].

3. Simulation design

We conducted a simulation study using custom script in R programming language [26]. In this simulation, we set up a hypothetical base population with 10 DM sires homozygous *mh/mh* and 1000 normal local dams assumed to have +/+ genotype. All sires and dams were mated at the age of two years old; after they were sexually matured. Two scenarios were made: 1) Discrete generations which followed the assumptions that the population only consisted of base generation and the first-generation offspring, and 2) overlapping generations; where the population also consisted individuals of the second generation as the offspring of the first generation. The second generation was therefore the offspring of the first-generation after they were sexually matured with the 10 DM sires in the base population while we avoid mating between related individuals. We run the simulation for five years or five consecutive mating cycles (batch) following **Figure 1**.

A

Y0	Y1	Y2	Y3	Y4	Y5
Base	G1 Batch 1	G1 Batch 2	G1 Batch 3	G1 Batch 4	G1 Batch 5

B

Y0	Y1	Y2	Y3	Y4	Y5
Base	G1 Batch 1	G1 Batch 2	G1 Batch 3	G1 Batch 4 G2 Batch 1	G1 Batch 5 G2 Batch 2

Figure 1. Simulation scenarios; A) discrete generations and B) overlapping generations

In the simulation, mating system were arranged to avoid inbreeding; dams' oestrus was synchronized, mating period was planned once a year resulting in calving at considerably close time range. We assume that sex ratio was 50:50 for male and female offspring and for both sexes, 90% of the calves survived until adulthood. Annual phenotypes, genotypes and allele frequencies were estimated and visualized.

4. Results and discussions

Population structure of the simulated population is presented in **Table 1**. Following the aforementioned scenarios, we can see the pattern of population expansion from the increasing numbers of individuals in the population over years. The number of individuals by sex differed. In the base population there were 10 sires and 1000 dams. As the mating cycle carried on and we limit the mortality by 10% of each cycle, the addition of male and female individuals became a linear function of the number of individuals from the base population.

We allow the simulation to run for five mating cycles and we record the frequency of every genotype (**Figure 2**). In **Figure 2A** we present the frequencies of the all three distinct genotypes under the discrete generation scenario, whereas in **Figure 2** we presented the same information under overlapping generations scenario. It is clearly visible that the frequency of normal (+/+) individuals decreased whilst the frequency of heterozygous individuals carrying *mh*-gene mutations increased overtime for both scenarios. The noticeable difference between discrete and overlapping scenario was concerning the frequency of the *mh/mh* genotype or the DM individuals. The number of the homozygous DM individuals was constant in the discrete scenario because there were only mating between DM sires and normal dams yielding in 100% heterozygous +/*mh* calf crop. On the other hand, in overlapping scenario, the heterozygous individuals in the first generation and first batch (Year 1; G1 Batch 1) was already matured at Year 3. The G1 Batch 1 females were then mated with unrelated DM sire. This scenario yielded 50% *mh/mh* calves and 50% +/*mh* calves and thus, the frequency of *mh/mh* genotype increased.

Table 1. The simulated population structures

Year	Sex	No. of Individuals	
		Discrete	Overlapped
0	F	1000	1000
0	M	10	10
1	F	1450	1450
1	M	460	460
2	F	1900	1900
2	M	910	910
3	F	2350	2350
3	M	1360	1360
4	F	2800	2552
4	M	1810	1562
5	F	3250	2956
5	M	2260	1966

The frequency of allele *mh* and + during the simulation is presented in **Figure 3A** for discrete generations scenario and **Figure 3B** for overlapping generations scenario. Due to the nature of the simulated population, the frequency of allele *mh* increased and the frequency of allele + decreased overtime in both cases. In the scenario of overlapping generations, the decrease in + allele and the increase in *mh*-allele frequencies happened more rapid compared to the earlier scenario due to the presence of new *mh/mh* genotyped individuals confirming the results in **Figure 2**.

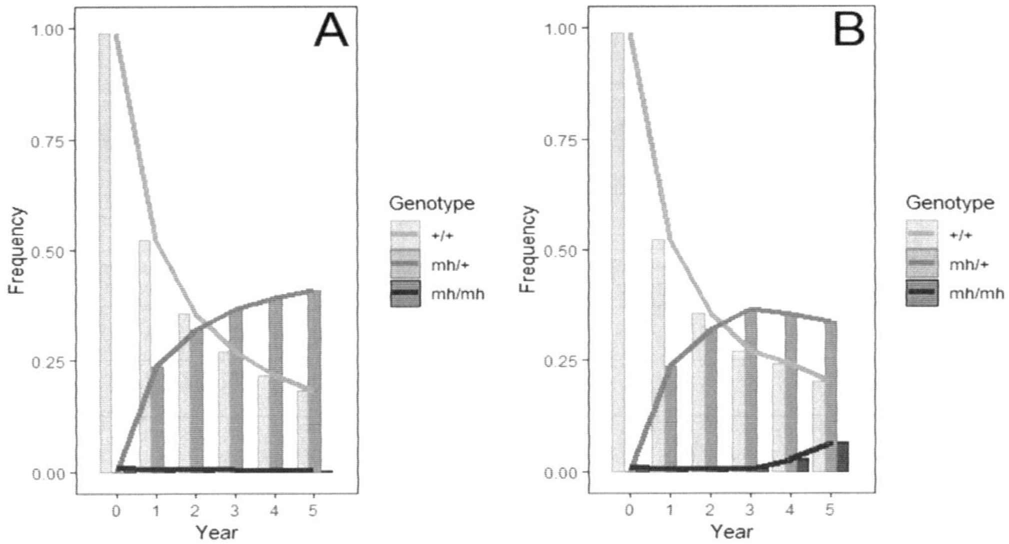

Figure 2. Genotype frequency in the simulated population: A) discrete generations and B) overlapping generations

Crossbreeding is mating between two genetically distinct breeds [20,21]. In this study we regarded the breeds as two populations: the DM population was homozygous for *mh*-allele whereas normal population only contained the + allele. The crossbreeding practice will create a new gene pool which contained both alleles and hence there will be three genotypes in the new admixed population.

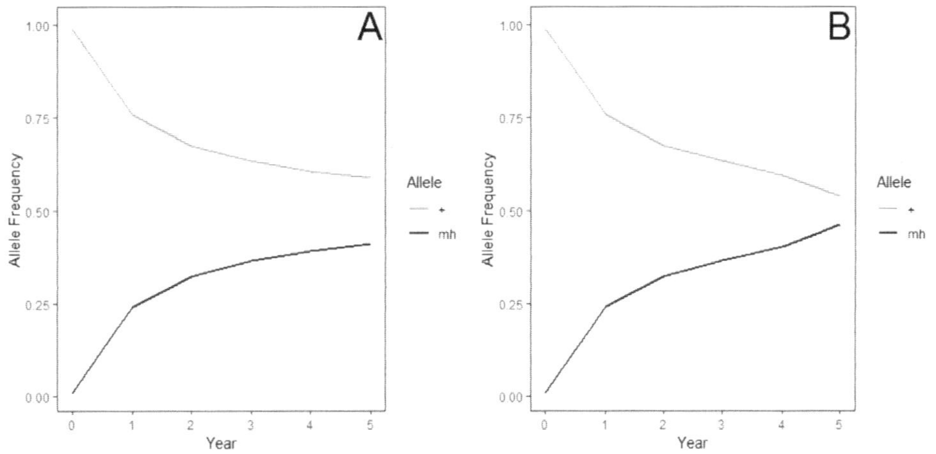

Figure 3. Allele frequency in the simulated population: A) discrete generations and B) overlapping generations

5. General Discussions

Cattle with DM phenotype produce considerably higher carcass percentage, more edible and premium cuts with lean and tender meat [12,14]. This merits however came with several physiological characteristics as summarized by Fiem et al [12] more specifically concerning the smaller skeletal systems, decreased internal organs and reduced skin weight compared to the normal cattle. These characteristics carried physiological consequences. DM cattle were prone to respiratory problems due to smaller lung and airway resistance. They have low endurance due to smaller heart and smaller digestive system [12,27].

From the reproductive aspects, DM cattle have narrower hip bone due to smaller skeletal system [12]; decreased pelvic area and pelvic opening compared to normal cattle and bigger foetal weight which often caused dystocia and perinatal mortality [12,19,28–30]. The reproduction problems were often solved by having caesarean surgery (C-section). Although the goal of C-section is to maintain future cow fertility and preserving calf life [31], C-section also brought problems to future reproductive efficiency of the cows. These problems including the prolonged calving interval as the result of longer interval form calving to first service and interval from first service to conception when compared to the cow with normal deliveries [32]. Higher calf mortality due to lower cow milk production also reported as the risk factors post C-section [33]. Previous reports mentioned that cows suffered more from post-surgery pain with side effects such as poor appetite, and uterine infection, thus reducing fertility as well as pregnancy rate [34]. Considering the poor sanitation and the fact that Indonesian smallholder farmers have less advanced relevant knowledge, post C-section risk as mentioned would be of a higher chance to occur. In addition to that, the cost for medicine and veterinary treatments increased. Therefore, instead of improving the reproductive aspects of DM cattle, this approach brought more problems such as longer calving interval, reduce in fertility, financially costly and laborious and posing welfare issues. In this regard, preserving cow fertility needs to be put as the highest priority for sustainable calf production.

From the simulation results, we can see that the crossbreeding between DM cattle and normal local cattle will produce homozygous offspring in generation 1 (G1). These cattle are potential as the meat producer because they carry the *mh*-gene mutation which is expected to increase the muscle mass and volume and in the end the carcass percentage and quality will be improved. Results also showed that in the discrete generation scenario, G1 individuals were not allowed to reproduce and thus there will be no addition of homozygous DM individuals in the population except the 10 DM sires in the base population. This way, we can guarantee that the *mh*-mutation will not spread in the dam population. However, if the G1 individuals more specifically females were allowed to reproduce, as in the scenario with overlapping generation, there will be an increase in the number of homozygous DM individuals in the population. Based on the simulated population, by the end of five years interval, the population would consist of 33.60% heterozygote individuals of each sex; whereas the frequency of individuals homozygous for *mh* allele were 6.20 and 6.40% for female and male respectively. Earlier study mentioned that in the mating between DM sires and normal dams, 42-45% of the offspring were DM [24,25]. The results suggested that there were 33.6% heterozygote females; and 42-45% out of them (which was around 14-16%) were double muscled. Double muscled female cattle are prone to dystocia and all of the carriers were potential agents in spreading the *mh*-mutation. The inability to control the spread of *mh*-gene in the breeding population will carry reproductive consequences as mentioned before.

Lastly, we fully aware of the policy makers' expectation when the attempt to introduce DM cattle was brought to surface. The consideration to create a composite cattle breed by attempting to utilize the DM phenotype superiority, however, is an absurd idea as the expression of DM phenotype was only affected by the number of *mh* allele present and **not** by how many percent of DM parental breed proportion existed in an individual.

6. Conclusion

Creating a composite breed utilizing the superiority of DM cattle is not at all feasible. Crossbreeding program with DM cattle, anyhow, seems promising. If DM cattle introduction in Indonesian cattle population is unavoidable, precautions are to be make regarding the reproduction of carrier cows. All of F1 have to be slaughtered and must not be allowed to reproduce. Otherwise the *mh*-gene mutation would spread in the population and disturb the calf production. DM introduction will pose a major threat in the Indonesian cattle population especially on the reproductive aspect.

References

[1] Kementrian Pertanian 2015 *Rencana Strategis 2015 - 2019* (Jakarta, Indonesia: Kementrian Pertanian Indonesia)
[2] Kementerian Pertanian 2010 *Pedoman umum program swasembada daging sapi tahun 2014* (Indonesia)
[3] Sutarno and Setyawan A D 2015 *Biodiversitas* **16** 327–54
[4] Widi T S M 2015 *Mapping the impact of crossbreeding in smallholder cattle systems in Indonesia* (Wageningen University and Research Centre)
[5] Agus A and Widi T S M 2018 *Asian-Australasian J. Anim. Sci.* **31** 1017–35
[6] Ditjennak 2017 Tingkatkan Produksi Daging Sapi Di Indonesia, Ditjen Pkh Kementan Kembangkan Sapi Belgian Blue
[7] Anonimous 2018 Serius Kembangkan Sapi Belgian Blue di Indonesia, Kementan Gandeng Perguruan Tinggi
[8] Grobet L, Martin L J R, Poncelet D, Pirottin D, Brouwers B, Riquet J, Schoeberlein A, Dunner S, Ḿnissier F, Massabanda J, Fries R, Hanset R and Georges M 1997 *Nat. Genet.* **17** 71–4
[9] Phocas F 2009 *J. Anim. Sci.* **87** 1865–71
[10] Van Soom A, Mijten P, Van Vlaenderen I, Van den Branden J, Mahmoudzadeh A R and de Kruif A 1994 *Theriogenology* **41** 855–67
[11] Arthur P 1995 *Aust. J. Agric. Res.* **46** 1493
[12] Fiems L O 2012 *Animals* **2** 472–506
[13] Felius M, Koolmees P A, Theunissen B and Lenstra J A 2011 *Diversity* **3** 660–92

[14] Kambadur R, Sharma M, Smith T P L and Bass J J 1997 *Genome Res.* **7** 910–6
[15] Dunner S, Miranda E, Amigues Y, Canon J, Georges M, Hanset R, WilliamsS J and Menissier F 2003 *Genet. Sel. Evol.* **35** 103–18
[16] Mota R R, Mayeres P, Bastin C, Glorieux G, Bertozzi C, Vanderick S, Hammami H, Colinet F G and Gengler N 2017 *J. Anim. Sci.* **95** 4288–99
[17] Hanset R 2004 *Belgian Blue Herd-Book.* 1–16
[18] Coopman F, Krafft A, Dewulf J, Van Zeveren A and Gengler N 2007 *J. Anim. Breed. Genet.* **124** 20–5
[19] Casas E, Thallman R M and Cundiff L V. 2011 *J. Anim. Sci.* **89** 979–87
[20] Falconer D S and Mackay T F C 1996 *Introduction to Quantitative Genetics* (Edinburgh: Prentice Hall)
[21] Bourdon R M 2000 *Understanding Animal Breeding* (Prentice Hall)
[22] Freetly H, Kuehn L A and Cundiff L V 2011 *J. Anim. Sci* **89** 2373
[23] Purchas R W, Morris S T and Grant D A 1992 *New Zeal. J. Agric. Res.* **35** 401–9
[24] Hanset R and Michaux C 1985 *Génétique, sélection, évolution* **17** 369–86
[25] Arthur P F, Makarechian M, Price M a and Berg R T 1989 Heterosis , Maternal and Direct Effects in Double-Muscled and Normal Cattle : I . Reproduction and Growth Traits P . F . Arthur , M . Makarechian , M . A . Price and R . T . Berg The online version of this article , along with updated information and serv 902–10
[26] R Core Team 2015 R: A language and environment for statistical computing
[27] Amory H, McEntee K, Linden A S, Desmecht D J, Beduin J M, D'Orio V and Lekeux P M 1995 *Can. J. Vet. Res.* **59** 135–41
[28] Fiems L O and Ampe B 2014 *Animal* **9** 94–103
[29] Arthur P F, Makarechian M and Price M a 1988 *Can. Vet. J.* **29** 163–7
[30] Mao W H, Albrecht E, Teuscher F, Yang Q, Zhao R Q and Wegner J 2008 *Asian-Australasian J. Anim. Sci.* **21** 640–7
[31] Newman K D and Anderson D E 2005 *Vet. Clin. North Am. - Food Anim. Pract.*
[32] Barkema H W, Schukken Y H and Guard C L 1992 *Theriogenology* **37** 489–506
[33] Barkema H W, Schukken Y H, Guard C L, Brand A and van der Weyden G C 1992 *Theriogenology* **38** 589–99
[34] Patterson D J, Bellows R A and Burfening P J 1981 *J. Anim. Sci.* **53** 916–21

The effects of rice hull supplementation or inclusion on performance and gastrointestinal weight of broiler chickens

S Hartini[13], DD Rahardjo[1] and H Sasongko[2]

[1]Faculty of Animal Husbandry, University of Papua, Manokwari, Indonesia
[2]Faculty of Animal Science, University of Gadjah Mada, Yogyakarta, Indonesia

[3]Email: sri.hartini8877@gmail.com

Abstract. The effects of rice hull inclusion or supplementation on the growth performance and gastrointestinal (GIT) weight of broiler chickens were investigated. Three diets, corn-soybean-based diet as control diet (CON), supplementation of 40 g/kg rice hulls on CON (SRH), inclusion of 40 g/kg rice hulls in the diets (IRH) were used for the experiment. The diets were fed to broiler chickens from 2 to 21 d of age. Each treatment was replicated 10 times (8 birds/ replicate or 80 birds/ treatment). From 14 to 21 d of age and across 2 to 21 d of age, broilers fed the IRH had a greater ($P \leq 0.05$) average daily gain (ADG) than those fed the SRH with birds fed the CON being intermediate. Feed intake (FI) followed a similar trend as ADG. From 14 to 21 d of age, birds fed the IRH had higher ($P \leq 0.05$) FI, followed by the CON, and the SRH as the lowest. Rice hull inclusion increased the intestinal content as a percentage of BW, with an effect similar to that of the CON ($P \leq 0.05$). Birds fed the IRH had a lower ($P \leq 0.05$) weight of empty ceca than those fed the SRH and the CON which were not different. The results indicated that the inclusion or supplementation of rice hull produces different effect on the growth performance and GIT weight of young broilers. Diet composition is suggested to be the factors that caused the differences. The diet composition due to rice hull inclusion affected the growth performance and GIT weight of broiler chicken more than the diet composition due to rice hull supplementation.

1. Introduction

The benefit of rice hulls for broiler performance is rarely to be discussed, in reality the use of rice hulls in poultry production is only popular as broilers litters or beddings. The chemical compositions of rice hulls are crude protein 2.4%, ether extract 0.5%, ash 12.9%, crude fibre 54.2%, total dietary fibre 74 %, insoluble fibre 72%, and soluble fibre 2% [1]. The high contents of crude fibre and insoluble fibre become the constraint on the use of rice hulls in poultry diets. However, after recent studies indicated that inclusion of insoluble fibre could improve i.e. utilization of other nutrients [2], gastrointestinal (GIT) health [3], growth performance [4] of broiler chickens and could reduce cannibalism mortality in laying hens [5], there is a growing interest in using the insoluble fibre sources in poultry diets. Rice hulls as one of insoluble fibre sources get more attention regarding their use in broilers diets. Hartini and Massora [6] showed that broilers could maintain BWG with the addition of 4 % rice hulls same as the control diet, but impaired BWG and FCR with the addition of 6% rice hulls. Sadeghi et al. [7] showed that inclusion of 3% rice hulls in the diets did not affect growth performance, carcass and gizzard weight, but shorten jejunum length and increased ileum weight of broiler at 42 days of age. Inclusion of 2.5% rice hulls in broiler diets has been reported to increase average daily gain dan feed to gain ratio better than the control diet [1]. In addition, an increase level of rice hulls from 2.5 to 5% in mash form did not improve ADG and F:G. Hartini and Purwaningsih [8] added 4% rice hulls with phytase in a commercial diet having low Ca and P found no effect on ADG, but an increase in percentage of carcass weight and a decrease in GIT weight.

Each fiber has each own physicochemical properties and its interaction with other component in the diet influence its effect on GIT [9]. Thus, it is hypothesised that the inclusion or supplementation of rice hull in the diet will produce different effects on broiler performance. The current study was designed to investigate the effects of rice hull inclusion or supplementation on growth performance and gastrointestinal weight of young broiler chickens.

2. Materials and Methods

2.1. Experimental diets, animals, and birds management

The diets used in the experiment were: 1) corn-soybean-based diet as a control diet (CON), 2) CON + 40 g/kg rice hulls (SRH), 3) inclusion 40 g/kg rice hulls in the diets (IRH). Rice hulls were ground through a hammer mill (2 mm screen) before being used. Composition and nutrient content of the treatment diets are shown in Table 1.

All diets were formulated isocaloric and isoprotein and met all nutrient recommendations of the NRC [10] for broilers starter. All experimental diets were provided as mash feed and were fed to broiler chickens from 2 to 21 d of age. A total of 240 unsexed broiler chicks (strain Lohman) at two days old (initial body weight of 53.6 ± 1.7 g) were randomly placed in 40 cages with 8 birds per cage and 10 cages (80 birds) per treatment. *Ad libitum* feed and water were provided during the experiment. The cages had continuous light.

2.2. Growth performance

Body weight (BW) and feed intake (FI) were weighed by cage at 2, 7, 14, and 21 d of age. From the data obtained, we determined the average of daily gain (ADG) (g/b/d), daily feed intake (ADFI) (g/b/d), and the average of gain to feed ratio (G:F) (g/g). Birds that died during the experiment were weighed and their remaining feeds were also weighed. The data obtained was included in the calculation of G:F.

2.3. Sample collection and analyses

At the end of experiment (21 d of age), all birds were weighed and one bird from each of ten replicates that had weight closed to the mean weight of birds per cage was chosen for dissection. Prior to dissection, the birds were not given any feed for about 8 hours. The birds were dissected at the jugular's vein. The dissection was done in compliance with Indonesia guide lines for animal handling and care [11].

2.4. Digestive organs weight

After the birds were killed, the body cavity was immediately opened, and the GIT was removed and weighed before and after removing the content. The gizzard with content and pancreas was also removed. The gizzard content was also collected. The weight of empty gizzard, GIT, including pancreas was expressed relative to 100 g live BW (without digesta) (g/100 g BW), whereas the weight of digestive content was expressed relative to 100 g live BW (g/100 g BW).

2.5. Statistical analysis

One-way analysis of variance [12] was used to analyse the data. When F test was significant, to inspect differences among group means we used Duncan's multiple range test. Statistical significance was determined at $P \leq 0.05$.

Table 1. Diet composition and nutrient content (as-fed basis).

Ingredient, g/kg	CON	SRH	IRH
Corn	527.40	527.40	355.70
Wheat bran	87.90	87.90	161.70
DDGS	78.90	78.90	76.30
SBM	128.50	128.50	120.90
PBPM	143.10	143.10	155.60
Palm oil	6.20	6.20	63.80
Rice hulls	-	-	40.00
NaCl	3.30	3.30	3.30
Lysin	0.30	0.30	0.10
DL-methionin	1.10	1.10	1.20
Mineral Mix	5.00	5.00	5.00
Ca_2PO_4	12.70	12.70	11.40
$CaCO_3$	5.60	5.60	5.00
Rice hulls	-	40.00	-
Nutrients (calculated), %			
ME, kcal/kg	3000.00	3154.00	3000.00
CP	23.00	23.00	23.00
EE	4.74	4.74	10.31
CF	3.09	5.09	5.53
Ca	1.00	1.00	0.94
Total P	0.90	0.90	0.98
Na	0.20	0.20	0.20
Lysin	1.10	1.10	1.10
Methionin	0.50	0.50	0.50

3. Results

The results are summarized in Table 2. The ADG and ADFI of birds from 2 to 7 d of age and from 7 to 14 d of age was not affected ($P > 0.05$) by the inclusion or supplementation of rice hulls. But from 14 to 21 d, broilers fed the IRH showed better ($P \leq 0.05$) ADG and ADFI than those fed the SRH and the CON. When we measured cumulatively, from 2 to 21 d of age, the IRH increased ($P \leq 0.05$) ADG more than the CON and the SRH, but the ADFI was similar among diets ($P > 0.05$). The effect of diets on GF in all periods was not significant ($P > 0.05$). Mortality was occurred during the first week of life (about 1 %) and it was not due to the treatment diet (data not shown).

Diets affected ($P \leq 0.05$) the weight of intestinal content and the weight of empty ceca, but the effects on the weight of empty gizzard was only tended to be significant ($P = 0.07$). Broilers fed the SRH had lower ($P \leq 0.05$) intestinal content than those fed the IRH and the CON which were not different. Birds fed the IRH had lower ($P \leq 0.05$) ceca weight and tended ($P = 0.07$) to have lower empty gizzard weight than those fed the SRH and the CON.

Table 2. Response of diets on growth performance and gastrointestinal weight.

Parameters measured	CON	SRH	IRH	SEM[1]	P-value
ADG, g/b/d					
2 to 7 d of age	15.65	15.90	16.07	0.163	0.59
7 to 14 d of age	26.10	27.04	26.97	0.491	0.70
14 to 21 d of age	42.06^{ab}	38.55^{a}	44.91^{b}	0.925	0.01
2 to 21 d of age	29.23^{ab}	28.35^{a}	30.71^{b}	0.375	0.03
ADFI, g/b/d					
2 to 7 d of age	19.90	20.17	20.41	0.211	0.63
7 to 14 d of age	46.18	47.04	45.26	0.837	0.70
14 to 21 d of age	64.10^{ab}	61.02^{a}	67.28^{b}	1.025	0.04
2 to 21 d of age	45.86	45.12	46.83	0.558	0.47
G:F, g/g					
2 to 7 d of age	0.788	0.788	0.789	0.007	0.99
7 to 14 d of age	0.577	0.574	0.596	0.012	0.75
14 to 21 d of age	0.658	0.635	0.673	0.015	0.59
2 to 21 d of age	0.640	0.629	0.657	0.080	0.37
Weight of empty digestive organs, % of BW (without digesta) at 21 d of age					
Gizzard	2.16	2.41	1.86	0.100	0.07
Small Intestine	3.26	3.49	2.80	0.141	0.12
Pancreas	0.37	0.33	0.32	0.017	0.49
Ceca	0.35^{b}	0.33^{b}	0.19^{a}	0.023	0.00
Weight of digestive content, % of live BW at 21 d of age					
Gizzard	1.07	0.95	1.00	0.050	0.61
Small Intestine	5.18^{b}	3.24^{a}	5.10^{b}	0.332	0.02
Ceca	0.46	0.38	0.31	0.039	0.30

[1]SEM: standard error of the mean
^{a,b}mean values within a row with different superscripts differ significantly ($P \leq 0.05$).

4. Discussion

Cumulatively, from 2 to 21 d of age, the inclusion of rice hulls increased ADG, but not G:F and ADFI, the ADFI only increased from 14 to 21 d of age. The inclusion of insoluble fiber sources at level up to 3% in low-fibre diet improved the growth performance of broilers [3]. Jiménez-Moreno et al. [1], including 2.5% of rice hulls (2mm in size) in broiler diets found an increase in average daily gain dan feed to gain ratio better than the control diet.

Sadeghi et al. [7] included 3% rice hulls (2 mm in size) in the diet did not find any significant effect on growth performance. We used 4% rice hulls (2 mm in size) in the present study and our diets contained an excess of fibre (about 5% CF in the diet). Based on the results in our present study, together with the information of Jiménez-Moreno et al. [1] and Sadeghi et al. [7], it suggests that the rice hull levels required to optimize growth performance of young broilers depend on diet composition. In our present study, the improvement of ADG observed from 14 to 21 d of age was consistent with an increase in ADFI. However, from 2 to 21 d of age, an increase in ADG on birds fed the IRH was not followed by an increase in ADFI, whereas those fed the SRH had the same ADG with the CON. So, it is likely that factors other than fibre was involved in increasing nutrient utilization of birds fed the IRH. The percentage of fat contained in the IRH is almost double than that contained in the SRH (10.31 % and 4.74 %, respectively). The higher dietary fat will reduce the rate of feed passage in the gut [13]. Presumably, the higher fat content in the IRH reduced digesta passage rate, allowing better nutrient utilization [13]. The fact that in our present study the intestinal content of birds fed the IRH was higher than those fed the SRH supported the suggestion.

The weight of empty ceca was lower in birds fed the IRH than those fed the SRH and the CON which were not different. If the nutrient utilization increased, substrates available for microorganisms might also reduce, consequently no much activity occurred in the ceca.

In our present study, inclusion of rice hulls tended to reduce the gizzard size, whereas the gizzard content was not different among diets. The results were differed from those reported by Hetland and Svihus [14] in their study with oat hulls. Different fibres have different physicochemical properties [9]. The insoluble fibre of rice hulls was almost 10% higher than that of oat hulls, whereas the soluble fibre was 59% lower than that of oat hulls [1]. The coarse oat hulls that was used in Hetland and Svihus [14] was ground through 1.6 mm sieve, whereas in our study the rice hulls were ground through 2 mm sieve. So, our data seem to indicate that the effects on size and content of the gizzard as a consequence of fibre inclusion were due to the physicochemical properties of the fibre rather than the particle size.

5. Conclusion

This study demonstrated that the inclusion or supplementation of rice hull did produce different effect on the growth performance and GIT weight of young broilers. However, diet composition rather than the individual fibre was suggested to be the factor that caused the differences. The diet composition due to rice hull inclusion affected the growth performance and GIT weight of broiler chickens more than the diet composition due to rice hull supplementation.

Acknowledgements

This work was supported by the Minister of Research and Technology of Higher Education Republic of Indonesia through contract No. 080/SP2H/L.T/DRPM/2018 under the National Competitive Grant, Competency-based Research. The support from PT. Japfa Comfeed Tbk in providing the research feedstuff is greatly appreciated.

References

[1] Jiménez-Moreno E, de Coca-Sinova A, González-Alvarado JM and Mateos GG 2016. *Poult. Sci.* **95**: 41-52
[2] Svihus B and Hetland H 2001. *Br. Poult. Sci.* **42**: 633 - 37
[3] Mateos GG, Jiménez-Moreno E, Serrano MP and Lázaro RP 2012. *J. Appl. Poult. Res.***21**: 156-74
[4] González-Alvarado JM, Jiménez-Moreno E and Mateos GG 2007. *Poult. Sci.* **86**: 1705-15
[5] Hartini S, Choct M, Hinch G, Kocher A and Nolan JV 2002. *J. Appl. Poult. Res.***11**:104-10
[6] Hartini S and Massora M 2014. *Proc. of the 14th European Poultry Conf.* (Stavanger, Norway) p. 561
[7] Sadeghi A, Toghyani M and Gheisari A 2015. *Poult.Sci.* **94**: 2734-43
[8] Hartini S and Purwaningsih P 2017. *Pak. J. Nutr.* **16** (4): 227-35
[9] Hartini S, Choct M, Hinch G and Nolan JV 2003. *Proc. Aust. Poult. Sci. Sym.* **15**: 135-138
[10] NRC 1994. 9th ed. (Washington, DC: National Academy Press).
[11] Decree of the Minister of Labor and Transmigration of the Republic of Indonesia 2014. No. 196
[12] SPSS 16.0 2007. *Command Syntax Reference, SPSS Inc.* (Chicago Ill.)
[13] Latshaw JD 2008. *Poult. Sci.* **87** (1): 89-95

[14] Hetland H and Svihus B 2001. *Br. Poult. Sci* **42**: 354-61

Supplementation of probiotic and prebiotic on the performance of broilers

M M D Utami[1] and N D Wahyono[2]

1 Department of Animal Science, Politeknik Negeri Jember, Jl. Mastrip PO Box 164, Jember, Indonesia
2 Department of Agricultural Production, Politeknik Negeri Jember, Jl. Mastrip PO Box 164, Jember, Indonesia

e-mail: merry.mdu@polije.ac.id

Abstract. The purpose of this research was to study the effect of probiotic-prebiotic on the performance of broiler. Two hundred one day broiler age were used in this study for 35 days. The chickens were randomly divided into five treatments, each treatment consists of four replications and each repetition consist of ten chickens. This research using completely randomized design, such as: T0: control, T1: probiotic 0.2% + prebiotic 0.2%, T2: probiotic 0.4% + prebiotic 0.2%, and T3: probiotic 0.2% + prebiotic 0.4%. The treatments started at 10 day-old chick. The parameters of research were feed consumption, body weight, and feed conversion. The results of significant analysis continued by Duncan's New Multiple Range Test. Two-hundred day old chicks divided into four treatments, each treatment has five replications, and every treatment has 10 chickens. The result showed supplementation probiotic 0,4% and prebiotic 0,2% increased body weight and decreased feed conversion ($P<0.05$) and had no significant difference in feed consumption.

1. Introduction

Feed costs in poultry farms are the largest production component reached 70%. Broilers need high-quality feed to sustain growth. Giving feed additive is proven to improve feed efficiency so that it provides profits for farmers. Problems arise when broilers are given high-quality feed and feed use additionally antibiotics. According to the study of [1] antibiotic resistance causes problems in animals and public health. Addition antibiotic in broiler is a problem in order to unsafe for consumption and cause drug-resistant drugs [2], residues in poultry carcass [3], and unbalanced microflora [4]. Antibiotic accumulation in the human body can kill microfloral.

Prebiotics, probiotics, symbiosis are used as feed additives to improve health and production performance. Feed conversion of broiler with probiotic-fed broilers was not different from those fed an antibiotic [5]. Probiotics are feed additives contains a number of bacteria (microbes) which give an effect which benefits of health because it can improve the balance of intestinal microflora, so it can provide advantages of protection, protection disease and improvement in digestibility feed. In addition, probiotics can also accelerate growth and increase body immunity from certain pathogenic diseases [6]. Broiler increase body weight significantly with probiotic and prebiotics [7].

Probiotics develop in the intestine and can benefit the host well directly or indirectly directly from the results of its metabolites [8]. Which bacteria contained in probiotics can change intestinal microecology such that microbes which can be profitable well developed [9]. Enzymes produced by microbes found in probiotics are amylase enzymes, protease and cellulose [10]. Using a prebiotic oligosaccharide dose of 0.4% is based on the results [11] observed using fructooligosaccharides 0.4% can significantly increase the population of Bifidobacterium bacteria spp and Lactobacillus sp. Lactobacillus spp. Lactobacillus spp. produce digestive enzymes, increase concentration intestinal digestive enzymes. According to [12] reported that the Lactobacillus strain shows amylase activity. Increased amylase activity in the small intestine in broilers given lactobacillus, but none effects on lipolytic and proteolytic activity [13]. Flora in intestinal bacteria has a role in digestion and absorption, such as carbohydrates, proteins, lipids, and minerals and in vitamin synthesis.

The effect of probiotics on growth is to prevent colon colonization by pathogenic bacteria, such as *Clostridium perfringens*, *Escherichia coli*, and *Salmonella spp.* through competitive exclusion mechanisms [14] [15], then [16] showed that *Bacillus subtilis* were able to increase immune modulation by producing antimicrobial factors against *Clostridium perfringens*. Reported by [17] inhibition of nutrient absorption in pathogenic bacteria by producing toxic metabolites that irritate the intestinal mucosa.

Prebiotics are food ingredients that cannot be digested and beneficial to the host by selectively stimulating growth and or the activity of one or a number of bacteria that are in the colon so that they can improve the health of the host. Prebiotics, in general, are carbohydrates which are not digested and is not absorbed especially by single-bellied cattle (monogastric like chicken and pork) usually in the form of oligosaccharides and dietary fiber (inulin) [18]. Some prebiotics such as fructooligosaccharides and inulin play a role in improving health by modifying the balance of intestinal microflora [19] and selectively stimulates bacterial growth beneficial such as Lactobacillus and Bifidobacteria [20].

Symbiotics are a combination of probiotics and prebiotics that have a positive effect on intestinal health, food digestion, and the life performance of broilers [21]. Symbiotic is defined as a combination of prebiotics and probiotics that benefit the host by increasing defense and implantation of feed supplements containing live microbes in the channel digestion by selectively triggering growth and or activating metabolism of a number of good bacteria so as to improve health the host. The advantage of this combination is to increase survival probiotic bacteria because specific substrates are available for fermentation so the body gets more perfect benefits from a combination of probiotics and prebiotics [20]. The effects of ratio probiotic and prebiotic in association to performance are still unclear. Therefore, This research was conducted to study the effects of supplementation probiotic and of prebiotics with a different ratio to improve the health and growth of broilers.

2. Materials and Methods

Two hundred one day broiler was obtained from a commercial hatchery were used in this study for 35 days. The chickens were randomly divided into five treatments, each treatment consists of four replications and each repetition consist of ten chickens. This experiment using completely randomized design, the treatments were: T0: control, T1: probiotic 0.2% + prebiotic 0.2%, T2: probiotic 0.4% + prebiotic 0.2%, and T3: probiotic 0.2% + prebiotic 0.4% diluted on water.

The ration used during the study was basal ration divide into two rations such as starter period (age 0-3 weeks) with 21-22% protein content and metabolic energy 2800-3000 Kcal/Kg and finisher period (age 3-5 week) with a protein content of 18-20% and metabolic energy 3000-3200 Kcal/Kg. The prebiotic and probiotic were obtained commercially. Chicks were given the experimental diets from 10 days of age.

Feed consumption and body weight were recorded weekly, and the feed conversion ratio was calculated. Data were analyzed based on a completely randomized design model using the SPSS software. For interpretation purposes, main effect means were used when the interaction term was not significant, and individual means were considered when the interaction term was significant. Means were considered as significantly different at $P<0.05$.

3. Results and Discussion

3.1 Feed Consumption

Supplementation of probiotic and prebiotic on productive performance of broiler, based on the result from the statistical test are shown in table 1. The effect of treatments on feed consumption of broiler showed results that were not significantly different (P>0.05). There is no significant difference in this effect may be because the chicken is given the same feed as the amount of protein and energy. The amount of feed consumption is largely determined by the energy content in the ration. If the energy content in the ration is high then feed consumption will decrease and low energy content makes increasing feed consumption.

This research has different result with [22] reported supplementation of probiotics causes an increase in feed consumption as much 2.6% higher than T0 and it is suspected that this difference will be significant if the number of chickens used was increased and rations were used not a commercial ration and [23] showed that supplementation of Lactobacillus culture in feed increase feed consumption of laying hens

Table 1. Effect of supplementation probiotic-probiotic on feed consumption

Treatments	Replication					Average
	1	2	3	4	5	
T0	2836	2864	2997	2913	2861	2894^{ns}
T1	2956	2998	2715	2918	2926	2902^{ns}
T2	3018	2883	3012	2910	2897	2944^{ns}
T3	2937	2915	2908	2914	2895	2914^{ns}

^{ns} Nonsignificant

3.2 Body Weight

The effect of treatments on body weight of broiler showed results that significantly different (P<0.05). the result from the statistical test is presented in table 2. The difference in body weight gain is closely related to more high feed consumption and possibly due to the increased digestibility of substances nutrition due to the provision of probiotics. Lipolytic, cellulolytic, and lignolytic microbes contained in probiotics are thought to have played a role active in increasing nutrient digestibility. This result different from [22] increased the dose of probiotics from 1 to 2 cc per liter of water drink did not have an effect (P> 0.05) on body weight gain chicken. It seems there are optimal limits on chickens in their tolerance to microbial populations in his digestive tract.

Table 2. Effect of supplementation probiotic-probiotic on body weight

Treatments	Replication					Average
	1	2	3	4	5	
T0	1694	1620	1637	1614	1693	1652^{a}
T1	1662	1667	1630	1624	1618	1640^{a}
T2	1782	1738	1797	1710	1712	1748^{b}
T3	1646	1641	1651	1646	1608	1638^{a}

$^{a,\,b}$ Means within each line and under the same factor with different superscripts are significantly different (P<0.05)

The result of body weight showed treatment of T2 has higher body weight than others treatment (P<0.05). Addition probiotics in broiler increase the growth of broiler [24] [25] [26], increasing body weight due to an increase in beneficial microbial populations [27], so cause improve digestive efficiency [28]. Supplementation from Lactobacilli mixture provides varied results. According to [29] chicken feed given *Lactobacillus sporogenes* and *Clostridium butyricum* found a significant weight gain, [30] reported weight gain increased significantly (P<0.05) in chickens containing feed *Lactobacillus acidophilus* and *Streptococcus faecium,* then [31] observed that feed efficiency and body weight increased significantly (P <0.05) compared to broiler with antibacterial products. From [32] observed weight gain can vary from 5% to 9% when chicken is added with probiotics containing a mixture of *Lactobacillus acidophilus*, *Lactobacillus casei*, *Bifidobacterium bifidum*, *Aspergillus oryzae*, and Torulopsis.

3.3 Feed Conversion

According to [33] in broiler management, performance production that must be observed includes life weight, weight gain, weekly ration consumption, ration conversion, and feed conversion every week. The effect of treatments on feed conversion of broiler showed results that significantly different (P<0.05). The result from the statistical test is presented in table 3. The lowest feed conversion was T2 (probiotic 0.4%+prebiotic 0.2%) was 1.67 according to [20] reported a combination of probiotics and prebiotics is called synbiotics, a mixture of probiotics and prebiotics that affect survival and implant microbial food supplements living in the digestive tract by activating the metabolism of one or a number of bacteria that promote health or selectively stimulate growth. combination of prebiotics and probiotics more effective when compared with addition separately [34]. As well as, [35] reported positive results from symbiotics to improve broiler performance. Probiotic-supplemented birds had a lower feed conversion rate (1.85) than control (1.89).

Many factors influenced feed conversion are feed consumption and body weight, as reported by [30] feed efficiency increased significantly (P<0.05) in chickens containing feed *Lactobacillus acidophilus* and *Streptococcus faecium.* Probiotics are feed additives in the form of living microorganisms given to livestock which has a positive effect on livestock consume. This concept of utilizing microflora balance is what becomes the basis for using probiotics to suppress the development of pathogenic bacteria, well in the digestive tract of the chicken or in the litter (environment in the cage livestock). Addition of probiotics to drinking water also functions to maintain the balance of the microflora ecosystem in the digestive tract and provides enzymes that can digest crude fiber, protein, fat and detoxify toxins or their metabolites [36].

Table 3. Effect of supplementation probiotic-probiotic on feed conversion

Treatments	Replication					Average
	1	2	3	4	5	
T0	1.67	1.77	1.83	1.80	1.69	1.75[a]
T1	1.78	1.79	1.66	1.79	1.81	1.76[a]
T2	1.69	1.61	1.67	1.70	1.69	1.67[b]
T3	1.78	1.78	1.76	1.77	1.80	1.78[a]

[a, b] Means within each line and under the same factor with different superscripts are significantly different (P<0.05)

Addition of probiotics containing Lactobacillus, *Bacillus thermophilum* and *Enterococcus faecium* on the broiler diet increases the height of the villi [37]. The size of villi and the crypt depth ratio directly affect the increase in the epithelial cell [38], longer villi are associated with active mitosis cells [39]. Then reported by [40] high mitosis cells cause acceleration of rapid replacement of villus crypt tissue thereby accelerating villus renewal. Using synbiotics caused histomorphological changes in broiler intestines, the changes were increasing in villi size and increasing the surface area of nutrient absorption [41], while short villi caused low nutrient absorption [42].

Supplementation probiotics and prebiotics in rations can increase the population microbes that are beneficial for livestock, prevent the development of microbes detrimental in the digestive tract so as to improve digestion food, thus giving probiotics can make consumption efficient feed.

4. Conclusion

The result showed supplementation probiotic 0,4% and prebiotic 0,2% increased body weight and decreased feed conversion (P<0.05) and had no significant difference in feed consumption

Acknowledgment

The authors really highly appreciate to Politeknik Negeri Jember for a grant to research and present this paper on the 1st International Conference on Food and Agriculture 2018 in Bali, Indonesia.

References

[1] Barton M D 2000 *Nutr. Res. Rev.* **13** 2 279–299
[2] Sørum H and Sunde M 2001 *Vet. Res.* **32** 3–4 227–241
[3] Burgat V 1991 *Rev. Prat.* **41** 11 985–990
[4] Pinto-Alphandary H, Andremont A and Couvreur P 2000 *Int. J. Antimicrob. Agents* **13** 3 155–168
[5] K. C. Mountzouris K C, Tsirtsikos P, Kalamara E, Nitsch S, Schatzmayr G and Fegeros K 2007 *Poult. Sci.* **86** 2 309–317
[6] Prangdimurti E 2001 Probiotik dan Efek Perlindungannya Terhadap Kanker Kolon," *Makal. Falsafah Sains. Inst. Pertan. Bogor*, 2001.
[7 Mateova S 2008 *Med. Weter* **64** 3 294–297
[8] Wizna H A, Rizal Y, Dharma A and Kompiang I P 2009 *Pakistan J. Nutr.* **8** 10 1636–1640
[9] Isolauri E, Kirjavainen P V and Salminen S 2002 *Gut* **50** 3
[10] Ziaei-Nejad S, Rezaei M H, Takami G A, Lovett D L, Mirvaghefi A R and Shakouri M 2006 *Aquaculture* **252** 2–4 516–524
[11] Moro G E 2003 *Acta Paediatr.* 92 77–79
[12] Ghadban G S 2002 *Arch. fur Geflugelkd.* **66** 2 49–58
[13] Jin L Z, Ho Y W, Abdullah N and Jalaludin S 2000 *Poult. Sci.* **79** 6 886–891
[14] Stern N J, Cox N A, Bailey J S, Berrang M E and Musgrove M T 2001 *Poult. Sci.* **80** 2 156–160
[15] Abudabos A M, Alyemni A H and Al Marshad B A 2013 *Int. J. Agric. Biol*, vol. 15, pp. 978–982, 2013.
[16] Yurong Y, Ruiping S, ShiMin Z and Yibao J 2005 *Arch. Anim. Nutr.* **59** 4 237–246
[17] Wilson J, Tice G, Brash M L and Hilaire S S 2005 *Worlds. Poult. Sci. J.* **61** 3 435–449
[18] Grizard D and Barthomeuf C 1999 *Reprod. Nutr. Dev.* **39** 5–6
[19] Mattila-Sandholm T, Myllärinen P, Crittenden R, Mogensen G, Fondén R and Saarela M 2002 *Int. Dairy J.* **12** 2–3 173–182
[20] Gibson G R and Roberfroid M B 1995 *J. Nutr.* **125** 6 1401–1412
[21] Patterson J A and Burkholder K M 2003 *Poult. Sci.* **82** 4 627–631
[22] Wiryawan K G and Dingle J G 1999 *Anim. Feed Sci. Technol.* **76** 3–4 185–193
[23] Nahashon S N, Nakaue H S and Mirosh L W 1994 *Poult. Sci.* **73** 11 1699–1711
[24] Cavazzoni V, Adami A and Castrovilli C 1998 *Br. Poult. Sci.* **39** 4 526–529
[25] Zulkifli I, Abdullah N, Azrin N M and Ho Y W 2000 *Br. Poult. Sci.* **41** 5 593–597
[26] Kabir S M L, Rahman M M, Rahman M B, Rahman M M and Ahmed S U 2004 *Int. J. Poult. Sci* **3** 5 361–364
[27] AFRCR F 1989 *J. Appl. Bacteriol.* **66** 5 365–378
[28] Nahashon S N, Nakaue H S and Mirosh L W 1993 *Poult. Sci.* **72** Suppl 1 87
[29] Jin L Z, Ho Y W, Abdullah N and Jalaludin S 1997 *Worlds. Poult. Sci. J.* **53** 4 351–368
[30] Tortuero F, Rodriguez M L and Barrera J 1989 *Arch. Zootec. Cordova* **38** 141 151–165
[31] Owings W J, Reynolds D L, Hasiak R J and Ferket P R 1990 *Poult. Sci.* **69** 8 1257–1264

[32] Awad W, Ghareeb K and Böhm J 2008 *Int. J. Mol. Sci.* **9** 11 2205–2216
[33] North M O and Bell DD 1990 *Commercial Chicken Production Manual Chapman and Hall. New York*
[34] Li X, Qiang L and Xu C 2008 *Arch. Anim. Breed.* **51** 1 64–70
[35] Awad W A, Ghareeb K, Abdel -RaheemS and Böhm J 2009 *Poult. Sci.* **88** 1 49–56
[36] Gaggìa F, Mattarelli P and Biavati B 2010 *Int. J. Food Microbiol.* **141** S15–S28
[37] Chichlowski M 2007 *Poult. Sci.* **86** 6 1121–1132
[38] Fan Y K 1997 *Poult. Sci.* **76** 12 1738–1745
[39] Samanya M and Yamauchi K 2002 *Comp. Biochem. Physiol. Part A Mol. Integr. Physiol.* **133** 1 95–104
[40] Yason C V, Summers B A and Schat K A 1987 *Am. J. Vet. Res.* **48** 6 927–938
[41] Caspary W F 1992 Oxford University Press
[42] Xu Z R, Hu C H, Xia M S, Zhan X A and Wang M Q 2003 *Poult. Sci.* **82** 6 1030–1036

Sustainability status of integrated rice-corn and beef cattle farming agriculture business in Jember regency

N B E Sulistyono[1], Z Fanani[2], M M D Utami[3]

1 Department of Agricultural Production, Politeknik Negeri Jember, Jl. Mastrip PO Box 164 Jember Indonesia
2 Department of Animal Science, Brawijaya University, Malang Indonesia.
3 Department of Animal Science, Politeknik Negeri Jember, Jember Indonesia.

e-mail: besjember@polije.ac.id

Abstract. The research was aimed to analyze the sustainability status of a particular form of integrated agriculture business system namely rice-corn pattern planting with beef cattle breeding. The selection of research location was done based on the Multistage Sampling Method, and Purpose Sampling Method. Jember Regency was selected as the research location, because information obtained from the preliminary survey shows that paddy fields with technical irrigation system were common in Jember Regency and thus several planting patterns can be used in a particular year, namely: Rice – Corn planting pattern. In addition, farmers are also used to breed cattle especially beef cattle as an additional activity to their main farming business. Data analysis was conducted using Sustainability Analysis with MDS Rap-SISPOTTA. Based on five sustainability dimensions used here namely ecological, economic, technology and infrastructure, sociocultural, as well as legal and institutional; results were obtained indicating that an integrated beef cattle and crops agriculture business system in the studied region has a sustainability index of 44.0. It indicates that the integrated agriculture business implemented is still less sustainable dimensions with the lowest index of sustainability and thus must seriously be given attention are technology and infrastructure, as well as legal and institution.

1. Introduction

Integration between food crops farming and beef cattle breeding is basically a mix of two commodities that can be developed simultaneously in a particular area, where the existence of a commodity is necessary for the other, and vice versa. Food crops will produce agricultural waste which can be utilized as animal feed. Meanwhile, beef cattle can produce potential manure to be utilized to fertilize the crops and improve soil physical condition. Through the fulfillment of any input facility especially animal feed for the beef cattle and organic fertilizer for the food crops, it is expected that the production cost will be reduced and production will increase. Ultimately, it will increase farmers' income and welfare as well as support food and meat self-sufficiency in Indonesia [1].

The implementation of a sustainable agriculture system can be used as a momentum to encourage the development of people economy. Basically, farmers are more than ready to accept a sustainable agriculture system because the necessary input is widely available in the surrounding environment. Even before being introduced to the agriculture intensification using fertilizer and chemical pesticide, farmers have been implementing environmentally-friendly sustainable agriculture system, for instance by using manure. Equipped with their traditional knowledge, farmers must be empowered to further increase their knowledge on sustainable agriculture, as well as to identify opportunities and market demands for high quality and environmentally-friendly products. Therefore, farmers can produce agriculture products with high economic value while also continue to preserve environmental function [2].

According to [3] sustainable agriculture is the management of agricultural resources to meet the changing needs of a human while continue to maintain or increase the quality of the environment and preserve natural resources. Also, purposes of managing a farming business for farmer households are productivity, security, continuity, and identity. The same was said by [4], there should be rearrangement in the form of integration towards ecosystem-oriented dual utilization. Although agro-ecosystem is a complex and complicated management, its most important specific characteristics are related to four principles. The four principles are equitability, sustainability, stability, and productivity.

2. Research Method

The research was conducted in Jember District in a purposive manner by considering that the region is one of the regions with quite wide technical irrigation fields and has the potential for beef cattle breeding development. The sample used in this research are farmers/breeders who implement plant cultivation system in 1 (one) year with Rice-Corn planting pattern model, and breed beef cattle for more than 1 (one) year, as well as have the experience in breeding beef cattle for more than three years.

The analysis method for sustainability status used the *Rap-SISPOTTA modified* method, namely a new multi-disciplinary rapid appraisal method to evaluate sustainability [5]; [6]. The assessment of sustainability was based on 3-5 dimensions or sustainable metrics aspects, namely economic, ecological, environmental, technical, institutional, and sociological.

Table 1 Amount of Attributes for Each Sustainability Dimension.

No.	Dimension	Amount of Attributes
1	Ecological	20
2	Economical	15
3	Technology and Infrastructure	12
4	Sociocultural	14
5	Legal and institutional	8
	Total	69

The value of each attribute for each dimension was analyzed using Multi-Dimensional Scaling (MDS). All values of each dimension's attributes were entered into Microsoft Excel program and then analyzed with MDS using Excel for Rap-SISPOTTA, a set of analysis can be simultaneously conducted including feasibility and significance analysis, sustainability status analysis, and sustainability model simulation analysis.

2.1. Feasibility and Significance Test for Sustainability Model.

Feasibility test of a sustainability model was done through the calculation of *goodness of fit* from the distance between an estimated point and original point through *S-stress* calculation. The *least square* technique was used to determine the goodness of fit based on the square root of *Euclidian distance (squared distance)* or the *algorithm of scale (ALSCAL)* method. ALSCAL method optimizes the squared distance on the original point squared data (W_{ijk}), thus *S-Stress* value was calculated using the following formula:

$$S = \sqrt{\frac{1}{m} \sum_{k=1}^{m} \left[\frac{\Sigma_i \Sigma_j \left(d^2_{ijk} - O^2_{ijk} \right)^2}{\Sigma_i \Sigma_j O^4_{ijk}} \right]}$$

The goodness of fit in MDS is the measure of how well one point may reflect the original data, which is determined by the *S-Stress* value resulted from the S value calculation. The goodness of fit in MDS analysis is determined by the S-Stress value resulted from a calculation where the low-stress value indicates a good-of-fit, while high S value indicates the opposite. In RAP-SISPOTTA modified, a good model is reflected by a *stress* value lower than 0.25, while a *stress* value higher than 0.25 means low good-of-fit. Rap-SISPOTTA modified also calculates uncertainty aspect using Monte Carlo analysis technique. Uncertainty is caused by the impact of error in scoring due to the lack of information, the impact of variation in scoring due to the difference in assessment, error in data entry, and high-*stress* level obtained from the ALSCAL algorithm. *Monte Carlo* analysis is also a simulation model to evaluate the impact of random error in the analysis conducted to the whole attributes of each dimension. In this case, *Monte Carlo* is not conducted using the scatter pilot method that shows the ordination of each dimension.

On [5] explained that with Rap-SISPOTTA modified, an object is observed based on two to five dimensions, thus the ordination technique in MDS is based on *n Euclidian Distance* in a particular dimension. For this research, five dimensions were used using the following formula:

$$d = \sqrt{\{(V_1-V_2)^2 + (W_1-W_2)^2 + (X_1-X_2)^2 + (Y_1-Y_2)^2 + (Z_1-Z_2)^2\}}$$

Significance test in the sustainability model was done based on the configuration or ordination of an object or point in MDS which is approximated by regressing the Euclidian distance (d), from the point I to point j with the original point (δ_y) in the following equation:

$$d_{ijk} = \alpha + \beta\delta_{ij} + \acute{\varepsilon}$$

The squared distance is the weighted Euclidian distance or can be written as:

$$d^2_{ijk} = \Sigma_{a=1}^{r} \left(X_{ia} - x_{ja} \right)^2$$

The determinant of significance can be seen from the processed RAP-SISPOTTA with the highest Root Means Square (RMS) value until the half value of each sustainability dimension, where [6] stated that the lever factor of significance is the attributes that sensitively influence the increase or decrease of sustainability status. Higher RMS value means greater significance or the greater role of an attribute towards the sensitivity of sustainability status.

2.2. Sustainability status assessment.

Assessment of the whole attributes of each sustainability dimension in integrated beef cattle and horticulture crops agribusiness system was categorized into good, pretty good, not good, and bad status. The assumption for performance was placed between 0 to 100% or bad to good. Between bad and good, performance interval was placed; enough or lacking, resulting to four performance levels i.e. bad, lacking, enough, and good, thus creating 25%, 50%, 75%, and 100% intervals.

Assessment result for the performance of each dimension's attributes was mapped in two reference points which is bad and good, while performance level was distinguished into 4 (four) categories, as stated by [5]. These categories are presented in the following Table 2.

Table 2. Categories of Sustainability Status

No.	Dimension Index	Category	Description
1	00,00 -24,99	Bad	Not Sustainable
2	25,00 – 49, 99	Lacking	Less Sustainable
3	50,00 – 74,99	Enough	Quite Sustainable
4	75,00 – 100,00	Good	Sustainable

Source: Kavanagh and Pitcher (2004).

3. Results And Discussion

The summary of MDS-Rap-SISPOTTA analysis result for the 5 (five) dimensions is presented in the following Table 3. The reference used as the basis to determine the analysis result's goodness of fit in MDS is *Stress* value less than 0.25 and an *R-square* value greater than 0.90. According to Table 3, it is shown that the *Stress* value of the five dimensions is less than 0.25. Similarly, the *R-square* value is also greater than 0.90. In conclusion, MDS analysis has met the Goodness-of-fit criteria and thus can be further discussed and analyzed

Table3. Summary of *MDS-Rap-SISPOTTA*analysis result.

Dimension	***Stress (S)***	***R-Square (R)***
Ecological	0.1569	0.9347
Economical	0.1596	0.9335
Sociocultural	0.1459	0.9429
Technology and Infrastructure	0.1584	0.9316
Legal and Institutional	0.1652	0.9412

3.1. Ecological Dimension

Based on the analysis result of MDS, sustainability status index for the ecological dimension is 58.53, which is in the 50-74.99 range indicating quite sustainable. In other words, the ecological condition of the studied region is quite supportive of integrated agribusiness system especially for Rice-Corn and beef cattle farming. The leverage of each attribute in the ecological dimension is illustrated in Tabel.4.

Table 4. Ecological Dimension Attribute's Leverage

No.	Attributes of Ecological Dimension	Leverage
1	Utilization of animal breeding waste for organic fertilizer	1.59
2	Utilization of agricultural waste for animal feed.	1.85
3	Beef cattle maintenance system	1.57
4	Land (soil fertility).	0.88
5	Level of land utilization for agriculture and animal breeding	6.00
6	Agro climate	4.34
7	Feed support capacity	12.43
8	Availability of Waste Water Processing Installation	16.2
9	Shed cleanliness	30.18
10	Availability of Animal Slaughterhouse.	28.87
11	Availability of slaughterhouse-waste management installation	29.48
12	The type of animal feed	5.40
13	Availability of land for animal feed (grass and corn leaves)	7.80
14	The quantity of animal breeding waste	7.94
15	The distance of breeding location to community settlement	6.01
16	Drought incidence	9.44
17	The frequency of flood incidence	5.95
18	Rainfall.	3.53
19	The condition of agricultural road infrastructure	16.68
20	The condition of village road infrastructure	2.35

Based on Tabel 4, attributes for the ecological dimension that highly influence the sustainability of this dimension are shed cleanliness, availability of slaughterhouse waste management facility, and availability of animal slaughterhouse. These are indicated by the leverage values of the three attributes that are higher than those of other attributes. Out of the three attributes, it can be said that to improve sustainability in ecological dimension there is a need for animal slaughterhouse that is closer in distance and managed well especially in terms of waste management and shed cleanliness.

3.2. Economical Dimension

Based on the analysis result of MDS, that sustainability status index of economic dimension is 52.44, which is inside the 50-74.99 range or indicates quite sustainable. This means that the existence of integrated rice-corn and beef cattle farming agribusiness in the studied region is quite beneficial from the economic perspective. The leverage of each economical is shown in Table 5. attributes with the highest influence towards the sustainability status of an economic dimension in integrated rice-corn and beef cattle farming agribusiness is the financial feasibility of beef cattle breeding, production market for animal breeding agroindustry, and changes in Regional Income of husbandry sector for the last five years.

Tabel 5. Economical Dimension Attribute's Leverage

No.	Attributes of Economical Dimension	Leverage
1	Average breeders' income relative to the minimum wage in Jember District.	2.77
2	Average breeders' income relative to total revenue.	3.55
3	The market for animal breeding agroindustry products.	4.88
4	Availability of animal market/agribusiness sub terminal.	2.18
5	The place where breeders sell their animals	2.55
6	Availability of animal feed industry	4.41
7	Changes in Regional Income value in husbandry (in the last 5 years)	4.84
8	Financial feasibility of beef cattle breeding business	5.07
9	Amount of subsidy	4.80
10	Percentage of poor population	2.42
11	Price of an animal commodity	4.09
12	Amount of agricultural workers	2.42
13	Type of key commodities	1.04

3.3. Sociocultural Dimension

MDS analysis result for sustainability status index of sociocultural dimension is 69.07. This index value is inside the 50-74.99 range, or in other words, indicates quite sustainable. This also means that the sociocultural condition of the studied region is quite supportive of integrated agribusiness system especially for Rice-Corn and beef cattle farming.

While the leverage of each sociocultural dimension's attribute is shown in Table 6. According to Table 6, attributes with the highest influence towards sustainability of the sociocultural dimension are time allocated for animal breeding agribusiness, the frequency of conflicts related to animal breeding, and the number of breeder households. The result of the analysis shows that time allocated and the number of households working in the animal breeding industry, as well as the number of conflicts – if not managed well and received serious attention – may prove to be a threat in the sustainability of integrated rice-corn and beef cattle farming agribusiness

IOP Conf. Series: Earth and Environmental Science **207** (2018) 012025 doi:10.1088/1755-1315/207/1/012025

Table 6. Sociocultural Dimension Attribute's Leverage

No.	Attributes of Sociocultural Dimension	Leverage
1	Work is done individually or collectively	48.23
2	Amount of animal breeder households	52.01
3	Annual growth of animal breeder households (2006-2011)	15.78
4	Environmental knowledge	51.05
5	Level of workforce absorption in animal breeding agroindustry	45.28
6	The frequency of conflicts related to animal breeding	52.31
7	Family participation in animal breeding agribusiness	49.97
8	Community role in animal breeding business	36.53
9	The frequency of counseling and training	5.557
10	Level of agricultural workforce absorption	3.16
11	Business alternative besides animal breeding	12.25
12	Amount of people working in animal breeding agroindustry	14.14
13	Time allocated for animal breeding agribusiness	53.40
14	Amount of village with people working in the animal breeding sector	13.60

3.4. Technology and Infrastructure Dimension

The leverage of each technology and infrastructure dimension's attribute has shown in Table 7.

Table 7. Technology and Infrastructure Dimension Attribute's Leverage

No.	Attributes of Technology and infrastructure Dimension	Leverage
1	Distribution of animal health facility or (*Poskeswan)*	2.86
2	Distribution of artificial insemination service facility	3.65
3	The use of vitamin and probiotic to boost animal growth	4.52
4	Animal feed technology	5.11
5	Animal breeding/agroindustry waste processing technology	3.63
6	Animal breeding product processing technology	3.66
7	Information and transportation technology	3.53
8	Availability of agribusiness facilities and infrastructures	4.01
9	Availability of public facilities and infrastructures	3.19
10	Level of animal breeding technology mastery	3.27
11	Availability of animal breeding information technology	2.19
12	Quality standardization for animal products	1.99

The output of MDS analysis for the sustainability status index of technology and infrastructure dimension is 18.41. This index value is in the 0.00-19.99 range which means not sustainable. The value was obtained because farmers or breeders mostly still use traditional methods in managing their business. Based on Table 7 below, attributes with the highest influence on the sustainability of technology and infrastructure dimension are animal feed industry and animal growth booster. This means that from a technology and infrastructure standpoint, these aspects do not support business sustainability due to the lack of innovation in feed technology and animal growth booster

3.5. Legal and Institutional Dimension

The result of MDS analysis for sustainability status index of legal and institutional dimension is 21.57. This index value is in the 20-24.99 range which means less sustainable. Meanwhile, attributes that highly influence the sustainability of this dimension is presented in Table 8. According to Table 8, out of eight attributes used to assess legal and institutional dimension, those with the greatest leverage or ones with the highest influence towards the sustainability of legal and institutional dimension are the availability of social institutions and policy synchronization between central and regional government.

Table 8. Legal and Institutional Dimension Attribute's Leverage

No.	Attributes of Legal and institutional Dimension	Leverage
1	Consultation and training center owned by farmers	4.55
2	Cooperation agreement with other regions in animal breeding	5.16
3	Policy synchronization between the central and regional government	7.03
4	Farming groups	6.66
5	Availability of social institutions	7.46
6	Microfinance institutions (bank/credit)	5.68
7	Agricultural extension office or (*BPP)*	2.87
8	Regional governing body	5.45

Figure 1 shows that a combination of five dimensions was used to measure the sustainability of integrated rice-corn and beef cattle farming agribusiness. In average, index value for the sustainability of integrated rice-corn and beef cattle farming agribusiness in Jember District is 44.00 or in the 25.00 – 49.99 range (less sustainable).

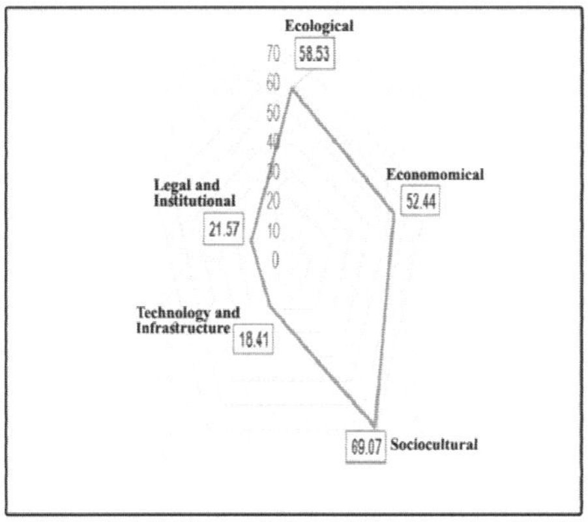

Figure 1. Sustainability Status Diagram

4. Conclusions

4.1. Conclusions

Based on five sustainability dimensions used in this research, the integrated rice-corn and beef cattle farming agribusiness in Jember District shows a sustainability index value of 44.00. This index value means that this particular agribusiness model is still less sustainable. Dimensions with the lowest sustainability index and must receive serious attention are technology and infrastructure dimension, as well as legal and institutional dimension, with each scored 18.41 dan 21.57 respectively.

4.2. Recommendations

In order to improve the sustainability status of aforementioned dimensions, serious attention must be given to improving the adequate provision of technology, infrastructure, legal, and institutional in the studied region so that the integration of rice-corn pattern planting and beef cattle breeding can be a sustainable business.

Refferences

[1] Kapa and Maximilian M J 2006. *Produktivitas Usaha tani dalam Sistem PertanianTerpadu: StudiKasus di Kecamatan Amarasi Kabupaten Kupang, Nusa Tenggara Timur. Proceedings of a Workshop to Identify Sustainable Rural Livelihoods, Held in Kupang, Indonesia*, 5–7 April (*ACIAR Proceedings No. 126*)

[2] Foot A.S., Banes S, Oge J A C G, Howkins J, Nielsen V C and Callaghan J R O 2006. *Studies on Farm Livestock Waste.* 1st ed. Agriculture Research Council, England.
[3] Pitcher T J 1999 *FAO Fisheries Circular No. FIRM/C:* **947** 47
[4] Ranaweera N, Dixon J N and Jodha N S 1999 *Journal of the Asian Farming Systems Association* **2** 1-15.
[5] Pitcher T J and Preikshot D B 2001 *Fisheries Research* **49**(3): 255-270.
[6] Kavanagh P. and Pitcher T J 2004 *Fisheries Centre Research Reports* 12 2.

IOP Conf. Series: Earth and Environmental Science **207** (2018) 012026 doi:10.1088/1755-1315/207/1/012026

An analysis of marketing channels on broiler carcass in Jember Regency

N D Wahyono[1] and M M D Utami[2]

1 Department of Agricultural Production, Politeknik Negeri Jember, Jl. Mastrip PO Box 164, Jember 68101, Indonesia
2 Department of Animal Science, Politeknik Negeri Jember, Jl. Mastrip PO Box 164, Jember 68101, Indonesia

e-mail: nanang_d_wahyono@polije.ac.id

Abstract. The aim of the study was to determine the pattern of marketing channels in the broiler carcass in Jember Regency. Data collected includes primary data and secondary data. Analysis of the data used by the marketing channel pattern. Primary data is collected from farmers selected as samples using a combination of interview methods, submitting a list of questions that have been prepared. Secondary data comes from literature, or scientific publications Descriptively discussed and tested the marketing margin for each marketing channel pattern tested with an independent sample t-test. The pattern 1 and 3 and the pattern 2 and 3 were a difference between each pattern. While pattern 1 and 2 was no difference between each pattern. The highest marketing margin was pattern 3 (IDR 12350).

1. Introduction

The poultry sector, especially broiler chickens is still the top priority meet animal protein needs. According to Rizaldi (2010), broiler meat contains a composition of good nutritional value and as a source of food ingredients contains animal protein. The consumption of broiler meat is increasing importance in Indonesia and per head consumption is still rising.The broiler is one of the livestock that has a good prospect on a small or large scale. The advantages of broilers compared to other types of livestock are the characteristics of broilers that have characteristics of rapid growth, the efficiency of ration use, short harvest periods, and soft fibrous meat. This makes the broiler chicken business to be one of the livestock businesses that have a significant role in the economy.

In Indonesia, broiler population in 2017 was 1698369, in 2015 was 1.528.329, in 2016 was 1.632.568 and the average of price of broiler carcass in 2014 about IDR 28976 per kg, in 2015 increased 3.47% to IDR 29981per kg. Furthermore, in 2016 increased 4.51% to IDR 31334 per kg. According to the East Java Animal Husbandry Office statistics in 2015, the production of broiler chickens in Jember Regency experienced a rapid increase from 7,689,080 in 2014 to 12,120,036 in 2015. But in 2016 broiler production decline to 11,932,763 and in 2017 decreased to 11,851,934. The phenomenon of decreasing production occurs cannot be separated from three factors that influence the success of broiler livestock businesses, such as day-old chick, feed, and management.

Broiler meat relatively easily to accept, because of the taste factor, production speed, and cost. Broiler meat is also almost there are no challenges from social or religious this product. So that the chicken cultivation business opportunity broiler is very broad and worthy doing. After the harvesting process, broilers are ready to be marketed. In a broiler farm business, marketing is defined as the activity of selling or bringing broiler chickens from farms to table. In Indonesia, broilers are usually marketed in the form of whole carcasses, namely the body parts of chickens slaughtered and without the visceral organs, legs, neck, head, feathers, and blood.

The broiler is one of the poultry commodities that contribute greatly to the food of the Indonesian people. Consumption of broiler meat increases every year because the price of chicken meat is affordable for all people. The economic characteristics of broilers are rapid growth, as meat producers with low feed conversion and ready to be cut at a relatively young age [1]. Broiler harvesting is carried out at the age of 4 to 6 weeks with a weight of 1.2 to 1.9 kg [2]. The growth of broiler is fairly fast around 30 days ready for harvest.

Broiler carcasses are the body parts of chickens slaughtered without the stomach, legs, neck, head, hair, blood. Percentage of the carcass is often used to assess meat livestock production [3]. According to [4], [5], and [6] based on the method of handling, broiler carcasses are divided into: fresh carcass (the carcass that has just been processed for no more than six hours and does not experience further treatment), fresh cold carcass (fresh carcass which is immediately cooled after processing so that the temperature in the meat becomes between 4-50°C), frozen carcass (carcass that has undergone a freezing process), then [7] explained carcass weight is strongly related to life weight and body weight gain. Body weight gain is directly caused by the availability of tissue-forming amino acids. The consumption of protein is directly related to the growth process, therefore the management of using feed ingredients that contain protein that is required for amino acid intake by the body [8]. The percentage of broiler carcasses varies between 65-75% of body weight, the heavier the chicken is cut, the higher the carcass [9].

Marketing according to [10] is a social process where both individuals and group involved in the process get what (product or service) they need and want by creating, offer, and exchange products or services value with other parties. Marketing is an activity carried out by individuals or groups to market, offering quality products to other parties in order to benefit [11]. Marketing can also be interpreted as an activity carried out to strive for a product to be accepted in the market and favored by the market [12]. Marketing channel is a series of interdependent and involved organizations in the process to make an item or service ready to be used or consumed [11]. Marketing channel is one of the important factors to increase sales volume which refers to the ease for consumers to access and get it.

A marketing channel is an intermediary group that deals with each other and which distributes products from producers to consumers. Marketing channels on both broiler and domestic chicken farms tend to belong. This is due to marketing channels starting with farmers, collectors, retailers, and then to consumers. Marketing margin is the price financed by the consumer minus the price received by the producer. High and low marketing margins are used to measure the efficiency of the marketing system. The greater the marketing margin, the more inefficient the marketing system [13]. For example, the price of broiler paid by consumers is IDR 23000/Kg, the producer receives IDR 20,000/Kg, the marketing margin is IDR 3000.

The price level that must be paid by consumers and accepted by producers depends on the prevailing form and structure, both competitive markets, monophonic markets, oligopsonistic markets, monopoly markets, and oligopoly markets. The length of the marketing channel also often leads to less efficient marketing. The marketing margin is high due to the small amount the producer receives. In the broiler carcass marketing process a marketing channel will be formed which requires marketing costs. Marketing costs affect prices at the consumer level. Marketing costs will be higher in line with the length of the marketing process. Marketing costs can be measured using marketing margins, namely the difference in prices paid by end consumers at prices received by producers [14]. Conducive marketing performance will provide incentives for farmers to produce and encourage the adoption of technology by farmers, and increase the share of prices received by farmers. A conducive policy can increase production, distribution, product development and proportional incentives for traders and welfare actors

An efficient trading system is the arrival of the product to the end consumer according to the place, time, and form desired by the consumer at the lowest cost and the fair distribution of the price paid by

IOP Conf. Series: Earth and Environmental Science **207** (2018) 012026 doi:10.1088/1755-1315/207/1/012026

the end consumer to all parties involved in the production and trading activities [15]. The size of the marketing margin depends on changes in marketing costs, the benefits of the intermediary trader, the price of the final consumer, and the price received by the producer. Based on the identification and formulation of the above problems, the purpose of this research studies carcass marketing distribution channels of broiler carcass in Jember Regency.

2. Materials and Methods

The study was conducted in the downtown of Jember Regency (Pasar Tanjung Market, Kepatihan Market, and Pagah Market). Selection this location is based on the consideration that in downtown is a production center of broiler carcass. Respondents as the subject of this research were 136 respondents. Respondents have sampled consists of farmers, retailers, wholesalers, and related parties broiler chicken farms. Various considerations like time, energy and constraints cost, then interview the element the sample is carried out in depth on respondents who really understand broiler carcass trading problems.

The data that will be used in this study is primary and secondary data. Primary data is collected from farmers selected as samples using a combination of interview methods, submitting a list of questions that have been prepared. Secondary data comes from literature, or scientific publications both private and government related to broiler chicken farms and reports from institutions or agencies that support, such as Department of Animal Husbandry, Central Bureau of Statistics and Department of Agriculture.

Data collected includes primary data and secondary data. Marketing channels are discussed descriptively and marketing margins are used as formulas following.

3. Results and Discussion

3.1. Marketing channels pattern

The efficiency of the business is one of the important components of creating a marketing system that can benefit many related parties in the broiler business, such as breeders, traders, and consumers. Through the implementation of efficient trading will ultimately affect the formation of price levels. The trading arrangements that occur in a commodity cannot be separated from the influence of the market structure that occurs.

Marketing channels are one aspect of marketing. Distribution can also be interpreted as a marketing activity that seeks to facilitate and facilitate the delivery of goods and services from producers to consumers. Marketing channel based on its intensity is divided into three such as intensive form that the type of channel that utilizes many traders big and small, a selective form, which is a type of distribution that only uses a few wholesalers and a small number of retailers, and exclusive form, which is a distribution channel that only involves one intermediaries in certain community environments to handle products. Channels that we currently encounter can be divided into two: direct and not direct channels. Direct channel from manufacturer to consumer. Usually, only a few items are marketed directly. Indirect channel, which is divided into two, from manufacturer to retailers to consumers and from producers to wholesalers to retailers to consumers.

Marketing channels play a role in marketing horticultural agricultural commodities such as farmers, traders, intermediaries or wholesalers and retailers [16]. Problems that arise in the marketing system of horticulture were marketing activities that have not run efficiently [17], in the sense that they have not been able to deliver agricultural products from producers to consumers at low cost and have not made fair distribution of remuneration from the overall price of the last consumer to all parties participating in the production and marketing activities of these agricultural commodities. Fair distribution in this context is the distribution of remuneration for marketing functions according to the contribution of each marketing institution that plays a role.

The marketing channel is every series of companies or people who take part in distributing goods and services from producers to consumer, distribution channels can be divided into several types, such as direct distribution channels, in this distribution entrepreneurs try to distribute goods that consumers buy directly where consumers live and indirect distribution channels, in this case, the employer uses outsiders to help to channel goods to consumers [18].

In addition, the trade in broiler carcass channels is influenced by the periodic products produced and the producers are relatively scattered. As a consequence, the price of chicken meat is strongly influenced by supply fluctuations. There are three channels in the broiler carcass marketing system such as Pattern 1 (farmers-consumers), Pattern 2 (farmers-retailers-consumers), and Pattern 3 (farmers-wholesalers-retails-consumers). Results of broiler carcass channels are presented in Table 1.

Table 1. Marketing channel of broiler carcass

Pattern	Total Farmers	Percentage of Total Farmer
Pattern I	14	10.29
Pattern II	94	69.11
Pattern III	28	20.58
Total	136	100

The results of the research on the marketing channel pattern in Jember regency were pattern 1 was 10.29%, pattern 2 was 10.29%I of 69.11%, and pattern 3 was 20.58. In Jember, the distribution channel of broiler carcass uses more medium marketing channels (pattern 2) because of a large number of intermediary traders involved so high prices issued by consumers. According to [10] reported that marketing channels are one of the determining factors for increase sales results, the marketing channel is a delivery channel product, if the channel can be used properly and optimally, then consumers can easily obtain the product. The result was consistent with the opinion of [19] more intermediary traders have involved in marketing channels the higher the price that consumers must pay.

3.2 Marketing Margin

Marketing margin is the difference in prices paid by the producer with the price paid by the end consumer [20], included in the marketing margin are all marketing costs incurred by the marketing costs and marketing profits, starting from the producer to the consumer. Marketing margins are divided into two: marketing costs and marketing charges. Marketing costs are related to the rate of return from factors of production, while marketing charges are related to how much is received by processors, collectors and trading institutions. The trading margin consists of three types, such as absolute, percentage and combination. Absolute marketing margins and percentages can be decreased, constant and increased with increasing quantity of marketed [21]. The relationship between demand elasticity at different levels of the trading chain provides several analytical uses. Results of the marketing margin are shown in Table 2.

Table 2. Marketing margin of broiler carcass

Pattern	Marketing Margin (Price, IDR)	Percentage of Total Marketing Margin (%)
Pattern 1	7450	25.13
Pattern 2	9850	33.22
Pattern 3	12350	41.65
Total	29650	100.00

The size of the marketing costs is depending on the size of the activities marketing institutions and facilities needed. Marketing costs also depend on length short of the marketing chain involved. The size of the marketing margin is influenced by changes in marketing costs, profits of traders intermediaries, prices paid by end consumers and the price received by the producer. Besides factors, the amount of marketing margin is also influenced by the distance of the producer area with the consumer and the nature items that will add to the whole marketing costs.

Marketing margins on this research results vary depending on the distribution channel. The difference in marketing margins in each marketing channel pattern is analyzed using Different test t independent sample t-test. The results of different tests t-test the results of the pattern (1 and 3) and the pattern (2 and 3) was a difference between each pattern. While pattern (1 and 2) was no difference between each pattern.

The process of distributing carcass by intermediary traders through several the distribution stage before reaching the hand end consumers namely transportation, payment of retribution, unloading, and loading livestock. When the distance between breeders and consumers close, then costs transport can be reduced. Getting farther distance and more and more intermediaries (marketing agency) involved in marketing, then marketing costs will be higher. According to [22] the length of the short chain of marketing and the number of marketing institutions will affect the amount of marketing margin. The longer the marketing chain, the lower the efficiency received by farmers, so that prices are paid by consumers The highest marketing margin was pattern 3 (12350), then [23] reported the longer marketing agencies are involved in the process marketing will increase the total margin and net profit obtained by the offender and the size of the marketing margin affected by changes in marketing costs, profits of intermediary traders, prices the final consumer is paid and the price received by the producer.

Conclusion

There were three patterns of distribution channels such as pattern 1 (farmers-consumers), pattern 2 (farmers-retailers-consumers), and pattern 3 (farmers-wholesalers-retails-consumers). The pattern (1 and 3) and the pattern (2 and 3) was a difference between each pattern. While pattern (1 and 2) was no difference between each pattern. The highest marketing margin was pattern 3 (IDR 12350).

Acknowledgment

The authors really appreciate to Politeknik Negeri Jember for a grant to research and to present this research on International Conference on Food and Agriculture.

References

[1] Dransfield E and Sosnicki A A 1999 *Poult. Sci.* **78** 743
[2] Robinson F E, Classen H L, Hanson J A and Onderka D K 1992 *J. Appl. Poult. Res.* **1** 33
[3] Havenstein G B, Ferket P R and Qureshi M A 2001 *Poult. Sci.* **82** 1509
[4] Georgsson F, Þorkelsson A E, Geirsdóttir M, Reiersen J and Stern N J 2006 *Food Microbiol.* **23** 677
[5] Mielnik M B, Dainty R H, Lundby F and Mielnik J 1999 *Poult. Sci..* **78** 1065
[6] Russell S M, Fletcher D L and Cox N A 1995 *Poult. Sci.* **74** 2041
[7] Feddes J J, Emmanuel E J and Zuidhoft M J 2002 *Poult. Sci.* **81** 774
[8] Sklan D and Plavnik L 2002 *Br. Poult. Sci.* **43** 442
[9] Jackson S, Summers J D and Leeson S 1982 *Poult. Sci.* **61** 2224
[10] Barich H and Kotler P 1991 *MIT Sloan Manag. Rev.* **32** 94
[11] Kotler P and Scheff J 1997 *Standing room only: Strategies for marketing the performing arts.* (Harvard Business School Press)
[12] Payne A F, Storbacka K and Frow P 2008 *J. Acad. Mark. Sci.* **36** 83
[13] Garleanu N and Pedersen L H 2011 *Rev. Financ. Stud.* **24** 1980
[14] Borden N H 1964 *J. Advert. Res.* **4** 2
[15] Lee H G and Clark T H 1996 *J. Manag. Inf. Syst.* **13** 113

IOP Conf. Series: Earth and Environmental Science **207** (2018) 012026 doi:10.1088/1755-1315/207/1/012026

[16] Kuma'at R 1992 *Sistem pemasaran sayuran dataran tinggi di provinsi Sulawesi Utara* (Thesis MS-FPS IPB, Bogor)
[17] Kader A A 2004 *International Postharvest Symposium 682* p 2169
[18] Jo Black N, Lockett A, Ennew C, Winklhofer H and McKechnie S 2002 *Int. J. Bank Mark.* **20** 161
[19] Sheth J N and Parvatiyar A 1995 *Int. Bus. Rev.* **4** 397
[20] Nagle T T and Müller G 2017 *The strategy and tactics of pricing: A guide to growing more profitably* (Routledge)
[21] Hardesty S D and Leff P 2010 *Renew. Agric. food Syst.* **25** 24
[22] Srivastava R K, Shervani T A and Fahey L 1999 *J. Mark.* p 168
[23] Butler R W 1980 *Can. Geogr. Géographe Can.* **24** 5

Current and future goat production in Jembrana Regency, Bali Province.

L Doloksaribu[1], B P McLachlan[2], R S Copland[2] and P J Murray[2]

[1] Faculty of Animal and Husbandry, Udayana University, Jimbaran Campus Badung-Bali, Indonesia

[2] School of Agriculture and Food Sciences, The University of Queensland, Gatton Campus, Queensland 4343, Australia

E-mail: lindawati_doloksaribu@hotmail.com

Abstract. This study was undertaken to establish a database for improving goat production in Bali. A survey was conducted from January to June 2014 on 68 farmers integrated with farm commodities, owning a total of 258 goats in Mendoyo District. Data on flock dynamics were 49 kidding occurrences in 2014 delivered 67 kids; 67% were single born, 29% were twins and 4% were triplets/quadruplets. The kids were a mixture of Gembrong, Benggala, Kacang, Etawah, and PE that had a 24% mortality rate. Litter size was on average 1.37 kids from does with 1st to 6th parity. Kids were housed in battery systems and fed fresh roughages with cut and carry feeding systems. The average bodyweights of female weaners 22.9±1.6 kg were significantly lower ($P<0.01$) than of male weaners 29.8±1.9 kg at day 135. This growth was supported by the abundance of available feed, and the housing used meant low incidence of parasites as indicated by FAMACHA© scores which were generally low ($P<0.01$). Fourteen goats were sold for IDR 22 M. Smallholder farmers in Mendoyo District reared goats in small flocks as the farmers had just re-started rearing goats again resulting in inefficient goat production, particularly when labour costs were included.

1. Introduction

Mendoyo District was selected as the sampling area as it had the largest goat population among the districts in Jembrana Regency [1]. Although Jembrana was designation regency for improvement of goat farming that was officially promulgated by the Indonesian Minister of Agriculture [2], this regency has had a large reduction in the number of goat farmers as well as goats, over the last 25 years. Jembrana Regency previously had 42,751 goats or about 45% of the total 95,430 goat population in Bali Province in 1990 [3]. In 1990, this was about double the current number of goats in Buleleng Regency i.e. 24,905 goats or triple that of Karangasem Regency i.e. 13,861 goats. Other villages in Jembrana Regency that previously had large numbers of goats and goat farmers [3] had been visited. However, these villages at the time of this study had none or very few goats and goat farmers. Available literature presented little information on the reasons some smallholder farmers had stopped farming goats years ago or they had just re-started farming goats recently, as well as the current situation of their goat rearing under smallholder production systems in Mendoyo District.

Most of the smallholder farmers studied in Mendoyo District apparently re-started rearing goats as they had received Simantri Programmes in October 2013 and April 2014 (Tunas, W. 2014, pers.

comm. 9 June) that motivated them to resume or improve their goat rearing management [4]. The smallholder goat farmers mostly had small flocks; therefore, they were increasing the number of goats in their flocks and apparently would not sell goats prior to Eid Qurban. A case study was conducted from the 3rd to 30th June 2014 to establish a database of current reproductive and productive efficiency of goat farming in Mendoyo District as they contributed about 2.4% of the total 841.8 km² size of Jembrana Regency, and as well they contributed about 11% of the total of 68,457 goats in Bali Province [1]. The objective of this study was to establish a database of the reproductive and productive efficiency of goat farming under smallholder production systems in Mendoyo District in Jembrana Regency, as well as a socioeconomic analysis of these systems. The database was used to set up development strategies for improving goat production, in Bali Province.

2. Materials and Methods

The goat farming studied in Jembrana Regency encompassed a village in Mendoyo District. Mendoyo District is located in West Bali National Park, the conservation forest of Jembrana Regency that is located in the western part of Bali. Jembrana Regency is situated 8^{0}9'58" to 8^{0}28'2" south and 114^{0}26'28" to 115^{0}51'28" east with average ambient temperatures 26.2 ^{0}C, and 83% annual average relative humidity, 1,721 mm rainfall and 3 knots wind velocity (www.bmkg.go.id). Most of the people lived in agricultural sectors and their income was derived primarily from soybean, coconut, cacao, coffee, and cloves plantations [1].

All 68 goat-owning families were interviewed, based on structured questionnaires, and 258 goats were observed to obtain data i.e. breed-type, sex, age, dentition status (I_0, I_1, I_2, I_3, I_4, toothless), FAMACHA© score, bodyweight, chest circumference, body length, height at withers and rump height, birth type (single and multiple), and parity were recorded over a month of direct animal observations. From the 258 goats, 67 kids were born during the data collection. Data was only collected in June 2014; therefore, data on goat reproductive performance was taken from the previous year. Most of the smallholder goat farmers had just re-started rearing goats and housed their goats only in battery housing systems; therefore, average daily gain of goats, kidding intervals as well as the effects of housing systems were not presented. The inputs included feeds, veterinary services, drugs, and labour cost i.e. labourers (from the family). The outputs obtained included sales of live animals, manure or goat products consumed at home which were then converted into cash (IDR million). IDR 1 million was equivalent to AUD$100.00.

Age of kids were estimated from their dentition status (I_0) while their birth dates were estimated by using the farmer's data and they were confirmed by looking at the condition and size of the dam's mammae or date of post-partum mating [5]. Kids were weaned, on average, at about 135 days of age. All 68 goat owning families completed questionnaires on a range of topics i.e. goat feeding and breeding management, income from goats and other sources and the demographics of people living on the farm. The relationships between liveweights and other factors were established using a linear regression analysis while correlations of the goats liveweight with other factors were computed by using SPSS version 22.

3. Results and Discussion

3.1 Household labourers and their profiles

Identifying the roles and the profiles of smallholder farmers involved in goat rearing under smallholder production systems in Mendoyo District could be used to improve their goat production. Ceasing goat rearing for quite some time, for various reasons, smallholder farmers in Mendoyo District re-started rearing goats again in October 2013 and April 2014 when they received Simantri Programmes (Tunas, W. 2014, pers. comm. 9 June). Smallholder farmers faced problems such as having traumatic experiences where their goats got sick and had low performance and/or died due to mouth and foot diseases; lack of capital to restart rearing goats again; low numbers of 2 ± 0.1 household labourers; and not being young anymore (47.3 ± 1.4 years) to manage their crops along with rearing goats at the same time. This indicated that the farmers were old and did not have much physical energy necessary to contribute to farming as would be expected from them. On the other hand, farmers who reared small flock sizes produced insufficient goat manure to fertilize the 0.9 ± 0.1 ha of their crops. Strangely farmers that had the highest education level tended to have the lowest labourer ratio, the smallest turn off rate and generated loss in both GM(A-B) IDR (4.661 ± 1.312) million and GM/doe IDR (2.518 ± 0.244) million.

Number of 0.2 ± 0.1 children per household also caused the reduction in the number of goats in Mendoyo District. Farmers' children went to study in Denpasar City or to work on cruise ships overseas. [6] and [7] reported that cruise ship work was an increasingly popular career choice for senior high school graduates in Bali Province. Limitations of the availability of family labourers resulted in a new critical threshold for farm growth strategies in Mendoyo District. Based on its low profitability and the amount of labour required, it is unlikely that the younger generations would be interested in taking up goat husbandry, so it is a trend that is bound to continue. This seemed to be a constraint to improving goat rearing in Mendoyo District.

However, 16% of the households in the present study were aged between 65 and 73 years with 9.2 ± 1.1 years of goat rearing experience and cultivated 0.9 ± 0.1 hectare of crop farms integrated with goat rearing although they were not young anymore. Smallholder goat farmers in Malang Regency, East Java, besides cultivating an average of 1.35 ha for crop farms, 32% respondents reared about 8 to 20 goats per household involving more family members as labourers i.e. 5-7 household labourers whose ages were younger than Indonesian manpower age standard i.e. 10 to above 65 years old [8].

This study revealed that the level of education of the household played an important role in dictating the best production parameters. The level of farmer education *per se*, could be expected to improve the efficiency of goat production, but it is hypothesised that farmers with a better education would implement improved husbandry practices. Although farmers' education level did not significantly affect their GM/doe, the farmers with the highest education level i.e. graduated from high school tended to have higher GM/doe IDR (2.419 ± 0.542) million. This indicated that the higher education level of the farmers, the better the decisions they make for efficient goat production such as having larger numbers of does per household, and having a larger ratio of household labourers to their goats. It may be assumed that increased education leads to a better understanding on how to have more efficient goat production [9] and [10]. Education and governance have major impacts on agricultural efficiency; therefore, farmers should have higher levels of education [11].

3.2 Goats and their profiles

Flock size, more importantly the number of does owned by households, was one of the factors that influenced the GM(A-B) or GM/doe gained by smallholder farmers in Bali Province. Smallholder farmers in Mendoyo District reared goats that were a mixture of Gembrong, Benggala, Kacang, Etawah Grade, PE and their crossbreds or backcrosses. All the body dimension measurements of preweaned, weaner and yearling males were significantly higher than of preweaned, weaner and yearling females ($P<0.05$). The average of 29.8 ± 1.9 kg bodyweights of male weaners in this study were significantly higher ($P<0.01$) than the 22.9 ± 1.6 kg of female weaners (Table 1). This was in agreement with [12] and [13] who reported that male kids grew faster and heavier than female kids. The reasonable growth rates of preweaned and weaned kids indicated that smallholder farmers in Mendoyo applied good kid rearing management, thus improving goat production.

Average height at withers of goats reared in Mendoyo District i.e. 66.0 ± 0.7 cm was categorized as a Large type goat [14] and [15] (Table 2). Based on the body dimension measurements, it appeared that goats were PE goats with longer ears. This indicated that larger size goats usually produced more milk and meat than smaller ones [15] and [16]. Female yearlings aged 0.5 – 1 year studied in Mendoyo District had an average of 26.2 ± 1.4 kg bodyweight, 66.1 ± 1.5 cm chest circumferences and 64.8 ± 1.3 cm height at withers. These body dimensions were within the range of the required physical standard of PE breeding stock by [17]. This indicated that it was a good time to expand the number of goats in Mendoyo District as good breeding stock was being farmed.

Table 1. Average chest circumference, bodyweight and FAMACHA© score of different physiological state from 258 recordings of goats reared in Mendoyo Districts Bali

Class of goat	Chest circumference (cm)		Bodyweight of goats (kg)		FAMACHA© score	
	No. of goat	Mean ± SEM	No. of goat	Mean ± SEM	No. of goat	Mean ± SEM
F preweaned	13	48.9 ± 2.0^{a}	13	10.5 ± 1.9^{a}	13	1.2 ± 0.2^{a}
F wcancr	18	61.1 ± 1.7^{b}	18	22.9 ± 1.6^{b}	18	1.5 + 0.2^{bijkl}
F yearling	22	66.1 ± 1.5^{cjk}	22	26.2 ± 1.4^{c}	22	1.3 ± 0.2^{aimn}
F pregnant	75	75.3 ± 0.8^{dlmn}	75	35.6 ± 0.8^{d}	75	1.3 ± 0.1^{cop}
F lactating	16	78.2 ± 1.8^{elop}	16	38.0 ± 1.7^{ej}	16	1.9 ± 0.2^{do}
F dry	48	74.4 ± 1.0^{fmoq}	48	33.4 ± 1.0^{fj}	48	1.6 ± 0.1^{ep}
M preweaned	18	48.3 ± 1.7^{a}	18	13.6 ± 1.6^{a}	18	1.7 ± 0.2^{a}
M weaner	13	68.9 ± 2.0^{gjr}	13	29.8 ± 1.9^{g}	13	1.1 ± 0.2^{fjm}
M yearling	11	69.8 ± 2.1^{hkqr}	11	28.7 ± 2.0^{h}	11	1.3 ± 0.2^{gk}
M buck	24	78.1 ± 1.5^{inp}	24	36.2 ± 1.4^{i}	24	1.7 ± 0.1^{hln}
All goats	258	66.9 ± 0.5	258	27.5 ± 0.5	258	1.5 ± 0.0

Class of goat	Chest circumference (cm)		Bodyweight of goats (kg)		FAMACHA© score	
	No. of goat	Mean ± SEM	No. of goat	Mean ± SEM	No. of goat	Mean ± SEM
All females	192	71.1 ± 0.8^{a}	192	31.3 ± 0.9	192	1.4 ± 0.0
All males	66	66.6 ± 1.4^{b}	66	27.5 ± 1.6	66	1.5 ± 0.1
All goats	258	68.8 ± 0.8	258	29.4 ± 0.9^{c}	258	1.5 ± 0.0^{c}

F=Female, M=Male. All preweaned=aged 0 – 4.5 months; All weaner=4.5 month – I_0; All yearling=I_1; F pregnant, F lactating, F dry, and buck had I_1 – toothless.
Means in a column with different superscripts differed significantly at the .05 level.

Table 2. Average body length, height at withers and rump height of different physiological state from 258 recordings of goats reared in Mendoyo Districts Bali

Class of goat	Body length (cm)		Height at withers (cm)		Rump height (cm)	
	No. of goat	Mean ± SEM	No. of goat	Mean ± SEM	No. of goat	Mean ± SEM
F preweaned	13	59.0 ± 2.8^{a}	13	49.1 ± 1.7^{a}	13	51.5 ± 1.7^{a}
F weaner	18	72.6 ± 2.3^{b}	18	59.0 ± 1.4^{b}	18	61.3 ± 1.5^{b}
F yearling	22	80.4 ± 2.1^{cjk}	22	64.8 ± 1.3^{cjk}	22	68.1 ± 1.3^{cjk}
F pregnant	75	91.5 ± 1.2^{dlmnov}	75	69.3 ± 0.7^{dlmn}	75	73.4 ± 0.7^{dlm}
F lactating	16	94.3 ± 2.5^{elqr}	16	72.9 ± 1.5^{eo}	16	75.6 ± 1.6^{eln}
F dry	48	91.6 ± 1.5^{fmqst}	48	68.8 ± 0.9^{flpq}	48	71.9 ± 0.9^{fmop}
M preweaned	18	58.7 ± 2.3^{a}	18	49.3 ± 1.4^{a}	18	51.8 ± 1.5^{a}
M weaner	13	80.4 ± 2.8^{gju}	13	65.8 ± 1.7^{gjmpr}	13	68.1 ± 1.7^{gjoq}
M yearling	11	85.4 ± 3.0^{hknsu}	11	64.4 ± 1.8^{hknqr}	11	69.2 ± 1.9^{hkpq}
M buck	24	93.7 ± 2.1^{iprtv}	24	75.1 ± 1.3^{io}	24	77.2 ± 1.3^{in}
All goats	258	80.8 ± 0.7	258	64.0 ± 0.4	258	66.8 ± 0.5

Class of goat	Body length (cm)		Height at withers (cm)		Rump height (cm)	
	No. of goat	Mean ± SEM	No. of goat	Mean ± SEM	No. of goat	Mean ± SEM
All females	192	86.4 ± 1.1^{a}	192	66.6 ± 0.7^{a}	192	69.9 ± 0.7^{a}
All males	66	79.9 ± 1.8^{b}	66	64.5 ± 1.2^{a}	66	67.0 ± 1.2^{b}
All goats	258	83.2 ± 1.1	258	65.5 ± 0.7	258	68.5 ± 0.7

F=Female, M=Male. All preweaned=aged 0 – 4.5 months; All weaner=4.5 month – I_0; All yearling=I_1; F pregnant, F lactating, F dry, and buck had I_1 – toothless.
Means in a column with different superscripts differed significantly at the .05 level.

Expanding quality-breeding stock could be one of the development strategies that will help in improving their goat production.

Overall, 49 parturitions produced 67 kids that were born in the first six months of 2014 in Mendoyo District. There were 75 pregnant does or 54% of the 139 does recorded in the 68 flocks in 2014 in Mendoyo District. This figure was larger than 25% of flock sizes recommended by [18]. This indicated that flock size could be expanded in two or three years to achieve a larger proportion of does, and bucks owned per household in Mendoyo District. Expanding flock size by households could be one of the development strategies that will help in improving their goat production with no goats sold in 2014.

The largest flock of 53 goats that included four bucks and 39 does, six female preweaned kids, a female weaner, two female yearlings and a male preweaned kid, had 37 pregnant does and was making a big effort to expand flock size in Mendoyo District. This was shown by selling zero goats and thus they had zero per cent annual turn off rate in 2014. Furthermore, the smallholder farmer had four bucks to mate the 39 does to ensure a high pregnancy rate i.e. 95% had been achieved, thus a high kidding rate. [17] required 34 ± 6 kg, 76 ± 7 cm and 71 ± 5 cm for breeding stock PE does aged 1-2 years while 49 ± 9 kg, 80 ± 8 cm and 67 ± 5 cm for breeding stock PE bucks aged 1-2 years for the three measurements, respectively.

This indicates that having a large flock size with the correct proportion of does and bucks as well as having three household labourers per household and a high standard of breeding stock could be one of the development strategies that will help in improving their goat production.

3.3 Socioeconomic analysis

Overall, total GM/doe of IDR (138.500) million with an average IDR (2.518 ± 0.174) million ranged from a loss of IDR 6.275 million to a profit of IDR 3.396 million. Of the 258 goats studied, 14 goats were sold in 2014 for the total price of IDR 22 million that contributed 48% to the total income of IDR 45.540. Total sale of manure contributed 52% or IDR 23.542 million and sale of goats contributed 48% or IDR 22 million to the total income of IDR 45.542 million for the flocks studied in Mendoyo District.

Although goat rearing contributed a high income i.e. GM (A-B) IDR 6.792 million and GM/doe IDR 3.396 million to the farmers, farmers in Mendoyo District preferred increasing the number of goats in their flocks. Their goats were quality-breeding stock as they achieved the required physical standard of [17]. Other farmers including the one who had the largest flock of 53 goats were expanding their flocks and apparently would not sell goats prior to Eid Qurban. Expanding flocks by smallholder farmers studied in Mendoyo District almost made a zero per cent of turn off rate and thus generated negative GM(A-B) and GM/doe. This indicated that in future, Mendoyo District could have more goats to produce organic fertilizer for their crops and to sell both fertilizer and live animals.

The average turn off rate recorded in 2014 was 5 ± 3% where only preweaned female and male kids and weaners female and male kids were sold in 2014. It was likely that smallholder farmers in Mendoyo District were not able to sell more goats prior to Eid Qurban in October 2014. Therefore, the turn off rate was likely to be stable and would not significantly affect GM(A-B) and GM/doe in all flock sizes. As members of the smallholder goat farmers association who received the Simantri programme, they had an agreement with the Bali Government that they would not sell their goats until they kidded. Overall, goat rearing in Mendoyo District, had average GM(A-B) and GM/doe of IDR (4.376 ± 0.565) million and IDR (2.518 ± 0.174) million, respectively, when they had a low turn off rate of 5 ± 3%. When all flocks were ranked based on GM/doe, only one flock had a positive GM/doe of IDR 3.396 million and this flocks consisted of four goats that included two does, and six goats were sold and had a 150% turn off rate. However, an assumption was made that owning the four goats with one buck, one female yearling and two does, in 2014, produced three female and three male kids (i.e. 2 does X 3 kids X 2 as twins per 2 years with 0% mortality). Therefore, when a female kid was kept for replacement, two female and three male kids and a culled doe could be sold with a 150% turn off rate.

Overall, goat rearing in Mendoyo District, had average GM(A-B) and GM/doe of IDR (4.376 ± 0.565) million and IDR (2.518 ± 0.174) million, respectively, when they had a low turn off rate of 5 ± 3%. When all flocks were ranked based on GM/doe, only one flock had a positive GM/doe of IDR 3.396 million and this flocks consisted of four goats that included two does, and six goats were sold and had a 150% turn off rate. However, an assumption was made that owning the four goats with one buck, one female yearling and two does, in 2014, produced three female and three male kids (i.e. 2

does X 3 kids X 2 as twins per 2 years with 0% mortality). Therefore, when a female kid was kept for replacement, two female and three male kids and a culled doe could be sold with a 150% turn off rate.

The bottom 20 households of GM/doe generated the largest loss of IDR 6.275 million when the farmer who had eight goats including one doe and sold no goats. Although this farmer did not generate the largest loss of GM(A-B), the farmer generated the largest loss of GM/doe as the GM(A-B) was divided by one doe. In contrast, the household that had the largest flock of 53 goats including 39 does had the largest ratio of household labourers to goats i.e. 17.7 goats per household had the largest total variable cost of IDR 42.271 million as well as the largest loss of GM(A-B) of IDR 37.435 million. However, this household was the 18th of the top 20 for GM/doe. This indicated that the number of does owned per household largely dictated the efficiency of goat production. The larger the number of does owned by the household, the greater opportunity they had efficient goat production and thus generated higher GM/doe ($P<0.05$).

Other factors that also contributed to these two production parameters had to be considered. Annual equivalent working hours, the labourer to goat ratio or farmer education level of flock size did not significantly ($P>0.05$) affect the GM/doe. This indicated that as long as the number of does owned per household did not generate positive values of GM(A-B) and GM/doe, other factors contributed to generate higher positive or negative values for GM(A-B) and GM/doe. That was as long as the number of does owned per household covered labourer, feed and health control costs, farmers generated positive GM(A-B) and GM/doe.

Goat marketing in Mendoyo District was not well managed by goat farmer associations as they had just re-started rearing goats again and had only sold a few goats. Goat brokers were common in Mendoyo District, and as such, there was a longer supply chain and this potentially reduced the possibility of gaining better selling prices for goat farmers in Mendoyo District. Most of the transactions between goat farmers and buyers occurred on goat farms. Goat buyers were categorised as four types: breeder, meat retailers, satay sellers and occasional buyers. As the goat population reduced in Jembrana Regency, farmers as well as larger satay sellers preferred buying goats from East Java Province that provided stable and lower prices. Smallholder farmers did not have any constraints to marketing their goats. In contrast, they faced constraints to fulfil the opportunities as well as the challenges to supply demand for goats within Bali Province, particularly for Eid Qurban.

3.4 Effects of managerial and environmental factors on production parameters

Although bodyweight is an important economic trait in meat type animals, all households interviewed in Mendoyo District were not sure how to predict the age of their goats and had never weighed their goats even when they sold them. Furthermore, farmers had never recorded the productive nor reproductive parameters of their goats. This is a constraint to rearing goats in Mendoyo District as it means farmers have no data on how well their animals are performing.

3.4.1 Objectives of goat keeping. All households interviewed in Mendoyo District had just re-started rearing goats again after they had experienced low goat production and great losses due to mouth and foot diseases in their goats. Therefore, their main priority of keeping goats was to increase the size of their flocks, as the amount of goat manure produced per household was insufficient to supply their organic fertilizer. This was in agreement with [19] who reported that to have high cacao production, one ha of cacao plantations needed goat manure produced by 25 to 28 goats. This indicated that the smallholder farmers in Mendoyo District have the opportunity to increase crop productivity with more goats; but expansion of their flocks may be constrained by their age and low numbers of children.

3.4.2 Feed and feeding management. Goats in this study were fed with fresh *Caliandra calothrysus*, *Gliricidia sepium*, *Sesbania sesban*, *Erythrina variegata*, *Artocarpus heterophyllus* or other bushes or field grasses at about 5 kg/goat/day regardless of their physiological state. This result was supported by [20] who reported that smallholder farmers in Bali, apart from growing crops as their main production, also practiced traditional silvipastoral systems by growing shrubs and trees and keeping livestock as a sideline.[14] stated that goats were great economic value ruminants for their efficient converters of low-quality forages into quality meat and milk products. This was supported by [21] who reported that feeding goats with various herbaceous roughages provided balanced proportions of nutrients in the goat's diet as they contained natural antioxidant rich diets that have become especially important for their growth, survival, maintenance and health status. This feeding management resulted in comparable bodyweights for all physiological states of goats reared in Mendoyo District (Table 1).

Smallholder farmers in this study also fed their goats with unfermented chopped cacao pods of about 100 g/goat/day. This was in agreement with [22] who suggested that goats be fed with un-processed cacao pods at a maximum amount of 100 g/goat/day, that increased daily weight gain of 70 g/goat/day. This was due to the un-processed cacao pods containing low protein, relatively anti nutritional compounds such as lignin, tannin and theobromine that limited the efficiency of its digestion [23]; [24] and [25].

Utilizing treated cacao pods or coffee pulp from their agriculture wastes, as goats feed that was mixed with cut and carry feeding, might be the best choice because of their limited number of household labourers or their old age. [8] reported that cut and carry feeding systems, such as looking for roughage, smallholder farmers in Malang Regency needed 36% of their daily labourer, while transporting the roughage, mixing the roughage with other feedstuffs and feeding the goats required about 12%, 5% and 21%, respectively. Cleaning goat housing and rearing kids, the farmers needed about 1% and 25% of daily labourer hour, respectively. Since feeding management was the most time consuming and costly for goat farming [26], by treating the cacao pods and coffee pulp, the farmers could minimise problems. Cacao pod contains crude protein from 6.8 to 13.8%; NDF from 55.3 to 73.9% and ADF from 38.31 to 58.98% was a good fibre source and could replace grass for goats [25]. Fermented cacao pods by using *P. chrysosporium* increased crude protein up to 13.8% from 8.7% of untreated cacao pods thus increased daily gain of goats from 102 g/goat/day [23] and 100 to 125 g/goat/day as well as milk production from 1.00 to 1.25 litre/goat/day [24]. Utilizing these agricultural wastes as goats feed rather than leaving them as mulch also helped to conserve the environment of Mendoyo District.

3.4.3 Health and disease control management. Although the farmers studied in Mendoyo District never injected their goats with Klosan200™ (an anthelmintic) or any other medications to control diseases or parasites, the incidence of anaemia as measured by the FAMACHA© technique in all classes of goats was nil. This was probably due to the housing management system being well maintained. Furthermore, the smallholder farmers in Mendoyo District had just received the package of Simantri Programme including the goat housings that apparently new.[27] suggested to house goats in individual battery housing systems for hygienic reasons as it was easy to clean the houses as well as to collect the goat manure. In addition, smallholder farmers had paid attention to the cutting height as well as the cutting time of roughages fed to their goats.

However, the mortality rate in the present study i.e. 24% of newly born kids was considered high as six kids died when they were born as twins and triplets. This was probably due to the kids born having low birth weights. Furthermore, most of the goats were housed quite a distance from farmers' houses, so the goat farmers paid less attention when the low birth weights newly born kids needed help to ensure they received sufficient quantities of colostrum as well as does' milk. This result was confirmed by [28] who reported that low birth weights of kids who needed help with suckling, does having little or no milk, and does abandoning their kids were the major contributors to high mortality of newly born kids. The distance to the goat housing also increased the occurrence of metabolic disturbances, toxicity, bloat and scabies in preweaned kids. High kid mortality as well as the distance to the goat housing were constraints to improving goat rearing in Mendoyo District.

4. Conclusion

In conclusion, smallholder farmers in Mendoyo District reared goats in small flocks as the farmers had just re-started rearing goats again resulting in inefficient goat production, particularly when labour costs were included.

Animal and Human Ethics Approvals

The University of Queensland Animal and Human Experimentation Ethics Committee, in agreement with the Australian National Health and Medical Research Council guidelines under approval numbers SAFS/A52/13 and SAFS/H13/19, respectively, approved the research conducted in this study.

Acknowledgements

The authors gratefully acknowledge support by The University of Queensland, Queensland, Australia for the research fund and the generosity of all goat farmers in Mendoyo District, Jembrana Regency, Bali Province. L. Doloksaribu also acknowledges support by the Directorate General of Higher Education and for her DIKTI scholarship.

References

[1] BPS-Bali., "Bali in Figures. Statistik Daerah Provinsi Bali," p. 70, 2015.

[2] Anonymous., *Promulgated Areas for Livestock in Jembrana and Tabanan Regencies, Bali Provinnce. Keputusan Menteri Pertanian RI Nomor: 43/kpts/PD.410/1/2015*. Indonesia: http://ditjennak.pertanian.go.id, 2015, pp. 1–6.

[3] Anonymous., "The role and improvement of livestock as the President Aid in Bali Province. Departement of Livestock and Veterinary of Bali Province, Denpasar," Denpasar, 1990.

[4] D. A. A. Elisabeth, "The Impact of Primatani Program in Accelerating Transfer of Agricultural Technology for Farmers and Local Government," *J. Ilm. Adm. Publik*, vol. 13, no. 2, pp. 278–90, 2012.

[5] T. H. Kunz, C. Wemmer, and V. Hayssen, "Sex, age, and reproductive condition of mammals," in *Smithsonian Institution Press*, Wilson, R. Cole, Nichols, and Foster, Eds. Smithsonian Press, 1996, pp. 279–90.

[6] L. P. Artini, P. Nilan, and S. Threadgold, "Young indonesian cruise workers, symbolic violence and international class relations," *Asian Soc. Sci.*, vol. 7, no. 6, pp. 3–14, 2011.

[7] P. Nilan and L. P. Artini, "Motivasi , Pengalaman , dan Harapan Kaum Muda Bali Bekerja di Kapal Pesiar," *J. Stud. Pemuda*, vol. 2, no. 1, pp. 75–89, 2013.

[8] B. Hartono, M. B. Hariyono, and F. Rochman, "Goat Raising as Source of the Household's Farming Income and Absorption of the family's labour alternative: Study at Tamansari Ampelgading Malang Regency East Java," *J. Indon. Trop. Anim. Agri.*, vol. 31, no. 2, pp. 99–104, 2006.

[9] A. Winter-Nation, "Robert McC. Netting. Smallholders, Householders: Farm Families and the Ecology of Intensive, Sustainable Agriculture. Stanford CA: Stanford University Press, 1993. xxii + 389 pp. Bibliography. Figures. Index. Photographs. Tables. $49.50. Cloth. $16.95. Pa," *Afr. Stud. Rev.*, vol. 38, no. 2, pp. 152–154, 1995.

[10] T. Shindina, Y. Lysenko, and N. Orlova, "Entrepreneurs' Training in Innovation-Oriented Society," *Procedia - Soc. Behav. Sci.*, vol. 214, no. June, pp. 1098–1108, 2015.

[11] N. Bayyurt and S. Yılmaz, "The Impacts of Governance and Education on Agricultural Efficiency: An International Analysis," *Procedia - Soc. Behav. Sci.*, vol. 58, no. 2012, pp. 1158–65, 2012.

[12] S. F. Souza *et al.*, "Comparison of growth curves in male and female goat kids of Saanen and of 7/8 Boer breeding during the suckling phase. Comparacdo das curvas de crescimento em caprinos jovens, machos e femeas, das racas Saanen e 7/8 Boer durante a fase de aleitamento," *Asoc. Latinoam. Prod. Anim.*, vol. 18, no. 1–2, pp. 27–32, 2010.

[13] L. F. D. Medeiros *et al.*, "Estudo Do Crescimento De Cabritos Mestiços Na Região Metropolitana No Estado Do Rio De Janeiro *," vol. 34, no. 1, pp. 35–46, 2012.

[14] C. Devendra and M. Burns, *Goat production in the tropics*. Commonwealth Bur. Animal Breeding Genet., 1970.

[15] C. Devendra and G. F. W. Haenlein, "Animals that Produce Dairy Foods | Goat Breeds," *Encycl. Dairy Sci.*, pp. 310–24, Jan. 2011.

[16] G. Abebe, G. Kannan, and A. L. Goetsch, "Effects of small ruminant species and origin (highland and lowland) and length of rest and feeding period on harvest measurements in Ethiopia," *African J. Agric. Res.*, vol. 5, no. 9, pp. 834–47, 2010.

[17] SNI., *Standar Nasional Indonesia-SNI 7325:2008 Bibit kambing Peranakan Etawah (PE).* Indonesia, 2008, pp. 1–11.

[18] P. Morand-Fehr *et al.*, "Strategy for goat farming in the 21st century," *Small Rumin. Res.*, vol. 51, no. 2, pp. 175–83, 2004.

[19] S. Guntoro, *Meramu Pakan Ternak dari Limbah Perkebunan*, 1st ed. Jakarta: Agro Media Pustaka, 2012.

[20] I. M. Nitis, "Silvipastoral Systems in Tropical Context," in *XVIII Intern. Grassland Congress 2000*, 1997, pp. 129–38.

[21] P. F. Surai, *Natural Antioxidants in Avian Nutrition and Reproduction*, 1st ed., vol. 1. Nottingham: Nottingham University Press, 2003.

[22] S. Prawirodigdo, T. Herawati, and B. Utomo, "Utilization of Coffee Pulp and Hull in the Diet for Sheep Fattening," in *Seminar Nasional Teknologi Peternakan dan Veteriner*, 2005, pp. 438–44.

[23] Suparjo., K. G. Wiryawan, E. B. Laconi, and D. Mangunwidjaja, "Goat Performance of given Fermented Cacao (Theobroma cacao). Performa Kambing yang Diberi Kulit Buah Kakao Terfermentasi," *Media Peternak.*, vol. 34, no. 1, pp. 35–41, 2011.

[24] J. Riyanto and C. Anam, "Integration of Etawah Crossbred Dairy Goat with Cocoa in East Java Province , Indonesia," 2012, no. April, pp. 193–5.

[25] Wisri P dan Susana IWR, "Potensi dan Pemanfaatan Kulit Buah Kakao sebagai Pakan Alternatif Ternak Ruminansia," *Wartazoa*, vol. 24, no. 3, pp. 151–9, 2014.

[26] FAOSTAT., *FAO Statistical Pocketbook 2015*. Rome, 2015.

[27] M. Martawidjaja, "Barn Design for Small-ruminant Production," in *New Technologies for Small Ruminant Production in Indonesia. A publication of the Small Ruminant-Collaborative Research Support Program and Research Institute for Animal Production Central Research and Development Indonesian Ministry of Agriculture*, no. September 1992, P. Ludgate and S. Scholz, Eds. Morrilton, Arkansas: Winrock International Institute for Agricultural Development, 1992, pp. 15–18.

[28] M. A. Snyman, "Factors affecting pre-weaning kid mortality in South African Angora goats," *South African J. Anim. Sci.*, vol. 40, no. 1, pp. 54–64, 2010.

Body dimensions of primiparous Bali cattle (*Bibos sondaicus*) supplemented with concentrate at their first three month of pregnancy

A A Oka, N N Suryani, M Dewantari, N P Sarini and N P Mariani

Faculty of Animal and Husbandry, Udayana University, Jimbaran Campus
Badung-Bali, Indonesia

E-mail: anakagungoka1961@gmail.com

Abstract. This study was undertaken to compare body dimensions between primiparous Bali cattle that were supplemented with concentrate or no concentrate at their first three month of pregnancy reared by smallholder farmers in Petang, Badung, Bali Province. Twenty primiparous Bali cattle that were fed with forage divided into two groups i.e. supplemented with concentrate (Rwc) and no concentrate (Rnc) from May to July 2018 using Completely Randomized Design. Body weight, body length, chest circumference, height at withers and rump width were measured every fortnight. Results showed that ADG of primiparous Bali cattle (Rwc) were significantly heavier ($P<0.01$) being 0.47±0.02 kg compared to 0.19±0.02 kg of primiparous Bali cattle (Rnc). Chest circumference, body length and rump width of those Rwc were also significantly higher ($P<0.01$) being 153.3±0.5 cm, 112.1±0.3 cm and 45.2±0.2 cm, respectively compared to those Rnc being 151.2±0.6 cm, 108.9±0.3 cm and 44.0±0.2 cm, respectively. High correlations ($P<0.01$) were found between chest circumference and body length of primiparous Bali cattle (Rwc), $R^2=0.911$, between chest circumference and body weight $R^2=0.648$ and between body length and body weight $R^2=0.591$. In summary, it is strongly suggested to feed concentrate particularly at the first three month of pregnancy of primiparous Bali cattle.

1. Introduction.

Bali cattle are native Indonesian germplasm that are regional and national superior assets having high genetic and economic potential to be developed as beef cattle. Although Bali cattle rearing played a very important role in the socioeconomic and social life of many Balinese smallholder farmers, they had them as side job with small flock size about 1-3 cattle per household (Suarna, I. W. 2018, pers. comm. 4 September). One characteristic of Balinese smallholder cattle farmers is that feeding cattle management is generally relied on roughage-based with cut and carry system. Roughage alone particularly field grass could not meet the livestock's requirement for energy or protein particularly in the early pregnancy of primiparous cattle. Feeding commercial concentrate to cattle is rare among smallholder cattle farmers in Bali. Smallholder cattle farmers sometimes fed their cattle with rice bran with small quantity resulted in various daily gain weights. Bali cattle reared by smallholder famers showed diverse responses to concentrate supplementation when cattle fed ration roughage based. Bali cattle gained bodyweight ranged from 0.13 to 0.24 kg/head/day when they were fed only with field grass. However, when they were fed with roughage and concentrate with 4.578 kcal gross energy and 16% crude protein contents they gained 0.76 kg/head/day [1]. Feeding adequate and balanced nutrients to Bali cattle based on their requirement is important to achieve their standard growth rate [2]. High genetic potential of livestock could be achieved through feeding management that met their requirements resulted in high productivity i.e. favorable growth rate and reproductive parameters.

Primiparous cattle are livestock of the prospective parents of the first time mated where they have had puberty so that they required special handling for preparing them as parent stocks. Pregnant cattle required sufficient nutrients so that their pregnancy process went safely until parturition. Bali cattle showed a diverse response to the improvement of the nutritional value of feed. Improvement of feeding management of early primiparous Bali cattle is paramount as they needed quality and balanced nutrients for their physiological status. The flow and distribution of nutrients resulted from metabolism of early pregnancy cattle is complex [3].

The study was undertaken to compare body dimensions between primiparous Bali cattle that were supplemented with and no concentrate at their first three month of pregnancy reared by smallholder farmers in Petang District, Badung Regency, Bali.

2. Materials and Methods.

Twenty primiparous Bali cattle aged between 18 – 24 months with average early bodyweight 202 - 247 kg reared by smallholder farmers in Belok Sidan Village, Petang District, Badung Regency, Bali Province were used in this study. The twenty primiparous Bali cattle that were fed with forage i.e. *Caliandra calothrysus*, *Sesbania sesban*, *Gliricidia sepium*, *Erythrina variegata*, *Pennisetum purpureum* and native grass divided into two groups. A group of 10 primiparous had concentrate supplementation (Rwc) i.e. 2 kg rice bran per head per day and another group of 10 primiparous had no concentrate supplementation (Rnc) during the 12 weeks of observation from May until July 2018 using Completely Randomized Design. Feeding of 2 kg of concentrate i.e. rice pollard and rice Weighing and measuring body dimension were done every fortnight. Body dimension measuring included body weight (kg), body length (cm), chest circumference (cm), height at withers (cm) and rump width (cm). Bodyweight (BW, kg) were measured in the morning before feeding cattle by using an Aflek© scale which had an accuracy to 1 kg. Body length (BL, cm) referred to the distance from the base of the ear to the base of the tail (where it joins the body). Extreme care was taken to ensure that the backbone was straight in both vertical and horizontal planes. All the measurements of body length were taken to the nearest 0.5 cm. Chest circumference (CC, cm) or hearth girth is a circumferential measurement taken around the chest just behind the front legs and withers. Chest depth (CD, cm) was measured from the backbone at the shoulder (standardized on one of the vertical processes of the thoracic vertebrate) to the brisket between the front legs. Height at withers (HW, cm) was the distance from the surface of a platform on which the animal stands to the withers. The measurement was made with a special measuring stick made with two arms one which was held vertical and the other at right angles to it sliding firmly up and down to record height. Rump height (RH, cm) is the distance from the surface of a platform on which the animal is standing to the rump using a measuring stick as described for height at withers. All measurement method was based on [4]. Data was analysed with T-test [5].

3. Results and Discussion.

Results showed that primiparous Bali cattle supplemented with concentrate 2 kg per head per day had during 12 weeks significantly larger chest circumference, body length, rump width and daily gain weight ($P<0.01$) compared to those of primiparous with no concentrate supplementation (Table 1). Height at withers and bodyweights of those primiparous that supplemented with concentrate also tended to be higher ($P>0.05$) than of the primiparous with no concentrate supplementation. Daily gain weight of primiparous supplemented concentrate gained 0.47 ± 0.2 kg was significantly heavier than 0.19 ± 0.02 kg of primiparous with no supplementation. This indicated that all the 20 primiparous Bali cattle still grew and they required energy and protein of nutrient to support their growth rate as well as the growth of foetus and also produce milk to their calves. [6] reported that an early pregnancy of beef cattle required dry matter intake (DMI) 5 kg, crude protein (CP) 10.2% or 0.55 kg, TDN 62% or 3.05 kg and Ca. 0.36% and P 0.2% to be able to gain about 500 g/head/day in early pregnancy beef cattle. This result was confirmed by [1] who reported that Bali cattle gained about 0.13 - 0.24 kg/head/day by feeding field grass. However, research conducted on Bali steer by [1], the steer gained 0.76 kg/head/day that was triple up to 5 times heavier that of steers with no concentrate supplementation. Furthermore, [7] reported that by increasing energy ration until 2300 kcal ME/kg significantly ($P<0.05$) increase energy intake and resulted in the peak of milk yield of 2179.83 ml/day with milk fat and lactose 8.56% and 4,76 %, respectively.

Table 1. Comparison between body dimensions between primiparous Bali cattle supplemented with and no concentrate (control)

Parameter	Control	Concentrate	Paired Samples Test		Correlation	
	Mean ± SEM	Mean ± SEM	Mean ± SEM	P	R^2	P
Chest circumference, cm	151.2 ± 0.6	153.3 ± 0.5	2.08 ± 0.59	0.001**	0.356	0.005
Body length, cm	108.9 ± 0.3	112.1 ± 0.3	3.18 ± 0.44	0.000**	-0.199	0.127
Rump width, cm	44.0 ± 0.2	45.2 ± 0.2	1.20 ± 0.27	0.000**	-0.148	0.260
Daily gain weight, kg	0.19 ± 0.02	0.47 ± 0.2	0.27 ± 0.03	0.000**	0.109	0.451
Height at withers, cm	112.5 ± 0.3	112.6 ± 0.3	0.18 ± 0.48	0.702	-0.362	0.004
Rump height, cm	113.9 ± 0.3	113.8 ± 0.2	-0.12 ± 0.31	0.701	0.148	0.263
Bodyweight, kg	239.0 ± 2.1	240.4 ± 2.3	1.450 ± 0.92	0.621	0.104	0.428

**. Correlation is significant at the 0.01 level (2-tailed).
*. Correlation is significant at the 0.05 level (2-tailed).

Table 2. Correlation relationship coefficient (R^2) of body dimension of primiparous Bali cattle supplemented with concentrate

Correlation			
Parameter 1	Parameter 2	R^2	P
Chest circumference, cm	Body length, cm	0.911**	0.000
Chest circumference, cm	Bodyweight, kg	0.648**	0.000
Chest circumference, cm	Rump height, cm	0.384**	0.002
Body length, cm	Height at withers, cm	0.363**	0.004
Chest circumference, cm	Rump width, cm	0.318*	0.013
Chest circumference, cm	Height at withers, cm	0.179	0.172

**. Correlation is significant at the 0.01 level (2-tailed).
*. Correlation is significant at the 0.05 level (2-tailed).

There are closed relationship between chest circumference and body length ($R^2 = 0.911$ and $P <0.01$) (Table 2). This indicated that primiparous Bali cattle supplemented with concentrate the growth of chest circumference was parallel with the growth of body length. When the chest circumference grew then the body length grew as well. The correlation between body dimensions was positive but they were not as strong as the correlation between chest circumferences with body length. The growth of Bali cattle particularly the circular body dimension started from chest circumference then it followed by abdomen circumference, anterior neck circumference and posterior neck circumference was the latest growth among the circumference dimension [8].

Measuring of body dimension i.e. body length, chest circumference, rump width were significantly higher ($P<0.01$) of those primiparous Bali cattle with concentrate supplementation than those of primiparous with no concentrate supplementation. This result was confirmed by [9] who reported that the condition of early pregnancy of primiparous, their body dimensions particularly chest circumference, and rump width were growing due to the growing of foetus as well as secretory cells to produce milk [7]. This indicated that the present study improved the growth rate of the primiparous Bali cattle. This result was supported by [6] that the present study provided sufficient energy and protein to primiparous cattle.

4. Conclusion

In summary, average daily gain weight of primiparous Bali cattle supplemented with 2 kg concentrate per head per day were significantly heavier (P<0.01) i.e. 0.47 kg/head/day than of those primiparous with no concentrate supplementation. It is strongly suggested to feed concentrate particularly at the first three month of pregnancy of primiparous Bali cattle.

Acknowledgements.

The authors gratefully acknowledge support by The Institute for Research and Community Service (LPPM) and Faculty of Animal and Husbandry, The University of Udayana, Bali-Indonesia for the research fund and the generosity of all Bali cattle farmers in Petang District, Badung Regency, Bali Province.

References

[1] A. A. Oka, I. N. T. Ariana, N. L. P. Sriyani, M. Dewantari and N. P. Sarini. Upaya Meningkatkan Produktivitas Sapi Bali Melalui Manipulasi Teknologi Pemberian Pakan Berbasis Hijauan. PROSIDING SEMINAR NASIONAL V HITPI, ISBN : 978-979-3660-42-4, pp:71-76, 2016

[2] Mariyono and E. Romajali. Petunjuk Teknis Teknologi Inovasi Pakan Murah Untuk Usaha Pembibitan Sapi Potong, Grati. Pp: vii + 29 h, 2007.

[3] Bali Cattle Research Center. Sapi Bali, *Sumberdaya Genetik Asli Indonesia*. Udayana University Press. Denpasar-Bali, 2012.

[4] I. B. Djagra, I. G. N Raka Haryana, I. G. M. Putra, I. B. Mantra dan A. A. Oka. *Ukuran Standar Tubuh Sapi Bali Bibit*. Laporan Hasil Penelitian Kejasama Bappeda Bali dan Fakultas Peternakan. Denpasar Bali, 2002.

[5] R. G. D. Steel and J. H. Torrie. *Prinsip dan Prosedur Statistika. Suatu Pendekatan Biometrik*. PT. Gramedia Jakarta, 1989.

[6] NRC. Nutrient Requirements for Beef Cattle. West Virginia University, 2013.

[7] N.N. Suryani, I. W. Suarna, N. P. Sarini, I. G. Mahardika, and M. A. P. Duarsa. Pemberian Ransum Berenergi Tinggi Memperbaiki Performans Induk dan Menambah Bobot Lahir Pedet Sapi Bali. Jurnal Veteriner. **18** No. 1. Pp 154-159, 2017.

[8] I. P. Sampurna and I. K. Suatha. 'The allometric growth of long and circular body dimension of Bali cattle', *Jurnal Veteriner*, **l. 11**, no. 1, pp. 46-51, 2010.

[9] I. B. Djagra. Pertumbuhan Sapi Bali: Sebuah Analisis Berdasarkan Dimensi Tubuh. *Majalah Ilmiah Peternakan*. Universitas Udayana, 1992.

Increasing antioxidant activity of quail (*cortunix-cortunix japonica*) eggs with the addition of sweet flag (*acorus calamus*) powder as a feed additive

Y F Nuningtyas[1] and E Widodo[1]

[1]Faculty of Animal Science, University Brawijaya, Malang 65145, Indonesia

E-mail: *fritanuningtyas@ub.ac.id*

Abstract. The aim of this research was to evaluate the addition of sweet flag powder on antioxidant activity and cholesterol activity of quails egg. Sweet flag (*Acorus calamus*) powder was analyzed using the DPPH and cholesterol activity. One hundred sixty (160) quails were divided into 5 treatments in which each treatment had 4 replications with 8 quails per replications. Quails were fed the experimental diets, namely basal diet, 0,1%, 0,2%, 0,3%, and 0,4% of sweet flag powder. Eggs were collected daily in every treatment. The analyze of egg were conducted starting from days 45 of experiments. The data was analyzed using Completely Randomized Design (RAL). The different level was analyzed using Duncan's Multiple Range Test. The addition of sweet flag powder as feed additive has a significant effect on the antioxidant activity of quails egg. The addition of 0,4 % of sweet flag powder improves antioxidant activity of quails egg. The addition of 0.4 % of sweet flag powder decreased cholesterol of quails egg yolk. The conclusion of this research that the addition of sweet flag powder as a feed additive up to 0,4 % had an effect on the antioxidant activity of quail eggs and decreased cholesterol of quails egg yolk.

1. Introduction

In Indonesia, the used of antibiotics as a growth promoter on feed was not allowed by the Government since 1st January 2017. Recently, researchers were looking for the natural bioactive compound as an alternative to changed antibiotic[5]. Antibiotic was used as a growth promoter in relatively small amount as a feed additive but improved feed efficiency and reproduction of livestock, therefore the used of the additive can increase the profit of the farmers. However, the accumulation residue from antibiotic will have a negative effect on the livestock product. The used of antibiotic should be restricted to prevent the problem of increasing resistance in bacteria of human and animal, especially resistance in gram-negative bacteria. Many natural herbs have a beneficial effect on the animal performance through as antioxidant activity [13]. The used of phytobiotic as a natural growth promotor (NGPs) has been identified as an effective alternative to changed antibiotic[8]. Phytobiotic as NGPs was developed as a feed additive, which is antimicrobial, antioxidant, to increase performance, immunity, and also improves a digestive tract of the poultry. Phytobiotics are found into various herbal plants, such as curcumin, ginger, garlic and the other rhizomes of plants[10]. Herbal plants have no chemical compounds where it means has no harmful effect on the health. In hence, the addition of herbal plants as a feed has a good effect on animals, as well as livestock production[21].

Quail is a type of bird that could not fly; the body size is relatively small; short-legged and can be pitted. The benefits of quail can be taken from eggs and meat. Eggs consist of four components namely egg yolks, egg whites, eggshell membranes and eggshells [16]. Male quail has a body weight of about 100-140 g while the female is slightly heavier which is between 120-160 g. Female quails will start laying eggs at the age of 35-42 days [1]. Quail meat contains 21.10% protein, while low-fat content is 7.70% [14].

Egg yolk of quails have high levels of fat and cholesterol, it can cause lowest consumers consumption, so an alternative product of quail egg yolks is low in cholesterol. To reduce cholesterol levels in egg yolks can be done by manipulating feed by adding feed additive consist of the natural bioactive compound.

Phytogenic is the result of plant secondary metabolites that contain nutritionally valuable, non-nutritious, or anti-nutritive compounds [8]. Phytogenic is categoryzed Generally Recognized as Safe (GRAS) relatively secure for human and animal. Phytogenic used includes herbal plants, spices, or extracts of active components contained in plants. The active component has been known to have anti-microbial activity, antifungal, and antioxidant activity [21]. Generally in a plant contains one or several active molecules [3]. At present, many studies have been carried out using plants or their extracts as feed additives to livestock because compared to synthetic antibiotics, phytogenic has proven to be more natural, free of residues, and more ideal to be used as a trigger for animal growth [7]. Some of the active phytogenic components include essential oils, flavonoids, saponins, and tannins as contained in the sweet flag (*Acorus calamus*).

Sweet flag (*Acorus calamus*) is a perennial wetland monocot plant from the *Acoraceae* family. The plant has been investigated has a number of chemical constituents from the rhizomes, leave and roots. The bioactive compound was identified on the *Acorus calamus* contain aromatic oil that used as a traditional medicine consist of *eugenol, asarilaldehida*, asarone, *kalameon, kalamidiol, isokalamendiol, akorenin, akonin, akroagermakron, akolamonin, isokolamin, sioburin, isosiobunin, epi-subunit, resin, amylum* and also *tannin* [9]. The major bioactive compound identified on the sweet flag are saponin and flavonoid [9], α- and β-asarones caryophyllene, isoasarone, methyl isoeugenol, and safrol. Hence, the used of herbal plants expected to provide health effect on livestock and also in the product. The purpose of this experiment was to evaluate the effect of the used sweet flag (*Acorus calamus*) powder as a feed additive on antioxidant activity and cholesterol of quails egg.

2. Materials and Methods

An experiment about the addition of sweet flag (*Acorus calamus*) powder as a feed additive on antioxidant and cholesterol of egg was conducted by two steps. The first step of the study was the test of the bioactive content of sweet flag powder. This test was carried out at the Airlangga University Laboratory to determine the flavonoid content. The variables observed were flavonoids and antioxidant activity content of the sweet flag.

The second step experiment was conducted at the Poultry Research Farm, Malang, Indonesia. One hundred sixty (160) quails 45 days old were divided into 5 treatments, in which treatment had 5 replication of 8 quails per replication. Five treatments consist T0 as a control (basal diet), T1 (basal diet + 0,1% of sweet flag powder), (basal diet + 0,2 % of sweet flag powder), (basal diet + 0,3% of sweet flag powder), and (basal diet + 0,4% of sweet flag powder). Quail is placed in 20 units of experimental cages with a size of (60 x 40 x 20) cm with a feeding and drinking area. The battery cage is placed in a cage room, which is equipped with two lighting.

2.1. Cholesterol analysis

Analysis of the chemical quality of quail eggs consists of egg yolk cholesterol content (mg / 100gr) using the *Libermann Burchard Color Reaction* method [11], where the procedure was egg yolk weighed 0.2 gram, then added 1 ml of KOH alcohol stirred until precipitate occurs. Then it was in a heater at 39 °C - 40 °C for 1 hour and added 2 ml of petroleum ether 40 °C -60 °C. After that 0.25 mL H2O was added and stirred for 1 minute and drying using the oven at room temperature. As much as 4 ml of acetate acid solution was added and stirred, then allowed for 35 minutes and read at 630 nm wavelengths with a gap of 0.5 nm using a spectrophotometer.

2.2. Antioxidant analysis

Preparing the solution of each sample by 100 ppm by dissolving 10 mg extract on 100 ml methanol and prepare a 50 ppm DPPH solution. DPPH stock solution was made by dissolving 5 mg DPPH solids into 100 ml of methanol. Then a comparison solution is prepared, which is a control solution containing 2 ml of methanol and 1 ml of 50 ppm DPPH solution. For the sample test, 2 ml sample solution and 2 ml solution were prepared for DPPH analysis. Then the solution incubated for 30 minutes at a 27 °C until there was a color changed from DPPH activity. All the samples were made Duplo. Analysis of the antioxidant activity of the DPPH method was done by looking at the respective color changes sample after incubation with DPPH. When the DPPH electrons pair with electrons in the sample extract will change the color of the sample starting from dark purple to bright yellow. Then sample absorbance value was measured using spectrophotometer at a wavelength of 517 nm.

Feed content of the basal diet was shown in Table 1. Data was analyzed using Completely Randomized Design (RAL). The differential of the data between treatment was to the analysis of variance followed by Duncan's Multiple Range Test.

Table 1. Feed content of basal diet on laying quails.

Composition	Nutrient (%)
Crude Protein	20,32
Crude Fat	3,92
Crude Fiber	3,50
Ash	10,06
Calsium	2,62
Sodium	0,16
Phosphor	0,67

Proximate analysis PT. JAPFA Comfeed Indonesia

3. Results and Discussion

Antioxidants are compounds that can inhibit reactive oxygen species or reactive nitrogen species (ROS / RNS) and free radicals so that antioxidants can prevent diseases associated with free radicals such as carcinogenesis, cardiovascular and aging [6]. Several epidemiological studies have shown that increased consumption of natural phenolic antioxidants found in plants and their products has great health benefits [4]. This is due to the presence of several vitamins (A, C, E, and folate), fiber, and other chemical substances such as flavonoids and polyphenols that are capable of binding free radicals. Analysis of bioactive compounds and antioxidant activity were taken to determine the chemical content of a substance. The raw material used in this study was sweet flag (*Acorus calamus*) powder where the results of the analysis can be seen in Table below.

Tabel 2. Bioactive compound and antioxidant activity of sweet flag powder.

Plants	Flavonoid (% b/b)	IC_{50} (mg/ml)
Sweet flag	33,76	0,22

Flavonoids were found in all parts of the plant including fruit, pollen, and roots. Flavonoids contribute to color in plant organs such as flowers, fruit, leaves, or color in pigments. Flavonoids are useful in attracting insects and other animals to help pollinate and spread seeds [18]. In the flavonoid test of the sweet flag with spectrophotometry method has a positive result about 33.76% b/b. Phenolic compounds consist of flavonoids and alkaloids have the potential as antioxidants which are polar compounds [19]. These polar compounds will be extracted in the methanol extract fraction because of the polar solvent methanol. The ethyl acetate extract fraction contained moderate polarity compounds and it was also possible to extract a small portion of polar compounds which had antioxidant activity.

Determination of IC50 values was carried out to determine the effectiveness of sweet flag powder to prevent 50% free radicals [10]. Based on the results of these calculations, it was obtained information that the sweet flag has 0.22 mg/ml was able to prevent free radicals by 50% in samples. The smaller IC50 value shows the higher antioxidant ability in counteracting free radicals [10]. The antioxidant activity of natural ingredients can be influenced by the levels of flavonoids [20]. Flavonoid compounds consist of phenol groups that are more complex with a higher degree of hydroxylation. The presence of hydroxyl groups in flavonoid compounds causes antioxidant activity. It can be caused by oxygen atoms in the hydroxyl group having free electron pairs that are sufficient to inhibit the reactivity of reactive atoms which make up free radical compounds [2].

The addition of sweet flag powder (*Acorus calamus*) as a feed additive on quail was shown in Table below:

Table 3. Effect of the addition sweet flag (*Acorus calamus*) powder on antioxidant activity and cholesterol on quails egg.

Treatment	Antioxidant Activity (IC50)	Cholesterol (mg/100 gram sample)
T0	199.17^{d}	604.25^{e}
T1	194.26^{c}	587.99^{d}
T2	191.67^{b}	549.76^{c}
T3	190.55^{b}	524.96^{b}
T4	180.09^{a}	511.76^{a}

$^{a\text{-}e}$ Means within a row with the different superscript were significantly different

Based on the data, the addition of sweet flag as a feed additive had a significant effect ($P < 0,01$) on antioxidant activity of quails egg. Antioxidant activity was measured by DPPH. The highest antioxidant activity of quails egg for each treatment shown respectively, T4 (180,09), T3 (190,55), T2 (191,67), T1 (194,26) and T0 (199,17). The results show that the addition of 0,4 % sweet flag had the highest antioxidant activity compared with other treatments. Essential oil of sweet flag had an antioxidant activity to increased quality of feed [12]. The total antioxidant activity of a bioactive compound was influenced by the structure of the test compound and the appropriate method was used for the test. The structure of the bioactive compound analysis was related to the presence of antioxidant groups. High antioxidant activity is obtained in compounds containing ortho-diphenol such as epicatechin, caffeic acid, epigalokatekingalat and others. The presence of phenolic OH groups that can be quickly abstracted by free radicals causes phenolic compounds to have the potential as antioxidants. In addition, antioxidant activity was also influenced by double bonds and carbonyl groups of heterocyclic rings from the core structure which can increase antioxidant activity by stabilizing phenolic radicals through electron conjugation and delocalization (Heim et al, 2002). Flavonoid levels have a positive relationship with antioxidant activity to reach IC50 where the higher the flavonoid content, the higher the IC50. In hence, it can be interpreted that an increased content of polyphenol compounds will increase the concentration value (ppm) of antioxidants to reach IC50, which means that the antioxidant abilities possessed stronger. Antioxidant activity has a correlation with the total components of phenolic compounds [23].

The results of the analysis cholesterol in quails egg with the different treatment had a significant effect ($P < 0.01$). The lowest cholesterol content on quails eggs shown in the treatment 4 with the addition of 0.04 % of sweet flag powder. Cholesterol content was decreased by 2.56 % compared with the basal diet (control). Decreased cholesterol content in quail egg may be due to the high content of flavonoid on the sweet flag powder [5]. Polyphenol compounds such as flavonoids can inhibit oxidation reactions through the mechanism of radical scavenging by donating one electron to unpaired electrons in free radicals so that the amount of free radicals becomes reduced. Flavonoids were strong inhibitors of lipid peroxidation, as binding of reactive oxygen or nitrogen species, also able to inhibit the activity of lipoxygenase and cyclooxygenase enzymes [6]. Alkaloid compounds have the ability to attach themselves between DNAs so that they interfere with DNA replication [22]. The mechanism of action is to interfere with the formation of a cross-crossing bridge of the constituent components of peptidoglycan in bacterial cells so that the cell wall layer is not formed intact and causes cell death [17]. Flavonoids have become a topic of learning and study because from various studies and existing data show that the function and use of flavonoids as anti-microbial, maintain oxidative stability, antioxidant, antimutagenic, anticarcinogenic, and have the ability to modify gene expression [15].

4. Conclusion

The used of sweet flag powder up to 0.4 % had increased the antioxidant activity and decreased cholesterol of quails egg yolk. The addition of sweet flag powder as feed additive has to increase antioxidant activity and decrease cholesterol on quails egg.

5. Acknowledgments

We would like to thank for the Faculty of Animal Science, the University of Brawijaya for the funding experiments.

References

[1] Anggorodi H R 1995 *Nutrisi Aneka Ternak Unggas*; Jakarta Gramedia Pustaka Utama

[2] Egwaikhide P A and C E Gimba 2007 Analysis of the phytochemical content and anti-microbial activity of *plectranthus glandulosis* whole plant *Middle-East J. Sci.Res* **2** 135-8

[3] Frankic T, M Voljc, J Salobir, and V Rezar 2009 Use of herbs and spices and their extracts in animal nutrition. *Acta Agri Slov* ***94*** *95-102*

[4] Ghiselli A, Nardini M, Baldi A and Scaccini C 1998 Antioxidant activity of different phenolics reactions separated from an italian red wine *J. Agric. Food. Chem.* 46 **2** 361-7

[5] Greathead H 2003 Plants and plant extracts for improving animal productivity *Proc. Nutr. Soc.* ***62*** *279-90*

[6] Halliwell B and Gutteridge J M C 2000 *Free Radicals in Biology and Medicine* Oxford University Press Oxford

[7] Hashemi S R, I Zulkifli, M Hair-Bejo, A Farida and M N Somchit 2008 Acute toxicity study and phytochemical screening of selected herbal aqueous extract in broiler chickens *Int J Pharmacol* ***4*** *352-60*

[8] Hashemi S R and Davoodi H 2011 Herbal plants and their derivatives as growth and health promoters in animal nutrition *Vet. Res. Commun.* ***35*** *169-80*

[9] Heim K E, Tagliaferro A R and Bobilya D J, 2002 Flavonoid antioxidant chemistry metabolism and structure-activity relationships *J. Nut. Bio* **13** 572-84

[10] Karthikumar S, Vigneswari K and Jegatheesan K 2007 Screening of antimicrobial and antioxidant activities of leaves of *Eclipta prostrata (L) J. Sci. Res. Ess 2* ***4*** *101-4*

[11] Kleiner I S and Dotti L B 1962 *Laboratory Instructions in Biochemistry, 6th Edition*: New York The CV Mosby Company

[12] Lee K W, Everts H and Beynen A C 2004 Essential oil in broiler nutrition *Int. J. Poult. Sci.* **3** 738-52

[13] Leeson S and Summers J D 2005 *Commercial Poultry Nutrition,3rd edition*: Nottingham University Press

[14] Listiyowati E and K Roospitasari 2000 *Tata Laksana Budi Daya Burung Puyuh Secara Komersial*: Jakarta Penebar Swadaya

[15] Marinova D, Ribarova and M Atanassova 2005 Total phenolics and total flavonoids in bulgarian fruits and vegetables *J. Univ. Chem. Tech. Metallurgy* 40 **3** 255-60.

[16] Roberts J R 2010 Factors affecting egg shell and internal egg quality *18th Annual ASAIM SE Asian Feed Technology and Nutrition Workshop* Cambodia 24-270510
[17] Robinson T 1995 *Kandungan Organik Tumbuhan Tingkat Tinggi*, *Edisi ke-6*: Transleted by K Padmawinata: Bandung Penerbit Institut Teknologi Bandung
[18] Sirait M 2007 *Penuntun Fitokimia Dalam Farmasi* Bandung: Penerbit ITB 158-9
[19] Suratmo 2009 Potensi Ekstrak Daun Sirih Merah (*Piper crocatum*) sebagai Antioksidan, *BSS* 205 **1** 1-5
[20] Takaya Y, Y Kondo, T Furukawa and M Niwa 2003 Antioxidant constituents of radish sprout (Kaiware-daikon) *Raphanus sativus L J. Agric. Food Chem,* **51** 8061-66.
[21] Windisch W, Schedele K, Plitzner C, Kroimayr A 2007 Use of phytogenic products as feed additives for swine and poultry *J. Anim. Sci.* ***86*** *140-8*
[22] Yildirim A, M Oktay and V Bulaloulu 2001 The antioxidant activity of the leaves of *Cydonia vulgaris Turk J Med Sci* ***31*** *23-27*
[23] Zheng W dan S Y Wang 2001 Antioxidant activity and phenolic compounds in selected herbs *J. Agri.c Food. Chem.* 49 **11** 51-65

Effects of Mount Agung eruption on botanical composition and nutritive value of ration fed and rumen performance of Bali cattle in evacuation zones

N N Suryani, I W Suarna and I G Mahardika

Faculty of Animal and Husbandry, Udayana University, Jimbaran Campus Badung-Bali, Indonesia

E-mail: mansuryanifapet@unud.ac.id

Abstract. Mount Agung eruption, August to December 2017, that is located in Karangasem Regency, Bali Province, caused Bali cattle to be evacuated. This study was undertaken to compare botanical composition and nutritive value of forage fed to Bali cattle and rumen performance before and after eruption in evacuation zones i.e. Sidemen and Nongan Villages. The first phase was a survey conducted in December 2017 on farmers to collect information on botanical composition fed before and after the eruption. The second phase was laboratory forage sample analysis from January to May 2018 for their nutritive value and rumen performance i.e. pH, NH_3 and total VFA, Digestible Organic Matter in Rumen (DOMR) and microbial protein synthesis (MPS). Data was analysed using Anova. Results showed that there was more composition of forage fed before eruption in both evacuation zones compared to those of after the eruption. In Nongan evacuation zone, DOMR was higher $4161.81^{b} \pm 188.18$ g/head/day vs. $3318.36^{a2)} \pm 236.89$ g/head/day but lower in Sidemen evacuation zones $4403.53^{a2)} \pm 143.81$ g/head/day vs. $4818.17^{a} \pm 279.77$ g/head/day. In summary, there was more botanical composition of forage fed to Bali cattle.

1. Introduction.

Bali cattle as one of the Indonesian beef cattle that plays important role in national meat supply. Not only have to pay attention on genetic potential of Bali cattle, but also have to pay attention on their nutrient intake to be able to improve their breeding stock. The improvement of Bali cattle have strategy goals such as to conserve and to improve the Bali germplasm. Furthermore, [1] urged that its impact-oriented benefits could directly sustain pro-poor initiatives to reduce poverty and hunger that are consistent with income growth, socioeconomic benefits, improved livelihoods and self-reliance.

The availability of forage through years plays important role in ruminant farming system. Forage provided to livestock consisted of native grass and legume that took over 40% to 80% from the total DM feed or about 1.5 to 3% from bodyweight of livestock. Forage is source of crude fiber, mineral and protein that supply the nutrient requirement of livestock [2]. The eruption of Mount Agung affected the availability of forage due to its ash contaminated the forage thus the growth of forage in Sidemen and Nongan conservation zones. As a result, the availability and the quality of forage provided to their livestock around the Mount Agung zone decreased. Furthermore, villagers and their livestock who lived around the Mount Agung zone had to evacuate to Sidemen and Nongan villages that were about 20 to 23 km from the Mount Agung zone. Therefore, smallholder farmers had to cut and carry forage around the both villages to provide to their livestock as well as donated concentrate by donators. As the villagers and their livestock stayed over three months in both evacuation zones, the availability of forage became crucial problem whether to keep or to sell their livestock with low prices. There more livestock were sold with low prices during the evacuation period in both zones that influenced the number of Bali cattle thus influenced the meat supply from Karangasem Regency. Karangasem Regency has the highest number of Bali cattle and was officially promulgated a designation regency for improvement of goat farming by the Indonesian Minister of Agriculture [3].

One of important limiting factors for animal performance is the daily amount of absorbable amino acids (AA) in their small intestine [4]. The amino acids reaching small intestine are supplied by microbial protein, the un-degraded feed protein, amino acids and peptides from feed which escape degradation, and endogenous secretions. The microbes that are produced in the rumen, and then pass down the digestive tract, may supply 60 to 80 per cent of the amino acids absorbed from the small intestine. The efficiency of microbial protein synthesis is a major factor affecting the overall amino acid requirement of ruminants [5]. Adequate dietary fiber is very important to maintain rumen health and microbial protein synthesis (MPS) by sustaining a stable environment in the rumen [6].

Based on these current phenomena, this study was undertaken to compare botanical composition and nutritive value of forage fed to Bali cattle and their rumen performance i.e. pH, NH_3 and VFA production, microbes protein synthesis before and after the Mount Agung eruption in evacuation zones i.e. Sidemen and Nongan Villages.

Based on these current phenomena, this study was undertaken to compare botanical composition and nutritive value of forage fed to Bali cattle and their rumen performance i.e. pH, NH_3 and VFA production, microbes protein synthesis before and after the Mount Agung eruption in evacuation zones i.e. Sidemen and Nongan Villages.

2. Materials and Methods.

Data collection was taken in two phases. The first phase was a survey conducted in December 2017 on farmers to collect information on botanical composition and nutritive value of forage fed before and after the eruption of Mount Agung in the Sidemen and Nongan evacuation zones. Nongan Village, Rendang District, Karangasem Regency and Talibeng Village, in Sidemen District in Karangasem Regency were about 20 and 23 km from Mount Agung zone. The second phase was laboratory forage sample analysis from January to May 2018 for their nutritive value as well as rumen performance of Bali cattle i.e. pH, NH_3 and total VFA of rumen fluids, Digestible Organic Matter in Rumen (DOMR) and microbial protein synthesis (MPS). Nutritive value of forages were analysed using Official Method of Analysis [7] and in vitro digestibility was analysed using modified In Vitro Technique by [8] NH_3 was analysed by Spectrophotometer according to [9], VFA was analysed by Gas Chromatography (GC) by [7] and MPS was analysed by [10]. Data was analysed by using descriptive statistics and Anova using multivariate using SPSS version 24.

3. Results and Discussion

3.1 Nongan evacuation zone

3.1.1 Botanical composition and their nutritive value. Twenty five smallholder farmers were surveyed who had flock size ranging between 2-14 Bali cattle of different physiological state. There were total of 163 Bali cattle consisted of 48 mature males, 67 mature females and 48 calves. Observation focused on mature males and females Bali cattle. Botanical composition and their nutritive values provided by smallholder farmers in Nongan evacuation zone before and after the Mount Agung eruption were presented in Table 1.

Botanical composition of forage provided by smallholder farmers before Mount Agung eruption was more varied with no concentrate supplementation. However, when smallholder farmers and their livestock were evacuated in Nongan Village, they provided less varied of botanical composition to their livestock with 1 kg of concentrate supplementation per head per day. This was due to the farmers found more difficult to cut and carry the forage as the availability and quality of forage decreased after Mount Agung eruption particularly after 3 months of the eruption. This resulted in lower crude protein.

Table 1. Botanical composition and nutritive values of feed provided in Nongan evacuation zone

(a) Before Mount Agung Eruption in Nongan Evacuation Zone							
Botanical Composition	Given (kg)	Dry Matter (%)	Crude Protein (%)	Crude Fiber (%)	Ether Extract (%)	Organic Matter (%)	Gross Energy (kkal/kg)
Elephant grass	25.00	61.21	6.34	18.85	4.17	50.63	2166.94
Calliandra	3.00	17.90	5.65	4.80	2.63	16.68	776.74
Gliricidia sepium	2.00	9.76	2.51	1.30	0.39	8.94	463.68
Jackfruit leaves	3.00	9.37	1.17	1.87	0.41	7.03	337.36
Cassava	0.30	1.76	0.06	0.09	0.01	1.70	87.85
Total	33.30	100.00	15.73	26.91	7.62	84.97	3832.57
(b) After Mount Agung Eruption in Nongan Evacuation Zone							
Botanical Composition	Given (kg)	Dry Matter (%)	Crude Protein (%)	Crude Fiber (%)	Ether Extract (%)	Organic Matter (%)	Gross Energy (kkal/kg)
Elephant grass	20.00	37.31	3.86	11.49	2.54	30.86	1320.93
Calliandra	5.00	22.73	7.18	6.10	3.33	21.18	986.43
Concentrate	1.00	14.07	2.30	1.53	1.21	13.99	591.36
Sugar cane shoot	2.00	25.89	0.97	0.57	1.52	23.37	991.47
Total	28.00	100.00	14.32	19.68	8.60	89.40	3890.19

3.1.2 Nutrient intake. Nutrient intake of Bali cattle before and after Mount Agung eruption was shown in Table 2. Intake of DM, CP, CF, EE, OM and GE of forage were significantly higher ($P<0.05$) when the cattle were in Nongan eruption zone after the Mount Agung eruption. This higher intake was due to the cattle provided with concentrate 1 kg per head per day. Concentrate had lower CF and higher digestibility compared to forage. This resulted in higher nutrient consumption.

3.1.3 Rumen fermentation in vitro. Rumen fermentation in vitro before and after Mount Agung eruption showed pH of rumen fluid of both rations was in normal range (Table 2). The differences of forage composition of ration caused pH in vitro was not significant ($P>0.05$). Rumen pH was one of factors that supported the growth of population and activity of rumen microbes. Optimal rumen pH for the optimal growth of population and activity of rumen microbes was 6.0 – 6.9 [11] and normal pH of rumen fluid was 6 – 7 [12].

Total VFA concentration rumen fermentation in vitro was also not statistically different ($P>0.05$). VFA was the products of carbohydrate fermentation as energy resources. Although energy consumption of Bali cattle was significantly higher ($P<0.05$) after the Mount Agung eruption, their VFA productions were not significantly different ($P>0.05$) before and after the eruption. However, NH_3 production was significantly higher i.e. 7.1156 mMol ($P<0.05$) after the Mount Agung eruption compared to 4.0698 mMol before the eruption. NH_3 was the product of the degradation of crude protein by microbial rumen. Protein content of feed after the eruption was lower although their consumption was higher i.e. 49.30%. The higher protein consumption resulted in higher NH_3 production. NH_3 concentration resulted in rumen fermentation in vitro of the ration that consisted of 70% forage and 30% concentrate ranged between 8.78 – 10.71 mMol [13]. An ideal N-NH_3 concentration that supported the growth of bacteria optimally was 6-21 mMol [14].

Table 2. Nutrient intake, rumen fermentation and microbial protein synthesis *in vitro* in Nongan evacuation zone

	Before Mount Agung eruption	After Mount Agung eruption
(a) Nutrient Intake :		
Dry Matter, kg	5.12	6.72
Crude Protein, g/d	805.67	1202.85
Crude Fiber, g/d	1378.52	1655.90
Ether Extract, g/d	390.07	509.60
Organic Matter, kg/d	4.35	5.67
Gross Energy, kkal/d	19630.82	26150.28
(b) Rumen fermentation *in vitro* :		
pH	6.8790	6.9760
Total VFA, mMol	62.9640	62.5400
N-NH_3, mMol	4.0698^{a}	7.1156^{b}
(c) Microbial protein synthesis *in vitro* :		
DOMR g	3318.36$^{a2)}$ ± 236.88608	4161.81^{b} ± 188.17950
Microbial N, g	106.19^{a} ± 7.58035	133.18^{b} ± 6.02174
MPS, g	663.67^{a} ± 47.37722	832.36^{b} ± 37.63590
Purine Absorption, mMol	146.06^{a} ± 10.42690	183.19^{b} ± 8.28300

Means in a column with different superscripts differed significantly at the .05 level.

3.1.4 Microbial protein synthesis in vitro. Digestible organic matter in the rumen (DOMR) was significantly higher ($P<0.05$) in ration of Bali cattle when they were in Nongan evacuation zone after the eruption (Table 2). The higher crude protein consumption resulted in higher NH_3 production. Microbial rumen required sufficient NH_3. The more production of NH_3 of the ration after the Mount Agung eruption was beneficial to the growth of rumen microbial and this was shown by the higher SPM i.e. 832.36^{b} ± 37.63590 g compared to those of before the eruption i.e. 663.67^{a} ± 47.37722g. DOMR and MPS of ration of Bali cattle in Nongan evacuation zone was significantly higher compared to DOMR and MPS before the eruption being 1843.08 g and 368.62 g [13].

3.2 Sidemen evacuation zone

3.2.1 Botanical composition and their nutritive value. Similarly to Nongan evacuation zone, king grass always dominated the composition of forage before and after the Mount Agung eruption (Table 3). Rearing Bali cattle by smallholder farmers was side job and done traditionally where the flock size ranged between 2 to 14, therefore, feeding their livestock was mostly forage and feeding concentrate was rare and costly. However, concentrate was fed to Bali cattle in both evacuation zones where in Sidemen zone it was 0.5 kg concentrate per head per day (Table 3). Feeding concentrate to Bali cattle after the Mount Agung eruption both in evacuation zones was due to donated by donators who were aware about livestock in evacuation zones.

3.2.2 Nutrient intake. Before the Mount Agung eruption, more botanical composition fed to Bali cattle in Sidemen evacuation zone. Forage fed as protein sources such as *Calliandra, Gliricidia sepium* and *Leucaena leucocephala*. In contrast, concentrate was the source of protein, so their protein content of ration was almost twice less than of the ration fed before the eruption being 8.85% and 17.94%, respectively (Table 3), and was the protein consumption (Table 4).

Table 3. Botanical composition and nutritive values of feed provided in Sidemen evacuation zone

(a) Before Mount Agung Eruption in Sidemen Evacuation Zone							
Botanical Composition	Weight (kg)	Dry Matter (%)	Crude Protein (%)	Crude Fiber (%)	Ether Extract (%)	Organic Matter (%)	Gross Energy (kkal/kg)
Elephant Grass	20.0	38.30	3.96	11.79	2.61	31.68	1355.74
Calliandra	2.0	9.33	2.95	2.50	1.37	8.69	404.97
Gliricidia sepium	4.0	15.27	3.92	2.03	0.61	13.99	725.25
L. leucocephala	4.0	17.77	5.30	3.49	0.93	16.68	835.48
Native grass	5.0	19.33	1.80	6.11	1.24	18.62	705.56
Total	35.0	100.00	17.94	25.92	6.77	89.67	4027.00
(b) After Mount Agung Eruption in Sidemen Evacuation Zone							
Botanical Composition	Weight (kg)	Dry Matter (%)	Crude Protein (%)	Crude Fiber (%)	Ether Extract (%)	Organic Matter (%)	Gross Energy (kkal/kg)
Elephant Grass	25.0	47.40	4.91	14.60	3.23	39.21	1678.03
Concentrate	0.5	7.15	1.17	0.78	0.61	7.11	300.49
Gliricidia sepium	2.0	26.31	0.99	0.57	1.54	24.10	1007.60
Native grass	5.0	19.14	1.78	6.05	1.23	18.44	804.48
Total	32.50	100.00	8.85	21.98	6.62	88.85	3790.60

3.2.3 Rumen fermentation in vitro. Nutrient consumption affected the rumen fermentation. The product of rumen fermentation in vitro showed that pH rumen fluid was in normal range (Table 4). Optimal pH for the growth of microbial rumen was 6 – 6.9 [15] and normal pH rumen fluid was 6 - 7 [12]. The concentration of total VFA and NH_3 was not significantly different ($P>0.05$) between before and after the Mount Agung eruption. Sufficient concentration of NH3 and total VFA was required for optimal growth and activity of microbial rumen being 50 mgNH_3/l and 70 – 130 mMol [16]. The level of VFA concentration was affected by basal ration, type of carbohydrate of feed, the physical form of the feed, rate of consumption, feeding frequency and the usage of chemical additive [16].

Table 4. Nutrient intake, rumen fermentation and microbial protein synthesis *in vitro* in Sidemen evacuation zone

	Before Mount Agung eruption	After Mount Agung eruption
(a) Nutrient Intake :		
Dry Matter, kg	6.55	6.61
Crude Protein, g/d	1174.77	1082.96
Crude Fiber, g/d	1697.42	1769.47
Ether Extract, g/d	443.12	405.34
Organic matter, kg/d	5.87	5.88
Gross Energy, kkal/d	26374.82	25072.86
(b) Rumen fermentation *in vitro* :		
pH	6.9330	6.8600
VFA, mMol	77.274	66.1440
N-NH_3, mMol	6.5051	5.5327
(c) Microbial Protein Synthesis *in vitro* :		
DOMR, g	$4818.17^{a} \pm 279.76791$	$4403.53^{a2)} \pm 143.80926$
Microbial N, g	$154.18^{a} \pm 8.95257$	$140.91^{a} \pm 4.60190$
MPS, g	$963.63^{a} \pm 55.95358$	$880.71^{a} \pm 28.76185$
Purine Absorption, mMol	$212.08^{a} \pm 12.31441$	$193.83^{a} \pm 6.32998$

3.2.4 Microbial protein synthesis in vitro. Microbial protein synthesis was affected by the level of DOMR. Before the Mount Agung eruption, DOMR was $4818.17^{a} \pm 279.76791$ g and after the eruption, the amount of DOMR was $4403.53^{a2)}$ ± 143.80926 g. The higher level of the DOMR before the eruption resulted in the higher level of SPM before the Mount Agung eruption (Table 4). Microbial protein synthesis was varied between livestock, as it affected by feed fed to the livestock and even the same quality of feed, their responses were also varied. Feeding forage and concentrate affected microbial protein synthesis variously 70-279 g MCP/kg DOMR [15].

4. Conclusion.

Botanical composition of forage provided to Bali cattle by smallholder farmers before the Mount Agung eruption was more varied compared to after Mount Agung eruption both in Sidemen and Nongan evacuation zones. Bali cattle in Nongan evacuation zone after Mount Agung eruption consumed more nutrient, produced more VFA, NH_3 and SPM ($P< 0.05$). In contrast, Bali cattle were in Sidemen evacuation zone after Mount Agung eruption consumed less nutrient, produced less VFA, Total NH_3 and SPM ($P<0.05$).

Acknowledgements.

The authors gratefully acknowledge support by Rector of The University of Udayana for the Udayana Research Fund allocated through the Institute for Research and Community Service (LPPM) and Faculty of Animal and Husbandry of the University of Udayana, Bali-Indonesia.

References

[1] B. Hartono and E. S. Rohaeni, "Contribution to income of traditional beef cattle farmer households in Tanah Laut Regency , South Kalimantan , Indonesia," vol. 26, no. 8, pp. 1–8, 2018.

[2] L. Abdullah, "Prospektif agronomi dan ekofisiologi Indigofera zollingeriana sebagai tanaman penghasil hijauan pakan berkualitas tinggi," *Pastura*, vol. 3, no. 2, pp. 79–83, 2014.

[3] Badan Pusat Statistik Propinsi Bali, *Provinsi Bali Dalam Angka 2017*. 2017.

[4] S. Tamminga, G. G. Brandsma, G. van Duinkerken, A. M. van Vuuren, and M. C. Blok, "Protein evaluation for ruminants : the DVE/OEB 2007-system," no. 070, 2007.

[5] U. Jasim, K. Z. Haque, K. Jasimuddin, and K. Mohammad, "Dynamics of microbial protein synthesis in the rumen - A Review," 2015.

[6] J. Xu *et al.*, "Effects of Forage Sources on Rumen Fermentation Characteristics , Performance , and Microbial Protein Synthesis in Midlactation Cows," *Asian Aust. J. Anim. Sci.*, vol. 27, no. 5, pp. 667–673, 2014.

[7] AOAC, "AOAC Official Methods of Analysis," *Assoc. Off. Agric. Chem. Washington, D.C.*, vol. 15th, no. Volume 1, pp. 136–138, 1990.

[8] D. Minson and M. N. McLeod, "The in vitro technique: Its modifications of estimating digestibility of large numbers of tropical pasture samples," CSIRO Melbourne, 1972.

[9] I. Solorzano, "Determination of ammonia in natural seawater by the phenol-hypochlorite method," *Limnol. Oceanogr.*, vol. 14, no. 5, pp. 799–801, 1969.

[10] X. B. Chen and M. J. Gomes, "Estimation of microbial protein supply to sheep and cattle basid on urinary excretion of purine derivatives-an overview of the technical details.," *Ocas. Publ. Buchsburnd, Aberdeen Ed. Rowett Res. Inst.*, no. September 1995, p. p.21, 1995.

[11] D. N. Kamra, "Rumen microbial ecosystem," *Curr. Sci.*, vol. 89, no. 1, pp. 124–135, 2005.

[12] L. Chiba, *Animal Nutrition Handbook Second Revision*. 2009.

[13] N. N. Suryani, I. K. M. Budiasa, and I. P. A. Astawa, "Suplementasi gamal sebagai rumen degradable protein (RDP) untuk meningkatkan kecernaan (In Vitro) ransum ternak ruminansia yang mengandung jerami padi," *Maj. Ilm. Peternak.*, vol. 16, no. 1, pp. 1–5, 2013.

[14] P. McDonald, R. a. Edwards, J. F. D. Greenhalgh, C. a. Morgan, L. a. Sinclair, and R. G. Wilkinson, "Animal Nutrition," *Anim. Nutr.*, p. 576, 2010.

[15] M. A. Karsli and J. R. Russell, "Effects of Some Dietary Factors on Ruminal Microbial Protein Synthesis," *Turkish J. Vet. Anim. Sci.*, 2001.

[16] J. France and J. Dijkstra, *Quantitative aspects of ruminant digestion and metabolism*. 2005.

IOP Conf. Series: Earth and Environmental Science **207** (2018) 012031 doi:10.1088/1755-1315/207/1/012031

Allele frequency estimation of BLAD (Bovine Leukocyte Adhesion Deficiency) in dairy cattle in Enrekang regency South Sulawesi Indonesia

M I A Dagong[1] , L Rahim[1], R R S R Aprilita Bugiwati[1] and Nurmulyaningsih[2]

[1] Animal Breeding and Genetic Laboratory, Faculty of Animal Science, Hasanuddin University

[2] Animal Production Study Program, Faculty of Animal Science, Hasanuddin University

E-mail: ihsandagong@gmail.com

Abstract. Bovine leukocyte adhesion deficiency (BLAD) is a genetic disorder in dairy cows that are generally lethal or premature in cattle that can have an impact on productivity and result in losses for farmers. The purpose of this study is to identify the distribution of the BLAD recessive allele in dairy cows raised in Enrekang dairy cattle development centers. Identification of recessive BLAD allele using the PCR-RFLP method. A total of 80 DNA samples in a collection of blood samples from FH dairy cows from Enrekang regency (50 heads from Cendana and 30 heads from Anggeraja district) were extracted and amplified by PCR, then the PCR product was cut with the *Taq*1 restriction enzymes. The identification of the BLAD carrier alleles is calculated according to the frequency of genotypes and alleles. The results obtained from this study is (+/-) heterozygous cows (BLAD carrier) found 1 head (with a frequency of the allele (-) (BLAD recessive) of 0.01) in the Cendana district of 50 samples, while the Bovine population in the district Anggeraja did not found any BLAD recessive allele. The results of this study concluded that there is a normal heterozygous dairy cows (BLAD carrier) in the dairy cows population in the Enrekang regency, although the proportion / frequency of alleles is very low (0.625%). This research can be used as a reference in selecting dairy cattle, especially those related to the spread of lethal alleles in dairy cattle development areas in South Sulawesi.

1. Introduction

Dairy cows have spread to several regions of Indonesia and, in general, farmers raise them on a small and large scale. Dairy cows have considerable potential to develop because, in addition to producing milk, they also produce meat. Genetic factors are one of the main factors that contribute to increase the productivity of dairy cattle. However, there are several genetic disorders that are lethal and often found in dairy cows. One genetic disorder in dairy cows that often occurs is bovine leukocyte adhesion deficiency (BLAD).

Bovine leukocyte adhesion deficiency (BLAD) or granulocytopathic syndrome is a genetic disease in cattle caused by the appearance of point mutations in exon 2 of the CD18 gene, this mutation causes β2 heterodimer molecules not expressed in the surface of neutrophils. In homozygous recessive conditions, this mutation causes lethal or premature death in cattle [1]. Bovine leukocyte adhesion deficiency (BLAD) is generally lethal in dairy cattle before reaching adulthood, thus affecting productivity and causing losses to farmers.

BLAD genetic disorders were first identified in North America [1] which then spread to various countries such as India [2], Turkey [3], Iran [4], Japan [5], Romania [6], Pakistan [7] and the Czech Republic [8].

Enrekang Regency is a dairy cattle development center in South Sulawesi. The development of dairy cattle population in this area is supported by the development of Dangke dairy products, a type of soft cheese made from milk [9]. There are two region for dairy cattle development in Enrekang, namely Cendana and Anggeraja districts. The purpose of this study was to identify the presence of recessive alleles from bovine leukocyte adhesion deficiency (BLAD) in dairy cows. With the knowledge of livestock carrying recessive alleles, breeding strategies can be carried out to reduce or eliminate BLAD recessive aleles.

2. Method

2.1. *Blood sample collection*

This study used about 80 DNA samples collected from blood samples from Fries Holland dairy cows on smallholder farms in Enrekang regency (50 heads from Cendana district and 30 heads from Anggeraja district). About 5 ml of blood sample were collected from the jugular vein of each dairy cow in a vacutainer tubes containing EDTA.

2.2. *DNA extraction and BLAD allele identification*

DNA was extracted and purified using DNA extraction kit by following the extraction protocol standards provided. The results of DNA extraction then continued for the PCR reaction process for the determination of BLAD alele. All procedur were carried out at the Integrated Biotechnology Laboratory of Animal Science Faculty, Hasanuddin University.

The PCR amplification was carried out in a total volume of 25 µL consisting of DNA template (~100 ng), 0.25 mM each BLAD forward and reverse primer (Table 1), 150 µM dNTP, 2.5 mM Mg2+, 1x buffer and 1 U/ µL Taq polymerase. The PCR condition including initial denaturing step of 94 °C for 4 min, followed by 34 cycles of 94 °C for 45 s, primer annealing at 64 °C for 45 s, and 72 °C for 1 min, with a final cycle at 72 °C for 5 min. The PCR products of BLAD gene was digested with T*aq*1 restriction endonuclease enzym according from manufacturer (Fermentas). Digested PCR products were electrophoresed on 2% agarose gel and then visualized under UV-transiluminator (gel documentation system).

Table 1. Primer sequence and restriction endonuclease enzyme.

Gene	Primer	Restriction enzyme	PCR size	Reference
BLAD	F: 5'- GAATAGGCATCCTGCATCATATC CACCA-3'	T*aq*1	357 bp	[10]
	R: 5'- CTTGGGGTTTCAGGGGAAGATGGA GTAG-3'			

Note : F = Forward, R = Reverse

2.3. *Data analysis*

Allele and genotype frequencies were calculated using PopGene program [11].

3. Results and Discussion

The amplified PCR product of the BLAD gene produced 357 bp fragment of amplicon. This result is in accordance with that reported by [10, 12]. Determination of the genotype of the BLAD gene in dairy cattle in this study using the PCR-RFLP (polymerase chain reaction restriction fragment length polymorphism) method with T*aq*1 as an enzyme restriction. The T*aq*1 enzyme recognizes the point mutation that changes adenine by guanine (A → G).

This is consistent with the report of [13] that the results of sequencing mutations of the BLAD allele known in base 383 in the CD18 gene contained a point mutation of adenine to guanine, which changes the function of the CD18 gene so that a genetic disorder known as BLAD appears. The DNA sequences in the BLAD gene based on the primer used and the T*aq*1 restriction sites shown on Figure 1.

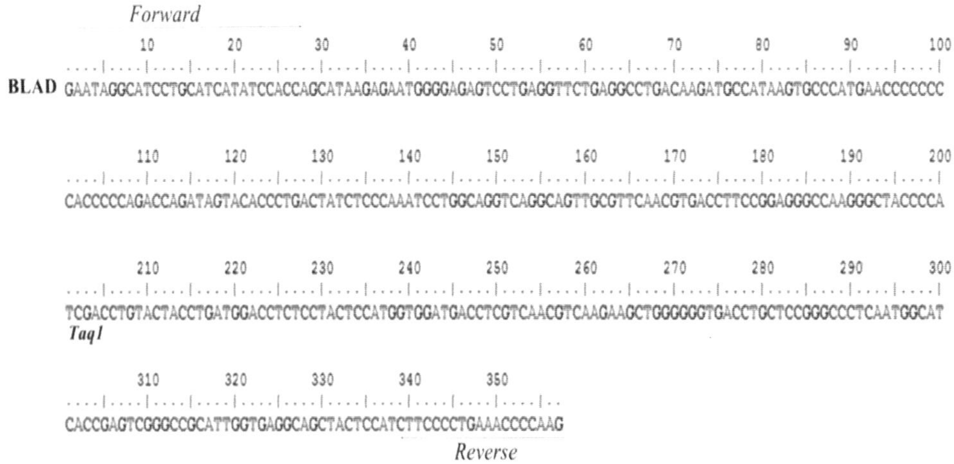

Figure 1. BLAD gene DNA sequences and the T*aq*1 enzyme restriction site.

The genotype of the BLAD gene was obtained by measuring the length of the BLAD gene fragment from the cut with the T*aq*1 enzyme. The visualization results show that the length of the fragments obtained is 357, 201 and 156 bp. The genotype found in dairy cattle can be seen in Figure 2.

The results show that each genotype can be distinguished based on the number of DNA bands that appear. Normal BLAD homozygous (+/+) were produced two fragments (156 and 201 bp), while heterozygous or BLAD carriers (+/-) with three fragments (156, 201 and 357 bp). This condition is in accordance with the research conducted by [10], who reported that the PCR product of BLAD gene which was restricted by T*aq*1 enzyme would distinguish normal alleles with two fragments (156 and 201 bp). Whereas BLAD carriers will produce three fragments 156, 201 and 357 bp.

Based on Table 2, it is known that the frequency of recessive BLAD genes in Enrekang Regency is very low (0.006), and is only found in Cendana district. Although only in a very small percentage, the presence of this recessive allele needs to be watched carefully considering the largest population of dairy cows in Enrekang Regency is in Cendana district. The possibility of the spread of this recessive alele comes from the origin of these dairy cattle which are mostly imported from Java, and the possibility of spread through artificial insemination. As is known the proportion of recessive alleles in Java is quite large as reported by [14], that the recessive BLAD frequency in dairy cows in Baturaden (Central Java) and Lembang (West Java) were 2.4 %.

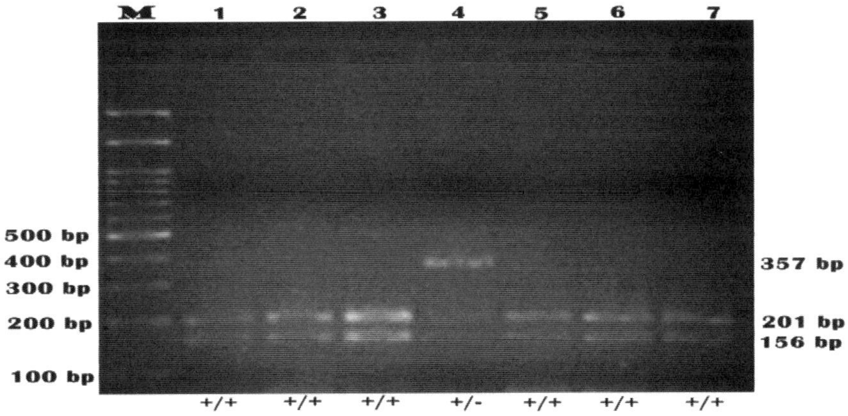

Figure 2. Visualization of BLAD gene PCR-RFLP, M: marker DNA ladder 100 bp, Line 1, 2, 3, 5, 6, 7 were normal BLAD homozygous (+/+), and Line 4 was heterozygous carrier (+/-) .

The results of genotype and allele frequencies analysis in BLAD gene from two dairy populations can be seen in Table 2.

Table 2. Alleles and genotype frequencies of BLAD gene

Dairy Population	Genotype Frequency +/+ (Normal)	Genotype Frequency +/- (Carrier)	Allele Frequency +	Allele Frequency -
Cendana	0.98	0.02	0.99	0.01
Angeraja	1	0	1	0
Total	0.987	0.013	0.994	0.006

The frequency of heterozygous dairy cattle (BLAD carriers) in some countries is quite high. [3] reported that in Turkey FH dairy cows found 8.4% heterozygous BLAD, whereas [10] reported that 3.5%. While in India (3.23%) [2], Iran (3.33%) [1] and Jepang (8.79%) [5].

BLAD is a genetic disorder that affects the low level of productivity of dairy cows that can cause premature death. Therefore, it is necessary to control the spread of BLAD cases by identifying heterozygous dairy cows (BLAD carriers). This is in accordance with research conducted by [15] that BLAD can cause large losses, namely if there is a mating between individual carriers who have 25% chance of having a premature death. According to [14] that in a program to reduce the negative effects of BLAD, what can be done is to identify heterozygous cows and prevent them from mating with a heterozygous parent.

4. Conclusion

There is a relatively small proportion of BLAD recessive alleles in Enrekang regency (0.006). Although in a small percentage, the existence of this allele needs to be watched out because it is in the center of dairy cattle development in Cendana district.

5. Acknowledgment

This research was supported by Institute of Research and Community Service of Hasanuddin University through the Internal Competitive Research Grant (Hibah Kompetitif Internal BOPTN 2015). We would also like to show our gratitude to the Enrekang dairy farmers in Cendana and Anggeraja district for cooperation in the implementation of this research.

References

[1] Shuster D E, Kehrli M E, Ackermann M R and Gilbert R O 1992 *Proc. Natl. Acad. Sci. USA.* **89** p 9225 - 9229
[2] Patel R K, Singh K M, Soni K J and Rao K R 2007 *J. Appl. Genet.* **48** p 153 - 155
[3] Akyus B and Ertugrul O 2006 *Acta Vet. Hung* **54** p 173 - 178
[4] Norouzy A, Nassiry M R, Shahrody F E, Javadmanesh A, Abadi M R M and Sulimova G E 2005 *Genetika* **41** p 1697 - 1701
[5] Nagahata H, Miura T, Tagaki K, Ohtake M, Noda H, Yasuda T and Nioka K 1997 *J. Vet. Med. Sci.* **59** p 233 - 238
[6] Vatasescu-Balcan R, Georgescu E S, Adina M M, Mariana R, Anca D and Marieta C 2006 *Roumanian Biotechnological Letters.* **11** p 2881 - 2884
[7] Nasreen F, Malik N A, Riaz M N and Qureshi J A 2009 *Hereditas.* **146** p 74 - 78
[8] Citek J, Rehout V, Hajkova J, and Pahkova J 2006 *Veterinarni Medicina* **51** p 333 - 339
[9] Baba S, Muktiani A, Ako A and Dagong M I A 2011 *Media Peternakan* **34** p 146 - 154
[10] Meydan H, Yidiz M A and Agerholm J S 2010 *Acta Veterinaria Scandinavica.* **52**:56
[11] Yeh F C, Yang R C and Boyle T 1999 POPGENE version 1.31 : Microsoft Window-based freeware for population genetic analysis. University of Alberta Canada. Edmonton, AB
[12] Meydan H, Ugurlu M and Yildiz M A 2012 *Tarim Bilimleri Dergisi–Journal of Agricultural Sciences.* **18** p 239 -245
[13] Kriegesmann B, Jansen S, Baumgartner B G and Brenig B 1997 *J Dairy Sci.* **80** p 2547 - 2549
[14] Farajallah A, Sumantri C and Muttaqin W N 2007 *Identifikasi alel pembawa bovine leucocyte adhesion deficiency (BLAD) pada sapi perah Friesian Holstein di Indonesia.* Seminar Nasional Teknologi Pertanian dan Veteriner. Departemen Biologi FMIPA Institut Pertanian Bogor.
[15] Herodita L U 2009. *Identifikasi bovine leukocyte adhesion deficiency (BLAD) pada peternakan sapi Friesian-Holstein di Jawa - Bali.* Departemen Biologi Fakultas Matematika dan Ilmu Pengetahuan Alam, Institut Pertanian Bogor, Bogor.

Identify the diversity of helminth parasites in cattle in Jember district (East Java - Indonesia)

A Awaludin[1*], Nurkholis[2] and S Nusantoro[3]

[1]Livestock Production Study Program, Department of Animal Husbandry, State Polytechnic of Jember, Jl. Mastrip PO BOX 164, Jember, East Java, Indonesia
[2]Livestock Production Study Program, Department of Animal Husbandry, State Polytechnic of Jember, Jl. Mastrip PO BOX 164, Jember, East Java, Indonesia
[3]Livestock Production Study Program, Department of Animal Husbandry, State Polytechnic of Jember, Jl. Mastrip PO BOX 164, Jember, East Java, Indonesia

*E-mail: aanawaludin@gmail.com

Abstract. Cattle are maintained by many rural communities, including in Jember district. Various health problems associated with parasitic infestations are common in cattle including worm parasitic infestations. This study aims to identify the diversity of worm parasites in cattle in Jember district. The benefit of this study is to obtain information on worm species so that it can be used as a database for strategies for prevention and treatment of helminthiasis in cattle in Jember district. The method used is to randomly take 314 cattle feces samples in the Jember area. Identification of worm species diversity by examining worm eggs found in fecal samples using the sedimentation method and the Whitlock method performed at BBVet (Balai Besar Veteriner) Wates. Based on the results of cattle feces examination in Jember, worm parasite eggs were found from the trematode class, *Fasciola* sp. and nematode class consisting of *Ostertagia* sp., *Trichostrongylus* sp., *Moniezia* sp., *Cooperia* sp., *Capillaria* sp., *Bunostomum* sp., *Strongyloides* sp. , *Oesophagostomum* sp., *Trichuris* sp., and *Toxocara* sp.

1. Introduction

A health problem in livestock is one of the obstacles faced in the development of livestock. Livestock productivity will be achieved optimally if the provision of adequate food and disease control is carried effectively. Diseases in livestock various, one of which is a disease caused by parasites, but parasitic diseases still lack the attention of farmers. Parasitic diseases usually do not cause in the death of livestock, but parasitic diseases can cause huge losses in the form of weight loss and animal productivity. Among parasitic diseases that are very detrimental are diseases caused by helminths. Losses caused by helminths parasites are delays in growth, especially in young cattle. The occurrence of parasitic helminth infections in the intestines of cattle will be pathogenic, especially if it coincides with poor conditions of animal feed [1].

The incidence of worm parasitic infestations is influenced by various things, one of which is the desert pasture environment contaminated by infective stage worm eggs or larvae. Disadvantages due to trematode infestation (fascioliasis) in the incidence of acute and sub-acute infections occur 2-3 weeks after infection occurs with marked anorexia, yellow conjunctiva, pale, abdominal pain, weight loss or sudden death. In chronic events, symptoms of anaemia and submandibular oedema are seen [2]. Parasitic infections are generally an important health problem and are most commonly found in grazing cattle. In grazing cattle, in Australia, these losses (nematodes and ectoparasites) reach around one billion dollars per year [3]. The spread of infective stage contaminants from the gastrointestinal worms in cattle came from the feces of infected livestock grazing in pasture fields. Worm eggs that come out with feces will be protected and stay alive for several months even with dry environmental conditions. The feces will be destroyed and scattered if exposed to rain so that it can spread and contaminate the pasture. The contaminated pasture will become an environmental factor for the spread of helminth infections and become an infectious medium if the grass is taken as food for other sensitive livestock. Seasonal factors and epidemiological conditions of several months (time) in a place are considerations in determining whether grazing or forage sources are safe for livestock that are kept especially in cattle with traditional maintenance systems [4]. The types of helminths that have parasites in ruminant animals are spread cosmopolitan, except for certain types that are only found in a particular geographical area. The incidence of intestinal worms in cattle is strongly influenced by geographical location and climate and seasons throughout the year. Parasitic helminths in cattle according to their morphology are divided into three classes: *nematodes*, *cestodes*, and *trematodes* that have different life cycles.

The fecal examination is one of the laboratory tests that have long been known to help diagnose a disease. Although at present there are various types of modern laboratory examinations, in some cases, fecal examination is still very necessary and cannot be replaced by other types of examinations [5]. The fecal examination is divided into two, qualitative and quantitative examination. Qualitative fecal examination, which is an examination based on finding worm eggs in each examination method without counting the number. A quantitative fecal examination is a fecal examination based on the number of worm eggs (the same type of worm eggs) found in each gram of feces [6]. The objective of this research is to identify the diversity of worm parasites in cattle in Jember district.

2. Methodology

This research was carried out from June 2018 to September 2018 in several locations which included field and laboratory activities. Preparation of equipment for sampling and preparation of samples is carried out at the Animal Production Laboratory, Department of Animal Husbandry, and State Polytechnic of Jember. Research samples (feces) were taken randomly from various types of cattle in Jember district. Cattle that are sampled come from folk farms by not differentiating the type of maintenance management in the community. Identification of worm eggs is carried out at BBVet (Balai Besar Veteriner) Wates, Yogyakarta. Data processing is done at the Department of Animal Husbandry, State Polytechnic of Jember.

The tools used for sample collection and preparation include gloves, long-sleeved plastic gloves (rectal exploration), masks, 5 grams of sample pots, drop pipettes, 33 L cool boxes, label paper, permanent board markers, refrigerators, and digital cameras.

The tools used for observing worm eggs include a 250 ml measuring glass, 100 ml glass beaker, binocular microscope, object glass, deck glass, scales, stirring rods, tube racks, centrifugal tubes, centrifuges, pipettes , wipes, and petri dishes.

The material used for the collection and preparation samples is 10% formalin, 1% eosin solution, and ice cubes.

Sample collection, samples taken was newly defecated cattle feces. The samples collected were 314 samples, 1 sample approximately 5 grams in the form of fresh feces or taken by rectal exploration. Then the feces are put into a sample pot plastic and mixed with 2-3 drops of 10% formalin and labelled. The number of feces samples that will be collected is 314 random samples from cattle in Jember district.

Identification of worm eggs, sedimentation test, feces samples were taken as much as 3 grams and then put into 100 ml beaker glass and added with 50 ml of distilled water, stirred with a stirring rod until the feces were destroyed and homogeneous. Homogeneous feces solution is taken using a pipette and put into a centrifuge tube up to 2/3 tubes. Then centrifuge is carried out at a speed of 2000 rpm for 5 minutes. The supernatant was discarded and the precipitate was added with distilled water to 2/3 tubes and centrifuged again at a speed of 2000 rpm for 5 minutes. This sedimentation is carried out until the supernatant appears clear then the supernatant is removed, the sediment formed is taken with a pipette and placed on an object glass added to the dye (1% eosin solution) then covered with a glass deck. Observation with a microscope 10 x 10 magnification and identification of worm eggs found.

3. Results and discussion

Total of cattle fecal samples collected from several areas in Jember district were 314 samples. The feces sample comes from various cattle with varying ages and does not differentiate between sexes. The feces sample was then labelled, sample codes 1-249 (249 feces samples) and sample codes 301-365 (65 samples).

An examination of fecal samples and identification of worm eggs were carried out at BBVet (Balai Besar Veteriner) of Wates, Yogyakarta. The examination carried out by qualitative test with sedimentation method, then carried out examination using a microscope and identification of worm eggs according to the morphology of worm eggs found.

The results of examination and identification of cow feces samples with sample codes 1 - 249 and 301 - 365 are listed in table 1.

From the 314 feces samples examined for worm eggs, 161 samples or about 51.3% were not found for worm eggs and 153 samples or about 48.7% were found for worm eggs. Worm eggs found were 113 fecal samples containing *Fasciola* sp. worm eggs, 48 fecal samples containing *Ostertagia* sp. worm eggs, 17 fecal samples containing *Trichostrongylus* sp. worm eggs, 9 fecal samples containing *Moniezia* sp. worm eggs, 13 fecal samples containing *Cooperia* sp. worm eggs, 13 fecal samples containing *Capillaria* sp. worm eggs, 2 fecal samples containing *Bunostomum* sp. worm eggs, 8 fecal samples containing *Strongyloides* sp. worm eggs, 1 fecal samples containing *Oesophagostomum* sp. worm eggs, 4 fecal samples containing *Trichuris* sp. worm eggs, and 2 fecal samples containing *Toxocara* sp. worm eggs.

The most commonly found worm eggs are species from the *nematode* class which is found as many as 10 types of worm eggs from the *nematode* class, the adult worms from the class of *nematodes* are parasitic in the intestine. This is related to the life cycle of *nematode* class worms does not require a reservoir host, so that the life cycle of this worm is easier [7]. Whereas for worm eggs from the class of *trematodes* found one type of species *Fasciola* sp. *Fasciola* sp. classified into the phylum Platyhelminthes, class Trematodes, order Digenea, family Fasciolidae, genus *Fasciola*.

Worm eggs can be identified by looking at the morphological form of worm eggs. *Nematode* class worm eggs in cows consist of several species. Species of *Ostertagia* sp. has an egg with 74-90µm in length, 38-44 µm in width, regular form of an ellipse, thin wall, chitinous shell with a smooth surface, the inside is covered with a thin yolk membrane. Species of *Trichostrongylus* sp. has egg with characteristics including medium-sized worm eggs, *Trichostrongylus axei*: 70-108 µm in length and 30-48 µm in width, *Trichostrongylus colubriformis*: 79-101 µm in length and 38-50 µm in width, *Trichostrongylus vitrinus*: 85-125 µm in length and 37-55 µm in width, irregular ellipse, thin wall, chitinous shell with smooth surface, the inside is covered by a thin yolk membrane. Species *Moniezia* sp. has eggs with characteristics including medium-sized worm eggs, more or less tri-quadrangular with 80-90 µm sizes, irregular round shape with round corners and bent walls, thick shell with a smooth surface, dark grey , contains an embryo surrounded by a piriform or pear-shaped apparatus. Species of *Cooperia* sp. has egg with characteristics included in medium-sized worm eggs, *Cooperia oncophora*: 74-95µm in length, 36-44 µm in width, *Cooperia punctata*: 69-83 µm in length, 29- 34 µm in width, regular small ellipse, thin chitinous shell with smooth surface, the inside is covered with a thin yolk membrane, very numerous blastomeres, hard to distinguish. Species of *Capillaria* sp. has an egg with features including small, lemon-shaped worm eggs, 45-50 µm in length, 22-25 µm in width, slightly protruding polar plugs, thick shell with a wrinkled surface, unsegmented, granular contents. Species of *Bunostomum* sp. has an egg with characteristics included in medium-sized worm eggs, 88-104 μm in length, 47-56 μm in width, irregular broad ellipse, thin chitinous shell with a

Table 1. Results of identification of worm eggs

No.	Sample Codes	Worm eggs
1.	1, 4, 6, 8, 9, 11, 15, 16, 24, 26, 29, 30, 31, 32, 34, 41, 42, 47, 52, 58, 63, 64, 84, 86, 88, 89, 90, 92, 93, 116, 117, 118, 121, 123, 137, 140, 151, 153, 155, 162, 185, 195, 203, 207, 212, 214, 216, 217, 218, 219, 221, 223, 225, 227, 228, 231, 232, 233, 235, 245, 304, 307, 309, 313, 322, 325, 329, 334, 341, 343, 347, 348, 355, 357, 363, 364	*Fasciola* sp.
2.	2, 50, 53, 100, 316, 353	*Fasciola* sp., *Trichostrongylus* sp.
3.	3, 56, 82, 85, 206, 349	*Fasciola* sp., *Moniezia* sp.
4.	5, 13, 17, 18, 19, 20, 21, 25, 28, 33, 36, 37, 38, 39, 44, 46, 48, 51, 54, 55, 57, 60, 61, 62, 67, 68, 69, 72, 75, 76, 78, 79, 80, 83, 87, 91, 95, 96, 97, 102, 104, 105, 107, 108, 109, 110, 111, 112, 113, 114, 115, 120, 122, 124, 126, 127, 128, 129, 131, 132, 134, 135, 138, 141, 142, 143, 145, 146, 147, 150, 152, 154, 158, 159, 160, 161, 163, 166, 167, 168, 169, 170, 171, 172, 173, 175, 176, 177, 178, 182, 184, 186, 187, 189, 190, 192, 196, 197, 199, 200, 201, 208, 209, 215, 222, 224, 226, 236, 238, 239, 242, 243, 244, 246, 247, 248, 249, 302, 305, 308, 311, 312, 314, 315, 317, 318, 320, 321, 323, 324, 326, 327, 331, 332, 333, 335, 337, 338, 339, 342, 344, 345, 346, 350, 351, 352, 354, 356, 358, 359, 360, 362, 365	Negative
5.	7, 22, 59, 65, 71, 74, 77, 94, 101, 103, 106, 119, 130, 144, 156, 194, 241, 301, 306, 328, 336, 340, 361	*Ostertagia* sp.
6.	10, 14, 35	*Fasciola* sp., *Cooperia* sp.
7.	12, 157	*Moniezia* sp.
8.	23, 45, 136, 193, 204, 213, 220, 229, 237, 310	*Fasciola* sp., *Ostertagia* sp.
9.	27, 66, 73, 81	*Fasciola* sp., *Capillaria* sp.
10.	40	*Fasciola* sp., *Ostertagia* sp., *Bunostomum* sp., *Strongyloides* sp.
11.	43	*Fasciola* sp., *Ostertagia* sp., *Trichostrongylus* sp., *Cooperia* sp.
12.	49	*Oesophagostomum* sp.
13.	70	*Strongyloides* sp., *Trichuris* sp.
14.	98, 148, 181	*Ostertagia* sp., *Trichostrongylus* sp.
15.	99, 139, 174, 179, 180, 319	*Capillaria* sp.
16.	125	*Fasciola* sp., *Ostertagia* sp., *Trichostrongylus* sp.
17.	133	*Toxocara* sp.
18.	149	*Ostertagia* sp., *Trichostrongylus* sp., *Strongyloides* sp., *Trichuris* sp., *Toxocara* sp.
19.	164	*Ostertagia* sp., *Trichostrongylus* sp., *Cooperia* sp., *Capillaria* sp.
20.	165, 183	*Cooperia* sp.
21.	188	*Ostertagia* sp., *Trichostrongylus* sp., *Cooperia* sp., *Trichuris* sp.
22.	191	*Ostertagia* sp., *Moniezia* sp.
23.	198	*Trichostrongylus* sp.
24.	202	*Ostertagia* sp., *Trichostrongylus* sp., *Strongyloides* sp.
25.	205	*Ostertagia* sp., *Capillaria* sp.
26.	210	*Ostertagia* sp., *Cooperia* sp., *Strongyloides* sp.
27.	211	*Cooperia* sp., *Trichuris* sp.
28.	230	*Fasciola* sp., *Cooperia* sp., *Strongyloides* sp.
29.	234	*Fasciola* sp., *Cooperia* sp., *Capillaria* sp.
30.	240	*Fasciola* sp., *Ostertagia* sp., *Strongyloides* sp.
31.	303	*Fasciola* sp., *Bunostomum* sp.
32.	330	*Fasciola* sp., *Ostertagia* sp., *Cooperia* sp., *Strongyloides* sp.

smooth surface, the inside is covered with a thin yolk membrane, there are 4 to 8 dark-stained blastomeres. *Strongyloides* sp species. has an egg with characteristics included in medium-sized worm eggs, 47-65 µm in length, 25-26 in width, regular broad ellipse, thin wall, colourless, chitinous shell with a smooth surface, embryonated. Species of *Oesophagostomum* sp. has an egg with characteristics included in medium-sized worm eggs, 75-98 µm in length, 46-54 in width, regular broad ellipse, thin wall, colourless, chitinous shell with a smooth surface, having 16 to 32 blastomeres. *Trichuris* sp. has an egg with characteristics including medium-sized worm eggs, 70-80 µm in length, 30-42 in width, lemon-shaped with two protruding and transparent polar plugs, coloured light or dark brown, thick wall, granular contents without blastomeres. Species of *Toxocara* sp. has an egg with characteristics included in medium-sized worm eggs, 69-95 µm in length, 60-77 µm in width, nearly spherical, thick wall, albuminous shell, granular contents, unsegmented. The worm species found in the *trematode* class is *Fasciola* sp. The *Fasciola* species has an egg with characteristics including large worm eggs, 130-145 µm in length, 70-90 µm in width, nearly regular ellipse, symmetrical, thin shell, granular, yellowish-brown whole egg filling contents, no blastomeres, has an operculum (pole lid) [8].

This parasitic worm (*Helminthiasis*) infection is very possible because of poor maintenance systems and worm medicine programs that have not been maximized. On folk farms, the management of cattle maintenance is still traditional, which the sanitation of the cage is still not too much attention so that the spread of these parasites becomes uncontrolled. Parasite (helminth) can continue to carry out their life cycle because the treatment of worms has not been routinely carried.

4. Conclusions and recommendations

Worm eggs identified from cow feces in Jember are from the *trematode* class, *Fasciola* sp., and from the *nematode* class, *Ostertagia* sp., *Trichostrongylus* sp., *Moniezia* sp., *Cooperia* sp., *Capillaria* sp., *Bunostomum* sp., *Strongyloides* sp., *Oesophagostomum* sp., *Trichuris* sp., and *Toxocara* sp. Research on the degree of infection of each worm parasite needs to be done to improve helminthiasis information in Jember district and be used to make policies in worm parasite eradication programs in Jember district.

5. Acknowledgements

The research team thanked profusely to the P3M Politeknik Negeri Jember who had provided research funding through PNPB State Polytechnic of Jember funding sources so that this research could be carried out.

6. References

[1] Koesdarto S 2001 *Jurnal Penelitian Medika Eksakta* **2** 114 - 122
[2] Love S C J and Hutchinson G W 2003 *Proc. 350 Post Graduate Foundation in Vet Scie.* **16** 11
[3] McLeod R S 1995 *International Journal for Parasitology* **25** 1363
[4] Hertzberg H, Figi R, Noto F and Heckendorn 2003 *Proc. the 2nd SAFO Workshop* 5 - 6
[5] Soejoto and Soebari 2002 *Parasitologi medik Jilid 3: protozoologi dan helmintologi* (Solo: EGC)
[6] Gandahusada S W, Pribadi and Hery D I 2000 *Parasitologi kedokteran* (Jakarta: Fakultas Kedokteran, Universitas Indonesia)
[7] Levine N D 1990 *Buku pelajaran parasitologi veteriner* (Yogyakarta: Gadjah Mada University Press)
[8] Thienpont D, Rochette F and Vanparijs O F J 1986 *Diagnosing helminthiasis by coprological examination* **2nd Edition** (Beerse, Belgium: Janssen Research Foundation) p 30 - 49

The blood haematological profile on laying hens that treated by different levels of yeast supplementation

D Pantaya [1] and M M D Utami [1]

[1]Animal Science Department, Politeknik Negeri Jember, Jl. Mastrip PO BOX 164, Jember, Indonesia

e-mail: dadieek@yahoo.com

Abstract: The study was conducted to evaluate the effect of different dose of yeast probiotics on hematological profile of hemoglobin, erythrocytes, hematocrit, MCV, MCH, MCHC, RDW- CV, leucocyte, eosinophil, basophil, neutrophil, lymphocytes in blood of laying hens before and after received a Newcastle disease (ND) vaccine. Total of 80 laying hens (18-week) were randomly divided into four treatments. Each treatments consisting 4 replicate with 5 hens each. The first treatment was fed a basal feed as a control, and the second, third and fourth treatment were basal feed that supplemented with the yeast that contains 1x 10^9 CFU/g at the level of 0.5, 1.0 and 1.5 g/per kg of feed during 2 weeks observation. Layers were vaccinated with ND vaccine strain LaSota on 7 days after yeast supplementation treatment. Blood samples were collected on 24 h before and 5 h after vaccination. Results indicated that treated with yeast showed not significant effect on MCV, MCH, MCHC, RDW-CV, leucocyte, basil, neutrophils, lymphocytes. Yeast supplementation had significantly ($p<0.05$) increase in the count of the hemoglobin, erythrocytes, and hematocrit, particularly at before vaccine administration. The result provides a new information that yeast supplementation has a beneficial effect on under low challenge or stress condition.

1. Introduction

Feed is the largest source of costs in poultry farming production systems, accounting for more than 70%. Several strategies can applicate to improved feed efficiency in a poultry farm. One of the strategies is to use antibiotics to improved productivity through reduced feed conversion and disease protection. Despite the well documented beneficial effects of antibiotics in increasing poultry growth, reducing mortality and increasing resistance to disease, but antibiotic use also is associated with several disadvantages and challenges.

Severals studies showed that there might a correlation between the use of antibiotics with resistance to certain microorganisms [1] [2]. Regarding the use of antibiotic massively as a driver of growth and demand for products antibiotics free from consumers, the alternative products need to be developed. Yeast is a feed additive material that has been widely used for ruminants and poultry as a natural growth promotor [3]. Yeast walls containing prebiotic compounds include mannan oligosaccharides, beta glucans, and amino acids [4] [5] which can function as agents to increase immunity to disease. In addition, yeast is a compound that can stimulate bacteria in the digestive tract [5] improve the health and immune system and as a bio control agent for reducing stress [3] [6].

The immune system in poultry can be identified from the hematological profile of livestock blood profiles. Hematological testing is not only used for diagnosis for medicinal purposes, but also for health monitoring of treatment responses [7]. In addition, the hematological analysis is used as a guide in diagnosing stress levels caused by vaccine administration that can cause stress and pressure on poultry. Therefore this study was aimed to determine the condition of hematology profiles in layer hens fed with the addition of yeast before and after vaccination treatment.

2. Materials and Methods

2.1. Animal and diets

The experimental was conducted in farm of Politeknik Negeri Jember, Indonesia. A total 80- 18 weeks-old Hi-sex Brown layer (body weight (BW)=1530±31 g) that provided from local pullet farm were allocated in the individual battery system for the layer. The hens were fed *ad-libitum* as a standard mixed of mash diet which has been formulating according to the manual guideline for layers (**Table 1**).

The chicken's experiments were divided into four treatments. Each treatments consisting 4 replicates with 5 hens in each replicate. The first treatment was fed a basal feed as a control, and the second, third and fourth treatment were fed the basal that supplemented with the yeast *Saccharomyces cerevisiae* that contains 1x 10^9 CFU/g at the dose of 0.5, 1.0 and 1.5 g/per kg of diets, respectively during 2 weeks observation. The feed was free from antibiotics and other feed supplements. Water was provided *ad- libitum* during the experiments periods. The Layer hens were vaccinated using ND (New castle disease) vaccine strain La-Sota via nasal drops on 7 d after yeast supplementation treatment.

2.2. Blood sampling and analysis

The blood sampled were taken in week 19 at 24 h before and 5 h after vaccine administration. The samples were collected as soon as possible from the wing vein (three/treatment) using a 1 mL syringe. One mL was carefully moved into heparinized tubes (BD Vacutainer). An automated hematology analyzer using XS series XS-1000i/XS-800i (Sysmex Corporation) according to ISO standard ISO 9001:2015 was used to analysis the concentrations (cells/μl) of WBC and RBC. The leukocytes and lymphocyte were determined by separation under an optical microscope.

The data were analyzed by one-way ANOVA using the General Linear Model (GLM) in MINITAB procedure (Minitab inc USA) with the following models $Y_{ij}=\mu+\alpha_i+\varepsilon_{ij}$. Where Y= dependent variables, μ: overall mean, αi: the effect of treatment, ε_{ij}: residual, The significant differences among the treatment groups were analyzed by Duncan's multiple range test at 0.05 significant level.

Table 1. Ingredients and chemical composition of the layers diets

Item	%

Ingredient (% of DM)	
Corn	49.5
Mix concentrate feed	34.5
Rice Bran	15.0
Top mix minerals [1]	0,94
DL-Methionine	0.05
L Lysin	0.01
Total	100
Calculated composition	
Metabolism energy (Kcal/kg of DM)	2917.0
Crude Protein (%)	17.2
Fiber (%)	3.9
Fat (%)	3.0
Calcium (%)	3.2
Phosphor (%)	0.7
Lysine (%)	1.13
Methionine (%)	0.56

[1])Minerals (%), P (0.25), Ca (2.0), Mg (0.45), Na (0.35); trace elements (mg/kg): Cu (15) vitamins (IU/kg): vitamin A (6,000), vitamin D3(1,250), and vitamin E (10 mg/kg).

3. Results and discussions

3.1 Feed consumption

Body weight and feed intake were measured weekly on week 18 to week 20. In the current study, the chicken reached of BW 1540 g at 18 weeks. This weight result was in accordance with the manual standard guideline of layers chickens. From the statistical analysis, there was no significant difference in feed consumption (Table 2), but yeast consumed by chickens increases according to the high level of treatment. The major yeast cell wall is oligosaccharide, glycoprotein, peptides, minerals, and vitamins [8] [9] which have been proposed to produce positively affect response the chickens that consume them in broiler chicks [10]. In agreement to [11] reported that yeast supplementation in feed can increase body weight as a growth promoter in starter chickens. However, the data from the addition of yeast level did not affect feed consumption. Another study reported that the addition of probiotics did not cause differences in feed consumption in broiler chickens [12]. Differences in animals response may be due to the formulation of yeast in the feed which is difficult in the comparison of yeast effects

Table 2. Effects of yeast supplementation on feed intake and body weight

	Yeast supplementation (g/kg of diets)			
Parameter	0	0.5	1	1.5
Live BW (g)	1542± 126	1547±90	1553 ± 96	1563±84
Average FI g /d/hen	108.5±3	106.3±8	107.3±5	102.4±8

BW= body weight, FI = feed intake

3.2. Blood hematology profile

This study aims to evaluate blood hematological profiles in daily conditions compared to high-stress condition (after vaccination treatment). Hematological blood profiles in the pre-vaccination and 5 h after vaccination are shown in Table 3.

In the pre-vaccination condition, statistical analysis showed that the addition of yeast did not show significant differences in parameters, MCV, MCH, MCHC, RDW CV, leucocyte, eosinophil, basophil, neutrophil, lymphocyte, except for hemoglobin (Hb), erythrocyte and hematocrit in sampling before vaccination. From the data shows that with increasing levels of consumption of yeast can increase Hb content. Comparison of laying hens fed with 1.5 g / kg yeast was higher than without addition of yeast on Hb content of (7.7 vs 6.7 g / dl), erythrocyte (2.56 vs 2.25) and Haematocrit (32.7 vs 29%). At the addition of 1.5 g / kg significantly increase the Hb content, the comparison of the results shows the influence of nutrients contained in yeast, with increasing yeast nutrients consumed. Blood Hb content is affected by Fe content, with more and more available elements of Fe will potentially increase hemoglobin. Hb content in yeast treatment 1.5 g / kg that is equal to 7.73 is still in the normal range [13]. According to reported that the normal Hb content in laying hens ranged from 7.3-10.9 g / dl.

Table 3. Effects of yeast supplementation on blood hematological profile

Parameter	Yeast supplementation (g/kg of diets)			
	0	0.5	1.0	1.5
Before (24 h) ND vaccine administration				
Haemoglobin (HGB) (g/dl)	6.76 ± 0.85^{a}	6.13±0.40^{a}	7.37±0.93^{ab}	7.73±0.61^{b}
Erytrocytes (RBC)(x 10^6)	2.25±0.11^{a}	2.25±0.04^{a}	2.53±0.17^{b}	2.56±0.06^{b}
Haematocrit (%)	29.00±0.95^{a}	28.63±0.25^{ab}	31.53±2.40^{b}	32.7±0.90^{b}
MCV (fl)	128.93±1.94	127.43±1.76	124.60±1.91	127.90±3.92
MCH	30.07±1.48	27.30±1.74	29.03±1.98	30,17±11.57
MCHC (g/dl)	23.30±1.15	21.40±1.54	23.30±1.55	23.60±1.39
RDWCV (%)	9.67±0.78	8.93±0.38	8.73±0.85	9.43±1.67
Leucocyt (WBC) (x 10^3/μl)	180.±11	186.±20	215.±23	202±22
Basophil (%)	1.23±0.61	3.07±2.89	0.97±0.57	0.50±0.10
Neutrophil (%)	82.63±3.10	72.20±15.59	84.07±5.63	82.10± 1.57
Lymphocyte (%)	14.87±2.50	23.43±13.18	14.30±5.27	16.43±1.19
Monocyte (%) *	1.27±0.15	1.27±0.46	0.67±0.21	0.97±0.31
Thrombosit (PLT)	1000±577	2000±1000	1333±577	2333±1527
After ***(5 h) ND vaccine administration***				
Hemoglobin (HGB) (g/dl)	6.57 ± 0.29	7.07±0.55	6.63±0.57	6.77±0.74
Erytrocytes (RBC (x 10^6)	2.24±0.08	2.40±0.17	2.37±0.14	2.32±0.16
Haematocrit (%)	29.30±0.70	30.50±2.34	29.37±1.86	30.00±2.07
MCV (fl)	131.10±5.38	126.90±2.79	123.73±2.59	129.11±3.07
MCH	29.33±0.47	29.40±0.66	27.9±1.03 2	29.10±1.31
MCHC (g/dl)	22.43±0.90	23.17±1.03	22.6±0.51	22.50±0.87
RDWCV (%)	9.63±1.1	8.93±0.68	8.4±0.64	9.43±1.64
Leucocyt (WBC) (x 10^3/μl)	185±11	203±17	184±19	180±16
Basophil (%)	1.37±0.21	1.70+0.61	1.67+1.11	1.03+0.55
Neutrophil (%)	78.27±2.23	79.23±3.21	83.73±7.9	81.73±4.46
Lymphocyte (%)	18.63±1.69	18.27±2.80	13.87±0.89	15.70±4.31
Monocit(%)	1.73±0.46	0.80±0.10	0.73±0.03	1.53±0.35
Thrombosit (PLT)	1333±577	4333±2516	1666±577	1666±577

[a] [m]eans in a row with no common superscript differ significantly ($p<0.05$).
MCV: mean corpuscular volume, MCH: mean corpuscular hemoglobin, MCHC: mean corpuscular hemoglobin concentration, RBC: red blood cell, WBC: white blood cell, MCV: Mean corpuscular volume, RDW-CV: red blood cell distribution width coefficient variation

Hemoglobin is a molecule and consists of four subunits. Each subunit contains one heme group which is conjugated by a polypeptide (globin). Heme is a porphyrin derivative containing iron [14]. Thus, it is suspected that the addition of yeast increases mineral supply, such as Fe, but is still in the normal amount of hemoglobin [9]. Hemoglobin functions to carry oxygen in red blood cells to be transported to all parts of the body. An increase in the amount of hemoglobin is influenced by oxygen levels and erythrocytes so that with the increase in the number of erythrocytes will be accompanied by an increase in the amount of Hb.

The value of erythrocyte in chickens given yeast was higher than the control. However, the amount of erythrocyte is still in normal condition. According to [15] reported the content of erythrocyte in chickens as big as 2.0-3.5x ($10^6/mm^3$). Increased of erythrocyte contributes to improving the body's defense system against disease [16]. Erythrocyte can act as a defense system for birds that produce cytokinin. The number of erythrocytes is strongly influenced by its formation. An important factor influencing the formation of erythrocytes is tissue oxygenation. If the amount of oxygen transported to the tissue decreases it will cause an increase in erythrocyte production. Haematocrit content also increased along with the addition of yeast.

Hemocytes value is influenced by erythrocyte so that if the erythrocyte increases the hematocrit value also increases. Hematocrit values will decrease if animals are under stress conditions, one of which is caused by high temperatures. In this study the hematocrit value increases with the addition of yeast, this indicates that the chicken is under lower pressure compared to without yeast.

Values, Hb, Erythrocyte and Hematocrit showed that they still had a positive effect on chicken performance. While the hematological profile of blood in chickens with the addition of yeast in the feed did not show a significant difference in hemoglobin parameters: erythrocyte, hematocrit MCV, MCH, MCHC, RDW CV, leucocyte, eosinophil, basophil, neutrophil, lymphocyte. However, the content of each parameter shows a normal value. Normally the value of blood constituents is influenced by several factors such as genotypical physiological conditions and macro and micro climatic conditions. No significant of blood constituents by the addition of yeast treatment is likely due to the body still trying to balance its metabolic processes due to stress during vaccine administration. This is because the time for blood collection is 5 hours after vaccination. According to [17] stress conditions cause unstable blood constituents, this can be seen in Table 3, the content of blood constituents shows a less consistent trend and blood sampling after vaccination was done 1 week after vaccination to determine the level of body titers from poultry.

4. Conclusion

It is concluded in the current study that the addition of yeast has a beneficial effect on under low challenge or stress condition but yeast supplementation failed to maintain stress pressure after vaccine administration.

5. Acknowledgments

This study supported by PNBP (Penerimaan Negara Bukan Pajak) grant under grant numbers 656/PL17.4/PL/2018 Politeknik Negeri Jember

6. References

[1] Quesada A, Reginatto G A, Ruiz Español A, Colantonio L D and Burrone M S 2016 *Rev. Peru. Med. Exp. Salud Publica* **33** 32–44
[2] Cosby D E, Cox N A, Harrison M A, Wilson J L, Buhr R J and Fedorka-Cray P J 2015 *J. Appl. Poult. Res.* **24** 3 408–426
[3] Gao J 2008 *Poult. Sci.* **87** 7 1377–1384
[4] Jouany J P 2006 *Anim. Reprod. Sci.* **96** 3–4 250–264
[5] Pantaya D 2016 *J. Dairy Sci.* **99 12** 9759–9767
[6] Sui Y, Wisniewski M, Droby S and Liu J 2015 *Appl. Environ. Microbiol.* **81** 9 2968–2975
[7] Chiang S H and Hsieh W M 1995 *Asian-Aust. J. Anim. Sci.* **8** 2 159–162
[8] Pantaya D, Wiryawan K G and Amirroenas D E 2016 *J. Vet.* **17** 1 143–154
[9] Kollár R 1997 *J. Biol. Chem.* **272** 28 17762–17775
[10] Kabir S 2009 *African J. Biotechnol.* **8** 15
[11] Fowler J, Kakani R, Haq A, Byrd J A and Bailey C A 2015 *J. Appl. Poult. Res.* **24** 1 66–72
[12] EO A, OMO I, AO O and OO O 2017 *Poult. Sci. J.* **5** 1 41–50

[13] Gaensly F, Picheth G, Brand D and Bonfim T 2014 *Brazilian J. Microbiol.* **45** 2 491–494
[14] Spiro T G, Stong J D and Stein P 1979 *J. Am. Chem. Soc.* **101** 10 2648–2655
[15] Londok J J M R, Manalu W, Wiryawan I K G and Sumiati S 2018 *J. Vet.* **19** 2 222–229
[16] Abbaspour N, Hurrell R and Kelishadi R 2014 *J. Res. Med. Sci. Off. J. Isfahan Univ. Med. Sci.* **19** 2 164
[17] Altan O, Altan A L I, Çabuk M and Bayraktar H 2000 *Turkish J. Vet. Anim. Sci.* **24** 2 145–148

The effect of adding probiotic *Saccharomyces cerevisiae* on dietary antibiotic-free on production performance and intestinal lactic acid bacteria growth of broiler chicken

S Wulandari[1*] and T M Syahniar[1]

[1]Study Program of Animal Production, Department of Animal Husbandry, State Polytechnic of Jember, Jl. Mastrip PO BOX 164, Jember, East Java, Indonesia

*E-mail: suci@polije.ac.id

Abstract. This study aims to determine the effectiveness of using yeast named *Saccharomyces cerevisiae* as a probiotic in the maintenance of broiler chickens that are free from antibiotic administration and to obtain the most efficient dose of *Saccharomyces cerevisiae* as a feed additive. The success of this study supports the government's program for the future that will prohibit antibiotics usage in poultry feed, especially broiler chickens. This study consisted of 5 treatments and 5 replications for each treatment. Each replication consisted of 10 heads of broiler chicken aged 14 days. Treatment I = formulation feed (control) without yeast addition (P0), treatment II = control plus yeast 0.5 g/kg of feed (0.05%) (P1), treatment III = control plus 1 g/kg of feed (0.1%) (P2), treatment IV = control plus yeast 1.5 g/kg of feed (0.15%) (P3), and treatment V = BR1 commercial feed (K). Observations were carried out since the chickens aged 14 days to harvested on aged 35 days. Data were analysed using one way analysis of completely randomized design (CRD) by IBM SPSS Statistics 23. Significantly different results were followed by Duncan's new multiple range test (DMRT) to determine differences between treatments. The results of this study indicate that the addition of *S. cerevisiae* up to 1.5 g/kg of feed (0.15%) on dietary free-antibiotics did not affect production performance (P1, P2, P3) compared to control (P0), however still lower than broilers fed BR1 commercial feed (K). Addition of *S. cerevisiae* as probiotic on dietary free-antibiotic had a greater role for improving health status by 0% mortality and tends to increase lactic acid bacteria in the duodenum intestinal tract.

1. Introduction

Chicken is a potential commodity that has economic value as the most popular source of animal protein that is quite cheap and affordable for society. Chicken consumers recently begin to avoid broiler chicken meat due to concerns about antibiotic residues. It was derived from chicken feed which used as growth promotor or medicine for preventing or treating broiler chicken health. This inspires nutritionist to provide quality rations and produce the product of broiler chicken meat that are free antibiotic by using probiotic as feed additive.

Probiotic is microbes or spores that live or develop in the intestine and be able to benefit their hosts from their metabolites directly or indirectly [1]. The characteristics of probiotic has not pathogenic and safe product when consumed [2], and improve the balance of intestinal microflora when entering the digestive tract [3]. The most probiotic that attracted the attention are the local yeast product named

Saccharomyces cerevisiae (*S. cerevisiae*). *S. cerevisiae* is safe and widely applied to the feed sector, especially in feed fermentation [4]. *S. cerevisiae* is used to improve animal health as a probiotic and immunostimulant which is administration of feed additives [5]. *S. cerevisiae* has the potential as a probiotic agent to replace the function of antibiotics as growth stimulants. The results of previous studies indicated that the use of *S. cerevisiae* culture can increase the amount of lactic acid bacteria (BAL) which will make the digestive tract environment become acidic, so that pathogenic bacteria cannot tolerate the acidic environment and cannot develop well. The success of this research strongly supports the government's program for the future which will prohibit the antibiotics usage in poultry feed, especially broiler chickens.

2. Methodology

The study was held in UPT Agricultural Production and the analysis was carried out at Feed Technology Laboratory, State Polytechnic of Jember. This study used 250 Day Old Chick broiler chickens. The local yeast was *Saccharomyces cerevisiae* of the Optimet @ Sc product which was a screening result from the previous study (a concentration of 10^9 CF/g). Dietary antibiotic-free contained crude protein (CP) 21% and metabolic energy (ME) 3020 kcal/kg, while for BR1 commercial feed contained CP= 20% and ME = 3100 Kcal/kg. This study consisted of 5 treatments and 5 replications for each treatment. Each replication consisted of 10 heads of broiler chicken aged 14 days. .

Treatment I = formulation feed (control) without yeast addition (P0), treatment II = control plus yeast 0.5 g/kg of feed (0.05%) (P1), treatment III = control plus 1 g/kg of feed (0.1%) (P2), treatment IV = control plus yeast 1.5 g/kg of feed (0.15%) (P3), and treatment V = BR1 commercial feed (K). Observations were carried out since the chickens aged 14 days to harvested on aged 35 days. Data were analysed using one way analysis of completely randomized design (CRD) by IBM SPSS Statistics 23. Significantly different results were followed by Duncan's new multiple range test (DMRT) to determine differences between treatments. Parameters observed were feed consumption, body weight gain, feed conversion ratio (FCR), amount of intestinal lactic acid bacteria (LAB) and *Escherichia coli* (*E coli*).

3. Results and discussion

The addition of probiotic *Saccharomyces cerevisiae* (*S. cerevisiae*) in feed did not give a significant effect for feed intake on P0 to P3. Feed intake of broiler chicken is shown at **Table 1**. This condition was caused the nutritional content sufficiently balanced for all formulation feed containing probiotic, and contained completely of nutritional requirement for broiler chicken. Energy content affects the level of consumption, broiler will stop eating when the energy content is required [6]. Formulation feed of P0 to P3 contained ME 3020 kcal/kg. Energy requirements for broiler finisher period is 2900 - 3200 kcal/kg [7].

Table 1. Feed intake of broiler chicken fed dietary antibiotic-free with and without the addition of *S. cerevisiae*

Weeks	Unit	P0	P1	P2	P3	K
3	g/head/d	70.31^a	69.44^a	71.86^a	72.60^a	81.80^b
4	g/head/d	104.90^a	103.02^a	105.96^a	108.48^a	123,01^b
5	g/head/d	132.07^a	128.88^a	131.19^a	139.08^a	158,42^b
3-5	g/head/d	102.42^a	100.45^a	103.00^a	106.72^a	121.08^b

^{abc} different superscript at the same row indicated significant differences ($p<0.05$)
P0 = formulation feed (control) without yeast addition, P1= control with yeast 0.5 g/kg of feed (0.05%), P2= control with yeast 1 g/kg of feed (0.1%), P3 = control with yeast 1.5 g/kg of feed (0.15%), and K = BR1 commercial feed.

Broiler chicken that were given commercial feed (K) had the highest level of consumption compared to the formulation feed (P0 to P3). There was possible form of feed produced from factory products was better than the formulation feed. Broiler chicken preferred granular feed rather than mass. Forms of formulation feed that were more destroyed easily than commercial feed granules. The results of previous studies showed that broilers fed with good quality pellet (88% pellet durability index) produced more consumption than low quality pellet (66% pellet durability index), thus the form of feed would produce lower production performance [6].

The effect of addition of yeast probiotic as an antibiotic substitute up to 1.5 g/kg of feed (0.15%) for 35 days of maintenance did not show an adverse effect on the weight gain. Weight gain of broiler chicken is shown at **Table 2**.

Table 2. Weight gain (WG) of broiler chickens fed with the addition of *S. cerevisiae* yeast

Weeks	Unit	P0	P1	P2	P3	K
2-3	g/head/d	227.23^{a}	216.18^{a}	234.25^{a}	243.99^{a}	361.72^{b}
3-4	g/head/d	358.61^{a}	344.12^{a}	335.24^{a}	359.75^{a}	440.09^{b}
4-5	g/head/d	441.85^{a}	417.35^{a}	418.97^{a}	449.6^{a}	631.36^{b}
3-5	g/head/d	1027.69^{a}	977.65^{a}	988.46^{a}	1053.34^{a}	1433.17^{b}

^{abc} different superscript at the same row indicated significant differences ($p<0.05$)
P0 = formulation feed (control) without yeast addition, P1= control with yeast 0.5 g/kg of feed (0.05%), P2= control with yeast 1 g/kg of feed (0.1%), P3 = control with yeast 1.5 g/kg of feed (0.15%), and K = BR1 commercial feed.

According to Mountzouris et al [8] that no effect of yeast administration on broiler chickens until the age of 35 days in unhealthy conditions, showed that it was necessary to increase yeast doses and longer maintenance periods from 35 days to 42 days. Shankar [9] also reported that yeast probiotic supplementation given up to 0.2% did not affect the performance of broiler chickens maintained for 42 days. It shows that the role of *S. cerevisiae* as probiotics and for antibiotic replacement is towards prefer as an improving health status to a growth promotor. Probiotic is a biological product which stimulate the immune system and increase its defense activity against pathogenic bacteria [10]. Mannanoligosaccharides (MOS) and 1,3/1,6 β-glucan are components of the yeast cell wall that act as immunostimulants [11], namely to increase the ability of T cells, B cells and macrophages to fight disease infections, as well as help repair damaged tissue in the body through a process of regeneration and recovery [12].

The addition of yeast probiotics in formulation feed on 0.05%, 0.1%, and 0.15% of dietary antibiotic-free (P1, P2, and P3) did not show a significant effect on feed efficiency as FCR value. The feed conversion ratio (FCR) of chicken broilers is shown at **Table 3**.

Table 3. Feed convertion ratio (FCR) of broiler chickens fed with the addition of *S. cerevisiae* yeast

Weeks	Unit	P0	P1	P2	P3	K
3	head/d	2.17^{b}	2.29^{b}	2.19^{b}	2.10^{b}	1.61^{a}
4	head/d	2.05^{ab}	2.1^{bc}	2.22^{c}	2.12^{bc}	1.96^{a}
5	head/d	2.10^{b}	2.18^{b}	2.23^{b}	2.18^{b}	1.76^{a}
3-5	head/d	2.09^{b}	2.17^{b}	2.20^{b}	2.14^{b}	1.78^{a}

^{abc} different superscript at the same row indicated significant differences ($p<0.05$)
P0 = formulation feed (control) without yeast addition, P1= control with yeast 0.5 g/kg of feed (0.05%), P2= control with yeast 1 g/kg of feed (0.1%), P3 = control with yeast 1.5 g/kg of feed (0.15%), and K = BR1 commercial feed.

The results showed a tendency for an increase in the number of intestinal lactic acid bacteria (LAB) by the addition of yeast as an feed additive of dietary antibiotic-free especially for P1 and P2, although not significantly different between treatments. The number of *E. coli* in the intestines was markedly decreased, as well as BR1 commercial feed. It is shown at **Table 4**.

Table 4 Amount of intestinal lactic acid bacteria (LAB) and intestinal *E coli* of broiler chickens fed with the addition of *S. cerevisiae* yeast

Parameter	Unit	P0	P1	P2	P3	K
LAB	CFU/ml	$(8.37x10^{7})^{a}$	$(4.15x10^{8})^{a}$	$(1.79x10^{9})^{a}$	$(9.97x10^{7})^{a}$	$(6.22x10^{8})^{a}$
E.coli	CFU/ml	$(3.89x10^{6})^{b}$	$(7.77x10^{5})^{a}$	$(4.04x10^{6})^{b}$	$(1.03x10^{6})^{a}$	$(4.77x10^{5})^{a}$

[abc] different superscript at the same row indicated significant differences ($p<0.05$)
P0 = formulation feed (control) without yeast addition, P1= control with yeast 0.5 g/kg of feed (0.05%), P2= control with yeast 1 g/kg of feed (0.1%), P3 = control with yeast 1.5 g/kg of feed (0.15%), and K = BR1 commercial feed.

S. cerevisiae was useful as a role for improving animal health as a probiotic and immunostimulant. *S. cerevisiae* as a probiotic was not killing microbes even increasing the number of beneficial microbes, in contrast to antibiotics which killed microbes both of harmful or beneficial for the body and showed resistance effect [12]. It was supported by 0% chicken mortality, even the broiler chicken fed dietary antibiotic-free and without any medicines usage during maintenance.

The addition of *S. cerevisiae* could significantly reduce the amount of *E.coli* in the duodenum, especially at doses of 0.05% and 0.15%. The development of LAB causes the condition of the digestive tract became acid that was not suitable for the growth of *E. coli*. In normal conditions, both microorganisms (pathogens and non-pathogens) are in balanced state. There are a number of interactions in the form of symbiosis and attachment between both of them. The balance of the intestinal microflora will be achieved when beneficial microbes can suppress harmful microbes by pushing out the pathogenic microbes from the digestive tract [13].

4. Conclusions and recommendations

Conclusions

Yeast administration of *S. cerevisiae* as a probiotic up to doses of 1.5 g/kg of feed (0.15%) in broiler chickens maintained without using antibiotics did not affect production performance compared to those without yeast. The maintenance of broiler chickens fed antibiotic-free with the addition of *S. cerevisiae* had a greater role for improving health status by 0% mortality and tends to increase lactic acid bacteria in the digestive tract.

Recommendations

The lack of a significant improvement between the yeast addition treatment and under no antibiotics condition, can imply that higher yeast inclusion levels up to 2% and longer experimental durations (i.e., 42 d instead 35 d) should be considered in the future studies. The consideration broiler chicken aged over 35 days are susceptible to disease.

5. Acknowledgements

The research team thanked profusely to the P3M of State Polytechnic of Jember for providing research funding through PNPB sources so that this research could be carried out.

6. References

[1] Kompiang I P 2009 *Pengembangan Inovasi Pertanian* **2** 3
[2] Prado F C Parada J L Pandey A Soccol. C R 2008. *Food Research Internaional.* **41** 2
[3] Dommels Y E M Kemperman R A Zebregs Y E M P Draaisma R B Jol A Wolvers D A W Vaughan E E and Albers R 2009 Appl Environ Microbiol **75** 19

[4] Pantaya D Morgavi D P Silberberg M Martin C Durand F Suryahadi K G W Boudra H 2016 *J Dairy Sci* **99** 12
[5] Ahmad R Z 2005 WARTAZOA **15** 1
[6] Dozier W A Behnke K C Gehring C K Branton S L 2010 *J Appl Poul Res* **19** 2
[7] Wahyu J 1992 Ilmu Nutrisi Unggas (Yogyakarta: Gadjah Mada University Press)
[8] Mountzouris K C Dalaka E Palamidi I Paraskeuas V Demey V Theodoropoulos G Fegeros K 2015 *Poultry Sci* **94** 10
[9] Shankar P A Premavalli K Omprakash A V John Kirubakaran J and Hudson G H. 2017. IJABR, **7** 2
[10] Priya B S and Babu S S. 2013 *IJPRAS* **3** 2
[11] Eltazi S M Mohamed K A and Mohamed M A 2014 *IJPRAS* **3** 2
[12] Anonimus 2004 Beta Glucan Research - *Saccharomyeces cerevisiae* [internet] available from http://www.betaglucan.org/ Accessed at 5 December 2017
[13] Rahmah A Suthama N and V D Yunianto VD 2013 AAJ **2** 3

IOP Conf. Series: Earth and Environmental Science **207** (2018) 012035 doi:10.1088/1755-1315/207/1/012035

Addition garlic extract in ration on fat deposition of broiler

M M D Utami[1], D Pantaya[1], N Nofida, N H D Larasati[1], and A Agus[2]

1 Department of Animal Science, Politeknik Negeri Jember, Jl. Mastrip PO Box 164, Jember 68101, Indonesia
2 Faculty of Animal Science, Universitas Gadjah Mada, Bulaksumur Yogyakarta 55281, Indonesia

e-mail: merry.mdu@polije.ac.id

Abstract. This research was conducted to determine the effect of garlic extract on the fat deposition of the broiler. The experimental design used Completely Randomized Design with four treatments, each treatment has of five replications and each replication consist of four broilers. The data using analysis of variance and when there were have significant differences, then further tested by Duncan New Multiple Range Test. The treatments were T0 (without garlic extract), T1 (2%), T2 (4), and T3 (6%). The treatments were started from 15 to 35 days of age. Parameters observed included: feed intake, weight gain, feed conversion, carcass weight, carcass percentage, abdominal fat weight, and abdominal fat percentage. The results showed that the addition of garlic extract did not affect (P>0.05) to feed consumption. The addition 2% of garlic extract in ration had a significant effect on body weight and feed conversion while addition 4% of garlic extract in ration decreased (P<0.01) abdominal fat weight, abdominal fat percentage, and increased carcass weight P<0.05) and percentage of the carcass (P<0.01).

1. Introduction

The Food Agriculture Organization estimates that demand for livestock production has doubled and produced more than half the total value of world agricultural production. Then by 2020, it is assumed that there will be a revolution in the field of animal husbandry (the next food revolution), namely 60% of meat production will be produced by developing countries, including Indonesia.

The broiler is one source of animal protein that is many consumed by Indonesian people. in the tropics area has high temperature and humidity, so broilers can be stress which the response of the environment adaptation. According to [1] broiler in the tropics can be stress which the responses of the environment. Adaptation responses can disturb the metabolism process. The energy consumed cannot be converted to be meat maximally and can impact of high abdominal fat. The addition of natural ingredients can be acted to increase the body weight, the feed conversion, and to decrease the abdominal fat.

The addition of natural ingredients can increase the body weight, the feed conversion, and to decrease the abdominal fat. Garlic (*Allium sativum* L) is used as an herbal supplement because based on research that has been conducted on experimental animals, the content of organosulfur compounds prevents cancer [2] [3]. Garlic supplements have been widely used in the world [4].

Garlic is a perennial herbaceous plant that forms a layer bulb, grows in clumps and stands up to 30 to 75 cm tall. The stem that appears on the ground is a false stem, the true stem is in the ground. From the base of the stem grow many fibers with a length of less than 10 cm. Garlic forms a layer bulb, each tuber consists of 8 to 35 cloves [5]. Each garlic bulbs contains 100 gram such as 4.5 grams of protein; 0.2 grams of fat; 23.1 grams of carbohydrates; calories around 95 calories; 0.22 mg of vitamin B1; 1.5 mg of vitamin C; calcium 42 mg; phosphorus 134 mg; iron 1 mg and water around 71 g. In percentage conversion, the nutrient content is as follows: carbohydrates 28%, protein 2%, crude fiber 1.5%, organosulfur compounds 2.3% and water content 65% [6].

Garlic contains more than 100 secondary metabolites [7]. Secondary metabolites contained in garlic form a complex chemical system and are a self-defense mechanism from damage caused by microorganisms and other external factors [8]. These secondary metabolite compounds contain sulfur and influence the taste and pharmacological properties of garlic [9]. The main organosulfur compounds with garlic tubers are non-volatile amino acids glutamyl-S-L-cysteine and essential oils of cysteine sulfoxide or alliin. The compound is a precursor to most other organosulfur compounds, reaching 82% of all organosulfur compounds [10]. The glutamyl-S-L-cysteine compound is formed from the amino acid biosynthesis pathway. The enzymatic reactions that occur from glutamyl-S-L-cysteine produce many derivative compounds, through two branches of reaction, namely S-all cysteine (SAC) forming pathways [11]. From the thiosulfinate formation pathway allicin compounds will be produced, then from this pathway will be formed allyl sulfide, dithiin, ajoene, and other sulfur compounds. Allicin is unstable [8] so it is easy to further reactions to organosulfur compounds [12]. Extraction of garlic with ethanol at a temperature below 0° C produces alliin, at 25° C produces allicin and at a temperature of 100° C by steam distillation method all the alliin content changes to allil sulfide [13]. Fresh garlic contains 2 to 6 mg per g of gamma-glutamyl-S-allyl cysteine or 0.2 to 0.6% of fresh weight and 6 to 14 mg per g of alliin or 0.6 to 1.4% of fresh weight. Furthermore, garlic extract contains 2500 to 4500 mcg/g of allicin from fresh weight before being cut, so that if made on average, one garlic contains 2 to 4 g of allicin [14].

Allicin can increase the conversion feeding so the broiler has a low fat abdominal and high meat production. Organosulfur is able to increase the meat production. Allicin and organosulfur can be found in garlic [15] and garlic contains a variety of organosulfur complexes, such as ajoene, S-allyl cysteine, S-allyl cysteine sulfoxide, diallyl disulfide, and S-methyl cysteine sulfoxide. Allicin as an antimicrobial agent. Allium sativum also contains approximately 100 sulfur compounds that are fundamentally pharmacological potential [16]. Scordinin helps the metabolism process and increase the sulfur of amino acid as methionine. Methionine is an essential amino acid to muscle tissue of broiler [17]. Garlic has to contain allicin, it was expected to improve the broiler performance.

The aim of this research to know the effect of garlic extract on broiler performance. The benefits of this research are giving the information about the effect garlic extract in the feed of broiler performance. Besides the use of fresh garlic, currently, there are some garlic preparations, namely garlic powder, garlic oil, and garlic extract. Garlic extract in ethanol with a concentration of 15 to 20% can be stored for about 20 months at room temperature and produce aged garlic extract [18]. In the form of extracts, garlic can be stored for up to 20 months. In garlic extract all allicin is converted to allyl sulfide, according to the percentage, diallyl trisulfide (DTS), diallyl disulfide (DDS) and diallyl sulfide (DS) [19].

2. Materials and Methods

2.1 Preparation of Garlic Extract

250 grams of garlic mashed using a juicer and then mixed with 500ml of 96% ethanol. The solution was filtered with 2 layers of gauzy, then filtered again with Whatman paper number 2 so obtained crude extract. The filtrate on the Erlenmeyer tube and placed in a water bath to evaporate ethanol until got 100 ml of garlic extract.

2.2 Birds and Treatments

The experimental arrangement was using a completely randomized design. Eighty Day Old Chick (DOC) were randomly divided into four treatments, each treatment consists of five replications and each replication consist of four chickens. Treatment T0 or control group received a basal feed with 21-

22% protein content and metabolic energy 2800-3000 Kcal/Kg without supplementation of garlic extract while T1, T2, and T3 were basal feed supplemented with 2%, 4%, and 6% garlic extract. The treatments were started from 15 to 35 days of age. The concentration of garlic extract through this process was 280 mg/mL. On the 1st day until the 14th days of age, the chicken was fed basal feed without supplementation of garlic extract. Feed supplemented with garlic extract was given at 3-5 weeks of age. Water and feed provided by ad-libitum.

2.3 Research Parameters

Feed intake was measured daily during the research. Feed consumption can be calculated by finding the difference between the amount of feed in one day with the rest of the feed that existed on the next morning. Weight gain was weighed every week during the research. The weight gain is measured by calculating the difference between the final weight and the initial weight. Feed conversion calculated weekly during the study. Conversion of feed is the ratio between the amount of feed consumed in one week with weight gain in one week. Carcass weight in this research comes from carcass without head, neck, fur, blood, legs and internal organ. The percentage of the carcass was calculated after the carcass weight obtained. The percentage of the carcass was calculated by weight of carcass divided by live weight and multiplied 100% [20]. The abdominal fat weight calculated by the fat found around the gizzard and the layers attached between the abdominal and bowel muscle. Abdominal fat percentage by calculating the weight of abdominal fat divided by body weight multiplied by 100%.

2.4 Data analysis

Data obtained from the results of this study were analyzed with analysis of variance (ANOVA) using Statistical Product for Service Solution program (SPSS 16.0). Results of analysis showing significant mean differences were continued by Duncan Multiple Range Test.

3. Results and Discussion

The research of the broiler performance from 3rd to 5th week included: feed consumption, body weight, feed conversion, carcass weight, carcass percentage, abdominal fat weight, the abdominal fat percentage can be seen in Table 1. Side effects and toxicity of garlic are not found so that they are safe for consumption. Safety of using garlic extract based on several studies has been recommended [8] there has been no reported toxicity of organosulfur compounds of garlic extract.

Table 1. Effect of garlic extract on broilers

Parameters	Treatments			
	T0	T1	T2	T3
Feed Consumption	3218,40^{ns}	3045,20^{ns}	3182,40^{ns}	3154,40^{ns}
Body Weight Gain	1443,00^{a}	1649,80^{c}	1596,00^{b}	1588,00^{b}
Feed Conversion	2,23^{c}	1,85^{a}	1,99^{b}	1,99^{b}
Carcass weight	1098,00^{a}	1179,00^{bc}	1202,00^{c}	1107,00^{ab}
Abdominal fat weight	29,40^{a}	27,55^{a}	24,00^{b}	28,25^{a}
Abdominal fat percentage	1,84^{a}	1,63^{bc}	1,47^{c}	1,78^{ab}
Carcass Percentage	68,60^{a}	69,95^{a}	73,21^{b}	70,49^{a}

^{ns} Nonsignificant

The analysis variance of the feed consumption showed that the addition on garlic extract in the feed broiler is not significantly (P> 0,05) on feed consumption. According to [1] showed that the addition on garlic extract is not significantly on feed consumption, it is because broiler needs energy, feed consumption will increase when the broiler consumed feed of low energy and consumption will decrease when the broiler consumed feed of high energy and report of [21] the broiler needed 3200 Kcal energy metabolism. The broiler consumed the same energy, so there is a no different influence in each treatment.

Addition of garlic extract in ration was significantly effected (P <0,01) on body weight of broiler. All compounds and enzymes contained in garlic were able to provide value to the metabolic system in the broiler. When the metabolism system in good condition, the absorption of feed consumption become the body weight can be optimal. Proteins have a function in building organs and tissues and providing amino acids. Although the research showed that protein did not give an effect on feed consumption, protein and energy given significantly on the body weight gain broiler, because of the content in garlic extract helped the process of feed absorption on the broiler [22].

The research showed that garlic extract had significantly effect on the feed conversion (P <0,01) because of energy and protein in the feed that contains extract of garlic is absorbed optimally. Allicin that showed antimicrobial is able to avoid the body from the pathogenic bacterial infection, methylatyltrisulfide prevent blood coagulation. Antimicrobial activity in garlic is caused by allicin and ajoene compounds [23]. Ajoene, an organosulfur compound found in onion oil, inhibits the growth of gram positive and negative bacteria [24]. A good metabolism process will give a good absorption process, so the value of feed conversion was low. The percentage of using garlic extract as much as 2% gives efficiently feed conversion than other treatments [25]. The percentage of addition garlic extract as much as 2% showed the best research on the feed conversion of the broiler. Addition garlic extract in the ration can increase the antibacterial activity in allicin, it can damage the intestinal mucosa and inhibit protein synthesis. When the intestinal mucosa is damaged and protein synthesis is inhibited it will decrease the metabolism system, the absorption of feed was not optimally, the body weight is low and the value of feed conversion ratio was high.

The result of supplementation of garlic extract in ration increased carcass weight. The result of the different test of carcass weight of T0 (control) was lowest (P <0,05) than T1, T2, and T3. Garlic extract contains bioactive diallyl sulfide (DAS) and diallyl disulfide (DADS). Diallyl sulfide (DAS) and diallyl disulfide (DADS) can increase the availability of amino acids containing sulfur, including methionine which is an essential amino acid that functions for muscle tissue growth. According to [26] methionine can increase the muscle mass of breast through the way of synthesizing creatine from methionine and then kept on muscle so that increased muscle mass. The increased muscle period will cause the flesh to increase so that the carcass weight was higher than control.

The percentage of carcasses in this research ranged from 68.60% to 73.21%. The result of variance analysis showed that the addition of garlic extract in ration increased the percentage of carcass significantly. The data is correlated with data on average carcass weight and abdominal fat weight data. The weight of carcass from highest to the lowest was T2, T1, T3, and T0, respectively, whereas abdominal fat weight from highest to lowest was T0, T3, T1, and T2, respectively. These data indicate that a high percentage of carcass obtained from high carcass weight and low abdominal fat weight.

Addition of garlic extract in ration significantly decreased abdominal fat. The result of the difference test of the mean of the lowest abdominal fat weight of T2 (P <0,01) compared to T0, T1, and T3. Garlic extract contains Allicin which has the binding properties of the functional part of Co-A while Co-A needs for cholesterol biosynthesis. Allicin in the gastrointestinal tract results in intestinal mucosal peeling, this condition triggers low digestive pH and suppresses cholesterol synthesis enzyme activity ie 3-hydroxy-3-methylglutaryl CoA reductase (HMGCo-A reductase) enzyme. This is in accordance with opinion [27] the addition of garlic significantly reduces the activity of HMG-A reductase. The decrease in HMG CoA reductase activity causes the amount of Co-A to decrease. Reduced Acetyl-CoA results in the obstruction of the lipogenesis process. According to [28] fat is formed from the conversion of intermediates, such as acetyl Co-A. Acetyl Co-A produced by metabolism process will enter into the citric acid cycle so that energy is produced, if energy requirement is sufficient, acetyl Co-A can experience lipogenesis to fatty acid and then esterified into triglyceride. The inhibited lipogenesis caused the transfused fats to the adipose tissue decreased so that abdominal fat decreases. Garlic extract supplementation in animal feed effectively decreases 15% cholesterol concentration and 30% LDL compared to controls [15]. Allicin is able to bind to the organosulfur group which is a functional part of co-enzyme A in the cholesterol formation process. Allicin and allyl mercaptan inhibit the HMGCo-A reductase enzyme and lanosterol accumulation [29].

Blocking activity of HMGCo-A reductase enzyme are indications of the absence of cholesterol synthesis and at the same time inhibit mevalonate [30]

The results of the analysis of variance performed showed that the addition of garlic extract in the ration decreased the abdominal fat percentage significantly. The average difference test of abdominal fat percentage on T0 (control) was highest (P <0,05) compared to T1, T2, and T3. The low percentage of abdominal fat in this study was obtained from the low abdominal fat weight. It shows the ability of bioactive allicin inhibits the process of lipogenesis in the body resulting in the decline in fat deposits. Less abdominal fat causes lower abdominal fat percentage. The abdominal fat percentage shows the number of parts of broiler that is not consumed by a human, so the greater the percentage of abdominal fat, the lower the quality of broiler.

4. Conclusion

Addition of garlic extract did not affect (P>0.05) to feed consumption, 2% of garlic extract had a significant effect on body weight and feed conversion while 4% of garlic extract decreased (P<0.01) abdominal fat weight, abdominal fat percentage, but increased carcass weight (P<0.05) and percentage of carcass (P <0.01).

Acknowledgment

The authors really highly appreciate to Politeknik Negeri Jember for a grant to research and present this paper on the 1st International Conference on Food and Agriculture in Bali, Indonesia.

References

[1] Utami M M D and Pantaya D 2016 *Penggunaan Ekstrak Bawang Putih dalam Pakan terhadap Performans Ayam Broiler Tropis Fase Starter* (*Prosiding*)
[2] Wargovich M J, Woods C, Hollis D M and Zander M E 2001 *J. Nutr.* **131** 113034S–3036S
[3] Serrano A, Ros G and Nieto G 2018 *Medicines* **5** 3 76
[4] Hossain M A, Akanda M R, Mostofa M and Awal M A 2014 *J. Adv. Vet. Anim. Res.* **1** 4 189–195
[5] Mangal J L, Singh R K, Yadav A C, Lal S and Pandey U C 1990 *J. Hortic. Sci.* **65** 6 657–658
[6] Rana S V, Pal R, Vaiphei K, Sharma S K and Ola R P 2011 *Nutr. Res. Rev.* **24** 1 60–71
[7] Kyo E 1997 *Phytomedicine* **4** 335–340
[8] Amagase H, Petesch B L, Matsuura H, Kasuga S and Itakura Y 2001 *J. Nutr.* **131** 3 955S–962S
[9] Vazquez-Barrios M E, Lopez-Echevarria G, Mercado-Silva E, Castano-Tostado E and Leon-Gonzalez F 2006 *Sci. Hortic.* **108** 2 127–132
[10] Li M 2002 *Carcinogenesis* **23** 4 573–579
[11] Santhosha S G, Jamuna P and Prabhavathi S N 2013 *Food Biosci.* **3** 59–74
[12] Cantwell M 2000 *Perishables Handl. Q* **102** 5–6
[13] Su F and Zhang B 2006 *J. Cereal. Oils.* **4** 17
[14] Lawson L D 1998 *Phytomedicines Eur. Chem. Biol. Act. Washington, DC Am. Chem. Soc.* 176–209
[15] Yalcın S, Onbaşılar E E, Reisli Z and Yalcın S 2006 *J. Sci. Food Agric.* **86** 9 1336–1339
[16] Singh V K and Singh D K 2008 *Annu. Rev. Biomed. Sci.* **10**
[17] Wafa Z 2008 *Pengaruh Penambahan Dl-Metionin terhadap Nilai Energimetabolis Ransum Ayam Broiler Starter Berbasis Jagung dan Bungkil Kedelai (Skripsi)*
[18] Banerjee S K and Maulik S K 2002 *J. Nutr.* **1** 1 4

[19] Lachance P A 1997 *Nutr. Press Trumbull, CT*
[20] Castellini C, Mugnai C and Dal Bosco A 2002 *Meat Sci.* **60** 3 219–225
[21] N. R. C. (US) 1984 *Nutrient requirements of poultry* No. 1. National Academies of Science
[22] Adibmoradi M, Navidshad B, Seifdavati J and Royan M 2006 *J. Poult. Sci.* **43** 4 378–383
[23] Kyo E, Uda N, Ushijima M, Kasuga S and Itakura Y 1999 *Phytomedicine* **6** 5 325–330
[24] Naganawa R, Iwata N, Ishikawa K, Fukuda H, Fujino T and Suzuki A 1996 *Appl. Environ. Microbiol.* **62** 11 4238–4242
[25] Horton G, M J, Fennell M J and Prasad B M 1991 *Can. J. Anim. Sci.* **71** 3 939–942
[26] Zhan X A, Li J X, Xu J R and Zhao R Q 2006 *Br. Poult. Sci.* **47** 5 576–580
[27] Bayan L, Koulivand P H and Gorji A 2014 *Avicenna J. phytomedicine* **4** 1
[28] Mattson F H, Benedict J H, Martin J B and Beck L W 1952 *J. Nutr.* **48** 3 335–344
[29] Gebhardt R and Beck H 1996 *Lipids* **31** 12 1269–1276
[30] Rasmussen P 1998 *Aust. J. Med. Herbal* **10** 3 94

Antimicrobial activity of essential oil from Indonesian medicinal plants against food-borne pathogens

T Budiati*[1], W Suryaningsih[1], S. Umaroh[2], B. Poerwanto[2], A Bakri[2] and E Kurniawati[1]

1 Food Engineering Department, Politeknik Negeri Jember, Mastrip PO Box 164 Jember 68101 East Java, Indonesia
2 Food Industrial Technology Department, Politeknik Negeri Jember, Mastrip PO Box 164 Jember 68101 East Java, Indonesia

*Email: titik_budiati@polije.ac.id

Abstract. The aim of the study was to evaluate antimicrobial activity of essential oil from local plants or plants cultivated in Indonesia against three food-borne pathogens, *Salmonella* Thypimurium, *Bacillus cereus* and *Staphylococcus aureus*. A total 6 essential oil extracted from ginger (*Zingiber officinale*), turmeric (*Curcuma longa*), candlenut (*Aleurites moluccanus*), lemongrass (*Cymbopogon citratus*), clove bud (*Syzygium aromaticum*), and galangal (*Alpinia galanga*). By using disc diffusion assay, the highest antimicrobial activity against *S.* Tyhimurium, *B. cereus* and *S. aureus* was shown by clove bud oil (3.67 $\pm$ 0.58 mm), lemongrass oil (3.58 $\pm$ 0.14 mm) and clove bud oil (5.25 $\pm$ 0.75 mm), respectively. By using Minimum Inhibitory Concentration (MIC), the concentration of essential oil to inhibit the growth of *S.* Tyhimurium, *B. cereus* and *S. aureus* was found on clove bud oil for 0.39 %, lemongrass oil for 1.56 % and clove bud oil for 0.78 %, respectively. Out of the essential oil tested, clove bud oil and lemongrass oil showed promising antibacterial activity against *S.* Thypimurium, *B. cereus* and *S. aureus*.

1. Introduction

In developed and developing countries, food-borne diseases were considered by pathogenic bacteria as the primary causes. Assuring food safety, the addition of chemical agents has been used in food industries. However, it becomes a concern due to the adverse effect on human health. The consumers' preference tends to choose the food products with the fewer chemicals and more natural preservatives. Essential oils were potential alternatives to assure food safety by inhibiting food-borne pathogenic bacteria in food. Essential oils has been studied to show antimicrobial activity on pathogenic bacteria [1]. The structure, functional groups, and composition of the essential oils play an important role in determining their antimicrobial activity [2; 3; 4]. However, the data regarding the antimicrobial activity of essential oils from Indonesian medicinal plants, such as ginger oil (*Zingiber officinale*), turmeric oil (*Curcuma longa*), candlenut oil (*Aleurites moluccanus*), lemongrass oil (*Cymbopogon citratus*), clove bud oil (*Syzygium aromaticum*), and galangal oil (*Alpinia galanga*) to against food-borne pathogens is still limited. Food-borne pathogens continue to be a major public health problem worldwide. These also contribute to negative economic impacts because of the cost of surveillance investigation, prevention and the treatment of illness [5]. The pathogenic bacteria which has been known to cause public health problem were *Salmonella* Thypimurium, *Bacillus cereus and Staphylococcus aureus*

S. Thypimurium is Gram-negative bacteria, rod-shape that causes salmonellosis. These pathogenic bacteria caused enteric fever and acute gastroenteritis within incubation period of 6-72 hours [6]. *Salmonella* is the most frequently isolated as foodborne pathogen, and is predominantly found in eggs, dairy products and poultry [5]. Other food-borne pathogen is by *B. cereus* which has resistant endospores as a cause of intoxication [7]. The spore may be present in raw and cooked foods that can survive in high temperature during food processing [8]. *B. cereus* is Gram-positive bacteria causing an emetic or diarrheal syndrome [6]. Other food-borne disease is Staphylococcal food-borne disease (SFD). SFD is one of the most common food-borne diseases worldwide which caused by the contamination of *S. aureus* enterotoxins in food [9]. *S. aureus* is Gram-positive bacteria that does not form spores but can cause the contamination during food preparation and processing [9]. This pathogenic bacteria can grow in the range of 7° to 48.5 ^oC, pH of 4.2 to 9.3 and sodium chloride concentration up to 15% NaCl [9]. The present study was conducted to evaluate antimicrobial activity of essential oil from local plants or plants cultivated in Indonesia against three food-borne pathogens, *S.* Thypimurium, *B. cereus* and *S. aureus*.

2. Material and Methods

2.1 Essential oil

A total 6 Indonesian medicinal plants extracted from ginger (*Zingiber officinale*), turmeric (*Curcuma longa*), candlenut (*Aleurites moluccanus*), lemongrass (*Cymbopogon citratus*), clove bud (*Syzygium aromaticum*), and galangal (*Alpinia galanga*). The pure essential oils were purchased from Nusaroma (Indonesia).

2.2The strains of microorganisms

Antimicrobial activity of the essential oil was investigated in 3 pathogenic bacteria *strains: S. enterica* serovar Typhimurium ATCC 14028, *B. cereus* ATCC 14579, *S. aureus ATCC 25923. S.* Typhimurium was cultured on Xylose Lysine Deoxycholate (XLD) agar (MerckKGaA, Damstads, Germany) for 24-48 hours at 37°C. Well colonies were appeared as red colonies with black centers and the agar will turn red. The colonies were purified by streaking onto tryptic soya agar (TSA) (MerckKGaA, Damstads, Germany). *Bacillus cereus* was cultivated on Brilliance *B. cereus* (BBC) agar (Oxoid, Baringstoke, Hampshire, United Kingdom) for 24 hours at 37°C. The colonies were appeared as blue colonies and were purified onto TSA bu streaking method. *S. aureus* was streaking on Manitol Salt Agar (MSA) (Himedia, India) and incubated at 37°C for 24-48 hours was appeared as yellow colonies. The colonies were purified onto TSA by using streaking methods.

2.3 Disc diffusion assay

Antimicrobial activity was evaluated using disc diffusion assay as Bauer [10]. Cultures were grown overnight in Triptic Soy Broth (TSB, Merck KGaA, Darmstadt, Germany) and incubated at 37 °C. The overnight cultures were diluted to a turbidity of 0.5 on McFarland scale. The cultures were streaked on Mueller Hinton Agar (Oxoid, Baringstoke, Hampshire, United Kingdom) plates using a cotton swab. After 30 min, 20 μL essential oil was dropped on the disc that has been placed on the plates and were incubated at 37 °C for 18–24 h. After the incubation period, the inhibition zones was measured in 4 different directions and recorded as milimeter.

2.4 Minimum Inhibitory Concentration (MIC)

Cultures were grown overnight in Triptic Soy Broth (TSB, Merck KGaA, Darmstadt, Germany) and incubated at 37 °C. The overnight cultures were diluted to a turbidity of 0.5 on McFarland scale. The cultures were dropped onto different concentration of essential oil. The first concentration was measured in 5 mL of essential oil. In the remaining concentration was added 2,5 mL Mueller Hinton

Broth (Oxoid, Baringstoke, Hampshire, United Kingdom) and was reduced gradually for 50%. Finally, each tube was added 20 μL of the bacterial suspension (10^6 CFU/mL). The tubes were then incubated for 24 hours at a temperature of 37 °C. The aliquot was dropped on Mueller Hinton Agar and incubated for 24 hours at 37 °C. The lowest concentration at which colonies grows occurred was taken as the MIC value (%). MIC values shown are mean values of three measurements. All experiments were performed in three replicates.

2.5 *Statistical Analysis*

The differences in MIC values of *S.* Typhimurium, *B. cereus*, *S. aureus* between different essential oils were determined by using one-way ANOVA (SPSS version 13.0) at a significance level of $P < 0.05$.

3. Results and Discussions

Minimum Inhibitory Concentration (MIC) of *S.* Typhimurium, *B. cereus* and *S. aureus* on six different essential oils was shown in table 1. By using disc diffusion assay, the three highest antimicrobial activity against *S.* Typhimurium was shown by clove leaf oil, lemongrass oil and ginger oil for 3.67 ± 0.58 mm, 2.92 ± 0.14 mm and 2.29 ± 0.36 mm, respectively. Statistical analysis showed significant difference ($P > 0.05$) between MIC value to inhibits the growth of each pathogenic bacteria.

Table 1. Inhibition zone (IZ) and Minimum Inhibitory Concentration (MIC) of food-borne pathogenic bacteria *S.* Typhimurium, *B. cereus* and *S. aureus* on six different essential oils

Food-borne pathogen bacteria	Ginger oil		Turmeric oil		Candlenut oil		Lemongrass oil		Clove bud oil		Galangal oil	
	IZ (mm)	MIC (%)	IZ (mm)	MIC (%)	IZ (mm)	MIC (%)	IZ (mm)	MIC (%)	IZ (mm)	MIC (%)	IZ (mm)	MIC (%)
Salmonella Typhimurium	2.29 ± 0.36	6.25 ± 0^b	1.75 ± 0	12.5 ± 0^c	No inhibiton	No inhibiton	2.92 ± 0.14	6.25 ± 0^b	3.67 ± 0.58	0.39 ± 0^a	1.17 ± 0	6.25 ± 0^b
B. cereus	1.92 ± 1.23	6.25 ± 0 ^{d}	1.63 ± 0.57	12.33 ± 0,29^e	No inhibiton	No inhibiton	3.58 ± 0.14	1.56 ± 0^c	3.25 ± 0.9	0.78 ± 0^b	0.75 ± 0.43	0.39 ± 0^a
S. aureus	4.33 ± 0.58	1.55 ± 0.05^b	1.88 ± 0.29	3.13 ± 0^c	0.5 ± 0	23.33 ± 2.89^e	1.75 ± 1.56	6.25 ± 0^d	5.25 ± 0.75	0.78 ± 0^a	1.92 ± 0.38	0.39 ± 0^a

Note : a,b = different alphabet means significant different at P<0.05 in the same row of MIC value

By using MIC, the highest antimicrobial activity of essential to inhibit the growth of *S.* Typhimurium was found on clove bud oil for 0.39 ± 0%. It was followed by three essential oils (lemongrass oil, ginger oil and galangal oil) that showed the same MIC value (6.25 ± 0%). The result was similar with Dorman and Deans [2]. Dorman and Deans reported that clove oil composed by volatile oil compounds which can inhibits *Salmonella* Pullorum and *S. aureus*. These were mainly carvacrol, δ-3-carene, eugenol and cis/trans citral. Other compounds were identified as carvacrol methyl ester and geraniol. There was only borneol which did not has microbial activity on *S.* Pullorum and carvacrol methyl ester on *S. aureus* [2]. Other study has reported that the component with phenolic structures, such as carvacrol and thymol have synergetic antimicrobial activity against major foodborne pathogens such as *S.* Typhimurium by attacking the bacterial membrane [11] and disrupting the cytoplasmic membrane of Gram-negative bacteria [12]. By altering the fatty acid composition of the bacterial membrane, carvacol might promote the sub-lethal injury to the cells [12]. However Moon reported that eugenol was not enough to show bactericidal effect to *S.* Typhimurium

[11]. Other compounds were reported has antimicrobial effect on *Salmonella* sp. and *S. aureus*. These were geranyl acetate, cis-hex-3-en-1-ol, menthone, α-pinene, β-pinene, terpinen-4-ol, α-terpineole [2].

By using disc diffusion assay, the highest antimicrobial activity against *B. cereus* was shown by lemongrass oil for 3.58 ± 0.14 mm. This was followed by clove bud oil and ginger oil for 3.25 ± 0.9 mm and 1.92 ± 1.23 mm, respectively. It was also observed on the minimum concentration inhibitory data as shown in table 1. Clove bud oil and lemongrass oil showed has high antimicrobial activity for 0.78 ± 0% and 1.56 ± 0% , respectively. The antimicrobial activity of the compounds in clove bud oil and lemongrass oil showed relatively higher on Gram-positive bacteria than those of Gram-negative bacteria. These results were similar to other studies. Moon [11] and Burt [3] reported that Gram-positive bacteria are more susceptible to carvacrol and thymol than Gram-negative bacteria. However α-terpineole reported to be more effective to inhibit Gram-negative (*S.* Typhimurium) than those Gram-positive bacteria (*S. aureus* and *B. cereus*) [3]. These compounds were composed by hydroxyl group in phenolic structure which having antibacterial properties to inhibit food-borne pathoges as reported by Dorman and Deans [2]. The mechanism of action generally was considered by disturbing the cytoplasmic membrane and the proton motive force (PMF), active transport, electron flow and coagulation of cell contents [3]. Futher, the alkyl substitution into the phenol nucleus might enhance the activity of antimicrobial by altering the distribution ratio between the nonaqueous and aqueous phases (including bacterial phases) which reduce the surface tension or altering the species selectivity [2]. Moreover, the hydrophobicity of essential oils enables to make the partition on the lipids of bacterial membrane and mitochondria which disturb the structure, render the membrane to be more permeable and promote the leakage of ions and other cell contents [3].

Lemongrass oil composed by terpenes (carvacrol, citral, linalool, and geraniol), phenolics (flavonoids and phenolic acids) and terpenes which have been effective as antimicrobial against pathogenic bacteria and deteriorative bacteria [13]. However the presence of citral and geraniol in lemongrass oil were more effective to inhibit *B. cereus* compared to those of *S. aureus* [14]. This study also observed the same results that antimicrobial activity of lemongrass oil was stronger on *B. cereus* than those of *S. aureus* as shown in table 1. The previous studi reported that the antibacterial activity of lemongrass oil components depended on the compositions, concentration, and cell target sites [14]. This study also observed that antimicrobial activity of components in lemongrass oil was less to against Gram-negative bacteria compared to those Gram-positive bacteria. In similar, Aiemsaard reported that citral more effective against Gram-positive bacteria (*S. aureus*, *S. agalactiae*, and *B. cereus*) than Gram-negative bacteria (such as *E. coli*) [14]. Citral and geraniol has been reported as the major bioactive components of lemongrass oil [15].

This present study found that inhibition zone of ginger oil on *S. aureus* was relatively larger than those of *S.* Typhimurium and *B. cereus* as shown in table 1. Inhibition zone on *S. aureus* was observed for 4.33 ± 0,58 mm. However the inhibition zone of. *S.* Typhimurium and *B. cereus* was found 2.29 ± 0.36 and 1.92 ± 0,36, respectively. Ginger oil has been reported to be composed by α-curcumene, α-zingiberene, β-bisabolene, β-sesquiphellandrene, cineole, 2,2-dimethyl-3-methylenenorbornane and rosefuran epoxide [8; 9]. Nader reported that extract ginger oil was contained by alkaloid, glicosides, terpenoid, flavonoid, phenolic [18]. Stoyanova [17] revealed that antimicrobial activity of the components in ginger oil was very weak on the Gram-negative and Gram-positive bacteria. Bassole and Juliani [4] observed that the low antimicrobial activity of ginger was promoted by the small amount of aldehydes and phenols constituents (3.804). However the large amount of terpene alcohols (32.508%) and terpene hidrocarbons (59.416%) was measured by previous study which may cause the weaker antimicrobial antivity [4].

Galangal oil was observed to be effective more on Gram-positive than those of Gram-negative. This present study found that the antimicrobial activity in galangal oil was less compared to clove bud oil, lemongrass oil, and ginger oil. This is in similar that previous study. Bassole and Juliani [4] revealed that the relatively low antimicrobial activity of galangal oil was related to their composition of chemical. Galangal oil was composed by cineole, 4-allylphenyl acetate, α-farnesene, (2,6-dimethylphenyl)borate and α-pinene [16]. This means that galangal oil composed by high terpende alcohols, ketones, esters and terpene hydricarbons but low aldehydes and phenols. The study revealed that ketones or esters, acetate had much weaker antimicrobial activity [4].

In similar, antimicrobial activity of turmeric oil was found to be weaker than those of clove bud oil, lemongrass oil dan ginger oil. Turmeric oil and galangal oil showed have the same antimicrobial activity to against *S.* Typhimurium, *B. cereus* and *S. aureus*. Turmeric oil was composed by the eucalyptol as the major component [19]. Eucalyptol was grouped as was ketones which had weak in antimicrobial activity [4]. Other compounds were α-pinene (1.50%), β-phellandrene (2.49%), β-pinene (3.57%), limonene (2.73%), 1,3,8-p-menthatriene (1.76%), ascaridole epoxide (1.452%), 2-methylisoborneol (2.92%), 5-isopropyl-6-methyl-hepta-3, dien-2-ol (2.07%) [19].

Candlenut oil showed low antimicrobial activity to inhibit *S. aureus* but it was not effective to inhibit *S.* Typhimurium and *B. cereus*. The previous study found that candlenut oil composed by flavonoids, cardiac glycosides, saponins, phenolic compounds, phytosteroids, tannins, triterpenoids, anthraquinones, and alkaloids [13; 14; 15; 16]. Mpala *et al.* [20] reported that candlenut (*A. moluccanus*) extracts to block microbial triggers of three bacteria (*Proteus* spp., *K. pneumonia, S. pyogenes*). In this present study, candlenut did not showed antimicrobial activity to inhibit *S.* Typhimurium and *B. cereus* which has not being reported by other researchers.

4. Conclusion

Clove bud oil showed the highest antimicrobial activity to inhibit the growth of S. Typhimurium, *B. cereus* and *S. aureus*. It was followed by lemongrass oil and ginger oil. These essential oils showed promising antimicrobial activity against *S.* Typhimurium, *B. cereus* and *S. aureus*.

5. Acknowledgment

Financial assistance provided by PNBP Politeknik Negeri Jember, Indonesia Ministry of Research and Higher Education is gratefully acknowledged.

6. References

[1] Rasool M H 2013 Antimicrobial activity of plant essential oils against the growth of *Escherichia coli IOSR J. Pharm.* **3** 1
[2] Dorman H J and Deans S G 2000 Antimicrobial agents from plants: antibacterial activity of plant volatile oils *J. Appl. Microbiol.* **88** 308
[3] Burt S 2004 Essential oils : their antibacterial properties and potential applications in foods-a review *Int. J. Food Microbiol.* **94** 223
[4] Bassolé I H N and Juliani H R 2012 Essential oils in combination and their antimicrobial properties *Molecules* **17** 3989
[5] Hendriksen R S, Vieira A R, Karlsmose S, Danilo M A, Wong L F, Jensen A B, Wegener H C, and Aarestrup F M 2011 Global Monitoring of *Salmonella* Serovar Distribution from the World Health Organization Global Foodborne Infections Network Country Data Bank: Results of Quality Assured Laboratories from 2001 to 2007 *Foodborne Pathog. Dis.* **8** 887
[6] Eng S K, Pusparajah P, Ab Mutalib N S, Ser H L, Chan K G and Lee L H 2015 *Salmonella*: A review on pathogenesis, epidemiology and antibiotic resistance *Front. Life Sci.* **8** 284
[7] Tewari A and Abdullah S 2015 *Bacillus cereus* food poisoning: international and Indian perspective *J. Food Sci. Technol.* **52** 2500
[8] Marollo R 2016 *Bacillus cereus* Food-Borne Disease *The Diverse Faces of Bacillus cereus* Ed V Savini (The Netherland: Academic Press) pp. 61–72.
[9] Kadariya J, Smith T C and Thapaliya D 2014 *Staphylococcus aureus* and staphylococcal food-borne disease: an ongoing challenge in public health *Biomed Res. Int.* 827965
[10] Bauer A, Kirby W, Sherris J and Turck M 1966 Antibiotic susceptibility testing by a standardized single disc method *Am. J. Clin. Pathol.* **45** 493
[11] Moon H and Rhee M S 2016 Synergism between carvacrol or thymol increases the antimicrobial efficacy of soy sauce with no sensory impact *Int. J. Food Microbiol.* **217** 35

[12] Luz I D S, de Melo A N F, Bezerra T K A, Madruga M S, Magnani M and de Souza E L 2014 Sublethal Amounts of *Origanum vulgare* L. Essential Oil and Carvacrol Cause Injury and Changes in Membrane Fatty Acid of *Salmonella* Typhimurium Cultivated in a Meat Broth *Foodborne Pathog. Dis.* **11** 357

[13] Balakrishnan B, Paramasivam S and Arulkumar A 2014 Evaluation of the lemongrass plant (*Cymbopogon citratus*) extracted in different solvents for antioxidant and antibacterial activity against human pathogens *Asian Pacific J. Trop. Dis.* **4** S134

[14] Aiemsaard J, Aiumlamai S, Aromdee C, Taweechaisupapong S and Khunkitti W 2011 The effect of lemongrass oil and its major components on clinical isolate mastitis pathogens and their mechanisms of action on Staphylococcus aureus DMST 4745 *Res. Vet. Sci.* **91** e31

[15] Moore-Neibel K, Gerber C, Patel J, Friedman M and Ravishankar S 2012 Antimicrobial activity of lemongrass oil against *Salmonella enterica* on organic leafy greens *J. Appl. Microbiol.* **112** 485

[16] Hamad A, Alifah A, Permadi A and Hartanti D 2016 Chemical constituents and antibacterial activities of crude extract and essential oils of *Alpinia galanga* and *Zingiber officinale Int. Food Res. J.* **23** 837

[17] Stoyanova A, Konakchiev A, Damyanova S, Stoilova I and Suu P T 2006 Composition and antimicrobial activity of ginger essential oil from Vietnam *J. Essent. Oil-Bearing Plants* **9** 93

[18] Nader M I, Ghanima K K, Ali S A and Azhar D A 2010 Antibacterial activity of ginger extracts and its essential oil on some of pathogenic bacteria *Baghdad Sci. J.* **7**

[19] Parveen Z, Nawaz S, Siddique S and Shahzad K 2013 Composition and Antimicrobial Activity of the Essential Oil from Leaves of *Curcuma longa* L. Kasur Variety *Indian J. Pharm. Sci.* 75 117

[20] Mpala L N, Chikowe G R and Cock I E 2017 *Aleurites moluccanus* (I.) Wilk. Extract Inhibit The Growth of Bacterial Triggers of Selected Autoimmune Inflamatory Diseases," *Pharmacogn. Commun.* **7** 83

[21] Arkhipov A, Sirdaarta J, Rayan P, Mc Donnel P A and Cook I E 2014 An examination of the antibacterial, antifungal, anti-Giardial and anticancer properties of *Kigelia africana* fruit extracts," *Pharmacogn. Commun.* **4** 62

[22] Kalt F R and Cock I 2014 Gas chromatography-mass spectroscopy analysis of bioactive petalostigma extracts: Toxicity, antibacterial and antiviral activities *Pharmacogn. Mag.* **10** 37

[23] Vesoul J and Cock I E 2012 The Potential of Bunya Nut Extracts as Antibacterial Functional Food Agents *Pharmacogn. Commun.* **2** 72

Increasing of fish quality using ice-sterofoam container for paseban fisherman at Jember Regency East Java, Indonesia

D E Rahmanto[1], **N Nurhayati**[2]

[1]Department of Engineering , Politeknik Negeri Jember, Mastrip PO BOX 164 Jember 68101 , Indonesia

[2]Pangan ASUH-Research Group, Department of Agricultural Product Technology – Faculty of Agricultural Technology – University of Jember, East-Java, Indonesia Jl. Kalimantan No. 37 FTP UNEJ Jember 68121

E-mail: nurhayati.ftp@unej.ac.id

Abstract. Using of fish storage containers is not always done by fishermen at Paseban Beach, Jember Regency East Java-Indonesia. This study evaluated the quality of fisherman catches fish based on population of microbial. The introduction of storage container was conducted without or using the ice cooler. There were three samples of fish treatment that evaluated i.e without storage container (on the boat floor), sterofoam container and ice-sterofoam container. Recall questionnaires conducted to evaluate the habits of Paseban fishermen in fish storing during fishing. There were 36% of fishermen no use the storage container for their fish during fishing, 14% of fishermen use the sterofoam container without ice cooler, and 50% of fishermen used ice-sterofoam container. The level of microbial contamination on the catch fish required SNI standard that reached 10^4 cfu/g. The storage of catch fish in the sterofoam container can decrease the level of contamination up to 2 log cycles (10^2 cfu/g). The quality of catches fish of Paseban fishermen can be classified as the SNI required (less than 5 x 10^5 cfu/g).

1. Introduction

Paseban Beach is located ± 50 km to the northwest of Jember city precisely in Paseban Village, Kencong Subdistrict, is a beach tourism object that stretches long and clean and can look towards the island of Nusa Barong. The fishermen on Paseban beach still use simple equipment to catch the fish. The fishing community on Paseban beach belongs to artisanal fishermen. Fishing communities are one group of people who are considered to be absolutely poor, even the poorest of the poor. Various studies have also shown that the condition of fishermen, especially fishermen of small-scale fisheries in Indonesia are at a marginal level.

Artisanal fishermen who are included as *small scale fisheries* are boat owners whose majority of their income depends on fishing activities in the sea, operating their own boats with a weight of the boat between 2.75-25 GT (or the size of the boat length of 5 meters to 15 meters, widths between 1.5 meters to 6 meters using simple fishing gear (such as gilnet, clown nets, *minitrawl*, fishing rods, stakes), using revenue sharing systems between owners and crew members, and selling fish catches within the local market limited[1].

There are can affect the quality of the end-product, if the fish raw material is bad. Choosing the appropriate fish containers is considered as one significant factor contributing to the fish freshness and quality.The changes of fish quality more easily be noticed and consist of changes the colour, odour or smell, taste, appearance and texture, that can affect the sensory preference[2].

The problem that arises is that the catching facilities and infrastructure are still simple and limited, so the number and quality of catch fish are also very low. Thus the selling price of Paseban beach fish is relatively very low. This has an impact on the level of welfare of fishermen who are low (poor) with the daily income of fishermen is very low around less than Rp. 20,000.00 per fisherman. Temperature control in containers can improve during fresh fish handling by using simple redesign and experimental testing. The design improvements can more stable temperature through sea freight and transport so get satisfactory improvements with simple and cost effective Procedures[3].

In general, the catch sea water fish is still quite feasible to consume, but to improve the quality of the caught fish, the vessels transformed the vessels that used to use ice coolers to ships using coolers *freezer*. The purpose of this study was to analyze the quality of caught-sea water fish by the various storage containers.

2. Materials and Methods

2.1 Materials

The media for determination of microbial contamination were PCA (*Plate Count Agar*), and the chromogenic media i.e HEA (*Hectoin Enteric Agar*) and SCA (*Salmonella Chromogenic Agar*). The tools used are autoclaves, laminar flow, a set of microbiology test kits.

2.2 Sampling of Sea Water Fish

Sampling of sea water fish was done on fresh caught fish with sterofoam storage containers, sterofoam containers with cube ice, and without storage containers. A total of 500g of sea water fish were dissolved in a sterile physiological solution contain 0.85% NaCl. Then the series dilutions up to 10^{-4} were 10^{0}, 10^{-1}, 10^{-2}, 10^{-3} and 10^{-4}. The last three series were inoculated on PCA, HEA and SCA media. Inoculation method used the pour plate on the last three series of dilution. The microbial population was calculated based on BAM standards 25 - 250 colonies/dishes. The contamination of *Salmonella* sp showed green colonies on HEA and purple media on SCA media. The contamination of *Eschericia coli* sp showed yellow colonies on HEA media and blue on SCA media[4].

2.3 Determination of Contaminant Microbes Population[5]

Determination of bacterial population was carried out using the BAM (Bacteriological Analytical Manual) namely:

$N = \{\Sigma C / [(1 \times n1) + (0.1 \times n2) \times (d)] \}$

N = number of colonies

ΣC = number of colonies in both dilution series used

n1 = number of plates used in the first dilution series

n2 = number of plates used in the second dilution series

d = series of lowest dilution used.

3. Results And Discussion

3.1 Fish Quality of Paseban Fishermen

Investigation of catches fish was carried out on the Paseban fisherman and local fish market (Tempat Pelelangan Ikan/TPI). The catch fish was shown in Figure 1.

Figure 1. Investigation of the catching fish quality of Paseban fisherman

Quality of fish without storage container was not neat and the chances of contamination was very high, both from the condition of the boat's facilities and the cross contamination by fishermen personnel. Contamination by fishermen can be caused by poor sanitation of personnel. The results of the questionnaire showed that more than 50% of fishermen respondents do not have a habit of washing hands, only 9% always do handwashing, and 36% seldom do handwashing (Figure 2).

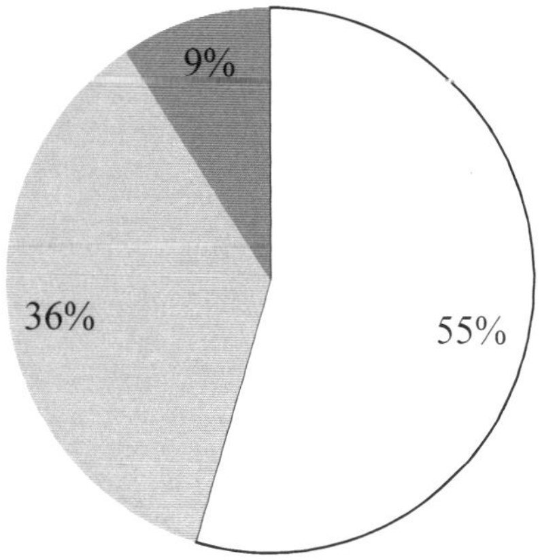

Figure 2. The handwashing habits of Paseban fishermen's: never (□), always (▩), seldom (■)

The fish is often handled in unhygienic conditions can causing spoilage by pathogenic bacteria contamination, and a loss of income as fish are sold for a low price. Fishermen may also lack knowledge, skills and the ability to invest in new equipment during fish storage[6].

3.2 Using of Fish Storage Container

Results of the investigation showed that there were differences in the effect of using containers with and without cube ice on the microbial population of contaminants. Sterofoam storage containers produce fewer microbial populations than without using containers. The catch storage in sterofoam containers equipped with ice cubes produces the least contaminated microbial population. There were still 36% of Paseban fishermen no used container to storage the cathes fish, 14 % are used sterofoam container, and 50% are used sterofoam container with cube ice (Figure 3).

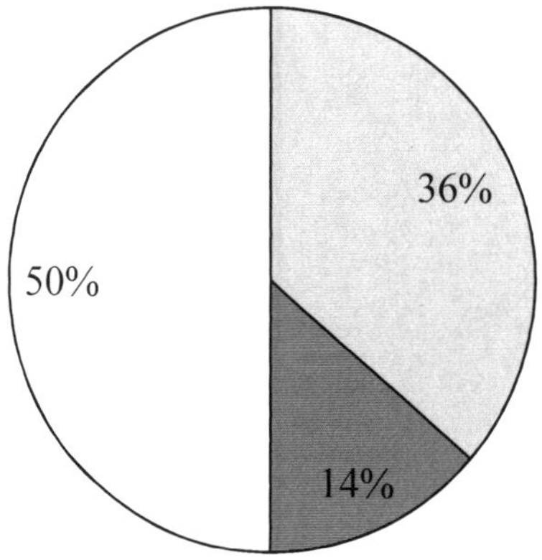

Figure 3. Using of fish storage containers by Paseban fishermen's: without container (□), sterofoam container (■),sterofoam container with cube ice (□)

With the limitations of fishing facilities, the Paseban fishermen have high hopes for the facilities, especially fish nets, and storage container. Figure 4 is the result of the fishermen's hope questionnaire. As many as 82% expect to get the aid of fishing gear in the form of nets, storage container and 18% expect the construction of a wave-reducing arm, considering the waves are big on Paseban Beach.

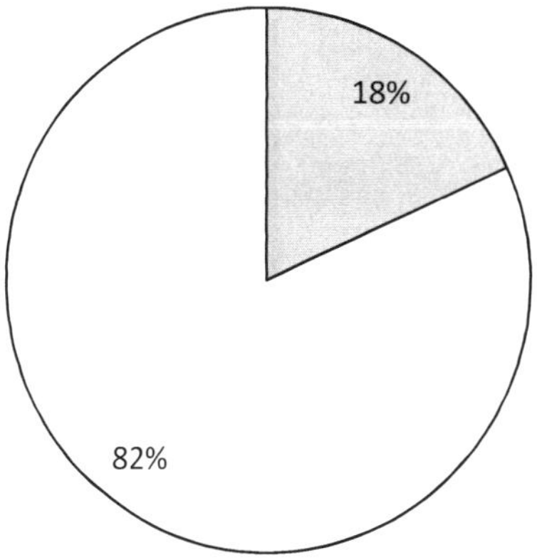

Figure 4. The hope of Paseban fishermen's to Government: a wave-reducing arm (■), fishing gear facilities (□)

3.3 Population of Microbial Contamination on Catches Fish of Paseban Fishermen Using Different Type of Container

The population of total microbial contamination on catches fish of Paseban fishermen up to 10^4 cfu/g, while the bacteria contamination in the range of 10 - 100 cfu/g or 1 - 2 log cfu/g cycles. The population of microbial contamination is presented in Table 1 and figure 5.

Table 1. Population of total microbe, *Salmonella* sp, *Eschericia coli*

Type of container	Population (cfu/ g)		
	Total microbe	*Salmonella* sp	*Eschericia coli*
Without container	1.14×10^4	0	1.6×10^1
Sterofoam container	3.73×10^2	0	1.4×10^1
Sterofoam container with block ice	1.45×10^1	0	2.0×10^0
Standard SNI (total plate count/ TPC)	$<10^5$	0	<3

Table 1 showed that the microbial population of contaminants is around 10^4 cfu/g. The treatment of storage containers showed significant results that storage using stereofoam could reduce the microbial population of contamination compared with no storage containers. Storage using container contain cube ice is the best method for improving the quality of catch fish. Storage methods with ice can reduce microbial and bacterial contamination populations up to 2 log cycles (Figure 5). Ice limits the growth of mesophilic bacteria. Bacteria are the fastest microbes in multiplying compared to other microbes (yeast and yeast and mold). Bacterial growth generally occurs at room temperature and most bacterial contamination populations are mesophilic bacteria[7]. The total difference in microbial population indicates the difference in the processing error [8]. This is suitable to be caused by a dirty, unhygienic and sterile environment that can trigger rapid microbial growth [9].

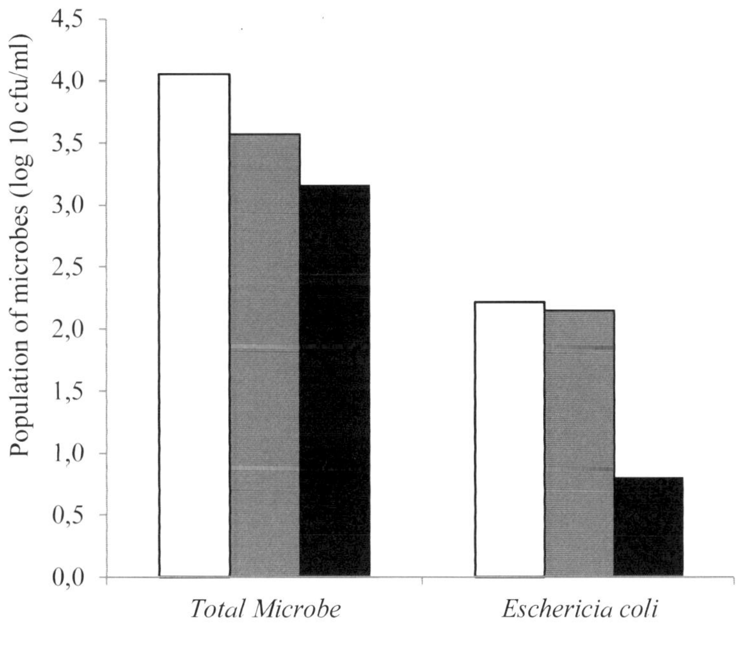

Figure 5. Population of Microbial Contamination on Paseban fishermen's: without container (☐), sterofoam container (▩), sterofoam container with cube ice (■)

Indonesian National Standard (SNI) No 2729: 2013 concerning microbiological requirements of fresh fish quality including total plate numbers, number of coliform bacteria, and identification of pathogenic bacteria. There were the maximum limit for total microbial contamination is less than 10^5 cfu/g, the contamination of *E. coli* bacteria is less than 3 and there should be no pathogenic microbes such as *Salmonella*, *Vibrio cholera* and *Vibrio parahaemolyticus*[10]. The caught sea water fish of Paseban fishermen contain 10^4 cfu/g of microbial total, 2 cfu/g of *E. coli* and no contain of *Salmobella* sp, if the fish storage in the container with cube ice. Thus the quality of catched fish from Paseban fishermen tested is in compliance with SNI standards of fresh fish that are permitted.

4. Conclusion

The habit of Paseban fishermen in storing catches is 36% of fishermen storage the sea water fish without using the storage containers. The level of microbial contamination in the catched fish still meets the required SNI standards which is less than 10^4 cfu/g. Storage of catched fish in a sterofom container was able to reduce the level of contamination by about 2 log cycles (10^2 cfu/g). The catching fish quality of Paseban fishermen was classified as meeting the required SNI standards (maximum 5 x 10^5 cfu/g).

5. Acknowledgments

Thank you to Kemristekdikti The Republic of Indonesia for funding through the Community Partnership Programme 2018. This work supported the scientific method of chromogenic media application for investigation of food safety by *Pangan ASUH*- research group.

References

[1] Charles TA. 2001. Sustainable Fishery System. London: Blackwell Science.

[2] Quang NH. 2006. Guidelines for handling and preservation of fresh fish for further processing in Vietnam. Fishers Training Programe. The United Nations University.

[3] Elíasson S, Margeirsson B, Arason S. 2013. Improved reefer container for fresh fish. Vinnsla, virðisaukning og eldi. Skýrsla Matís 01-13 Janúar 2013 ISSN 1670-7192

[4] Nurhayati N, Oktavianto A, Suswati E, Rahmanto DE. 2018. Effectiveness of low thermal destruction on drinking water contain enteropathogenic bacteria isolated from wellspring at Mojo VillageLumajang Regency-Indonesia. J-Sustain. 6 (1).

[5] Jackson, G.J., Merker, R.I., Bandler, R. 2001. Bacteriological analytical manual, 8th Edition, Revision A, Office of Special Research Skills, CFSAN

[6] Ward A, Beyens Y. Fish handling, quality and processing: training and community trainers manual. Smartfish Working Papers. 001. Commision del'Ocean Indien.

[7] Sampels, S. (2015). The effects of storage and preservation technologies on the quality of fish products: A review. *Journal of food processing and preservation*, *39*(6), 1206-1215.

[8] Nurani TW, Haluan J, Sudirman S, Lubis E. 2010. Analysis of fishing port to support the development of tuna fisheries in the South Coast of Java. *Indonesia Fisheries Research Journal*. 16 (2): 69-78.

[9] Poli, B. M., Messini, A., Parisi, G., Scappini, F., Vigiani, V., Giorgi, G., & Vincenzini, M. (2006). Sensory, physical, chemical and microbiological changes in European sea bass (*Dicentrarchus labrax*) fillets packed under modified atmosphere/air or prepared from whole fish stored in ice. *International journal of food science & technology*, *41*(4), 444-454.

[10] (SNI) Standar Nasional Indonesia. Fresh Fish (Ikan Segar). ICS 67.120.30 Badan Standardisasi Nasional (Indonesian)

IOP Conf. Series: Earth and Environmental Science **207** (2018) 012038 doi:10.1088/1755-1315/207/1/012038

The Potency of betacyanins extract from a peel of dragon fruits as a source of colourimetric indicator to develop intelligent packaging for fish freshness monitoring

Ardiyansyah[1], Mulia W Apriliyanti[1], Wahyono A[1], Fatoni M[1], Poerwanto B[1], Suryaningsih W[1]

[1] Department of Food Technology, Politeknik Negeri Jember, Jalan Mastrip POBOX 164 Jember 68101, Indonesia

Email : ardi@polije.ac.id[1]

Abstract. A Famous freshwater fish in Indonesia is *Osphronemus gouramy* (O. gouramy). This fish has a thick meat and savory taste. Recently, many researches have focused on developing the intelligent packaging for detecting spoilage fish The aims for this research is to develop smart packaging through immobilizing betacyanin into glucomannan-Polyvinyl alcohol matrix used to monitor fish freshness. The color stability of matrix films was monitored using camera. Furthermore, the colourimetric films were used to observe freshness based on a TVB-N level of gourami (O. gouramy) on chilling temperature. The incubation result of betacyanin extracts with variation pH shows that the color change of betacyanin extracts at pH 10 and pH 12 are purple and yellow. Stability of colorimetric films color is reported that the colorimetric films are more stable after incubation on chilling temperature until 7th day. Evaluation of fish freshness of *Osphronemus gouramy* (O. gouramy) fillet was conducted at chilling temperature. Transformation of a colorimetric film becomes yellow after gouramy fillet is, where it's simultaneously with increase of the level of TVBN for *O.gouramy* fillet is 39,74 mg/100g. Thus, it can be concluded this color change corresponded with TVB-N level are increased.

1. Introduction

Fish is one of the natural food which has a short-term of shelf life because it has heavy protein and water content. Good handling practice is needed to reduce the fish spoilage during supply chain. TVB-N level is one of the parameters to measure Fish freshness. Generally, the analysis of TVB-N level needs the technician's advanced skill to handle the instrumentation such as kjehdal intrumentations [1].

Intelligent packaging is a novel packaging which has ability to monitor food quality with the real-time conditions. Intelligent packaging such as one package containing matrix films cross-linked with a colorimetric indicator to monitor pH, temperature or freshness of some products. Fish spoilage was monitored with smart packaging used chemosynthetic as colour agents are reported by Kuswandi, et al [2] and Paquet et al [3]. Recently, a source of a colorimetric agent from a natural pigment to develop a smart packaging was investigated such as anthocyanin. Some advantages of the anthocyanin use are most obtainable in flesh, peel, leaf, or of fruits, and water soluble. Choi et al [4] and Ma & wang [5] suggested that colorimetric films used anthocyanin were important source to evaluate fish freshness based-on ammonia gases production during fish spoilage process. It is widely recognized that anthocyanin is a safe ingredient to blend with food or beverage.

Currently, dragon fruits are most commonly found in Indonesia, Malaysia, Taiwan, Australia, Colombia, Vietnam, and Mexico. It has been reported that the antioxidant activity of dragon fruit was 22.4 ± 0.29 measured by DPPH method at EC_{50}, the vitamin C equivalent was 118 ± 4.12 μmol per g of peel dried and flesh extract, and flesh has total phenolic contents was 42.4 ± 0.04 mg of gallic acid equivalents (GAE)/100 g of fresh weight. Moreover, betacyanins (betanin equivalents) content of flesh and peel were 10.3 ± 0.22 and 13.8 ± 0.85 mg per 100 g of fresh weight, respectively [6].

The aims for this research is to develop smart packaging through immobilizing betacyanin into glucomannan-Polyvinyl alcohol matrix. The micrograph of colorimetric films was recorded using scanning electron microscope. The color stability of matrix films was monitored. Furthermore, the colourimetric films were used to observe freshness based on a TVB-N level of gourami (O. gouramy) on chilling temperature.

2. MATERIAL AND METHODS

2.1. Materials

Dragon fruits were harvested from farms of Jember Residence, Indonesia. A fully mature and no physical damage of dragon fruits were selected for study. Gouramys were obtained from local pool of Jember Residence, Indonesia. The chemicals used in this study were NaOH (Merck, Indonesia), HCl (Merck, Indonesia), glucomannan (CV. Nura jaya, Indonesia), polyvinyl alcohol and (sigma-aldrich, Singapore).

2.2. Methods

2.2.1 Extraction of betacyanins

Solution was prepared by extracted a 50 gr of fresh red pitaya peel with 50 mL boiled-distilled water until 5 minutes, after that this solution was filtered.

2.2.2 Quantification of Betacyanin Content

After preparing extraction solutions, the betacyanin content was calculated according to with : [BC (mg/100 g) = (A × DF × MW × V × 100)/(ε × L × W)], where MW is molecular weights of betanin (550 g/mol), A is absorbance at 538 nm and corrected at 600 nm, DF is the dilution factor, L is the path length of the cuvette, V is sample volume (ml), W is the weight of dry matter in gram, and ε is the molar extinction coefficients of betanin (60,000 L mol^{-1} cm^{-1}) [7].

2.2.3 Preparation of fish freshness film

1 g of PVA and 2 g of glucomannan mixed with 70 mL distilled water, and the mixture was boiled on 100 °C and blended used a magnetic stirrer until a formed homogeneous mixture. after that adjust pH for mixture until pH 8.5, 30 ml of betacyanin extract solutions was then added. Furthemore, the film was created by casting solution into a pan with a dimension of 5 x 10 cm. A pan was dried at 40°C for 36h in oven dryin (modified Zhai et al. [1]).

2.2.4 Color stability of colorimetric films evaluation

Color stability of colorimetric film was evaluated visually during 7th days based on L, a, b values using MiniScan EZ 4500L, Hunter lab, virginia, USA (Hunt et al, [8]).

2.2.5 Betacyanin Extracts spectral

betacyanin solution incubated at different pH (2, 4, 6, 8, 10, and 12), after that evaluate the spectra using a UV-Vis Spectrophotometer (Ultra Spec 2100 pro, amersham biosciences, US).

2.2.6 Surface Profile of Colorimetric Films

Micrograph of the colorimetric films was recorded by SEM (TM-3030, Japan)

2.2.7 Trial of Fish Spoilage

100 g of fish fillet was immediately displaced into a plastic box. The matix films were installed into headspace of the plastic box (25 x 25 cm). The plastic box was incubated on a chilling temperature (4°C) until 8th days. Using camera to evaluate the color change of the colorimetric films every 24 h [1].

2.2.8 Quantification of Total Volatile Basic Nitrogen (TVB-N)

Determination of TVB-N value has used automatic distillation methods [9]. Where, 20 g of fish meal mixed with 40 ml of 7.5% TCA solutions, 5 ml filtrate was collected into the distillation tube. automatic distillation used to produduce steam (Vapodest50s, Germany), and gasses were exposed with boric acid (0.1 N). TVB-N was calculated following by :

$$\text{TVB (mg/100 g)} = \frac{(\text{Vsample} - \text{Vblanko}) \times \text{N HCl} \times 14{,}007 \times 100}{\text{Weight sample} \times \text{fp}} \quad (1)$$

3. Results and Discussion

3.1 Spectral and Color of Betacyanins

UV-vis spectra of betacyanin solutions compared with the color changes is illustrated in Figure. 1. That result shows that betacyanin solutions color was purple on pH 2 until pH 10, the color change of betacyanin extract to yellow after incubated at pH of 12.

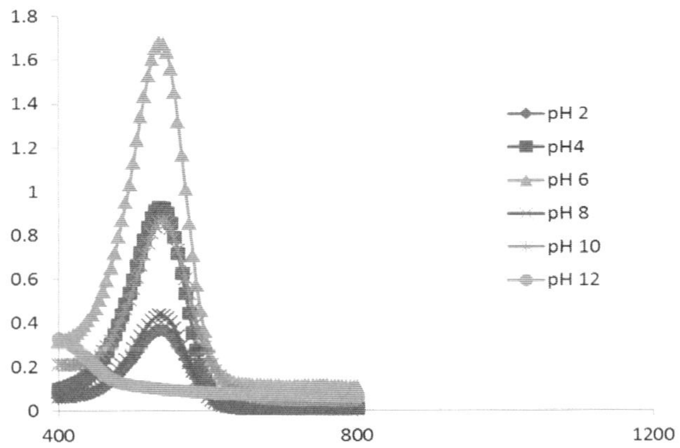

(a)

(b)

Figure 1. (a) UV-vis Color and (b) Spectral of betacyanins solution at pH 2, 4,6, 8, 10 and 12.

On the pH 6 it shows that the maximum absorbances was 538 nm. This was in according to the study of Fernanda Robert de Mello et al. [10] who reported that betalain extract solution from red pitaya peel had maximum absorption peak at 540 nm. Additionally, it was found that the maximum absorption peak altered to 400 nm (yellow color) after the pH value of 12. Previously, Herbach et al. [11] reported that the betalains turned into yellow products such as betalamic acid (424 nm) after exposed to high temperature. The mechanism reaction of color change of betacyanin after incubated at base condition showed in fig. 2.

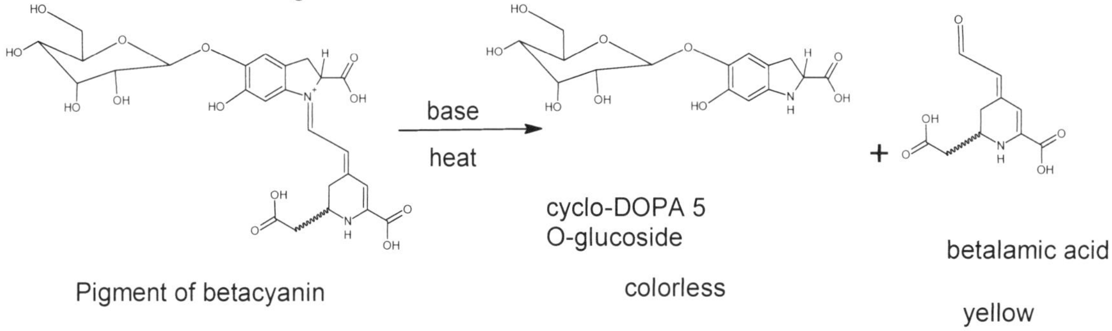

Figure 2. Mechanism reactions of the color change of betacyanin [12].

3.2 SEM Micrographs Analysis

Before preparation of colorimetric films, the level of betacyanin extracts for indicator solution was 119+0.71 mg/100 g. Then, this was used for developing glucomannan-PVA-betacyanin. The micromorphology of a glucomannan-PVA-betacyanin film and Glucomannan-PVA films is showed in Figure 3. This result showed that the glucomannan-PVA film forms a two-phase structure due to separation of glucomannan and PVA. Similarly, Cano et al, [13] , and Zhai et al, [1] reported that a two-phase structure formed due to the degree of immiscibility of PVA. In the contrast, in presence of glacial acetic acid, starch/PVA shows a homogeneous mixtures or a single-phase. Starch/PVA films became elastic after incorporating a plasticizer such as CaCl2 [14]. After that, the micrograph result showed that betacyanins extracts were successfully trapped into polymer matrix.

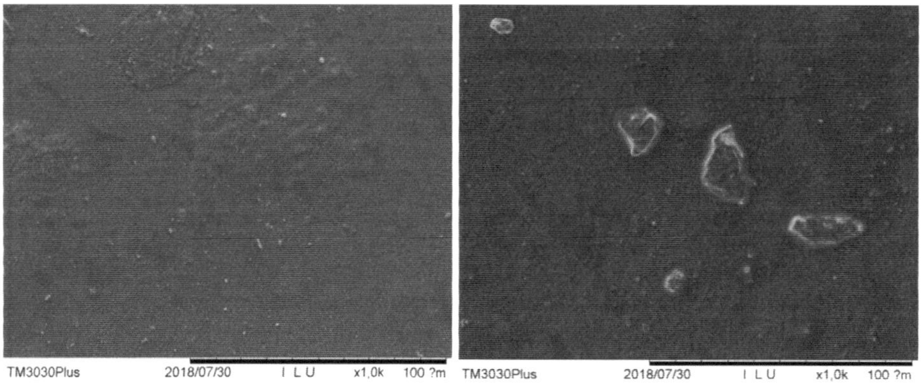

(a) (b)

Figure 3. Micromorphology of (a) glucomannan-PVA-betacyanin film, (b glucomanan-PVA film).

3.3 Stability of the colorimetric films color

Mostly, natural pigment has short shelf life is extract of betacyanin. This depends on the various factors such as pH, medium extractions and temperature. It has been reported that betacyanin extract in water corresponding to the first-order kinetic model with its self-life (t1/2) was 90% shelf life of 76 hours and 23 hours at 25 °C [15]. However, starch-PVA matrix probably inhibited oxidation process of the betacyanins [16]. The color stabilty of betacyanin-glucomannan-PVA is presented in Table 1. This result suggested that there was no significant color change on film during incubation at 4°C. On the other hand, Zhai et al,[1] reported that anthocyanin-glucomannan-pva matrix stored into chilling temperature at 4 °C had great color stability.

Table. 1. Value of L, a, b, after incubation until 7th days at 4°C.

Day	L	a	b
0	40.935 ±0.22	67.12 ±0.41	-33.22 ± 0.056
1	41.02±2.72	43.98 ±2.80	-14.815 ± 0.01
2	41.895 ±1.12	44.875 ± 0.78	-15.16 ± 0.06
3	42.045 ±0.42	43.98 ± 0.077	-14.315 ±0.01
4	39.725 ±0.06	44.265 ±1.08	-13.175 ± 0.04
5	39.225 ±0.14	41.885 ± 1.08	-13.945 ±0.38
6	39.305 ± 0.14	40.255 ±0.06	-12.225 ±0.134
7	39.38 ±0.05	39.93 ± 0.084	-12.225 ±0.134

3.4 Evaluation Fish Freshness Using Colorimteric Film

The transformation of indicator films color are shown in Figure 3. In the beginning, the color of glucomannan-PVA-betacyanin film is purple. Furtheremore, the color became yellow after 8 days. This color changed due to the increasing TVB-N level from gouramy. The color change due to interaction or the film exposured with gases such as NH_3, H_2S, where is before explained. Initially, TVBN value of gouramy was 2.4 mg/100 g. After eight days, TVB-N exceeded the standard values (39.74 mg/100 g). On the contrast, it has been reported that anthocyanins from roselle [1], purple sweet potato [4], and red cabbage [17], Zhang, Lu, and Chen [18] correspondingly with pH changes or the level of freshness matter. Thus, the results are suggested that the glucomannan-PVA-betacyanin can be used to evaluate fish freshness.

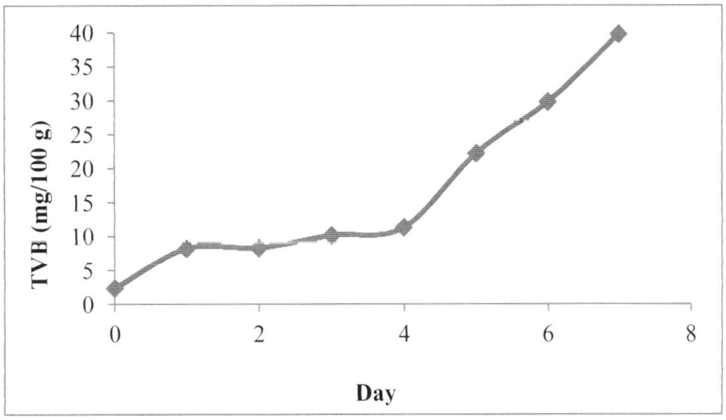

(a)

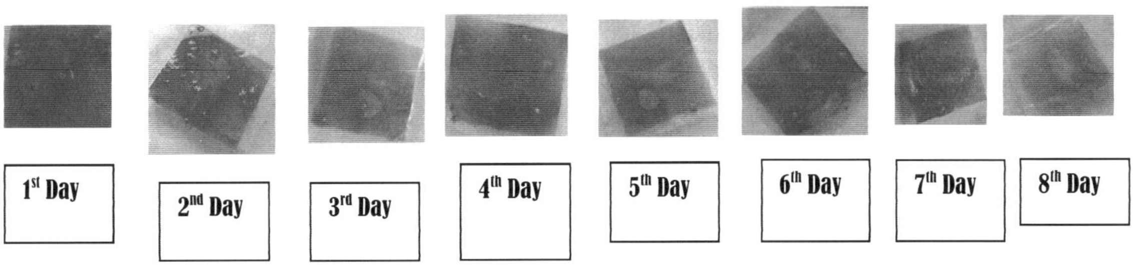

(b)

Figure 4. (a) TVB-N value of gouramy after stored on 4 °C until 8 days and (b) transformation of the colorimetric films.

Table 2. L, a, b, value of colorimteric films after applicated on monitoring fish fressnesh until 8th days at 4°C.

Day	L	a	b
0	40.935±0.22	67.18± 0.48	-33.245±0.09
1	39.16± 0.08	60.615±0.7	-27.565±0.64
2	39.075±0.035	54.875 ±0.78	-15.095+0.106
3	34.825±3.17	43.98±2.8	-9.535±0.57
4	38.625±1.49	44.195±0.23	-13.205±0.06
5	46.955±0..035	29.85±0.46	-0.065±0.68
6	46.405±0.926	12.02±0.056	12.225±0.13
7	44.185± 0.035	24.02±2.77	13.34±0.084
8	45.295± 2.97	6.075±0.46	24.52±1.35

4. Conclusion

The Result showed that the color stability of colorimetric films was considerably stable until 7 days at 4 °C. This result exhibited that the glucomannan-PVA film showed not homogen due to the separation of glucomannan and PVA. The fish freshness monitoring using film indicator was conducted at chilling temperature. The color of colorimetric film became yellow until 8th days. Thus, it can be concluded this color change corresponded with TVB-N level are increased.

5. Acknowledgement

The authors thank for all who help to the completion this paper.

References

[1] Zhai Xiaodong 2017 *Food Hydrocolloids*, Vol. 69, pp. **308-317**.
[2] Kuswandi B 2012 *Food Control,* 25(1) : **184-189**.
[3] Pacquit A 2007 *Food Chemistry*, Vol. 102 (2): **466-470**.
[4] Choi I, Lee J, Lacroix M, and Han J 2017 Food Chemistry, 218, **122-128**.
[5] Ma Q and Wang L 2016. *Sensors and Actuators B: Chemical,* Vol.235, pp. **401-407**.
[6] Wu L C, Hsu HW, Chen Y C, Chiu C C, Lin Y I, and Ho J A 2006 Food Chemistry 95: **2319-327**.
[7] Jamilah B, Ee S C, Kharidah M, Dzulkifly, M A. and Noranizan A 2013 Food and Bioprocess Technology: An International Journal 6: **1332-1342**.
[8] Hunt, R.W.G. 1991. Measuring colour. 2nd ed, UK: Ellis Horwood Ltd, Chichester
[9] Malle P, 1989. Journal of Food Protection.52:**419–423**.
[10] Mello fernanda robert de 2015, vol.45, n.2, pp.323-328, Epub.
[11] Herbach K M 2006 Journal of FoodScience. v.71, p.**41–50**.
[12] Florian C 2004 vol 15 **19–3**.
[13] Cano A 2015 Journal of Food Engineering, 167, **59-64**.
[14] Jiang X 2012 Carbohydrate Polymers, 90(4), **1677-1684**.
[15] S Priatni 2015 Procedia Chemistry 16, **438-444.**
[16] Sui X 2016 Food Chemistry, 192, **516-524**.

[17] Pereira V A 2015 *Food Hydrocolloids*, Vol. 43, **180-188**.
[18] Zhang X 2014 *Sensors and Actuators B: Chemical*, Vol. 198, **268-273**.

IOP Conf. Series: Earth and Environmental Science **207** (2018) 012039 doi:10.1088/1755-1315/207/1/012039

Utilization of edamame soybean (*Glycine max (L) Merril*) as modified of enteral formula high calories

A. Widiyawati[1*] and Y. Susindra[1]
[1] Clinical Nutrition Study Program, Politeknik Negeri of Jember, Indonesia

*Email : agatha@polije.ac.id

Abstract. Malnourished patients may experience a decrease in nutrient intake due to anorexia, nausea, and vomiting which can lead to further worsening the patient's condition. Insufficient nutritional intake causes undernutrition so it is necessary to administer a high calorie enteral formula to improve the nutritional status of the patient. The source of a grain-based protein based on dietary requirements is edamame. Edamame has a high protein content which reaches 36% higher than other types of soybeans. This study aims to analyze physical quality viscosity, determine the best treatment formulation, as well as the bioavailability of the best-treated protein with a standard hospital formula and analyze acceptability. Methods this study uses experimental laboratory. Added edamame flour treatment was 55%, 65%, 75%, 85%, and 95%. Conclusion the treatment of adding edamame flour as a modification material for high calorie formula gives a very real influence (Sig. < 0,01), the best treatment was obtained in the formula of high calorie modification of the third treatment edamame flour (P3) with the percentage addition of edamame flour 75%, and the formula for high calorie modification of edamame flour was not significantly different from the standard formula of the Hospital.

1. Introduction

Malnutrition in hospitals describes deficiencies, excess or imbalance of nutrients that affect inpatients due to complex interactions between disease, food and nutrition. The prevalence of malnutrition in Indonesia is still very large, namely in 2003 as many as 75% and in 2004 as much as 59.6%, the main cause of malnutrition in hospitals is because they do not get enough energy and protein intake. Adequate intake of nutrients for patients is needed to prevent the decline in nutritional status during hospitalization. Malnutrition can also occur before entering the hospital because of illness or insufficient nutritional intake, malnutrition can also occur after being hospitalized. The results of previous research studies showed that there was a decrease in nutritional status for 75% of patients during hospitalization. Nutritional therapy is an important part of the treatment and recovery process and can shorten the stay [Nurmala et al., 2014[1]].

One of the complaints that patients often feel at home is a disorder of swallowing or dysphagia that is related to the patient's ability to eat, which in turn can result in a decline in nutritional status.The intervention recommended to overcome this situation is through the provision of a special diet. This special diet is an enteral formula given to patients with swallowing disorders to avoid malnutrition due to inadequate nutritional intake of patients due to decreased awareness and difficulty in swallowing. Food ingredients that can be used as a substitute for soybean flour in a high-calorie diet formula are edamame soy flour. Edamame is a food source of vegetable protein that has more protein

content than other types of plant foods. Edamame is a functional food that is very potential because it contains bioactive components including bioactive peptides, omega-3 fatty acids, isoflavones, sterols, and saponins, as well as high food fiber content in edamame soybeans which are proven to reduce LDL cholesterol and nutrients contained in more edamame easily digested by the body than yellow soybeans. [Almatsier, 2010[2]; Muaris, 2013[3]; Winarti, 2010[4]].
Based on the description above, the researcher analyzed the difference in making high-calorie formula modified edamame flour with standard hospital formula.

2. Materials and methods

This study uses experimental laboratory using a completely randomized design study. Added edamame flour treatment was 55%, 65%, 75%, 85%, and 95%. Each treatment was repeated 5 times so that there were 25 experimental units.

Table 1. Comparison of material composition at each treatment level

Food material	Unit	P1	P2	P3	P4	P5
Edamame flour	g	25,85	30,55	35,25	39,95	44,65
Skimmed milk powder	g	21,15	16,45	11,75	7,05	2,35
Canola oil	g	3	3	3	3	3
Coconut oil	g	2	2	2	2	2
Sugar Flour	g	6	6	6	6	6
Maltodextrin	g	4	4	4	4	4
Total	g	62	62	62	62	62

Description: Each treatment is dissolved with water until it reaches 200 ml

Data were analyzed by One Way Anova statistic test at P = 0,01 and continued with Duncan's Multiple Range Test (DMRT) test with real level 1% and processed with SPSS 18.0 for Windows. The best treatment determination test uses the Effectiveness Index test (De Garmo).

3. Results

Viscosity test results were analyzed statistically using One Way Anova test to determine whether there was an effect of adding edamame flour to the viscosity value of each treatment. Before the Anova test is performed, the data is first tested for normality to determine whether the data is normally distributed or not.

Table 2. The results of one way ANOVA test for the viscosity of high-calorie formula modified edamame flour

	Sum of Squares	df	Mean Square	F		Sig.
				F-count	F-Table	
Between Groups	4035,22	5	807,05			
Within Groups	945,76	24	39,41	20,48	2,77	,000
Total	4980,99	29				

Anova test results in Table 2. show that the treatment of adding edamame flour as a modified ingredient in high calorie formula has a very real effec (F-count > F-table) reinforced with the value of

Sig. <0.01.This shows that the percentage comparison of the amount of edamame flour and skim milk affects the viscosity of the formula for high calorie modification. So, the six treatments have real viscosity differences. From Table 2. then continued with Duncan's test to find out the comparison between treatments.

Table 3. Test results of the duncan viscosity formula of high-calorie formula modified edamame flour

Treatment			Subset for alpha = 0,01			
		N	1	2	3	4
Dimension 1	K	5	2,18			
	P1	5	5,48			
	P2	5	9,36	9,36		
	P3	5		18,32	18,32	
	P4	5			23,91	
	P5	5				35,84
	Sig.		,10	,03	,17	1,00

Means showing for groups in a homogeneous set.
a. Uses Harmonic Mean Sample Size = 5,00.

The high calorie formula modification of the best edamame flour from the six formulas (P1 - P5) was determined from the results of the effectiveness index test. The results of ranking the importance of variables on product quality and the weight of each variable are obtained from the opinions of 30 respondents. After knowing the ranking of each variable, then the variable weight and the normal weight of each variable can be determined. Next is to determine the best and worst value of each variable for each treatment. Then the best treatment can be determined by calculating the value of effectiveness (Ne) and the result value (Nh) of each treatment. Determination of the best treatment based on the highest number of results (Nh).

Table 4. Determination of the best treatment

Treatment	P1	P2	P3	P4	P5
Number of Nh	0.19	0.22	0.25*	0.04	0
Rating	III	II	I	IV	V

Remarks *: Best Treatment

The results of the determination of the best treatment is the formula for high calorie modification of the first treatment of edamame flour (P3) with the percentage addition of edamame flour 75% then carried out a proximate test to determine the nutrient content of the formula consisting of water content, ash content, fat, protein, and carbohydrates, and testing of food fiber content. The results of the nutrient level test for the formula of high calorie modified edamame flour 75% (P3) can be seen in Table 5.

Table 5. Nutritional composition of high-calorie formula modified edamame flour per 100 g

No	Nutritional Composition	Value
1	Energy (kcal)	465,53
2	Protein (gram)	38,50
3	Fat (gram)	19,57
4	Carbohydrates (gram)	33,85
5	Water (gram)	4,13
6	Ash (gram)	3,94
7	Food Fiber (gram)	3,27

Table 6. Value of physical quality and chemical quality of standard hospital formulas and high calorie formula modified edamame flour (per 100 g)

No.	Parameter Mutu	Formula Diet Standar RS	Formula tinggi kalori Modifikasi Tepung Edamame
Mutu Kimia:			
1	Energi (kkal)	421,88	465,53
2	Protein (gram)	16,88	38,50
3	Lemak (gram)	14,38	19,57
4	Karbohidrat (gram)	57,19	33,85
Mutu Fisik:			
5	Viskositas (cp)	2,2	18,3

Table 7. Test results T formula of the standard hospital diet with a high calorie formula modified edamame flour

	Paired Differences			t	df	Sig. (2-tailed)
	Mean	Std. Deviation	Std. Error Mean			
Hospital Standard Diet Formula	102,50	179,73	117,68	,095	8	,902
Formula for High Calorie Modification of Edamame Flour	113,73	192,20				

T test is known the results of different tests between the standard formula of the Hospital with the formula of high calorie modification of edamame flour,obtained results that the value of physical quality (viscosity) and chemical quality values (energy, protein, fat, carbohydrates, and dietary fiber,) between the two formulas are not significantly different indicated by the value of Sig. > 0.01. So the formula for high calorie modification of edamame flour is not significantly different from the standard formula of the Hospital.

4. Discussion

Edamame is a food source of vegetable protein that has more protein content than other types of plant foods. Edamame is a functional food that is very potential because it contains bioactive components including bioactive peptides, omega-3 fatty acids, isoflavones, sterols, and saponins, as well as high food fiber content in edamame soybeans which are proven to reduce LDL cholesterol [3;4].

Enteral nutrition formula is a nutrient given to patients who cannot meet thei r nutritional needs through the oral route. Enteral nutrition is the therapy of giving nutrients through the gastrointestinal tract using a special tube / feeding tube. The method of administration can be through nasogastric route or nasal intestine (nasoduodenal or nasojejunal route). Giving nutrients can also be done by bolus or infusion through an enteral infusion pump. Enteral nutrition is a method of feeding that is simpler, cheaper, and rarely causes complications. Early enteral nutrition can reduce catabolic responses, reduce infectious complications, improve patient tolerance, maintain intestinal integrity, maintain an immunological response, be more physiological, and provide the right source of energy for the intestine at the time of illness. Proper administration of enteral nutrition will provide the patient with nutrients in a form that can be used by his body's metabolism without causing gastrointestinal disorders such as intestinal cramps or diarrhea [5; 6]. Energy is needed by the body for basal metabolism, especially for patients who experience hypermetabolism, which requires high energy to prevent malnutrition. The energy value in food depends on the amount of carbohydrates, proteins, and fats in these foods. The energy value of the high-clorical formula modification of edamame flour was higher than the standard RS diet formula, because the energy value of the high-calorie formula was modified by the greater edamame flour, which was 465.86 kcal / 100 g (Laboratory Test Results). So that the formula for high calorie modification of edamame flour meets the energy requirements of high calorie formula. The protein value of the formula for high calorie modification of edamame flour was higher than the standard formula protein RS. This is because the high-calorie formula protein value modified by edamame flour is greater than 38.50 g / 100 g (Laboratory Test Results). The high protein content of the formula of high calorie modification of edamame flour meets the requirements of high calorie formula protein, and is appropriate if given to patients who are not allowed to eat solids, patients before and after surgery, infectious patients, breastfeeding mothers and in adolescents during growth, and to increase body weight.

The fat contained in the formula for high calorie modification of edamame flour is higher than the fat in the standard hospital formula. This is due to the fat value of the formula with high calorie modification of edamame flour which is greater than 19.57 g / 100 g (Laboratory Test Results). However, despite its high fat content, a source of fat in a high-calorie formula, modified edamame flour is an essential fatty acid that is good for the body, namely omega-3 fatty acids found in edamame soybeans and omega-9 fatty acids found in canola oil and coconut oil . According to Winarti [4], soybeans are low in saturated fatty acid content, soybean fat contains 15% saturated fatty acids, while around 60% is unsaturated fat. Omega-3 fatty acids found in soy can reduce bad cholesterol or LDL and can also increase good cholesterol in the body, HDL. Based on research, omega-9 has a protective power that can reduce LDL blood cholesterol, increase HDL cholesterol greater than omega-3 and omega-6. Diets based on canola oil have been shown to reduce plasma cholesterol levels compared to diets containing higher levels of saturated fatty acids [7]. Canola oil consumption also affects biological functions that affect the risk of disease. Canola oil contains lower levels of saturated fatty acids (about 6%) than other oils and contains high levels of unsaturated fatty acids. Canola oil (*Brasicca napus* L.) has high nutritional value and vitamin E content is useful for skin care. Coconut oil has a natural sweetness from coconut and contains 92% saturated fatty acids (in the form of triglycerides), most (about 70%) are lower saturated chain fatty acids known as medium chain fatty acids (MCFAs). MCFAs are not common to different vegetable oils with lauric acid at 45-56%. Various fractions of coconut oil have medium chain triglycerides and are excellent solvent for flavours, essences, emulsifiers etc. Its metabolism is different from that of the normal vegetable oils containing long chain fatty acids [8].

The carbohydrate formula high in calories modified by edamame flour is lower than carbohydrates in the standard formula of the hospital. This is because the high carbohydrate formula modified carbohydrate value of edamame flour is 33.85 g / 100 g (Laboratory Test Results). However, the high-calorie carbohydrate formula modification of edamame flour is not less than 10% of the standard hospital formula energy. This means that the formula for high calorie modification of edamame flour still meets the requirements of the carbohydrate formula for stroke diet.

The high-calorie formula modified edamame flour has met the requirements of enteral formula according to [AsDI, 20059] as follows:

a. Formula energy content is high in calories, modification of edamame flour 1 kcal / 1 cc liquid.
b. The nutrient content of the high calorie modification formula has been balanced,
c. The raw material for edamame flour is easily absorbed, because food in the form of flour is easier to digest compared to fresh ingredients.
d. Contains dietary fiber which is 3.27 g.
e. The high-calorie formula is modified by lactose-low edamame flour because it contains only a little milk.
f. High-calorie formula modified edamame flour does not contain cholesterol
g. The high-calorie formula modified with edamame flour contains balanced nutritional value, enough vitamins and minerals.

The high-calorie formula modified with edamame flour is the enteral formula given by NGT to patients who are not allowed to eat solids, patients before and after surgery, infectious patients, breastfeeding mothers and in adolescents during growth, and to increase body weight ,also to meet daily needs and to prevent malnutrition. The suggestion of presenting this formula to meet nutritional needs in a day is given as much as 62 g, and in a day given as much as 5 meals to get energy as much as 1500 kcal. The presentation method is 62 g of formula added with warm water at a temperature of 70°C until it reaches a volume of 200 ml, and obtained an energy density of 1 kcal / 1 ml enteral formula according to the requirements of enteral formula.

5. Conclusion

The conclusion of the treatment of adding edamame flour as a modified ingredient in a high calorie formula gives a very real influence Sig. <0.01, the best treatment was obtained in the formula of high calorie modification of the third treatment edamame flour (P3) with the percentage addition of edamame flour 75%, and the formula for high calorie modification of edamame flour was not significantly different from the standard formula of the Hospital. Based on the conclusions above, For further research, it is necessary to test the osmolarity, Serum amyloid A (SAA) proteins, and to be given to patients who are allowed to eat solid food, before and after surgery, infectious patients, nursing mothers and adolescents during growth, and to increase body weight.

Acknowlement

Researchers are grateful to the Non-Tax State Revenue (PNBP) which has provided research funding assistance with contract number 648 / PL17.4 / PL / 2018.

References

[1] Nurmala et al. 2014 *Jurnal Gizi dan Dietetik Indonesia* **2** 14-22.
[2] Almatsier, S. 2010. *Penuntun Diet Edisi Baru*. Jakarta: PT. Gramedia Pustaka Utama.
[3] Muaris, H. J. 2013. *Khasiat Edamame Untuk Kestabilan Kesehatan*. Jakarta: PT. Gramedia Pustaka Utama.
[4] Winarti, S. 2010. *Makanan Fungsional*. Yogyakarta: Graha Ilmu.
[5] Hartono, A. 2006. *Terapi Gizi & Diet Rumah Sakit Edisi 2*. Jakarta: Penerbit Buku Kedokteran, EGC.
[6] Beck, M. E. 2011. *Ilmu Gizi dan Diet Hubungannya Dengan Penyakit-Penyakit Untuk Perawat dan Dokter*. Yogyakarta: Yayasan Essentia Medica.

[7] Gopaka KAG, Gauray R, Bhatnagar AS, Prasant KPK and Preeti C 2009. *Coconut Oil : Chemistry, Production and Its Applications – A Review.* Department of Lipid Science & Traditional Foods, Central Food Technological Research Institute (CSIR), Mysore – 57002.
[8] Lin L, Hanja A, Angela D, Lisa C, Shaunda D, Alvin B and Peter JHJ 2013. Evidence of health benefits of canola oil. *Nutrition Reviews*. *71*: 370-385.
[9] AsDI. 2005. *Panduan Pemberian Makanan Enteral*. Jakarta: CV Jaya Pratama.

Full Trap Method in Handling Warehouse Pests in Ledokombo, Jember

I Erdiansyah[1], F Mayasari[1], S U Putri[1], V Kartikasari[1], Eliyatiningsih[1]

[1]Departement of Agriculture Production, Politeknik Negeri Jember, Jalan Mastrip POBOX 164 Jember 68101, Indonesia
[1]iqbal@polije.ac.id

Abstract. Rice is the main source of food in Indonesia which its needs is increasing each year. The minimum availability of rice in the market is caused by the scarcity of rice stock in rice storage warehouses due to the existence of warehouse pests. Beside decreasing the quantity and quality of rice, they also affect financial losses by farmers. The declining quality of rice causes the price offered by collecting merchants to be very cheap and far below the Floor Price. The Team of Community service provides counseling and training of warehouse pest control using full trap to farmer groups in Jember. The Team of Community service also conducts counseling about warehouse sanitation in order to keeping the hygiene of warehouse properly. To increase farmers' income, the Team of Community service also conducts marketing management training. Full trap method is a method combination of physical and chemical control. The management was applied by purple lights and chemical pesticides, humidifiers with colorful lights and insect pheromones, and mouse traps from buckets. The results showed wet pests that were caught in full trap media are *Sitophilus oryzae pests, Rhyzopertha dominica, Cryptolestes ferrugineus, Necrobia rufipes, Tribolium castaneum, Sitophilus granaries, Periplaneta amiricana, Solenopsis fugax, Rattus rattus diardi, Spodoptera litura*. As the population of warehouse pests decrease, while rice quality and packaging increase.

1. Introduction

Rice is one of the staple foodstuffs of the Indonesian people and almost 90% of Indonesians consume rice as their staple food [1]. The need for rice increases every year based on data from the Central Statistics Agency for the past 10 years, Indonesia's rice imports in 2015 reached 861,601 tons [2]. Constraints in national rice cultivation, not only in the aspects of production and land area, but also in the aspect of post-harvest rice. Post harvest includes harvesting and processing of grain. Storage is one of the important stages because rice experiences a process of decreasing quality and quantity. Agricultural development policies in 2015-2019 include self-sufficiency policies, competitive product development, and strengthening of systems and institution. In connection with this matter, the government has also set grain quality standards and supervised the processing, transportation, and storage suggestions so that grain quality can be maintained and the level of damage can be reduced so that the economic value remains high.

The presence of warehouse pest attacks results in a decrease in the quality and quantity of grain which is certainly detrimental to farmers financially. Several factors that influence product quality during storage include biotic and antibiotic factors. Biotic factors include the presence of insects, mites, mice, molds, and warehouse pests from the Coleoptera order, namely Sitophilus oryzac, Sitophilus granarius, Rhizopertha dominica, Tribolium castaneum and Cryptolestes ferrugineus. Damage due to Sitophilus oryzae, for example, can cause weight loss of 6.29%, water content of 15.9%, broken rice 24.40%, rice flowing 10.08%, rice being musty and cloudy white [3].

Several factors that are likely to affect the population and development of warehouse pests (insects) are hardness, humidity, density of seeds, and calcifying grains [4]. Climate of micro warehouses with an average temperature of 34 °C, 59% humidity and good warehouse aeration still allow the presence and development of warehouse pest insects. During the 1.5 month period of storage of rice in the warehouse, there were several changes in the quality and quantity of rice. Post-harvest methods carried out by farmers in several areas in Jember Regency are still relatively simple, ranging from drying of grain and grain storage in the warehouse lacks assistance [5].

Bulog is a corporation namely a food institution in Indonesia that deals with the rice trading system. Bulog carries out maintenance and control of integrated warehouse pests, including the hermetic method of CO2 stack technique and the use of plastic Cocoon [6]. The growth of pest population is influenced by environmental factors including availability of oxygen, humidity and temperature, so that by manipulating the environment it is expected to be able to physically control pests. Innovations in grain drying, packaging, storage processes, and the application of full trap techniques in warehouse pest control need to be carried out by rice farmers in Jember, this is aimed to maintaining the quality and quantity of grain / rice sold. The aim of the study was to determine the diversity of species of pest insects that attacked rice grain commodities in the storage warehouse of Suren Jaya 2 farmer groups and Jaya Abadi farmer groups by using the Full Trap method.

2. Method

The research was carried out in the rice milling storage warehouse owned by Suren Jaya 2 farmer group in Suren Village and Tani Jaya Abadi Group in Lembengan Village, Jember Regency, East Java Province. The study used qualitative methods and counseling of two Jember district farmer groups, counseling on warehouse sanitation so that the cleanliness of rice storage warehouses was well maintained. To increase farmers' income, the implementing team also conducts marketing management training. Retrieval of observation data is done by counting the insects caught in the 'Full Trap' trap. Observation data retrieval was carried out for 4 times for 4 months with an interval of 1 month observation. Each insect caught in the 'Full Trap' trap is then taken to the laboratory to be observed. The things that were observed were: 1. Type of pests, pests were sorted / separated according to the type and location of the harvest. Pest samples were collected in 70% alcohol to be identified. Identification of insect pests was carried out at the Plant Protection Laboratory of the Department of Agriculture in the State Polytechnic of Jember. Identification of insect pests using the material library [7], [8], [9], [10], [11], [12], [13] *dalam* [14]. 2. Insect populations, Insects found to be sorted / separated then manually counted under a digital / multimedia microscope.

3. Result and Discussion

The application of full trap technology that has been carried out on both partners includes rice warehouse sanitation counseling, training and installation of full traps in rice storage warehouses, and rice marketing counseling. Through counseling and training, it is expected that partners will be able to improve the quality and quantity of grain / rice they have. The extension of rice warehouse sanitation aims to enable partners to maintain the cleanliness of the rice warehouse. A clean rice warehouse is a good determinant of rice quality. Dirty warehouses will cause the growth of fungi, insects, rodents, and even attacks of rice lice. Therefore, one way to maintain rice in good condition before being sold is the right post-harvest treatment, including storage in a clean and representative rice warehouse.

The second activity after warehouse sanitation being carried out is the installation of a full trap or trap for various warehouse pests. Full traps installed include Light Trap for insect traps and rice lice, and

Bucked Trap for warehouse rodent pest traps. Light trap uses a purple LED light, which has been assembled with a solution (water, detergent, and sodium benzoate). The solution serves to kill insects and reduce mobility when they enter the solution. To attract insects, methyl eugenol is used which functions as pheromone sex in insects. This fluid attracts the attention of male insect pests, thus inhibiting breeding and population growth. This liquid is applied and assembled with automatic chamber sprayers. Bucked Trap application is done using a bucket that is assembled with bottles and boards which functions as a trap for warehouse rat pests.

Evaluation of full trap installation activities, it is known that some pests managed to enter the pest trap. Some pests include *Sitophilus oryzae, Rhyzopertha dominica, Cryptolestes ferrugineus, Necrobia rufipes, Tribolium castaneum, Sitophilus granaries, Periplaneta amiricana, Solenopsis fugax, Rattus rattus diardi, Spodoptera litura.* Installation of traps and trapped pests in a trap is expected to reduce the number of pests that damage the quality and quantity of rice stored in a warehouse.

Types of Pests found in Farmer Groups in Jember Regency, The results showed that there were 10 types of pests in grain storage warehouses in Suren Jaya Farmer Group 2 Suren Village and Tani Jaya Abadi Group Lembengan Village including *Sitophilus oryzae pests, Rhyzopertha dominica, Cryptolestes ferrugineus, Necrobia rufipes, Tribolium castaneum, Sitophilus granaries, Periplaneta amiricana, Solenopsis fugax, Rattus rattus diardi, Spodoptera litura* consisting of 6 types of Coleoptera order, 1 type of Bllatodea order, 1 type of order of Hymenoptera, 1 type of order Rodentia, and 1 type of order Lepidoptera (Figure 1).

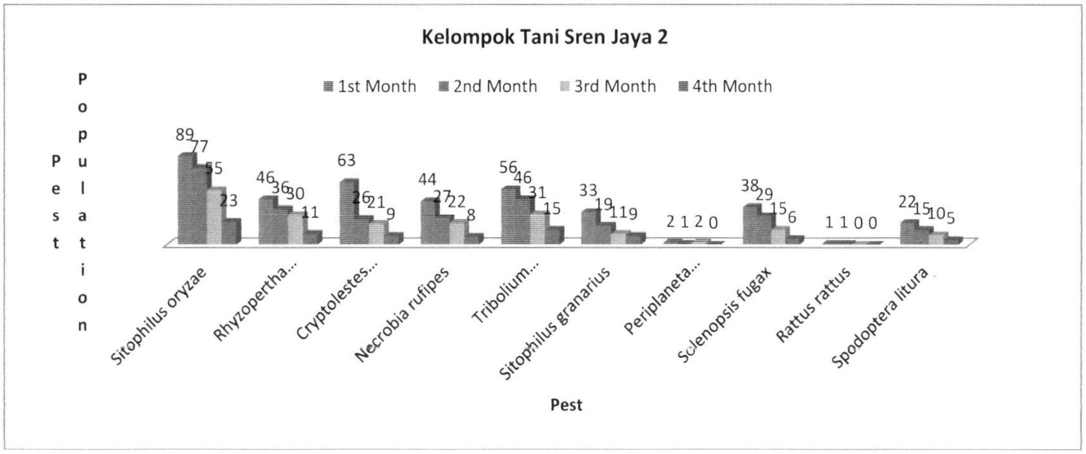

Figure 1. Warehouse pest population in the Suren Jaya Farmer Group 2

The results of identification based on features seen from both body size, color, compound eyes, limbs, wings and antennae indicate that these insects are *Sitophilus oryzae* includes reddish brown Imago and has reddish brown spots. On the front wing, there are four reddish yellow spots that form a distinctive pattern. Imago's body is oval and has a length ranging from 2 - 3.5 mm and a width of 1.1 - 1.3 mm. Having a snout and antenna shaped in the form of Lamellate (Figure 2). This pest is classified into Phylum Arthropods, Subfilum Mandibulata, Klas Insecta. Pterygota class, Order Coleoptera, Family Curculionidae, genus Sitophilus and species of S. oryzae [14].

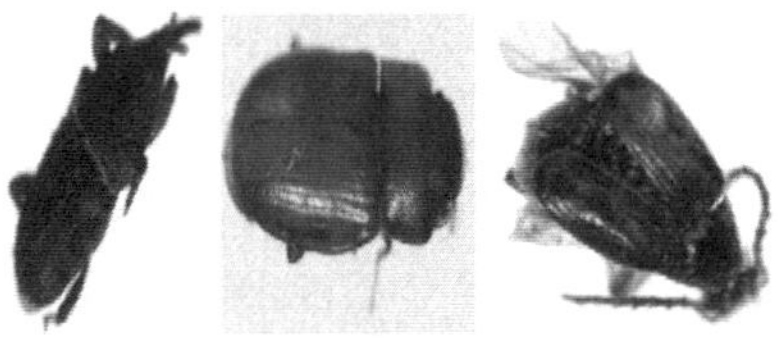

Figure 2: (a) *S. oryzae*, (b) *T. castaneum*, (c) *C. cepalonica Imago*

Tribolium castaneum has the characteristics of Imago's reddish brownish red color, which is about 5-6.5 mm long and 2 mm wide. The clavate-shaped antenna resembles mace, the segments enlarge regularly from the base to the tip. Imago has a mace-shaped antenna extending towards the tip regularly. These pests are found in each sample location. These insects are classified into phylum Arthropoda, Sub-phyllum Mandibulata, Class Insecta, Pterygota sub-class, Coleoptera Order, Tenebrionidae Family, Genus Tribolium and T. castaneum species [14].

Rhyzopertha dominica has a characteristic appearance: Adult beetles are 2-3mm long. Red-brown or brown-black. Life cycle: Females produce 200-500 eggs. At an ideal temperature of 34 ° C and 70% relative humidity (RH), cocoon formation takes 3 days. At an ideal temperature of 34 ° C and Relative Humidity (RH), the life cycle lasts 3-4 weeks. Habit: Eggs are placed in cracks. Larvae will be in grain to continue growing. Larvae develop faster in grain than in flour. Grow up after they appear. Commonly found in grain storage warehouses [14].

Ferrugineus cryptolestes has the following characteristics: Very flat, light brown, hair-like antenna, sometimes very long. A very common species in Australia is *C. ferrugineus*. Important pests in grains and their processed products. Life cycle. Optimum: 23 days at 33 0C, 70% RH. Range: 18-42.5 0C, 40-90% RH, resistant to cold conditions. The maximum population growth rate per month: 55 times. Biology. Eggs: placed randomly. Larva: move freely, not confined. Adults: long-lived, active eating, able to fly, walking fast, able to enter food packed through small gaps [15].

Necrobia rufipes has a morphological characteristic consisting of an antenna, caput, compound eyes, abdomen, thorax, front legs, legs and a pair of wings. Adult body size is about 4-5 mm. The upper surface of the body is metallic bluish and shiny green. The lower surface of the abdomen is dark blue. Their feet are light reddish or orange. A reddish-brown antenna with a dark brown or black tip. In the female beetle, it has an ovipositor patch, has a pair of ovaries, 8 or 9 abdominal segments, a system of oviducts that are channeled out where to lay eggs. Whereas the male beetle, on the 10th abdominal segment has a genitals in the form of a penis, has an outer clamping organ and a median part of the piercing organ [15].

Sitophilus granaries have traits. In the front head there is a snout, *S. Zeamay* / *S. oryzae* has two orange spots on the front wing, the rear wing appears below elitra, *S. granarius* does not have the wing characteristics. Is a major pest in cereals and hard cereal products such as pasta. Life cycle. Optimum: 25 days at 30 0C, 80% RH. Range: 14-35 0C, minimum 40% RH. Maximum population growth rate per month: 25 times. Biology. Eggs: Put on a hole that is prepared in advance on the seeds and covered with a waxy plug. Larvae: Does not move, and develops entirely in grain. Adults: long life, active eating and can fly [15].

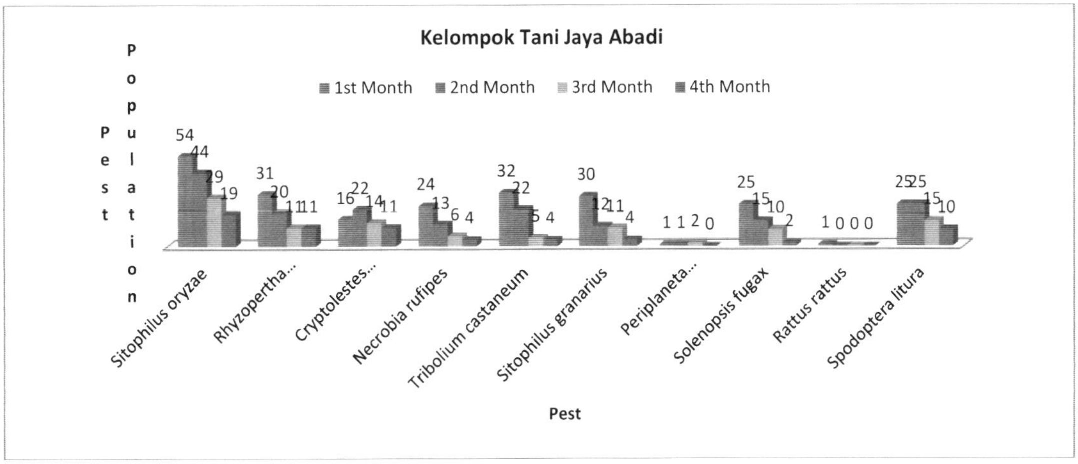

Figure 3. Warehouse pest population in the Lembengan Village Tani jaya Abadi

Tani Jaya Abadi Group Lembengan Village including *Sitophilus oryzae* pests, *Rhyzopertha dominica, Cryptolestes ferrugineus, Necrobia rufipes, Tribolium castaneum, Sitophilus granaries, Periplaneta amiricana, Solenopsis fugax, Rattus rattus diardi, Spodoptera litura* consisting of 6 types of Coleoptera order, 1 type of Bllatodea order, 1 type of order of Hymenoptera, 1 type of order Rodentia, and 1 type of order Lepidoptera (Figure 3).

4. Conclusion

The application of full trap technology that has been carried out on both partners includes rice warehouse sanitation counseling, full trap installation activities which include light trap and bucked trap, as well as counseling on grain storage storage facilities. The results of the evaluation of the full trap installation showed that several warehouse pests trapped were *Sitophilus oryzae, Rhyzopertha dominica, Cryptolestes ferrugineus, Necrobia rufipes, Tribolium castaneum, Sitophilus granaries, Periplaneta amiricana, Solenopsis fugax, Rattus rattus diardi, Spodoptera litura.*

Acknowledgment

The Authors would like to acknowledge the financial support from Directorate of Research and Community Service, Directorate General of Research and Development Strengthening, Ministry of Research, Technology and Higher Education, In accordance with the Community Service Program Implementation Funding Agreement Number: 071 / SP2H / PPM / DRPM / 2018, March 9, 2018.

Reference

[1] Sinaga, I.W. 2010. *Peramalan Kebutuhan Konsumsi Beras di Propinsi Sumatera Utara Tahun 2008-2010 dengan Menggunakan Metode Kuadrat Terkecil.* Tugas Akhir. Universitas Sumatera Utara

[2] Badan Pusat Statistik. 2017. Impor Beras Menurut Negara Asal Utama, 2000-2015. https://www.bps.go.id/LinkTabelStatis/view/id/1043. Diakses tanggal 7 April 2017.

[3] Siregar, A.Z., Tobing, M.C., Pinde, Lumongga. 2010. Pengendalian *Sitophilus oryzae* (Coleoptera: Curculionidae) dan *Tribolium castaneum.*

[4] Zulfahnur. 2010. Kajian Resistensi Lima Jenis Beras Varietas Lokal Terhadap Serangan Sitophilus Zeamais Motsch. Skripsi. Institut Pertanian Bogor.

[5] Dinas Pertanian Tanaman Pangan. 2014. Strategi pengembangan agribisnis komoditas padi dalam meningkatkan ketahanan pangan Kabupaten Jember. Jember
[6] Badan Urusan Logisik. 2012. Perawatan dan Pengendalian Hama. http://www.bulog.co.id/phgt.php. Dikases tanggal 7 April 2017.
[7] Kamel, A. H. and Hassaneim, M. H, 1967.Biological studies on Corcyra cephalonica Stainton. Bulletin de la Societé Entomologique d'Egypte, Cairo, 51: 175-96.
[8] Hinton, H. E. A A. S. Corbet. 1975. Common insects pests of stored products (A guide to their ideentification). Trustees of fhi British Museum (Natural History). London.
[9] Banks, H.J. 1979. Identification of Stored Product Cryptolestes spp (Col: Cucujidae); A Rapid Techique for Preparation of Suitale Mount. Journal of the Australian Entomology Society, Vol. 18.
[10] Halstead, D. G. H, 1986. Key For The Identification of Beatles Associated With Stored Products. Journal of Stored Product Research, Vol 22. No.4 1986.
[11] Haines, C.P. 1991. Insect and Arachinids of Tropical Stored Product Their Biology and Identification . Natural Resource Institute, Central Avenue, Chatam Maritime, Kent Mey 4 TB, United Kingdom.
[12] Brower, J. 2003. Stored Product Management.Oklahoma Cooperative Extension Service Division of Agricultural Sciences and Natural Resources Oklahoma State University. www.okstate.edu/ag/aged.
[13] Buntin, G. D., S. P. Keith., M.J. Weiss, and James A. Webster, 2003. Handbook of Small Grain Insects. photographs, Maps, and Identification Keys. Entomological Societyof America and APS PRESS.
[14] Ilato, J. Dien. M. F., Rante. C.S. 2012. Jenis dan Populasi Hama Beras Di Gudang Tradisional dan Modern Di Provinsi Gorontalo. Jurnal Eugenia Vol 18 No. 2.
[15] Suparjo. 2010. Teknik Penyimpanan Pakan: Kerusakan bahan Pakan Selama Penyimpanan. https://jajo66.files.wordpress.com/2010/12/penyimpanan2010.pdf. Universitas Jambi.

Performance of Resistance of Rice Varieties Recommendation of Jember Regency to Brown Planthopper Pest (*Nilaparvata lugens* Stal.)

I Erdiansyah[1] and Damanhuri[1]

[1]Departement of Agriculture Production, Politeknik Negeri Jember, Jalan Mastrip POBOX 164 68101, Indonesia
[1] iqbal@polije.ac.id

Abstract. This study aims to determine the susceptibility and resistance response of Jember Regency suggested rice varieties from brown planthopper pest (*Nilaparvata lugens* Stal). The research used randomized block design (RAK). Differences between treatments were tested by DMRT at 5% level. The treatments in this study are using seven varieties of rice such as: treatment A (Sintanur), B (Cibogo), C (Ciherang), D (Inpari 13), E (Cigeulis), F (IR-66) and G (IR-64), each with three replicates. The results showed that the highest planthopper pest populations was on Sintanur variety with an average of 7.10 insects and intensity of attacks in the last observation by 34.07%. This parameter affects the number of productive tillers, grain weight and percentage of dry milled grain hollow. During the study, brown planthoppers preferred a young aged rice, whereas in the older age of the plant population, brown planthoppers were rarely found. The most resistant varieties of this research was the variety IR66 which has a low intensity of the attack on the last observation by 1.11% with a population of brown planthopper pest average of 6.36 insects. Thus, to inhibit the development of brown planthoppers population, Sintanur varieties are not recommended while other varieties can be planted in Jember regency.

1. Introduction

Rice is an agricultural commodity that is needed in Indonesia, because almost 90% are used as a main food source. The main obstacle in rice cultivation in Indonesia is the problem of brown planthopper pests which causes low rice production to cause crop failure [1]. Brown planthopper (Nilaparvata lugens) is a type of insect that has the potential as the most dominant pest attacking rice plants in Indonesia and Asia [2]. Planting resistant varieties can reduce the development of brown planthopper pests [3]. The causes of brown planthopper population explosion include the use of nitrogen fertilizer exceeding the dose, and the non-alternating planting period and insecticides [4].

Until now, farmers still rely on synthetic insecticides to overcome the problems caused by brown planthopper pests. According to [5], the use of synthetic insecticides can harm both the environment and human health. Therefore, other integrated control techniques are needed, both using natural methods such as the use or planting of resistant varieties [6].

Rice varieties that are often cultivated by farmers in Jember are Sintanur, Cibogo, Ciherang, Inpari 13, Cigeulis, IR-64, and IR-66. Among these varieties, resistance brown planthopper biotypes has not been identified that attacked the research site or in the field. Therefore, this research needs to be done to determine the resistance of rice varieties tested against brown planthoppers that attack at the study site.

2. Method

This research was conducted in Lembengan Village, Ledokombo District, Jember Regency. The research took place from October 2016 to February 2017. The materials used were rice varieties: Sintanur, Cibogo, Ciherang, Inpari 13, Cigeulis, IR-64, and IR-66. The tools used are: insect nets, petri dishes, tweezers, brushes, loops, keys to insect determination, microscope, glass beaker, raffia rope, meter, electric scales.

The study used a randomized block design (RBD) method consisting of seven rice varieties, namely: A = Sintanur, B = Cibogo, C = Ciherang, D = Inpari 13, E = Cigelius, F = IR-66 and = IR-64 (control). Each treatment was repeated three times. Planting is carried out on paddy fields with a length of 21 m and a width of 9 m. This rice field is divided into 21 plots, each plot measuring 2x2 meters. Planting distance of 20x20 cm, distance between plots measuring 75 cm. Seed removal is done at the age of 25 days, each planting hole is filled with one plant. Observation data will be analyzed and to distinguish differences between treatments data were analyzed by variance (ANOVA) while to distinguish the mean between treatments DMRT test was conducted at the level of 5%.

The parameters observed are:

1. The population of brown plant hopper (Nimpha population and Imago brown planthopper.) Observed at the age of the plant after moving from the nursery (0 days), observation interval once a week.

2. Percentage of damage to plant leaves (%) by systematic random sampling on rice sample units. Determination of attack intensity (IS) is indirectly done by calculating the score of damage to plant parts by using a formula (Digilib, 2003):

$$\text{P} = \frac{\sum (nxv)}{ZxN} x100\%$$

Information :
P = Attack / Damage intensity
n = Number of examples observed for each damage category
v = Score score for each damage category
Z = The highest damage category score
N = Number of sample units observed

Observation of score scores for the intensity of attacks caused by brown plant hopper pests on each clump of rice plants.

(a) Score of 0: No damage to the leaves of the plants observed
(b) Score 1: Very little damage
(c) Score 3: Most of the first and second leaves of the plant turn yellow in part
(d) Score 5: Yellow and dwarf plants or about 10-25% wilted plants
(e) Score of 7: More than half of the plants wither or die and the remaining plants are very stunted or dry
(f) Score of 9: All plants die (IRRI, 1985).

3. Rice Crop Performance by looking at the morphology of the leaves and stems of the tested rice plants under a microscope.

3. Result and Discussion

3.1 Population of Brown Planthopper Pests Nilaparvata lugens. Stal on some rice varieties

The highest pest population at the age of 52 days was found in Sintanur variety rice plants, namely 13.00 head and the lowest was in Inpari 13 variety, which was 2.67. Furthermore, the highest population of planthopper pests at the age of 66 days was found in Sintanur 6.0 insects and Ciherang 6.67 insects while the lowest population was in Cibogo variety which was 0.67 per clump. At the age of 73 hst equal to 66 hst, the highest population of planthopper pests was found in Sintanur 3.0 insects and Ciherang 3.33 insects while the lowest population was in Cibogo variety which was 0.33 per clump.

Population increase from 10 days to 24 days is very rapid, because brown plant hopper prefers rice plants in the vegetative phase or young age [7], brown planthopper is abundant in young plants because the tissue is still soft. In old age plants, the brown planthopper population is rarely found. This was presumed because the availability of food for brown plant hopper in rice plantations at generative age was less, so brown plant hopper moved or migrated to other fields looking for rice crops that were still young. In line with age, the population of brown planthopper pests which initially increased, gradually decreased in six varieties of rice plants.

The peak of population development in rice varieties observed, occurred at various ages, between 24 - 31 days. Base The peak of population development in rice varieties observed, occurred at various ages, between 24 - 31 days. Based on [8], the economic threshold for brown plant hopper that has been determined is an average of 2 head per clump for rice plants aged less than 40 days and 5 per clump for rice plants more than 40 days after planting [9]. Thus, the leafhopper population in this study has exceeded the economic threshold at the age of 17-52 days, because the average population at the age of 17 days was 4.71 and up to 52 years old the population ranged above 7.38 heads. Cibogo variety (6.67 head) and Inpari 13 (5.67 head), had a population below the economic threshold at the age of 45 days, then IR64 (6.67 head) and Ciherang (5.33 insect) decreased at age 52 hst, IR 66 variety (4.33 insects) and Cigeulis (4.33 insects) at the age of 59 hst, while the Sintanur variety (6.67 insects) the new population fell at the age of 66 hst.

Thus, in this study, varieties that have brown planthopper populations below the shortest economic threshold are Cibogo and Inpari 13, namely 28 days, then IR 64 and Ciherang 35 days, then IR 66 and Cigeulis 42 days, the longest are Sintanur varieties namely 49 days. Thus, to inhibit the development of brown planthopper populations, Sintanur varieties are not recommended for planting.

3.2 The intensity of the attack of the Brown Planthopper Pest Nilaparvata lugens. Stal

The treatment of IR64 variety has a low attack intensity of 3.70% on the last day of observation. The highest intensity of brown planthopper pest attack was on sintanur varieties with an attack intensity of 34.07%, and the lowest intensity of brown planthopper pest was IR66 with 1.11%. It is clear that sintanur varieties are varieties that are susceptible to brown planthopper pests. In addition to sintanur varieties, the Cigeulius variety has a criterion susceptible to brown planthopper pests, namely 31.48%. [10], the response of plants to pests and diseases caused by the most destructive brown planthopper pest damage to rice plants, according to the resistance assessment more than half of the plants wither and the plants become stunted or die, according to the symptoms caused in the field. research land. According to [11], damage to the rice plant network is caused by most of the reovirus in the planthopper is pathogenic and causes various symptoms in the host plant.

IR64 is more resistant than sintanur varieties, but more susceptible when compared to IR66 varieties. It was seen in the observation of 101 hst plant age, the intensity of the attack on IR66 varieties was the lowest with symptoms of leaves per clump of yellowing rice plants at the time of observation. This is also the tolerance of rice plants on brown plant hopper with its ability to recover or process the healing of rice plants after being damaged due to the attack of brown planthopper pests [12]. In this case, pests are present in rice plants, but the damage and losses can be minimized because of the ability of these varieties to repair (repair) and replace (damage) damage caused by brown planthopper pests, so that the plant can continue to grow again.

IOP Conf. Series: Earth and Environmental Science **207** (2018) 012041 doi:10.1088/1755-1315/207/1/012041

Table 1. The Effect of Several Rice Varieties on the Population of Brown Planthopper Pests Nilaparvata lugens. Stal

Treatment	Brown planthopper pest population														Average
	10 hst	17 hst	24 hst	31 hst	38 hst	45 hst	52 hst	59 hst	66 hst	73hst	80 hst	87 hst	94 hst	101 hst	
Ciherang	1.33 a	5.00 a	14.33 a	15.67 a	15.00 a	9.67 a	5.33 bcd	3.33 a	6.00 a	3.33 a	2.33 a	1.33 a	0.33 a	0.00 a	5.93
Cigeulis	1.00 a	4.00 a	16.67 a	16.33 a	11.33 a	7.00 a	9.00 abc	4.33 a	4.00 ab	2.67 ab	2.33 a	2.33 a	1.00 a	0.00 a	5.86
Sintanur	0.33 a	6.67 a	21.33 a	11.00 a	10.67 a	15.33 a	13.00 a	7.67 a	6.67 a	3.00 a	1.67 a	1.67 a	0.33 a	0.00 a	7.10
IR66	1.67 a	4.00 a	18.33 a	20.00 a	12.33 a	11.33 a	10.67 ab	4.33 a	3.33 abc	1.33 abc	0.33 a	0.67 a	0.67 a	0.00 a	6.36
Cibogo	0.33 a	2.00 a	17.67 a	11.00 a	6.00 a	6.67 a	4.33 cd	1.67 a	0.67 c	0.33 c	1.00 a	0.33 a	0.00 a	0.00 a	3.71
IR64	1.67 a	6.67 a	15.33 a	12.67 a	13.67 a	8.67 a	6.67 abcd	2.67 a	1.00 bc	1.00 bc	1.00 a	1.00 a	0.00 a	0.00 a	5.14
Inpari 13	0.67 a	4.67 a	14.00 a	16.33 a	9.00 a	5.67 a	2.67 d	4.67 a	1.00 bc	1.67 abc	1.33 a	1.00 a	0.00 a	0.00 a	4.48
Rata-rata	1.00	4.71	16.81	14.71	11.14	9.19	7.38	4.10	3.24	1.90	1.43	1.19	0.33	0.00	

Remarks: Numbers followed by the same letter in the same column are not significantly different at the 5% level on the Duncan Multiple Range Test (DMRT).

Table 2. Effect of Several Rice Varieties on the Intensity of Brown Planthopper Pests Nilaparvata lugens. Stal

Treatment	intensity of brown planthopper attacks														Average
	10 hst	17 hst	24 hst	31 hst	38 hst	45 hst	52 hst	59 hst	66 hst	73 hst	80 hst	87 hst	94 hst	101 hst	
Ciherang	0.00 a	1.85 a	7.40 a	9.99 a	14.81 a	12.22 a	11.18 a	7.11 abc	8.52 ab	11.11 abc	12.59 ab	13.33 ab	19.26 ab	20.74 ab	10.72
Cigeulius	0.00 a	0.37 a	5.56 a	10.73 a	11.85 a	8.14 a	5.92 a	11.48 ab	14.81 a	15.92 ab	22.22 a	22.59 a	30.00 a	31.48 a	13.65
Sintanur	0.00 a	1.11 a	6.66 a	8.14 a	10.73 a	13.70 a	12.96 a	15.18 a	14.07 a	19.62 a	22.22 a	23.70 a	28.87 a	34.07 a	15.08
IR66	0.00 a	0.74 a	3.70 a	6.85 a	7.77 a	8.88 a	7.03 a	2.59 c	0.74 b	1.11 c	1.11 b	1.11 b	1.11 b	1.11 b	3.13
Cibogo	0.00 a	0.00 a	2.22 a	5.55 a	4.07 a	6.66 a	4.44 a	4.07 bc	2.96 ab	2.96 bc	3.70 b	5.18 b	3.70 b	3.70 b	3.52
IR64	0.00 a	1.85 a	8.51 a	10.00 a	12.22 a	7.77 a	5.92 a	2.59 bc	0.37 b	0.37 c	1.85 b	2.96 b	3.70 b	3.70 b	4.42
Inpari 13	0.00 a	1.11 a	8.51 a	8.88 a	7.40 a	7.40 a	5.92 a	9.25 abc	1.48 b	7.03 abc	8.51 ab	3.33 b	4.81 b	4.81 b	5.60

Remarks: Numbers followed by the same letter in the same column are not significantly different at the 5% level on the Duncan Multiple Range Test (DMRT).

Cibogo and IR64 varieties have the same low intensity of attack that is 3.703%, and the Inpari 13 variety is 4.813%. According to the initial description, the variety is a variety of rice plants with a rather resistant criteria with the characteristics of the first yellow and first few leaves when observed on the ground. So this level of resistance becomes resistant [13]. Because IR64 is more resistant compared to Sintanur varieties, it is more susceptible when compared IR66.

Commonly, varieties that are sensitive to brown plant hopper are which hair or hair on the torso are rare or few but rather long and between long hairs there are also small, relatively dense small feathers. In resistant rice cultivars have a variety of feathers or hair but are generally longer and more abundant in the leaves and stems of rice plants [14]. Coarse fiber on the leaf and stem surface of rice varieties tested in accordance with the description of varieties according to [15] is that each variety has a rough leaf and stem surface on resistant varieties while in varieties susceptible to leaf surfaces and rather rough rod. In IR66 varieties tend to have coarse stem walls and this is also the same in IR64 varieties, whereas in Sintanur varieties have rather coarse stems and large stems and the presence of cavities in the stem.

3.3 Performance of Resistance to Seven Test

Table 3. Performance of Tested Resistance of Rice Varieties

Varietas Padi	The average population of brown planthopper pests (ekor)	Average Intensity of Brown Planthopper Attack (%)	Heavy Grain Dry Grain (g)	Morphological Performance	Conclusion of Resilience
Ciherang	5.93	10.72	425.00 c	The surface of the stem and leaves is rough	rather vulnerable
Cigeulis	5.86	13.65	483.33 c	The surface of the stem and leaves is rough	susceptible
Sintanur	7.10	15.08	375.00 c	There is no stem hair or rarely seen with a microscope 10x magnification	vulnerable or sensitive
IR66	6.36	3.13	1475.0 a	Stem hairs do not exist or are rarely seen with a microscope 10x magnification	Tolerant
Cibogo	3.71	3.52	1216. 7 a	The surface of the stem and leaves is rough	Antixenoksis
IR64	5.14	4.42	1266.7 a	There are rod hairs seen with a microscope 10x magnification and rough surface of the stem and leaves	Antixenoksis
Inpari 13	4.48	5.60	833.33 b	The surface of the stem and leaves is rough	Antixenoksis

Remarks: Numbers followed by the same letter in the same column are not significantly different at the 5% level on the Duncan Multiple Range Test (DMRT)

In rice varieties that are resistant to brown planthopper biotypes, three have the highest silica content and in varieties that are resistant to planthopper biotypes, one and two has little silica content meaning that the most silica content in rice plants has hard and stocky stems causing activity, especially feeding brown planthopper is disturbed and resistant to brown plant hopper [16].

In accordance with the description of varieties about pest resistance, it can be inferred that Sintanur varieties are sensitive to brown planthopper biotype 3, so that the population and intensity of brown planthopper pest attack is the largest in Sintanur variety, Cigeulis variety which is pest resistant ie brown plant hopper biotype 3 and ciherang variety which is somewhat resistant against biotype 3 planthopper, while in IR66 and inpari 13 varieties, pest resistance was resistant to brown planthopper biotypes 1, 2 and 3 so that the population and intensity of brown planthopper pest was the lowest. The population and intensity of brown planthopper pests in all rice varieties tested affected rice growth and production.

Plant varieties that are resistant to pests and diseases have a group of factors contained in plants and obtained naturally, while their nature is to reject, prevent or tolerate attacks by pests and diseases. Factors that control the nature of resistance include physical, chemical, anatomical, physiological and genetic factors [16]. Antixenoxis is the insect's preference for mechanical factors derived from the physical structure and surface properties of plants. The structure and physical properties of the surface of the plant include, thickness of the skin, length and dense hairs on the surface of the leaf, the size of the stomata and the thickness of the cuticle layer. The insect's preference for mechanical factors is closely related to the structure of the tools and the way to take food and lay eggs held by insects.

The mechanism of resistance in the seven varieties of rice plants tested showed that the Inpari 13 variety, Cibogo and IR 64 showed antixinoxic resistance which means that the three varieties contained morphological resistance which did not benefit the brown planthopper pests in the test plants, namely the presence of long and hard fur and the surface of the stem and coarse leaves that inhibit the way of eating, the feathers are found on the leaves and stems that are seen using a 10x magnification microscope on the IR 64 variety, and can be seen in the population, the intensity of the brown planthopper pest attack is low.

In Sintanur, Cigeulis and Ciherang varieties showed the population and intensity of large brown planthopper pest attacks and also the yield of milled dry grain weight per low experimental plot, this was due to the morphological resistance of the three varieties, Sintanur, Cigeulis and Ciherang, which were very low. Sintanur varieties which have no or rarely found hair on the stem seen with a 10x magnification microscope as well as on the rather rough surface of the stem and leaves.

In IR 66 varieties which showed a high population of brown planthopper pests, the lowest intensity of brown planthopper pests and the yield of milled dry grains per experimental plot were the highest compared to the production of other varieties. This is because the resistance of IR 66 varieties shows tolerant resistance which means tolerant is one of the properties possessed by plants that can heal themselves from damage to pest attacks, although the number of pests that attack amounts to the same as those that attack IR 66 varieties, the presence of feathers on the stems and leaves and the rough surface of the stem and leaves (table 3).

4. Conclusion

Treatment of IR66 and IR64 (control) varieties had low attack intensity and *Sintanur* varieties had high attack intensity, so it was known that sintanur varieties were susceptible varieties when compared to controls. 2. High in the number of productive tillers, the weight of dry milled grain and the percentage of empty grain on Sintanur varieties is lower when compared to other varieties and also control.

Acknowledgment

The Authors would like to acknowledge the support from Reasearch and Development Agency of Politeknik Negeri Jember, East Java Province, Indonesia.

References

[1] Badan Meteorologi dan Geofisika, 2011. Produksi Beras Diperkirakan Turun 5 %. Available At: http://api.or.id/2011/02/11/produksi-padi-tahun-ini-turun/. Dalam Koran Tempo 2016.Diakses 17 Oktober 2016.

[2] Zhu, L. L., Zhu, C. L., Weng, Q. M., Huang, Z., He, G. C. 2004. *Reaserch Progress On Brown Plant Hopper Resistence Genes In Rice (in Chines)*. Hubei Agric. Sc:1:19-24

[3] Bahagia, M. A. G., J. Pritchard and B. Ford Lloyd. 2011. *Brown Planthopper (N. lugens.Stal) Feeding Behavior on Rice Germplasm As An Indications Of Resistance. Plos One*. www.plosone.org. July 2016/Volume 6/issue 7/e 22137.

[4] Yuniarti, A. 2004. Uji Ketahanan Beberapa Varietas Kedelai (*Glicine max*. L. Merril) Terhadap Lalat Bibit Ophimyia phaseoli Tyron (Diptera:Agromyzidae). Karya Tulis Ilmiah. Faperta. Universitas Jember.Astuti, P. 2008. Karakterisasi Fenotipe Pada Beberapa Kultifar Padi Dalam Hubungan Tingkat Ketahanan Terhadap Wereng Coklat (Nilaparvata lugens Stal). Universitas Sebelas Maret.

[5] Sogawa, K., Liu, G. J., Shen, J. H. 2003. *A Review On The Hyper-Susceptibility of Chines Hybrid Rice To Insect Pests. Chin. J.* Rice Sci., 17:23-30.

[6] Marwoto. 2006. Teknik Budidaya Tanaman Jilid 2. Direktorat Pembinaan Sekolah Menengah Kejuruan, Direktorat Jendral Managemen Pendidikan Dasar dan Menengah. Departemen Pendidikan Nasional. Jakarta.

[7] Untung, K. 2001. *Pengantar Pengelolaan Hama Terpadu*. Gadjah Mada University Press. Yogyakarta.

[8] Baihaki. 2011. *Strategi Fundamental Pengendalian Hama Wereng Batang Coklat Dalam Pengamanan Produksi Nasional*. . Pengembangan Inovasi Pertanian 4(1), 2011: 3-75.

[9] IRRI. 1985. *Genetic Evaluation For Insect Resistance In Rice*. Los Banos. Laguna. Manila Philippines.

[10] Heinrichs, E.A., F.G. Medrano, and H.R. Rapusas. 1995. Genetic evaluation for resistance in rice. IRRI. Los-Banos, Philippines. 356 p.

[11] Nakashima, N. and Nota, H. 1994. *Nucleotide Sequense of Nilaparvata lugens Reovirus Genome Segment S8 Coding for The Major Other Capsid Protein. Journal of General Virologi* (1994), 75, 2803-2806.

[12] Suwito, T. dan I. Hanarida, 1993. Peningkatan ketahanan varietas padi terhadap wereng coklat (*Nilaparvata lugens* Stal.). *Seminar hasil penelitian Tanaman Pangan*, Balittan Bogor.

[13] Badan Besar Penelitian Tanaman Padi dan Badan Penelitian dan Pengembangan Pertanian, 2010. *Deskripsi Varietas Tanaman Padi*. Departemen Pertanian.

[14] Astuti, P. 2008. Karakterisasi Fenotipe Pada Beberapa Kultifar Padi Dalam Hubungan Tingkat Ketahanan Terhadap Wereng Coklat (Nilaparvata lugens Stal). Universitas Sebelas Maret.

[15] Suprihatno, B., A. A. Daradjat, Satoto., Baehaki, S.E., Suprihanto., A. Setyono., Indrasari, Wardana dan Hasil Sembiring. 2010. *Deskripsi Varietas Padi*. BBPTP, BPPP Deptan. Indonesia

[16] Sodiq, M. 2009. *Ketahanan Tanaman Terhadap Hama*. UPN Press. ISBN 978-979-3100-53-1. Surabaya.

Exploring the potency of gathotan and gathot as diabetes functional food: resistant starch analysis

D E Puspaningtyas[1], P M Sari[1] and R J Kusuma[2]

[1] Nutrition Science Program, Health Science Faculty, Universitas Respati Yogyakarta, Jalan Raya Tajem KM 1.5, Maguwoharjo, Depok, Sleman, Daerah Istimewa Yogyakarta 55282, Indonesia

[2] Department of Health and Nutrition, Faculty of Medicine Public Health and Nurse, Universitas Gadjah Mada, Jalan Farmako, Sekip Utara, Daerah Istimewa Yogyakarta 55281, Indonesia

E-mail: puspaningtyas.desty@gmail.com

Abstract. Indonesian traditional fermented food made of cassava, gathotan and gathot, was considered as diabetes functional food due to higher dietary fiber and lower carbohydrate content compared to cassava. Fermentation processes in gathotan and gathot involving lactic acid bacteria can potentially hydrolyze some natural starch into undigested resistant starch correlated with the value of glycemic index. Nonetheless, there has been no study on the gathotan and gathot potency as resistant starch sources. This study aimed to examine resistant starch content in cassava, gathotan, and gathot as diabetes functional food. Cassava was obtained from the traditional market in Yogyakarta. Gathotan was made through spontaneous fermentation while gathot was made by overnight soaking then steaming the gathotan. Resistant starch was analyzed by using enzymatic process through glucooxydase method. ANOVA test was used to analyze the resistant starch content difference in cassava, gathotan, and gathot at 5% significance level. The highest resistant starch content was found in cassava (24.44%), followed by gathotan (22.5%), and gathot (21.85%) ($p=0.025$). Fermentation process of cassava to gathotan and gathot decrease resistant starch content. Significant differences of resistant starch content were seen between cassava and gathot.

1. Introduction

Diabetes mellitus, a serious chronic disease, is the largest health problem in the world caused by insulin function decrease [1, 2]. The prevalence of diabetes mellitus increases year by year, especially for the past three decades and is growing rapidly in low- and middle-income countries [1, 3]. The worldwide prevalence of diabetes mellitus among adults has risen from 4.7% (1990) to 8.5% (2014) [2]. The global prevalence was expected to increase to 639 million by 2045 compared to the global prevalence in 2017 (451 million) [4]. Basic health research conducted by ministry of health showed that there was an increase in diabetes prevalence from 2007 (1.6%) to 2013 (2.1%) [5, 6]. Meanwhile, survey conducted by World Health Organization and International Diabetes Federation in 2016 and 2017 described the prevalence of diabetes mellitus in Indonesia was 6.6% and 6.7% respectively [7, 8].

Considering the increase of diabetes prevalence, deaths attributable to diabetes and high healthcare expenditure because of diabetes [4] a comprehensive strategy is needed to prevent and treat diabetes mellitus. Dietary management can be used to control blood sugar levels and other metabolic profile in diabetic patients [1]. Diet with high soluble and insoluble dietary fiber, high resistant starch and low glycaemic index can improve diabetes condition, especially controlling postprandial blood glucose levels and utilisation of blood glucose and fatty acid uptake to the muscle [9–12].

Indonesian traditional fermented food made of cassava, gathotan and gathot, was considered as diabetes functional food due to higher soluble and insoluble dietary fiber and lower carbohydrate content compared to cassava [13]. Recent intervention study showed that the most important characteristic of fiber associated with improved insulin sensitivity was the fermentability of the substrate. Fiber that is lost to caecum then potentially fermented by intestinal microflora to produce short chain fatty acid (SCFA), circulate free fatty acids (FFA) and regulate some physiological mediators and inflammatory markers so that it can influence health and decrease the risk of getting diabetes mellitus [14].

Not only does fermentation processes involving lactic acid bacteria in gathotan and gathot increase total dietary fiber and decrease carbohydrate content [13], it can potentially hydrolyze some natural starch into undigested resistant starch correlated with the value of glycemic index. Furthermore, the previous study described that cassava fermentation to gathotan and gathot can lower aflatoxin content and produced some antioxidant compounds [15, 16]. As a result, gathotan and gathot can be potentially used as diabetes functional food. To the best our knowledge, there has been no study on the potency of gathotan and gathot as resistant starch sources. The aim of this study is to examine resistant starch content in cassava, gathotan, and gathot as a preliminary study of diabetes functional food development.

2. Methods

This was an observational laboratory study analyzing resistant starch content in cassava flour, gathotan flour, and gathot flour. This study was conducted in Dietetic and Culinary Laboratory of Universitas Respati Yogyakarta on May until September 2017.

2.1. Materials

Pepsin from gastric porcine mucosa (Sigma Aldrich, Singapore), porcine pancreatic α-amylase (Sigma Aldrich, Singapore), Amyloglucosidase from *Aspergillus niger* (Sigma Aldrich, Singapore) and Glucose FS (Diasys, Germany) were purchased and used in this study.

2.2. Development of gathotan and gathot

Samples used in this study were cassava, gathotan, and gathot. Cassava obtained from local market in Yogyakarta was used to make fermented cassava; Gathotan and Gathot. In a brief, cassava was peeled, washed and dried for five days by sun drying. Dried cassava then was fermented for two weeks by laying dried cassava in a sack for spontaneous fermentation to occur. The cassava will turn into blackish color after two weeks of fermentation and will be called gathotan. Gathot was made by soaking gathotan for 2x24 hours then steaming for 30 minutes. The next step, cassava, gathotan, and gathot were made into flour by drying them in cabinet dryer, grinding with grinder, and sieving with 40 mesh size.

2.3. Resistant starch analysis

Resistant starch analysis was analyzed according to Goni et al. (1996). In brief, protein in 100 mg samples was hydrolyzed using 20 mg of pepsin at 40° C for 60 minutes at pH 2 prior to resistant starch analysis. The mixture then cooled at room temperature and the pH was adjusted back to 6.9 for hydrolysis of digestible starch using 40 mg of pancreatic α-amylase at 37° C for 16 hours. Hydrolysis was performed in magnetic stirrer and solution was continuously stirred during starch hydrolysis. The resulting solution was centrifuged for 15 minutes at 3000 rpm and pellet was collected. The pellets were solubilized in 2 M KOH and digested further with amyloglucosidase at 60°C and pH 4.74 for 45 minutes to hydrolyze resistant starch. The mixture was centrifuged at 3000 rpm for 15 minutes and glucose was assayed in the supernatant using glucose FS kit according to manufacturer protocol [17, 18]. All analysis was performed in triplicates.

2.4. Statistical analysis

Data were presented as mean ± standard deviation. Shapiro wilk test was used to test normality of the data. Levene's test was also performed for homogeneity of variance test. Because of both condition met, one way ANOVA continued by multiple comparisons Tukey HSD was performed to analysis the data. Significance level was set at 5%. All statistical analysis was performed in SPSS version 16.

3. Result and discussion

Analysis of resistant starch content showed that the highest resistant starch content was found in cassava (24.44%), followed by gathotan (22.50%), and gathot (21.85%) (Figure 1). From the statistical analysis result using one way ANOVA, there were significant differences in resistant starch content among the samples with p=0.025 (Table 1).

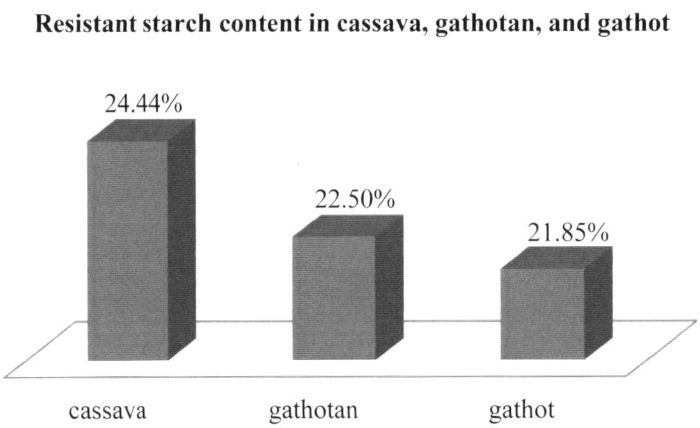

Figure 1. Resistant starch content in cassava, gathotan, and gathot.

Table 1. Differences in resistant starch content among cassava, gathotan, and gathot.

Products	% Resistant starch content Mean±SD	*p*
Cassava	24.44±0.85^{a}	0.025^{c}
Gathotan	22.50±1.09^{ab}	
Gathot	21.85±1.37^{b}	

Different letter notations indicate differences between groups (a, b)
^{c}significant (p<0.05)

Gathotan is fungal fermented cassava product from Gunung Kidul, Yogyakarta, Indonesia. It is mainly colonized by *Botryodiplodia theobromae* Pat. Further fermentation of gathotan into gathot can change the texture and color of gathot. Gathotan has black color inside it as important characteristic of gathotan, and gathot has chewy texture [15, 16, 19].

Many processes in making cassava to gathotan and gathot potentially change the resistant starch levels. According to the previous research, resistant starch classified into some categories which are a non-digestible matrix (RS1), ungelatinized starch (RS2), retrograded starch (RS3), and chemically modified starch (RS4). Processing method, storage condition and time had an effect on resistant starch formation. Sample with steam cooking process at atmospheric pressure had higher resistant starch than the samples steam cooked at higher pressure [20].

Figure 1 showed that fermentation of cassava into gathotan and fermentation gathotan to gathot can reduce resistant starch levels of the sample. Moreover, the resistant starch levels difference was looked between cassava and gathot. There was no resistant starch levels difference between cassava–gathotan and gathotan–gathot (Table 2).

This results is similar with the previous study evaluating effect of natural fermentation on resistant starch of legume based fermented foods. That study stated that fermentation significantly decreased the resistant starch content in all legume fermented products. The reduce levels of resistant starch in fermented food can be assumed as the effect of degradation of the complex starch into simple sugars and oligosaccharides by the enzymes produced by fungi or bacteria in fermented food and to a loss in the structural integrity of the starch granules [21]. Previous study stated that fermentation of cassava into gathotan may cause starch degradation [15, 22]. Texture changes, a chewy texture, of gathot may relate to the change in starch nature due to the enzymes or acid produced by fungi that can degrade starch fraction [19, 22].

Another study investigating the effect of soaking and autoclaving on resistant starch (RS3) content of tapioca pearls showed that autoclaving treatments increased RS3 content of tapioca pearls higher than soaking treatment. It may be due to decrease in intramolecular and intermolecular hydrogen bonds and or increase the crystallinity and amylose to amylopectin ratio of tapioca pearl starch [23]. Analysis of amylose to amylopectin ratio in cassava, gathotan, and gathot may potentially be used to describe the resistant starch content difference among them. As observed before, as the amylase content increases, resistant starch also increases [24].

Although there are differences in the levels of resistant starch between cassava and gathot, cassava and processed cassava products are food sources that contain resistant starch which can be beneficial products to improve and maintain health [24]. Substitution of gathotan flour to gathotan noodle increased blood glucose level quicker but also it reduced blood glucose level to fasting blood glucose quicker than white bread as control meal due to the resistant starch [15]. Resistant starch is good for reducing blood glucose level and enhances glycemic control. Study conducted to seventeen individuals with type 2 diabetes mellitus for 12 weeks supplementation with 40 g resistant starch each day can lower postprandial glucose levels, lower fasting non-esterified fatty acid (NEFA) concentrations, increase muscle uptake of fatty acids and increase soleus intramyocellular lipid (S-IMCL), decrease TNF alpha [25]. Study carried out to normal and overweight women evinced that there was reduction in glycemic response by combining soluble fiber and resistant starch. Consumption of food containing moderate amount of soluble fiber and resistant starch can improve glucose metabolism [26]. The optimal intake of resistant starch to give positive effect for health is up to 45 gram/day [27].

Resistant starch, as a prebiotic and symbiotic, is non-digestible starch that can escape digestion from the small intestine and pass into the colon then is fermented by intestinal microflora to form short chain fatty acids (mainly acetate, propionate, butyrate) [28–32]. The increase of short chain fatty acid can give stimulus to enteroendocrine L-cells, adipose tissue, liver, and hypothalamus. The increase of peptide tyrosine tyrosine (PYY) and glucagon-like peptide 1 (GLP-1) is the response of enteroendocrine L-cells. This effect leads to a decrease body weight and fat mass. The hypothalamus gives response in the form of anorexigenic signaling increasing, appetite decreasing and food intake decreasing. Liver will compensate by decreasing glucose output and inflammation and also improving insulin sensitivity. Furthermore, adipose tissue decreases free fatty acid output and inflammation and increases adipocyte differentiation and insulin sensitivity [28, 33].

Besides the positive effect on appetite and nutrient intake changes, resistant starch can improve glucose metabolism, optimize energy metabolism, and balance the gut microbiota [34]. Microflora in human intestine ferments resistant starch and non-starch polysaccharides to short chain fatty acids. Then, colonic blood flow, fluid and electrolyte uptake will be stimulated by the short chain fatty acids.

Moreover, resistant starch acts as prebiotic which beneficially affect the host by enhancing the growth and activity of bacteria in the gastrointestinal tract selectively [29].

4. Conclusion and recommendation

Fermentation process of cassava to gathotan and gathot decrease resistant starch content. Significant differences of resistant starch content were seen between cassava and gathot. There was no resistant starch difference between cassava–gathotan and gathotan–gathot.

The next study can further examine the type of resistant starch formed from cassava fermentation process into gathotan and gathot and test the concentration of amylose and amylopectin considering the level of amylose to amylopectin ratio has an effect to resistant starch level. Moreover, the further research can analyze the effect of gathotan and gathot administration to metabolic mechanism and prebiotic profile of diabetes mellitus, such as to influence appetite and nutrient intake, improve glucose metabolism, improve lipid profile, optimize energy metabolism, balancing the gut microbiota, produce short chain fatty acids, stimulate the secretion of peptide tyrosine tyrosine (PYY) and glucagon-like peptide 1 (GLP-1), etc.

5. Acknowledgments

We would like to thank the Minister of Research, Technology and Higher Education of the Republic of Indonesia for funding support through DIPA Kopertis V Yogyakarta Grant 2017. The contract number of this study is 226/BPDIPA/VI/2017.

We are also grateful to the head of research and community service centre of Universitas Respati Yogyakarta for the support. All authors declare no competing interests.

References

[1] World Health Organization 2016 http://apps.who.int/iris/bitstream/handle/10665/204871/9789241565257_eng.pdf;jsessionid=7D22CD964925F7616D0569A706FD9BAA?sequence=1
[2] World Health Organization 2017 http://www.who.int/news-room/fact-sheets/detail/diabetes
[3] Wild S, Roglic G, Green A, Sicree R and King H 2004 *Diabetes care* **27** 1047–53
[4] Cho N H, Shaw J E, Karuranga S, Huang Y, da Rocha Fernandes J D, Ohlrogge A W et al 2018 *Diabetes Research and Clinical Practice* **138** 271–81
[5] Badan Penelitian dan Pengembangan Kesehatan Kementerian Kesehatan Republik Indonesia 2007 *Riset kesehatan dasar: RISKESDAS* (Jakarta: Kementerian Kesehatan Republik Indonesia)
[6] Badan Penelitian dan Pengembangan Kesehatan Kementerian Kesehatan Republik Indonesia 2013 *Riset kesehatan dasar: RISKESDAS* (Jakarta: Kementerian Kesehatan Republik Indonesia)
[7] World Health Organization 2016 http://www.who.int/diabetes/country-profiles/idn_en.pdf
[8] International Diabetes Federation 2018 https://www.idf.org/our-network/regions-members/western-pacific/members/104-indonesia.html
[9] Yulianto W A, Susiati A M and Andhini H A N 2017 *IOP Conf. Ser.: Earth Environ. Sci.* **102** 012101
[10] Bodinham C L, Smith L, Thomas E L, Bell J D, Swann J R, Costabile A et al 2014 *Endocrine Connections* **3** 75–84
[11] Lattimer J M and Haub M D 2010 *Nutrients* **2** 1266–89
[12] Witwer R S 2013 *Agro Food Industry Hi Tech* **24** 3 34–36
[13] Sari P M, Puspaningtyas D E and Kusuma R J 2017 *Indonesian Journal of Nutrition and Dietetics* **5** 2 88–92
[14] Maki K C and Rains T M 2012 http://www.intechopen.com
[15] Purwandari U, Tristiana G R and Hidayati D 2014 *International Food Research Journal* **21** 4 1629–34

[16] Oktaviana R S and Pangesthi L T 2014 *e-journal Boga* **03** 3 141–50
[17] Goni I, Garcia-Diz L, Manas E and Saura-Calixto F 1996 *Food Chem* **56** 333–37
[18] Vatanasuchart N, Niyomwit B and Wongkrajang K 2009 *Kasetsart J. (Nat. Sci.)* **43** 178–86
[19] Purwandari U, Nava N and Hidayati D 2014 *Microbiol Indones* **8** 3 112–20
[20] Abioye V F, Adeyemi I A, Akinwande B A, Kulakow P and Maziya-Dixon B 2017 *Cogent Food & Agriculture* **3** 1296401
[21] Yadav B S, Sharma A and Yadav R B 2007 *Journal of Agricultural Technology* **3** 1 21–27
[22] Prabawati S, Richana N and Suismono 2011 http://www.litbang.pertanian.go.id/download/one/104/file/Manfaat-Singkong.pdf
[23] Kasote D M, Oak M D, Nilegaonkar S S and Agte V V 2018 *International Food Research Journal* **25** 1 1–8
[24] Pereira B L B and Leonel M 2014 *Food Sci. Technol* **34** 2 298–302
[25] Bodinham C L, Smith L, Thomas E L, Bell J D, Swann J R, Costabile A et al 2014 *Endocrine Connections* **3** 75–84
[26] Behall K M, Scholfield D J, Hallfrisch J G and Liljeberg-Elmstahl H G M 2006 *Diabetes Care* **29** 976–81
[27] Maziarz M P 2013 *Diabetes Spectrum* **26** 1 35–39
[28] Chambers E, Guess N, Viardot A and Frost G 2011 *Diabetes Manage* **1** 5 521–28
[29] Garg N K, Singh A and Chaudhary D P 2017 *Int.J.Curr.Microbiol.App.Sci* **6** 5 2046–57
[30] Jyoshna E and Hymavathi T V 2017 *Journal of Pharmacognosy and Phytochemistry* **6** 3 55–58
[31] Topping D L, Fukushima M, Bird A R 2003 *Proceedings of the Nutrition Society* **62** 171–76
[32] BIrt D F, Boylston T, Hendrich S, Jane J-L, Hollis J, Li L 2013 *Adv. Nutr.* **4** 587–601
[33] Bindels L B, Munoz R R S, Gomes-Neto J C, Mutemberezi V, Martinez I, Salazar N et al 2017 *Microbiome* **5** 12 1–16
[34] Lei Z, Ting L H, Li S, Chen F Q, Ling Q L, Ping J W 2015 *Biomed Environ Sci* **28** 4 291–97

IOP Conf. Series: Earth and Environmental Science **207** (2018) 012043 doi:10.1088/1755-1315/207/1/012043

Purple Yam Flour (*Dioscorea alata* L.) Processing Effect on Anthocyanin and Antioxidant Capacity in Traditional Cake "Bolu Cukke" Making

R. Larief[1*], A. Dirpan[1] and Theresia[1]

[1] Department of Agricultural Technology, Hasanuddin University, Indonesia

*Email : rindamias04@yahoo.com

Abstract. Purple yam (*Dioscorea alata* L) is a food that has the potential as a source of natural antioxidants, because of the anthocyanin component. Anthocyanin can be damaged from high temperatures (heating) during processing and manufacturing processed products. The purpose of this study was to analyze the content of anthocyanin compounds and the antioxidant activity in purple yam flour before and after processing. The determination of anthocyanin test was done by using pH spectrophotometry method and DPPH method for antioxidant test. The main analysis parameters wereanthocyanin content and antioxidant activity. The results showed that the anthocyanin content in the processing of purple yam to flour and sponge cake was decreased. The highest anthocyanin content of 38.12 mg/ 100g is found in purple yam and the lowest is 2.99 mg/100 g in cake "bolu cukke" with 10 grams of purple yam flour. The highest antioxidant activity was 79.08% in purple yam and the lowest was 22.64% in cake bolu cukke with purple yam gram 10 gram addition. The decrease in antioxidant activity is directly proportional to the decrease in anthocyanin levels.

1. Introduction

Indonesia is a tropical country, therefore Indonesia is rich in biodiversity. One of the many crops produced is tubers such as purple yam. This plant grows well from the lowlands to an altitude of 800 m above sea level, but can also grow at an altitude of 2700 m above sea level [2]. In the dry season purple uwi bulbs experience a period of rest. To avoid damage easily, purple yam bulbs are stored in a dry place. Towards the rainy season this tuber will sprout. Bulbs that have sprouted are used as seeds. Harvesting can be done after 7-9 months after planting, marked by yellowing of the leaves, which means the tubers are ripe [9].

Purple yam are one of the varieties of tubers that have the potential as a source of local food ingredients that contain carbohydrates, proteins, fats, fiber, vitamins, minerals, and contain antioxidants [10]. Purple yam plants grow in the Soppeng area, but the people do not use these plants, even though purples can be processed into flour to extend shelf life and can be used as an alternative to reduce wheat flour in meeting consumer needs. The advantages of this processed flour can be seen from its form, among others, easily stored ingredients, small volume, easy to transport, and can be used for various processed food products [17]. Another interesting thing from the Dioscorea plant is that it contains anthocyanin compounds and antioxidant activity.

According to [5], the purple color found in local varieties of tuber indicates that there is an anthocyanin which is an antioxidant and can counteract free radicals in the body. Anthocyanin content has colored compounds which are mostly red, blue and purple. Strong purple color shows anthocyanin levels and has high antioxidant activity [14]. Antioxidants are compounds that serve as a neutralizing increase in free radicals, protecting cells from the toxic effects produced and contributing to disease prevention. Anthocyanin acts as an antioxidant that can counteract free radicals so as to prevent aging, cancer and other degenerative diseases. Anthocyanin content also functions as antimutagenic, antocarcinogenic and can reduce blood sugar levels and antihypertension [13]. Changes in the content of anthocyanin levels and antioxidant activity in purple yam, purple yam flour and cukke sponge products made from purple yam flour during processing will occur. So this research was carried out by analyzing the anthocyanin compound content and activity on purple yam bulbs, purple yam flour and cukke sponge products made from purple yam flour and it is hoped that this research could become information to the public about the benefits of anthocyanin compounds and antioxidant activity in purple yam.

2. Research Methodology

2.1. Tools

The tools used in this study are analytic scales, measuring cups, erlenmeyers, test tubes, vortices, propipets, measuring pipettes, measuring cups, wathman filter no. 42 and UV-Vis Genesys 20 spectrophotometers, and cuvettes.

2.2. Ingredients

The ingredients used are uwi tuber, uwi tuber flour, cukke sponge products made from purple yam flour, 98% methanol and 96% ethanol, 0.2 mM DPPH solution, acetate buffer, potassium chloride (pH 1), buffer sodium acetate (pH 4.5) and aluminum foil.

2.3. Research Procedure

The initial stage of preparation of samples, namely purple yam obtained from Soppeng district, South Sulawesi, then cleaned with running water to remove the sticky soil. Then the purple yam is peeled to separate the tuber meat and skin. Furthermore, anthocyanin levels and antioxidant activity were tested. Furthermore, the process of making purple uwi flour begins with purple yam washing from the soil and other impurities, then theprocess is carried out *blanching* by soaking it in hot water at $80^{\circ}C$ for 1 minute until the entire purple yam is submerged in water, then anthocyanin levels and antioxidant activity are carried out. Then stripping purple yam skin using a knife. The next stage is slicing with a *slicer* with a thickness of ± 1-2 mm, then the tubers are soaked in 5% salt solution for 2 hours, then anthocyanin levels and antioxidant activity are tested. Then washing with running water was carried out, then anthocyanin levels and antioxidant activity were tested. Then drying with a *cabinet dryer* at $60^{\circ}C$ for 6-8 hours, then testing anthocyanin levels and antioxidant activity. Then the dried uwi was crushed with a *grinder* and sieved with 80 mesh size and tested for anthocyanin levels and antioxidant activity. How to make cukke sponge by using purple yam flour, flour and rice flour, first of all is purple yam flour, flour and mixed rice flour according to the treatment then roasted. Then the eggs and brown sugar are stirred until blended. Then added tbm, and baking soda. After thickening, the mixture of purple yam flour, flour and roasted rice flour was added, little by little. Then added cooking oil and stirred again until smooth. After that, prepare the cake mold "cukke sponge" then spread the cooking oil into the cake mold. Then put the mixture into the mold and then baked until cooked. Furthermore, anthocyanin levels and antioxidant activity were tested.

2.3.1. Anthocyanin Extraction [3]

Samples of purple yam, purple yam flour and cukke sponge 'made from purple yam flour, weighed 1 g then put into a test tube. In the test tube a 1 ml acetate buffer was added, then the purple yam was incubated at $40^{0}C$ for 1 hour with a shaking of 170 rpm using a centrifuge. After the education is complete, into the tube is added 10 mL of acidified 70% ethanol (pH 3.5). The tube is then inserted into the waterbath at 80^0C for 1 hour. After the incubation in the water bath is complete, the tube is centrifuged so that the extract and residue are separated. Anthocyanin extract is separated from the residue by decantir. Anthocyanin extracts are stored in tubes that have been coated with aluminum.

2.3.2. Determination of Total Anthocyanin Concentration Using Differential pH Method of Spectrophotometry

Determination of total anthocyanin content by differential pH method [8]. The sample was dissolved in two different buffer solutions, 1 ml, the sample was diluted with potassium chloride buffer (pH 1) until the volume was 10 ml, and the sample was diluted with sodium acetate buffer (pH 4.5). Then it was allowed to stand for 15 minutes, then the absorbance measurements were carried out using spectrophotometry. Absorbance measurements are carried out using each wavelength read at a length of 516 and 700 nm.

$$\text{antosianin content (mg/L)} = \frac{A\ x\ MW\ x\ DF\ x\ 1000}{\varepsilon\ x\ L}$$

Where:
A = Absorbance [A516 - A700] pH 1 - [A516 - A700] pH 4.5,
ε = Koefisien ekstingsi molar extension (Cyanidin-3-glycoside: 26900 L / mol cm)
L = width of cuvette (1 cm)
MW = molecular weight of cyaniding-3-glycoside 448.8 g / mol,
DF = factor of dilution.

2.3.3. Measurement of Antioxidant Activity with DPPH Method

As much as 1 ml the extract of the sample was dissolved in 98% ethanol as much as 5 ml and then vortexed until homogeneous. Then it was centrifuged at 4000 rpm for 10 minutes, then the supernatant was pipetted as much as 4 ml, then 1 ml of DPPH 0,2 mM was added. After that, leave it in a dark place for 30 minutes. Then absorbance is measured with a wavelength of 517 nm. The amount of the percent (%) value of antioxidant activity can be calculated using the equation:

$$A = \frac{(\textbf{\textit{Absorbansi blanko}} - \textbf{\textit{absorbansi sampel}})}{\textbf{absorbansi blanko}}\ x\ 100\%$$

Where:
Akt. antioxidant: The amount of the percentage (%) antioxidant activity
Abs / Absorbance: Absorption value (a) blank and sample

3. Results and Discussion

The results of anthocyanin analysis on the purple yam during the sowing process to become cukke sponge cake ranged between 2.99 - 38.12 mg / 100g. Figure 01 shows that the lowest anthocyanin level in cukke sponge cake treatment with the addition of 10 grams of purple tuber flour is 2.99 mg / 100 g. While the highest anthocyanin content is found in purple yam with a yield of 38.12 mg / 100g. The results of the variance analysis showed that the purple yam processing during the tasting process to become a cukke sponge cake had a significant effect on the level of 5%, so that further testing was carried out. Duncan's test results showed that the purple yam treatment had a significant effect on drying treatment, flour, and the addition of purple yam flour 10, 20 and 30 grams. But thetreatment *blanching* and washingdoes not have a real effect.

The total anthocyanin content obtained in purple yam is 38.12 mg / 100g. After processing into flour, there is a decrease in anthocyanin levels. The anthocyanin content with treatment *blanching* decreased anthocyanin levels, ie 36.73 mg / 100g. The decrease in anthocyanin levels that occur occurs because in theprocess *blanching* can damage the anthocyanin caused by contact with heating and moisture, which can affect the damage to anthocyanins caused by enzymatic damage and heating treatment resulting in anthocyanin degradation. This is in accordance with [11] reduced anthocyanin levels due to the presence of two, namely the steam process (heating with water vapor), anthocyanins damaged by enzymatic damage and heating treatment.

Table 1. Total anthocyanin content and antioxidant activity in purple uwi during processing into flour and used as substitute for making sponge cake

Ingredients and processing	Anthocyanin content of mg / 100 g	Antioxidant Activity %
Purple yam	38.12	79.08
Blanching	36.73	61.75
Washing	32.63	40.75
Drying	29.29	32.16
Flour	27.27	30.01
Cukke sponge cake + 10 gram purple yam flour	2.99	22.64
Cukke sponge cake + 20 grams of purple yam flour	8.04	24.06
Cukke sponge cake + 20 grams of purple yam flour	20.71	26.9

3.1. *Anthocyanin Content*

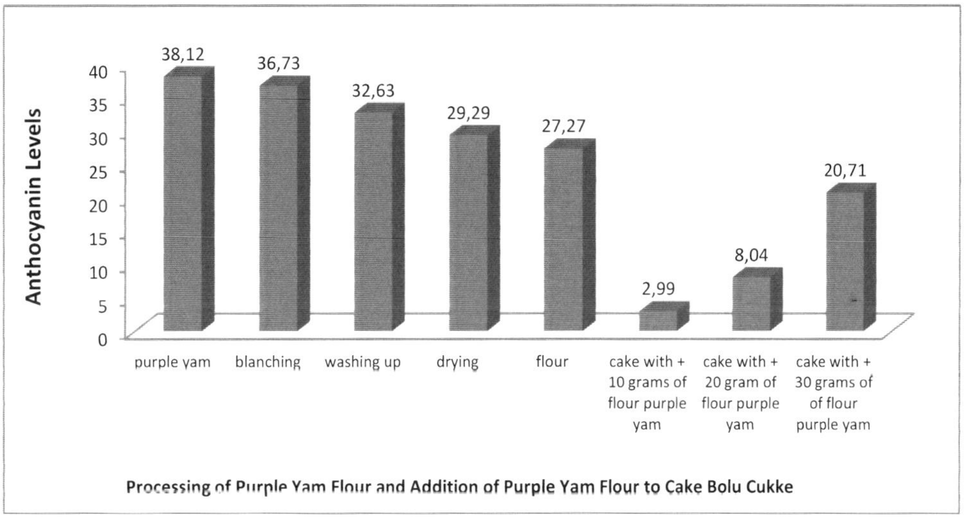

Figure 1. Total content anthocyanin and in purple yam during processing into flour and cukke sponge cake.

Washing treatment also decreased anthocyanin levels, where after purple yam was *blanched* then purple yam was washed. This washing process has experienced a decrease in anthocainin levels of 32.63 mg / 100g. The loss of anthocyanin is caused by the dissolution of anthocyanin compounds in the immersion of purples yam slices before drying, because anthocyanins are soluble in water. This is consistent with [16], who stated that anthocyanins are a group of pigments that cause reddish color, located in cell fluids that are soluble in water.

Drying treatment also decreased anthocyanin levels by treatment after purple yam washing, then drying was done using a dryer cabinet. After testing anthocyanin levels in this treatment, anthocyanin levels were produced which were 29.29 mg / 100 g. The decrease in anthocyanin levels in this treatment is due to the drying process of anthocyanin pigments which is easily damaged because anthocyanins are not stable at high temperatures. In making purple yam flour, drying is carried out at a temperature of 60^{0}C and requires a long time and contact with oxygen which can cause anthocyanin oxidation. High heat causes the anthocyanins contained in the material to be damaged, but this is also affected by the heating temperature, heating time, and the size of the treated material. This is consistent with [20], on high heating, the stability and resistance of anthocyanin dyes changes and results in anthocyanin damage. This is also confirmed by [6] anthocyanins begin to experience damage above 60°C. Drying is carried out at a temperature of 60°C but requires a long time and contact with oxygen which causes anthocyanin oxidation. The decrease in anthocyanin in purple yam flour produced is 27.27 mg / 100g. The loss of anthocyanin because it has been treated with repeated heating and washing, because anthocyanins are heat-stable and water-soluble anthocyanins. This is in accordance with [21], which states that the color of a product is determined based on the natural pigments present in food ingredients, besides the presence of heating factors can reduce the content of natural pigments contained in food.

3.2. Antioxidant Activity

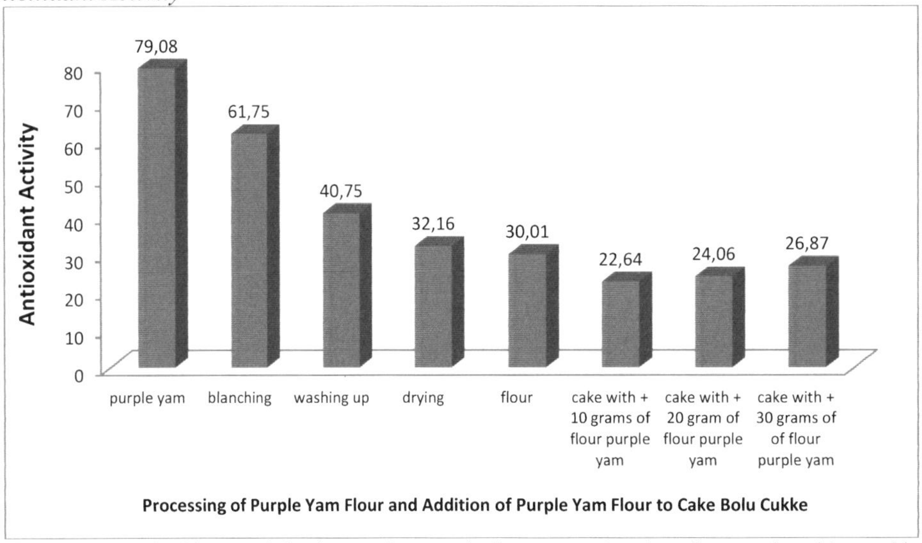

Figure 02. Antioxidant activity in purple yam during processing into flour and making cukke sponge cake Antioxidant

Reduced levels of anthocyanin in the cukke sponge cake treatment with the addition of purple yam flour 10 grams, 20 grams and 30 grams, due to theprocess *blanching* and drying before making flour, thus damaging the purple yam color of anthocyanin content and influencing purple yam flour is produced and ultimately affects purple yam flour. The decrease in the stability of anthocyanins is also caused by the added sugar which is brown sugar which affects the color of the resulting cukke sponge turning brownish. In addition, from several types of sugar that have been tested (sucrose, fructose, glucose, and xylose) which apparently can increase anthocyanin degradation with a mechanism of formation to form polymer pigments and browning and due to the different levels of purple uwi flour substitution that produce different levels of anthocyanin. The higher the substitution of purple yam flour, the higher the anthocyanin content produced. And the roasting temperature used can increase anthocyanin degradation. This is in accordance with [1], which states that the color of the cake turns brown can be caused by sugar. Sugar can reduce anthocyan stability. In addition, from several types of sugar that have been tested (sucrose, fructose, glucose, and xylose) it was found to increase the degradation of anthocyanins by a mechanism of formation to form polymer pigments and browning. This also occurs in research that has been carried out by [7] that the variation of the substitution of purple sweet potato flour in the manufacture of syringes can affect anthocyanin levels. Anthocyanin increased in keu syringe by 6.25466 mg compared to control (wheat flour). Besides that, there was also a study conducted by [4] that the variation of mixing flour and brown rice flour in the manufacture of steamed sponge increased compared to control (100 percent flour) could affect the levels of anthocyanin in the steamed sponge produced.

Antioxidant activity in purple yam and during the siege process to become cukke sponge cake ranged between 22.64 - 79.08%. Figure 02 shows that antioxidant activity decreases during the siege process to become a cukke sponge cake, the highest antioxidant activity is obtained in purple yam treatment which is 78.08% and the lowest in cukke sponge cake treatment with 10 gram purple uwi flour substitution of 22.64% . The results of the variance analysis showed that the purple yam processing during the tasting process to become a cukke sponge cake had a significant effect on the level of 5% on the content of the purple yam antioxidant activity during the siege process to become a cukke sponge cake so that further testing was carried out. Duncan's test results showed that the purple yam treatment had a significant effect on washing, flour and cukke sponge cake treatment with the addition of purple yam flour 10, 20 and 30 grams. But the blaching and washing treatment has no real effect.

Antioxidant activity experienced a different decline after undergoing the siege process to become a keuku cukke. Figure 02 shows that the highest to lowest antioxidant activity in the order is purple yam treatment, *blanching*, washing, drying, purple uwi flour, cukke sponge cake with 30 grams of purple yam flour substitution, cukke sponge cake with 20 grams of yam purple flour substitution and finally cukke sponge cake with 10 grams of purple flour flour. This is due to the process of making purple uwi flour through a process of *blanching* and drying, which causes a decrease in antioxidant compounds found in purple uwi flour, because the antioxidant component is not heat resistant. The difference in antioxidant activity produced is due to the resulting anthocyanin levels. Where anthocyanin levels are directly proportional to antioxidant activity. This is consistent with [18], which states that the decrease in antioxidant activity is directly proportional to the decrease in total phenolic and anthocyanin levels and the process of processed products. This is also confirmed by [15], which states that phenolic and anthocyanin compounds are closely related to the ability of ingredients as a source of antioxidants.

4. Conclusion

The anthocyanin content in the processing of yam purple becomes flour and sponge cake has decreased. The highest anthocyanin content of 38.12 mg / 100 g is found in purple yam and the lowest is 2.99 mg / 100 g in cukke sponge cake with 10 grams of purple yam flour. The highest antioxidant activity was 79.08% in purple yam and the lowest was 22.64% in cukke sponge cake with purple gram 10 gram addition. The decrease in antioxidant activity is directly proportional to the decrease in anthocyanin levels.

Kuhn T 1998 Density matrix theory of coherent ultrafast dynamics *Theory of Transport Properties of Semiconductor Nanostructures (Electronic Materials* vol 4*)* ed E Schöll (London: Chapman and Hall) chapter 6 pp 173–214

5. References

[1] Ayu K, Rachmawanti A, and Sigit B *2014 Jurnal Teknosains Pangan* **3** 1

[2] Balitbang Pertanian Departemen Pertanian 2005 *Mengenal Plasma Nutfah Tanaman Pangan* . http : //biogen.litbang.deptan.go.id./mengenal_plasmanutfah.
[3] Bridgers EN, Chinn MS and Truong VD 2010 *Journal Industrial Crops and Products* 613-620.
[4] Devillya PD, Wijanarka A, Febriana N 2016 *Jurnal Medika Respati* **11** 34-43
[5] Fang Z, Wu D, Yü D, Ye X , Liu D and Chen J. 2011 *Food Chemistry* **128** 943-948.
[6] Fennema RO 1985. *Food Chemistry second edition.* Revised and Expanded. Academic Press. New York.
[7] Gartika IR 2015 *Kajian Penggunaan Tepung Ubi Jalar Ungu Pada Produk Kue Semprit Sebagai Unggulan Yang Berpotensi Sebagi Makanan Fungsional.* State University of Yogyakarta. Yogyakarta.
[8] Giusti MM dan Worlstad RE 2001. *Charecterirization and Measurement of Anthocyanin by UV-Visible Spectroccopy.* Oregon State University.
[9] Heyne K 1987 *Tumbuhan Berguna Indonesia Jilid I* Balitbang Kehutanan Dephut p. 542 – 544.
[10] Indrastuti E, Harijono dan Susilo B 2012 *Jurnal Teknologi Pertanian* **13** 169-176
[11] Iversen CK 1999 *Jurnal of Food Science* **64** 37-41.
[12] Jiao Y, Jiang Y, Zhai W and Yang Z 2012 *Afr J Biotechnol* 11: 7046-7054
[13] Jusuf M, Rahayuningsih SA and Ginting E 2008. *Warta Penelitian dan Pengembangan Pertanian* **30** 4
[14] Krisnawati. 2009. *Homemade Food : Olahan ubi untuk bayi dan Balita.* Gramedia. Jakarta.
[15] Leo L, Leone A, Longo C, Lombardi DA, Raimo F, Zacheo G 2008 *Journal of Agricultural and Food Chemistry* **56**: 4154–4163
[16] Nollet LML 1996 *Handbook of Food Analysis: Physical Characterization and Nutrient Analysis.* Marcell Dekker Inc, New York.
[17] Tejasari S, Hartanti, Herlina BH, Purnomo 2001. *Laporan penelitian kajian tepung umbi-umbian local bahan pangan olahan* Jember University p. 121
[18] Salim M, Dharma A, Mardiah E and Oktoriza G 2017 *Jurnal Zarah* **5** 7-12
[19] Utami R, Esti W and Annisadyah ARD 2013. *Jurnal Teknosains Pangan* **2** 3
[20] Winarno FG 2002 *Kimia Pangan dan Gizi* PT Gramedia Pustaka Utama Jakarta.
[21] Winarno FG 2004 *Kimia Pangan dan Gizi* PT Gramedia Pustaka Utama.Jakarta

Antioxidant Activity, Total Phenol, and SensoryProperties of Melinjo Peel Tea with Pre-Treatment

M W Apriliyanti[1], M Ardiyansyah[1], A M Handayani[1]

[1]Agriculture Technology Department, Politeknik Negeri Jember, Indonesia

* Email : mulia_apriliyanti@polije.ac.id

Abstract. Melinjo (*Gnetum gnemon*) is a melinjo fruit producing plant that has many benefits. Melinjo peel contains ascorbic acid, tocopherol, and polyphenols which have activity as antioxidantand potential as xanthine oxidase inhibitors. Melinjo peel has the potential as an herbal tea because of its antioxidant properties. This study aims to determine antioxidant activity, total phenol, and sensory properties of melinjo peel tea with pre-treatment variation (concentrations of citric acid 0%, 0.05%, 0.1% and the blanching process for 5 and 10 minutes). The total phenol content and antioxidant activity were analyzed to support the potential of melinjo tea as an herbal tea. The results showed that melinjo peel herbal tea can be accepted by panelis and the most preferred one is the tea made fromthe blanching treatment duration of 5 minutes and 0% citric acid concentration.The steeping water of melinjo peel herbal tea has a clear yellow color, a rather distinctive taste and a slightly typical aroma of melinjo peel. Melinjo peel tea contains total phenol between 2.02 - 2.52 mg/ g and antioxidant activity 32.47 - 48.47%.

1. Introduction

The seeds of melinjo (*Gnetum gnemon*) have been known to have various benefits. It is very well known throughout Indonesia because almost all parts of the melinjo tree are very useful, e.g. as foods or food ingredients (the flowers, seeds, and leaves) and as household appliances (the stem). The old melinjo trees can produce melinjo seeds which have high economic value, i.e. as the main ingredient for making melinjo crackers [1].

Melinjo trees are commonly planted for their leaves, seeds, and flowers. One of the products of melinjo which has been widely known is *emping melinjo* (a type of cracker). This product is typical Indonesian food which is usually consumed as a snack or can be eaten together with rice. To make the crackers, the peel of melinjo seeds is often separated from the fruits and discarded as agricultural waste.

Melinjo peel contains ascorbic acid, tocopherols, and polyphenols which have antioxidant activity as well as the potential to serve as xanthine oxidase inhibitors. Xanthine oxidase plays an important role in the process of formation of uric acid by successively catalyzing hypoxanthine into xanthine and then uric acid [2]. According to Cos [3], some antioxidant compounds have the potential to serve as xanthine oxidase inhibitors because they are able to capture electrons. Flavonoids in

flavones and flavonols have higher inhibitory power than other flavonoids because the position of the hydroxyl group is easier to capture electrons from the active side of xanthine oxidase. Other compounds such as polyphenols and saponins also have the potential to serve as xanthine oxidase inhibitors because they have hydroxyl groups serving as electron acceptors from xanthine oxidase.
Based on the studies conducted by Ardiyansyah & Apriliyanti [4] and Khoiriyah [5], melinjo peel tea with oxidation and non-oxidation treatment as well as withering duration produced a total4 phenol of 25.30-79.67% (through the titration method) and antioxidant activity of 48.23-53.16%. Nevertheless, the resulting melinjo peel tea has a deficiency, i.e. its organoleptic properties have not been well received by consumers. Furthermore, Pujimulyani's research [6] suggested that white turmeric blanched in 0.05% citric acid medium, 100°C for 5 minutes had total phenol content, total flavonoids, condensed tannins, DPPH values, and FRAP significantly higher than fresh white turmeric extracted with 6 types of solvents. In addition, the increased phenolic content of white turmeric correlated significantly with the increased antioxidant activity of white turmeric after undergoing blanching than that in a fresh condition.

This study, therefore, was aimed at determining the antioxidant activity, total phenol, and sensory properties of melinjo peel tea with pretreatment. Melinjo peel tea with pretreatment performed using the concentration of citric acid and the blanching process was expected to increase the antioxidant activity and total phenol, as well as the sensory properties accepted by consumers. Melinjo peel tea processing is an effort to utilize waste from an Indonesian local plant that has the potential to contain bioactive compounds. This study is also expected to get a pretty patent regarding the processing of melinjo peel tea which is beneficial for health.

2. Material and Methode

2.1 Materials and Tools

The raw materials needed were peel of melinjo (*Gnetum gnemon*), distilled water, folin ciocalteu, Na_2CO_3, Gallic acid, 1,1-diphenyl-2-pycrilhidrazil (DPPH), and methanol. The tools used were a blender, basins, glass cups, Erlenmeyer, measuring flask, measuring cup, glass bottles, cups, gauze, filter paper, analytic balance, oven, hot plate, volume pipette, micropipette, and spectrophotometer.

2.2 Processing of Melinjo Peel Tea

The selected melinjo peel was old/ripe, dark orange to red, and still fresh. In addition, the peel must be clean and not damaged. Melinjo peel was separated from dirt and washed with running water. Melinjo peel was weighed according to the necessary treatments. The pretreatment was done by blanching the peel using citric acid media and setting the duration of the blanching process. The concentration used was 0; 0.05%; and 0.1% and the duration was 5 and 10 minutes. The drying process was performed at a temperature of 105 ^{0}C for 30 minutes in an electric oven so that the drying can be spread evenly and quickly. After being dried, the peel was ground with a blender until turning into powder. Afterward, the powder was packaged using a tea bag. Parameter analysis was performed on each treatment of the produced melinjo peel tea. The parameters observed in the tea included the analysis of bioactive compounds, i.e. the content of total phenol, antioxidant activity, and functional groups, as well as sensory properties.

2.3 *Research Design*

This study employed a factorial completely randomized design which consists of two factors with 3 replications. Factor 1 with 3 levels included the concentration of citric acid media which is 0; 0.05; and 0.1%, while factor 2 with 2 levels included the duration of the blanching process which is 5 and 10 minutes.

2.4. Analysis Parameters

2.4.1 Determination of antioxidant activity.
The antioxidant activity testing was performed using a sample of 0.1 ml and then added with 5 ml of methanol, vortexed, and taken as much as 4 ml. A 1 ml of 20 ppm DPPH was then added. The solution of 1,1-diphenyl-2-pycrilhidrazil (DPPH) in methanol was vortexed and then incubated in a dark space for 30 minutes and absorbed at a wavelength of 517 nm (modified from Brand-Williams [7]). The antioxidant activity was calculated by the following formula:

$$[(\text{Abs control} - \text{Abs sample})/ \text{Abs control}]*100$$

where:
Abs control: the absorbance of DPPH before reacting with the sample
Abs sample: the absorbance of DPPH after reacting with the sample

2.4.2 Determination of total phenol content.
The analysis of the total polyphenol content was performed spectrophotometrically using the modified folin ciocalteu method [8]. The sample extract with a certain volume was put into a test tube, then added with distilled water to a volume of 5 mL. Afterward, a 0.5 mL of follin ciocalteau was added into the test tube, then vortexed, and allowed to stand for 5 minutes. Then added Na_2CO_3 (7% as much as 1 mL., Distorted, and allowed to stand for 60 minutes in a dark space. The absorbance value was measured using the spectrophotometer at a wavelength of 765 nm. The total polyphenol content in the extracted sample was calculated using a standard curve made from Gallic acid (GA) in some concentrations. The total polyphenol content in the material was expressed as mg GAE/g of sample.

2.4.3 Sensory test.
The sensory or organoleptic test was performed by presenting the samples and giving an organoleptic questionnaire to each panelist. The method of serving samples was done by brewing melinjo peel tea in a tea bag into hot water. One tea bag contains two grams of melinjo peel tea powder. The brewing was done with 150 ml of hot water for 10 minutes. This sensory test was performed by 20 trained panelists.The hedonic test was performed to determine the level of panelists' preference for the tested products, which were three formulation treatments without and using foam and bubble effects and the best formulation. The test was done by giving a score on each sample according to each panelist's favorite level. The rating scale used for the hedonic test ranged from 1 - 5 (dislike - extremely like).

2.5 Data Analysis
The observation data was analyzed using the Analysis of Variance (ANOVA) method. If the test results were different, the analysis was then proceeded with the Tukey HSD (Honestly Significant Difference) test with 5% and 1% to see the difference between treatments.

3. Results and Discussion

3.1 Antioxidant Activity
The DPPH radical (1,1-diphenyl-2-picryhydrazyl) is a stable nitrogen organic radical, which has a purple effect. The antioxidant activity test using DPPH method is based on the measurement of the reduction ability of DPPH radicals. The measurement can be done by measuring the decrease in absorbance [9]. The purple DPPH solution is a collection of free radicals and will be bound by H ions from antioxidant compounds so that the purple intensity will decrease. The decrease in purple intensity can be measured at a wavelength of 517 nm [7]. The capacity of radical capture is usually measured by using a DPPH (1,1-diphenyl-2-picryhydrazyl) radical compound, which is stable and able to receive electrons or hydrogen radicals to become a diamagnetically stable

compound (Soares *et al.*, 1997). Furthermore, Duh *et al.*, (1999) stated that the ability of DPPH radicals can be reduced or stabilized by antioxidants at a wavelength of 517 nm. Therefore, DPPH is commonly used to assess radical capture capacity.

The variance analysis results showed that the duration of blanching had no significant effect, while the concentration of citric acid had a very significant effect. This is because the duration of blanching has a short time span, i.e. only 5 minutes. The higher the concentration of citric acid, the higher the antioxidant activity for citric acid is a preservative which can be used to increase antioxidant activity. The interaction of the two treatments was very significantly different. The average test results of antioxidant activity of melinjo peel herbal tea can be seen in Table 1.

Table 1. Antioxidant Activity

Citric Acid Concentration (%)	Blanching Duration (minutes)	Antioxidant activity (%)
0	5	35.32^{a}
	10	32.37^{bc}
0.05	5	38.95^{ab}
	10	33.70^{c}
0.1	5	40.14^{d}
	10	48.47^{d}

Note: Figures followed by the same letters indicate 'not significantly different' with Duncan's test 5%.

3.2 Total Phenol

Phenol is a polar compound with the highest solubility in polar solvents. Phenolic compounds have antioxidant activity because they are able to donate H atoms from hydroxyl groups to radical compounds. Phenolic compounds include a variety of compounds derived from plants that have the same characteristic, i.e. an aromatic ring containing one or two OH- groups. Phenolic compounds in nature are very wide, diverse in structures, and easily found in all plants, leaves, flowers, and fruits. Thousands of natural phenolic compounds which have been known in terms of their structures include flavonoids, simple monocyclic phenols, phenyl propanoids, polyphenols (lignin, melanin, tannin), and phenolic quinones (Rohyami, 2008).

The results of variance analysis showed that the duration of blanching had a significant effect and the concentration of citric acid had a significant effect with a downward trend. The interaction of the two treatments had a significant effect. The total phenol value of melinjo peel herbal tea can be seen in Table 2. The highest total phenol was obtained from blanching treatment for 5 minutes and a citric acid concentration of 0% (2.51 mg/g).

Table 2. Total Phenol (mg of gallic acid/ 100g)

Citric Acid Concentration (%)	Blanching Duration (minutes)	Total Phenol (mg/ 100 g)
0	5	2.51^{a}
	10	2.41^{d}
0.05	5	2.19^{ab}
	10	2.15^{c}
0.1	5	2.08^{bc}
	10	2.02^{cd}

Note: Figures followed by the same letters indicate 'not significantly different' with Duncan's test 5%.

3.3 Sensory Test

The sensory test of melinjo peel herbal tea was performed using the hedonic method and the test of color quality, taste and aroma parameters. The analysis results of quality parameter sensory test towards the color of melinjo peel tea can be seen in Table 3. The results of variance (ANOVA) analysis indicated that the variation of blanching duration significantly affected the color of melinjo peel tea while the treatment of citric acid concentration had no significant effect. The interaction between the two treatments had a significant effect on the color quality of melinjo peel tea. The color of brewed melinjo peel tea was recorded from 2.07 to 3.97 quality scale (i.e. from clear yellow to almost solid yellow). Blanching is a preliminary process of food for further processing and can improve the color and texture of food ingredients for it can activate the polyphenol oxidase enzyme.

The blanching process is not only intended for preservation methods but as a preliminary treatment which is generally conducted between raw material preparation and final treatment such as heat sterilization, drying, and cooling. Blanching method is performed by soaking a material in hot water (boiling) or in steam (steam blanching). Factors that influence the duration of blanching include the type of fruit and vegetables, size and unit, blanching temperature, and heating method (Winarno, 1991).

Table 3. Parameters of Melinjo Peel Tea Color Quality

Citric Acid Concentration (%)	Blanching Duration (minutes)	Color Quality
0	5	2.07 [ab]
	10	3.97 [ef]
0.05	5	2.97 [bc]
	10	3.97 [de]
0.1	5	3.40 [bcd]
	10	3.28[a]

Note: Figures followed by the same letters indicate 'not significantly different' with Duncan's test 5%.

The mean sensory test results of the quality parameters for the taste of melinjo peel tea can be seen in Table 4. The results of variance (ANOVA) analysis showed that the variation of blanching duration significantly affected the taste of melinjo peel herbal tea while the treatment of citric acid concentration had no significant effect. The interaction of the two treatments significantly affected the quality of melinjo peel tea taste. The blanching process can improve the quality of melinjo peel tea taste compared to the previous studies that suggested the taste of melinjo peel tea is still very typical of melinjo peel. Thus, the results of this study can reduce the typical taste of melinjo peel with a quality value of 2 to 3, which means the taste of the tea ranges from not typical to slightly typical melinjo peel.

Table 4. Parameters of Melinjo Peel Tea Taste Quality

Citric Acid Concentration (%)	Blanching Duration (minutes)	Taste Quality
0	5	2.98 [ab]
	10	2.12 [ef]
0.05	5	2.78 [bc]
	10	2.42 [de]
0.1	5	2.78 [bcd]
	10	3.02[a]

Note: Figures followed by the same letters indicate 'not significantly different' with Duncan's test 5%.

The mean sensory test results of the quality parameters for the melinjo peel herbal tea aroma can be seen in Table 5. The results of variance (ANOVA) analysis showed that the variation of blanching duration significantly affected the aroma of melinjo peel herbal tea while the treatment of citric acid concentration had no significant effect. The interaction of the two treatments had no significant effect on the aroma quality of melinjo peel tea. The aroma of the melinjo peel tea was recorded on a quality scale of 2.60 to 3.43, which means the aroma ranges from not typical to slightly typical melinjo peel. Fatah [14] explained that blanching aims to remove unpleasant odors, help remove tannin compounds, strengthen tissue so that the shape or texture of the fruit remains stable despite various processing processes, and activate enzymes in the fruit.

Table 5. Parameters of Melinjo Peel Tea Aroma Quality

Citric Acid Concentration (%)	Blanching Duration (minutes)	Aroma Quality
0	5	3.43 [ab]
	10	2.60 [def]
0.05	5	2.92 [bc]
	10	2.68 [bcde]
0.1	5	2.90 [bcd]
	10	2.73[a]

Note: Figures followed by the same letters indicate 'not significantly different' with Duncan's test 5%.

The sensory test results of melinjo peel herbal tea with the hedonic method can be seen in Figure 1. Based on the data from Figure 1, the most liked color, taste, and aroma of brewed melinjo peel tea are those made by blanching the peel for 5 minutes with a citric acid concentration of 0%. When being correlated with the test results of the brewed melinjo peel tea aroma quality attributes, the best treatment to produce tea with an aroma not typical of melinjo peel is blanching the peel for 5 minutes with a citric acid concentration of 0% with a quality value of 2.67. The study of melinjo peel herbal tea with blanching treatment and citric acid concentration can reduce the typical aroma of melinjo peel, which is a quality scale of 3.43, and enhance the taste of melinjo peel herbal tea with a slightly typical taste of melinjo peel with a quality scale of 2.02 to 3.02.

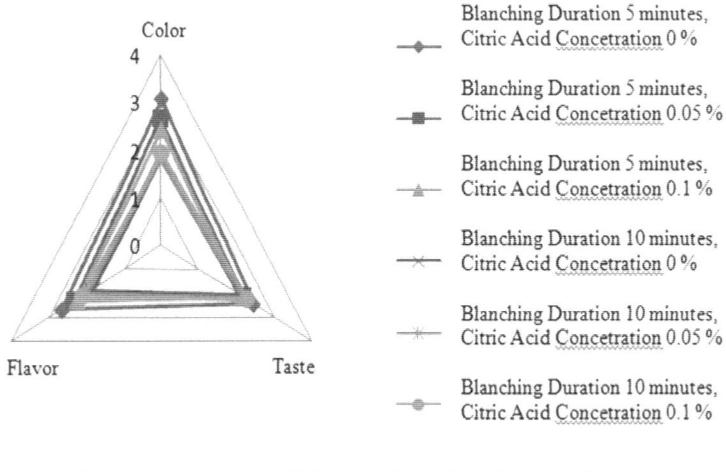

Figure 1. Results of Melinjo Peel Tea Hedonic Test

4. Conclusions

The melinjo peel tea has the potential to serve as herbal tea due to its antioxidant activity and bioactive compounds, i.e. total phenol. Melinjo peel herbal tea contains a total phenol of 2.02 - 2.52 mg/ 100g with the antioxidant activity of 32.47 - 48.47%. The brewed water of melinjo peel tea which is liked and accepted by panelists is that made by blanching the peel for 5 minutes with a citric acid concentration of 0%. The most-liked brewed melinjo peel herbal tea has the following characteristics: clear yellow, slight taste of melinjo peel, and slight aroma of the typical melinjo peel.

Acknowledgments

We would like to express our gratitude to the State Polytechnic of Jember, in particular, the Center for Research and Community Service (P3M) that has organized and funded this study through the PNBP Fund Sources Research Program in 2018.

References

[1] Enda I 2007 *Physical Characterization and pH Test of Melinjo Peel Soaking Solution and Melinjo Peel Violence* Minor Thesis (Bogor: Bogor Agricultural University) p.1-35
[2] Santoso M, Naka Y, Angkawidjaja C, Yamaguchi T, Matoba T, and Takamura H 2010 *Journal Food Science and Technology* **16** (6) 549-556
[3] Cos P, Ying L, Hu C J P, Cimanga K, Poel B V, Pieters L, Vlietinck A J, and Berghe D V 1998 *J. Nat. Prod* Vol. **61** (1): 71-76. (http://www. pharmanet.com.br/pdf/np970237h.pdf)
[4] Ardiyansyah M and Apriliyanti M 2016 *Jurnal Inovasi* **16** (2) 89-92
[5] Khoiriyah F 2017 Minor Thesis (Jember: State Polytechnic of Jember). (unpublished) p.40-45
[6] Pujimulyani D, Raharjo S, Y Marsono, and Santoso U 2010 *Journal of Agritech* **30** (2) : 68-74
[7] Brand-Williams Cuvelier M E and Berset C 1995 *Food Sci. Technol* **28** 25–30 (https://doi.org/10.1016/S0023-6438(95)80008-5)
[8] Slinkard K and Singleton V L 1977 *J. Enol. Vitic.* **28** 49-55
[9] Prior R L, Wu X, and Schaich K 2005 Journal *of Agricultural and Food Chemistry* **53** 4290-4302
[10] Soares J R, T C P Dins, A P Cunha, and L M Ameida 1997 *Free Rad. Res.* **26** 469-478
[11] Duh P, Y Tu, and G Yen 1999Antioxidant activity of water extract of Harng Iyur (Chrysanthemum morifolium Ramat) *Lebensm Wiss U Techno* **32** 269-277.
[12] Rohyami, Y 2008 Determination of Flavonoid Content of Mahkota Dewa Fruit Meat Methanol Extract (*Phaleria macrocarpa Scheff Boerl*) *Journal of Logic* **5** (1). P.233-240
[13] Winarno F G 1991 Kimia Pangan dan Gizi. (Jakarta : PT. Gramedia Pustaka Utama)
[14] Fatah A, Memet and Bachtiar Y 2004 *Making Various Candied Fruits* . PT. Agromedia Library, Bogor

IOP Conf. Series: Earth and Environmental Science **207** (2018) 012045 doi:10.1088/1755-1315/207/1/012045

The effect of blanching and drying temperature upon proximate composition of okra flour

R Rindiani[1*], H Warsito[1], N Maria Rosiana[1]

[1]Clinical Nutrition Study Program, Politeknik Negeri Jember
Jln. Mastrip, Jember 68101, Indonesia

*Email : rindiani@polije.ac.id

Abstract. Okra is one of the flagship products of Jember with its potential as a source of functional food due to its components. Okra's components comprise protein and fat. However, Okra has been used in limited chores, take dishes for instance, through boiling, grilling and frying, not to mention that okra is a perishable food. Therefore, the methods to process Okra to become various types of food need to be developed such as transforming it into form of flour. This study aims to determine the effect of processing on proximate okra flour. It used Randomized Block Design with two factors. Blanching process was taken as the first factor, followed by drying temperature, 50°, 60° and 70° C. The results showed no significant interaction between blanching process and drying temperature on the precise content of Okra flour. Blanching process did not indicate a significant difference in the precise content of Okra flour. Boiling temperature will significantly reduce the moisture and carbohydrate of the flour content, and will also increase the ash, protein, and fat of the content of okra flour.

1. Introduction

Okra was firstly developed by PT Mitra Tani 27 in the district of Jember. So far, Okra has merely been used for daily needs such as soup, dried, grilled, and fried food. This way, the use of Okra becomes very limited. In additon, Okra belongs to perishable vegetable. Therefore, the process of Okra to be developed into various types of food such as flour needs to be taken into consideration. Furthermore, Okra flour will be more easily processed and / or developed into various types of some other food such as cookies, noodles, and bread.

The flour process goes through a high temperature drying method known as food preservation method by reducing the content of water so that it lives even longer. Decreased water content results in the decrease of microorganism activity. Drying Okra makes use of hot air (70°C), air velocity (1.5 m/s), freezing dryer (-30 ± 2°C), drying with the use of microwave (70 kPa), a combination between hot air drying and microwave drying (70°C for 2 hours with 1.5 m/s air flow velocity), the combination of freeze drying by vacuum drying using microwave, freeze dried for 10 hours then vacuum with a pressure of 70 kPa (1).

Instead of drying, the process of okra flour also uses blanching. This way aims to reduce microbes and anti-nutrient and increase the brightness of the product. Drying and blanching can affect the proximate content of Okra as it contains components with physiological functions so that Okra can potentially become the source of functional food. Okra also consists of protein up to 16.27%, fiber 11.63%, fat 10.57% and ash 9.63% (2). Moreover, Okra provides important sources of vitamin, calcium, potassium and other minerals. The research aims to identify the effect of Okra flour processing on its proximate content.

2. Methods

2.1 Materials
Okra was obtained from PT Mitra Tani 27 in Jember Regency. It had similar age, maturity, size, color, and had no defects, then sent to the laboratory, washed, drained and cut crosswise.

2.2 The Preparation of Flour
The Okra was blanched on hot air (95±2^{0}C) for 10 minutes before being dried. The drying was conducted on the dehydrator at 50°C, 60°C and 70°C until a fixed moisture content was reached. Whereas, some other Okra were not blanched but dried immediately, then the dried okra were ground to release 80 mesh sieve. The flour was packed and sealed in polypropylene plastic and stored in dry condition until used (3).

2.3 Proximate analysis
Okra flour was then analyzed for moisture content using oven method, ash and fat content using soxhlet method, protein content used Kjedahl method (4) and carbohydrate content was calculated by difference. All analyzes were conducted using duplication.

2.4 Statistics analysis
Determination was carried out in four replications and the error was reported as standard deviation. Data were subjected to analysis of variance (ANOVA). Significance was accepted at $p<0,05$.

3. Result and Discussion

The process of Okra flour was divided into two steps: blanching and drying. ANOVA analysis results showed no interaction between blanching and drying temperature on the proximate content of Okra flour. Therefore, it was examined through each factor to determine the effect between treatments.

3.1 Effect of blanching on the proximate content of okra flour
Blanching is a short process of mild heat treatment for texture modification, enzymes deactivation, number of microbes reduction, as flavour preservation, nutritional value and flavour (5). Blanching uses hot water in temperatures of 7-100°C with alloted time based on the use of materials. Compared to other blanching methods, it is an easier, cheaper and more efficient method. Nevertheless, the use of hot water can cause changes in the components and product sensory due to the leaching or damage of nutrient saved in hot water (6). Comparison between the blanched and proximate content of Okra flour and that of non-blanched one is displayed in Table 1.

Table 1. The effect of blanching on proximate content of okra flour (% wb)

Proximate content	Without Blanching	With blanching
Moisture	9,01 ± 0,20	9,08 ± 0,17
Ash	2,48 ± 0,06	2,49 ± 0,07
Protein	11,56 ± 0,12	11,61 ± 0,13
Fat	2,51 ± 0,11	2,57 ± 0,18
Carbohydrate	74,42 ± 0,39	74,26± 0,23

Table 1 shows no significant effect indicated by blanching on the moisture, ash, protein, fat and carbohydrate content of okra flour. The Moisture content without blanching was 9.01% while with blanching showed 9.08%. This is also indicated by other studies that moisture is not affected by blanching on plantain flour (7,8). Low moisture makes Okra material lives longer. According to (8), products with moisture above 12% have shorter life.

The ash of non-blanching Okra flour (2.48%) was not significantly different from that of blanching process (2.49%). Ash shows stable mineral of Okra during the blanching process. Blanching process also has no effect on plaintain flour ash (8). Ash in Okra flour (2.48-2.49%) was higher than that in sweet potato flour (0.42-0.49%) (5), raw plantain flour (1.66-2.00%) (9), and on plantain flour (2.80-2.97%) (8). Carbohydrate in Okra flour only ranges from 74.26 to 74.42% and were not affected by blanching. Carbohydrate in Okra flour is lower than dried sweet potato flour (78-83%) (10), plantain flour (81.63-82.40%) (8). Blanching process on okra flour did not give a significant effect on changes in proximate content of okra flour. Therefore, blanching process can be applied for Okra flour. Blanching is an important step of fruit and vegetable processing. Blanching is aimed to reduce anti-nutrient content and number of microbes, to deactivate the enzymes and to increase the brightness of Okra flour. In addition, some other studies showed that blanching provide many advantages as it can reduce processing cost for plantain and banana flour through the increase of efficiency of drying. (5)

3.2 Effect of drying temperature on the proximate content of okra flour

Drying through dehydrator aims to extend the life of the product by reducing the product moisture. It aims to inhibit microbial growth and enzyme activity. As indicated in blanching, the use of heat during the drying process can cause changes in the components and sequences of the product. The comparison of drying temperature difference to Okra flour components is displayed in Table 2.

Table 2. The effect of drying temperature differences on the proximate content of Okra flour (%wb)

Proximate content	Drying temperature (°C)		
	50	60	70
Moisture	9,20 ± 0,03[a]	9,09 ± 0,03[b]	8,84 ± 0,08[c]
Ash	2,41 ± 0,05[a]	2,51 ± 0,05[b]	2,54 ± 0,04[b]
Protein	11,44 ± 0,03[a]	11,59 ± 0,05[b]	11,72 ± 0,13[c]
Fat	2,38 ± 0,04[a]	2,45 ± 0,05[b]	2,78 ± 0,04[c]
Carbohydrate	74,56 ± 0,10[a]	74,35 ± 0,13[b]	74,11 ± 0,17[c]

Table 2 shows that the drying temperature has a significant effect on moisture, ash, protein, fat and carbohydrate of Okra flour. Heat used in drying causes more water to evaporate from the flour so the moisture will likely decrease (9.20%, 9.09%, 8.84%). Blanching comes before drying. During the blanching process, starch gelatinization occurs, then starch granules will surely swell. The swelling of the starch on Okra flour will increase the permeable properties of the cell wall so that much water is absorbed and evaporated. When moisture in the product decreases other components such as ash content, protein content and fat content (shown in Table 2) will increase. However, increased levels of ash, protein and fat will reduce carbohydrate content in Okra flour.

Okra drying process occurs for 30 hours at a temperature of 50°C, 25 hours at a drying temperature of 60°C and 20 hours at a drying temperature of 70°C. The moisture in Okra is a type of water bound to the material. The water has a binding energy sufficiently strong to be released from the material so that it requires a long drying time. In addition, the moisture of Okra is adequately high so that it takes a long time to reach constant moisture.

3. Conclusion

Blanching has no significant effect on proximate Okra flour. Whereas, drying temperature has a significant effect on water, protein, fat, carbohydrate of Okra flour. Further research needs to be conducted focusing on functional properties of Okra flour applied in food products.

IOP Conf. Series: Earth and Environmental Science **207** (2018) 012045 doi:10.1088/1755-1315/207/1/012045

4. Acknowledgement
Special thanks to DIPA Politeknik Negeri Jember (State Polytechnic of Jember) for the research funding.

5. References
1. Jiang N, Liu C, Li D, Zhang Z, Liu C and Wang D 2017. LWT - *Food Sci Technol.* **82** 216–26
2. Adetuyi FO and Ibrahim TA. 2014. *Niger Food J.* **32** 128–37
3. Pendre NK, Nema PK, Sharma HP, Rathore SS and Kushwah SS. 2012. *J Food Sci Technol.* **49** 378–81.
4. AOAC. 1995. *Official Method of Analysis of The Association Analitycal Chemist.* Inc. Washington, DC
5. Harijono, Estiasih T, Saputri DS and Kusnadi J. 2013. *Adv J Food Sci Technol.* **5** 1342–1350
6. Mala KS, Kurian AE and Srinivasulu K. 2016. *Int J Innov Stud Sci Eng Technol.* **2** 17–24
7. Adeniji, Adebayo T, Tenkouano and Abdou 2008 *Agro-Science* **7** 99–105
8. Falola AO, Olatidoye OP, Adesala SO and Daramola O. 2015. *Effect of Different Processing Methods on Proximate , Functional and Pasting Properties of Plantain Flour* **03** 522–528
9. Fagbemi TN 1999 *Plant Foods Hum Nutr.* **54** 261–269
10. Fashina AB, Adejori EA and Akande FB 2017 *Int Food Res J.* **24** 1349–1352

Surimi from Freshwater Fish with Cryoprotectant Sucrose, Sorbitol, and Sodium Tripolyphosphate

Fahrizal[1*], N Arpi[1] , S Rohaya[1], R Febriani[1]
[1] Agricultural Product Technology Department, Faculty of Agriculture, Syiah Kuala University, Jl. Teuke Hasan Krueng Kalee No.3, Kopelma Darussalam, Banda Aceh, 23111 Indonesia

*Corresponding author's e-mail: fahrizal.z@unsyiah.ac.id

Abstract. Research has been conducted to produce surimi from freshwater fish (tilapia and snakehead fish). In general, this study aims to determine the process of making surimi from freshwater fish so that the concept of fishery product diversification can be applied. The specific purpose is to study the manufacture of surimi from freshwater fish and to investigate the addition of cryoprotectants consisting of sucrose, sorbitol and sodium tripolyphosphate (STPP) with different concentrations on the quality of frozen surimi. A factorial randomized block design consisting of two factors was used. The first factor was the type of fish, namely tilapia and snakehead fish. The second factor was a combination of cryoprotectants which consists of four levels namely K1 = (2% sucrose, 2% sorbitol, and 0.1% STPP), K2 = (3% sucrose, 3% sorbitol, and STPP 0.2%), K3 = (4% sucrose, 4% sorbitol, and 0.3% STPP), K4 = (5% sucrose, 5% sorbitol, and 0.4% STPP). Parameters analyzed were yield, moisture content, ash content, protein content, pH, water holding capacity (WHC) and folding test. The results showed that surimi from snakehead fish species had high yield, moisture content, protein content, and folding test, but had a low WHC value. The best surimi was obtained from snakehead fish which was added cryoprotectant combination of 5% sucrose, 5% sorbitol, and 0.4% STPP. The surimi has a moisture content of 78.91%, protein content of 19.84%, pH 6.38, WHC 2.08%, and folding test of 3.66 (not cracked after the first folding).

1. Introduction

Surimi is deboned fish meat that has been subjected to leaching and dewatering processes to remove sarcoplasmic proteins and concentrate myofibrillar proteins [1]. In production of surimi, fresh fish is skinned, deboned, and washed. Cleaned fish flesh is then minced, washed in iced water, dewatered by filtration and pressing, blended with cryoprotectant (such as sucrose, sorbitol, and STPP), wrapped, and stored in freezer.

Cryoprotectant protect surimi from protein dehydration, so as to prevent protein freeze-denaturation of myofibrillar proteins during frozen storage, hence preserving the gel-forming ability of surimi. Sucrose, sorbitol, and sodium tripolyphosphate (STPP) are food additives often used with different concentrations in the production of surimi [2][3].

This study used freshwater fish, tilapia (*Orheochromis niloticus*) and snakehead fish (*Channa striatus*). In Aceh, these two types of fish are not common cultivated and underutilized. In harvest season, fish are abundant, however fish price and fish utilization are low. Therefore, to increase the use of fish that is rarely used, it is necessary to make a diversification of freshwater fish into processed products that are economically valuable such as surimi.The production of surimi employs underutilized fish, result in product diversification of freshwater fish, and generate value-added product.

The purpose of this research is to study the process of making freshwater fish surimi, so that the concept of product diversification by utilizing raw materials for fishery products can be applied. In addition, this study also aims to understand the function of cryoprotectant addition which consists of sucrose, sorbitol and sodium tripolyphosphate (STPP) with different concentrations so that it can produce high quality frozen surimi.

2. Method

Fresh tilapia (*Orheochromis niloticus*) and snakehead fish (*Channa striatus*) were obtained from local fish farms. They were deboned directly after gutting and washing with water. Fish flesh that has been weighed and added ice were minced by using chopper. Fish meat that has been minced was washed 5 times using ice water with a ratio of fish and water 1: 5. During the last washing, salt (NaCl) was added as much as 0.30% of the weight of the minced fish. Then the washed minced fish was wrapped using a filter cloth and pressed by using a hydraulic press with a pressure of 50 kg.f/cm^2 for 5 minutes. Minced fish then mixed with combination of cryoprotectants i.e, K1 = 2% sucrose, 2% sorbitol, and 0.1% STPP, K2 = 3% sucrose, 3% sorbitol, and 0.2% STPP, K3 = 4% sucrose, 4% sorbitol, 0.3% STPP, K4 = 5% sucrose, 5% sorbitol, and 0.4% STPP. Each surimi sample was subdivided into 2-kg blocks, packed into low density polyethylene bags and freeze until the internal temperature reached –22 °C, then-stored at –20 °C until analysis. The surimi was analyzed for yield, moisture, ash and protein contents, folding test (folding test), measurement of pH (degree of acidity), and water holding capacity (WHC).

Data were collected and analyzed using the completely randomized block design. Analysis of variance was performed on SPSS version 17.0 for Windows.

3. Results and Discussion

3.1 Yield

Surimi yield of tilapia and snakehead fish ranged from 24.76% - 35.23% with an average of 27.73%. The effect of fish species on surimi yield can be seen in Figure 1.

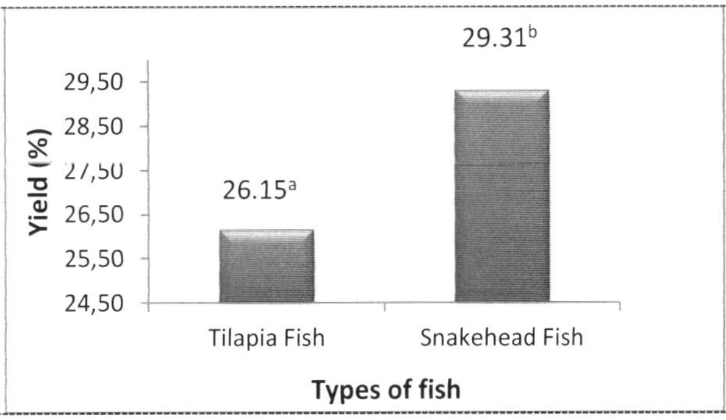

Figure 1. Surimi yield based on fish species.

Tilapia surimi (N1) had lower yield compared to snakehead fish surimi (N2). The yield of tilapia surimi was 26.15%, lower than the yield of snakehead surimi which was 29.31%. This is presumably because the water and protein contents of snakehead fish surimi were higher than the moisture content of tilapia surimi (Figures 2 and 3).

3.2 Moisture Content

Water is an important component in food ingredients because water can affect the appearance, texture, and taste of food [4]. Water contents of tilapia and snakehead fish surimi ranged from 74.69 to 81.82% with an average of 78.54%. Study showed that the moisture content obtained from some surimi had met the quality requirements of SNI frozen surimi. According to [5], the moisture content of fish surimi is between 80-82%.

Based on analysis of variance, the fish species had a very significant effect ($P \leq 0.01$) on the water content of surimi, while the combination of cryoprotectant and interaction between fish species and cryoprotectant combination had no significant effect ($P > 0.05$) on the moisture content of surimi . The effect of fish species on water content of surimi can be seen in Figure 2.

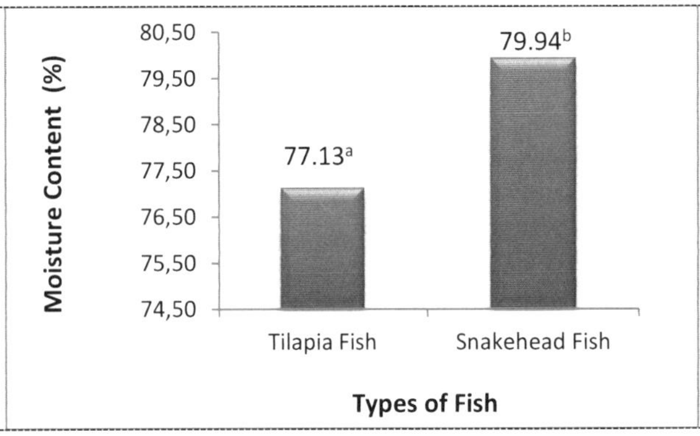

Figure 2. surimi moisture content as affected by fish species.

Figure 2 shows that tilapia surimi (N1) has lower moisture content than snakehead surimi (N2). The moisture content of tilapia surimi was 77.13% while that of snakehead surimi was 79.94%. This is presumably because the water content of each type of fish is affected by habitat and species of fish. Tilapia has water content of 77.8% whereas snakehead fish has a water content of 89%.

3.3 Ash Content

According to [6], ash content reflects the amount of non-flammable minerals contained in volatile substances. Surimi ash content produced from tilapia and snakehead fish ranged from 0.33% - 0.73% with an average of 0.60%.

Based on the analysis of variance, fish species, the combination of cryoprotectant and their interaction had no significant effect ($P > 0.05$) on the ash content of surimi.

3.4 Protein Content

Protein contents in foodstuffs generally determined the quality of the foodstuff [7]. Protein levels of tilapia and snakehead fish surimi ranged from 8.75 to 26.26% with an average of 16.26%. Fish species had a very significant effect ($P \leq 0.01$) on protein content of surimi produced. Whereas the cryoprotectant combinations, and the interaction between fish species and cryoprotectant combinations had no significant effect ($P > 0.05$) on protein content of surimi produced. Figure 3 shows the effect of fish species on protein content of surimi.

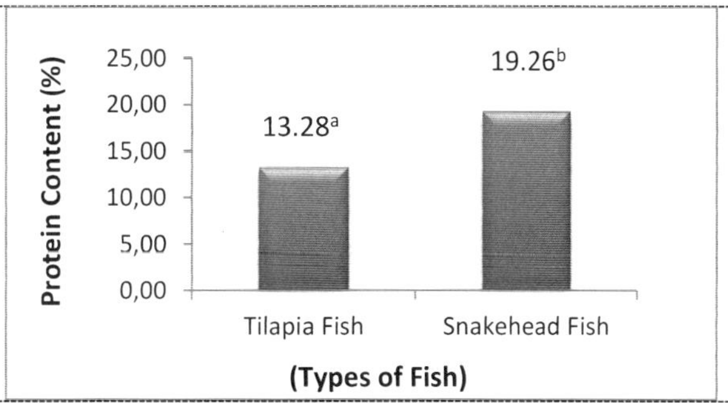

Figure 3. Effect of fish species on surimi protein content.

Figure 3 shows that the protein content of tilapia surimi lower than that of snakehead fish surimi. Tilapia surimi had protein content of 13.27% while snakehead surimi had 19.26%. This is presumably because fish protein contents vary based on the fish species. Tilapia had protein content of 20.08% while snakehead fish had protein content of 25.2%. In the surimi process production, not all proteins are desirable. The important protein in surimi production is the miofibril protein, therefore some water soluble protein will be wasted during the washing process. Sarcopoplasmic protein is one of the unwanted water soluble proteins which can inhibit gel formation of surimi.

3.5 pH

The pH is one factors that influence surimi gel formation. The pH of surimi produced in this study ranged from 4.44 to 6.70 with an average of 5.27. Based on the analysis of variance, fish species, combination of cryoprotectant and their interaction on the pH of surimi were significant. The effect of fish species, combination of cryoprotectant and interaction between species of fish with cryoprotectant combinations can be seen in Figure 4.

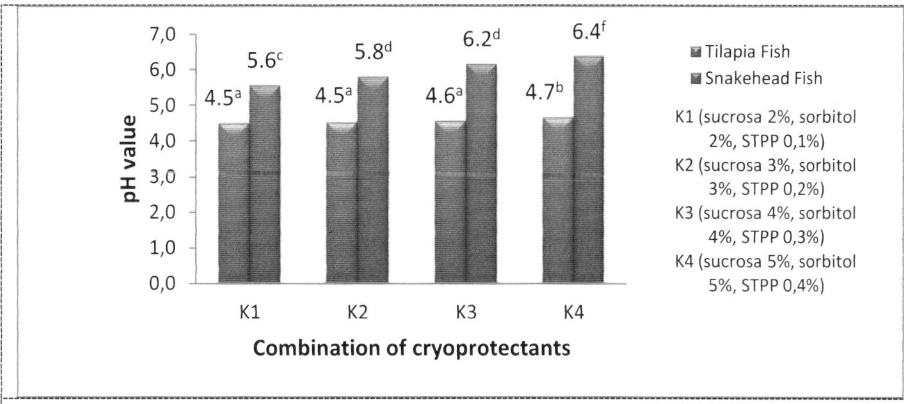

Figure 4. Effect of fish species, cryoprotectant combination and their interaction on acidity (pH) of surimi.

Based on Figure 4, the pH of tilapia surimi tended not to increase, while the pH of snakehead fish surimi increased along with the increase of cryoprotectant concentration. This is presumably because the concentration of cryoprotectant can increase the pH, so that it can produce a chewy texture. This can be seen in the snakehead fish surimi which had a high folding test compared to tilapia surimi. According to [8], the characteristics of a good quality surimi have a pH of 6.5 - 7.0. If the pH value of a product is lower than 6.0, the fish gel will not be formed.

3.6 Water Holding Capacity

Water holding capacity (WHC) is the ability of meat protein to bind water both from the meat itself and from outside. WHC is an important factor in gel formation [9] and is closely related to free water released. The results of surimi WHC analysis from tilapia and snakehead fish ranged from 0.40% - 8.477% with an average of 3.71%. Analysis of variance showed that the fish species and the combination of cryoprotectant had significant effect (P≤0.05) on the WHC value, while the interaction between fish species with cryoprotectant combinations had no significant effect (P> 0.05) on the WHC. The effect of fish species on WHC of surimi can be seen in Figure 5.

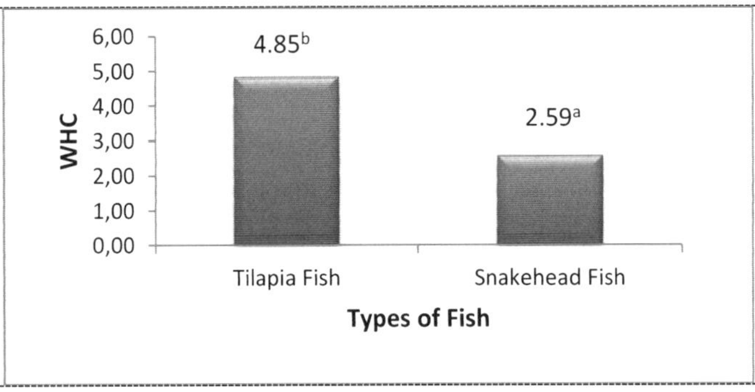

Figure 5. The effect of fish species on surimi water holding capacity.

Based on analysis of variance, surimi water holding capacity of tilapia higher than that of snakehead fish, with WHC of 4.84% and 2.59%, respectively. This is might due to the lower water content of tilapia surimi. Similarly, snakehead fish have a low WHC due to the higher protein and water contents of the surimi. The effect of cryoprotectant combination on WHC of surimi can be seen in Figure 6.

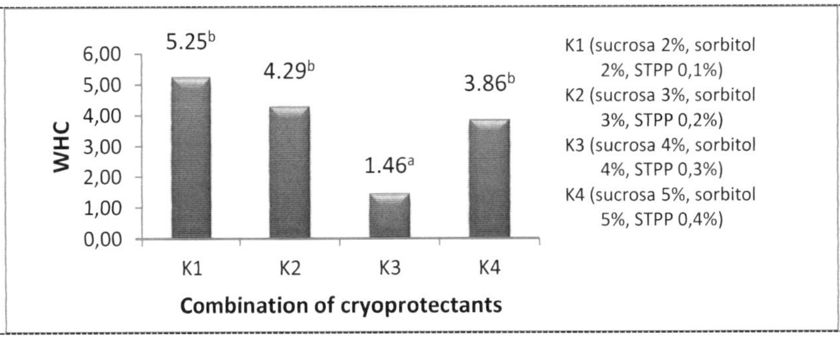

Figure 6. The effect of cryoprotectant combination on water holding capacity (WHC) of surimi.

Figure 6 shows that all surimi containing cryoprotectant combination had the same WHC except for K3 (4% sucrose, 4% sorbitol, and 0.3% STPP). This findings contrary to the results of [10] which showed that a cryoprotectant combination of 4% sucrose, 4% sorbitol, and 0.3% STPP produced the best surimi with WHC of 64.66%. In the process of making surimi, cryoprotectant serves to improve the ability of water as a binding energy, preventing the exchange of water molecules from proteins so that it can stabilize the proteins. If the proteins damaged, the ability to bind water will be reduced and thus cause the WHC to decrease.

3.7 Folding Test

Folding test aims to determine the level of surimi elasticity subjectively. Folding test of tilapia and snakehead fish surimi ranged from 1-5 (cracked when pressed with the finger to not crack after two folds) with an average of 2.25. Based on analysis of variance, fish species had a very significant effect ($P \leq 0.01$) on surimi folding test. While the interaction between combination of cryoprotectant and fish species had no significant effect ($P > 0.05$) on the surimi folding test.

Surimi folding test showed that tilapia surimi had lower folding test of 1.5 (cracked when pressed with the finger to crack immediately after the first folding) compared to that of snakehead fish surimi which had value of 3 (gradually cracked after the first fold). The effect of fish species on surimi folding test can be seen in Figure 7.

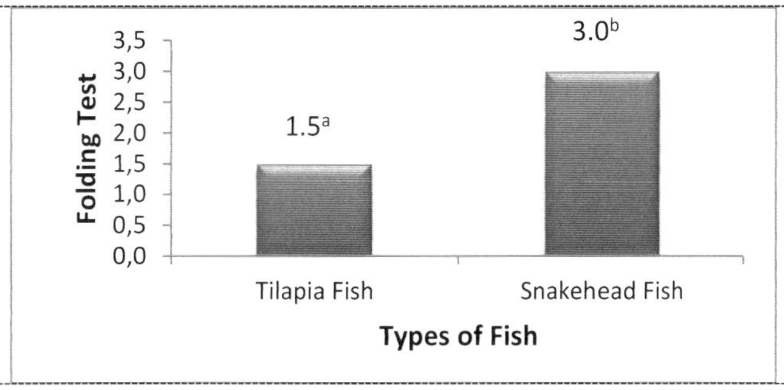

Figure 7. The effect of fish species on surimi folding test.

The strength applied during surimi folding test depends on the protein contained in the fish. According to [5], fish gel products that have high gel strength will produce a higher folding test and bite test, with a range from 4-5 (AA grade) for folding test and 7-10 for bite test. The higher the folding test, the better the quality of the surimi gel will be.

4. Conclusions

Surimi from the snakehead fish had high moisture and protein content, also had high folding test, but had low WHC. The higher the cryoprotectant concentration, the higher the pH. However, combination of cryoprotectant concentration added gives the same WHC value except for K3 (4% sucrose, 4% sorbitol, 0.3% STPP) which had the lowest WHC. The best treatment was obtained from surimi using snakeheadfish with cryoprotectant combination of 5% sucrose, 5% sorbitol, and 0.4% sodium tripolyphosphate (N2K4). The surimi had moisture content of 78.91%, protein content of 19.84%, pH 6.38, WHC 2.08, and a folding test of 3.66 (did not crack after the first folding).

5. Acknowledgment

We thank Syiah Kuala University, Banda Aceh, Indonesia and Ministry of Technology, Research and Higher Education of the Republic of Indonesia for providing the financial support.

Reference

[1] Lee CM 1984 *Food Technol.* **38**: 69–80.
[2] Pipatsattayanuwong S, Park JW and Morrissey MT 1995 *J. Food Sci.* **60**: 1241-1244
[3] Park JW and Lanier TC 1987 *J. Food Sci.* **52**: 1509–13.
4] Jin SK, Kim IS, Kim SJ and Jeong KJ 2007 *J. Food Eng* **81**: 618-623.
[5] Indonesian National Standards 1992 *National Standardization Agency* Jakarta
[6] Sediaoetama AD 1985 *Science of Nutrition Vol I* Dian Rakyat Jakarta.
[7] Winarno FG, Fardiaz S and Fardiaz D 1980 *Introduction to Food Technology* PT Gramedia Jakarta.

[8] Trondsen T 1998 *Blue Whiting Surimi: New Perspectives on Market Value* Fisheries Research.
[9] Muchtadi TR and Sugiyono 1992 *PAU Food and Nutrion*, IPB Press Bogor.
[10] Peranginangin R, Wibowo S, Nuri Y and Fawza 1999 *Slipi Marine Fisheries Research Installation*, Indonesia Marine Fisheries Research Institute, Jakarta.

IOP Conf. Series: Earth and Environmental Science **207** (2018) 012047 doi:10.1088/1755-1315/207/1/012047

Determination of Water Content Of Intact Sapodilla Using Near Infrared Spectroscopy

Kusumiyati [1*], Yuda Hadiwijaya [2], Ine Elisa Putri [2]
[1,]Lecturer of Agronomy Studies Program, Agriculture Faculty, Padjadjaran University
[2]Alumni of Agronomy Studies Program, Agriculture Faculty, Padjadjaran University

*Corresponding Author: kusumiyati@unpad.ac.id

Abstract. Water content is one of components that affects the fruit firmness of sapodilla. Fruit firmness is used as an indicator of physical fruit quality, which determines consumer acceptance. Generally, determination of water content is done by drying the fruit itself. However, this method damages the fruit, therefore needed a non-destructive method which able to determine the water content of sapodilla. Non-destructive method is also very useful to reduce physical damage to fruit (bruses) caused by the touch of the consumers hands to assess the fruit firmness. Near infrared spectroscopy is a non-destructive measurement that has been widely used to determine various components of fruit quality, including water content. The purpose of this study was to determine water content of intact sapodilla non-destructively using near infrared spectroscopy. This research used experimental method with multivariate data analysis by modelling, continued by data processing using ISIS software (Integrated Spectronics Pty, Ltd, Australia and Unscrambler multivariate data software (version 9.7, CAMO, Oslo, Norway). Further research resulted value has the same accuracy as the value of non-destructive measurements. The result showed that near infrared spectroscopy was able to determine water content of intact sapodilla non-destructively.

1. Introduction

Most sapodilla fruit farmers determines the appropriate time to harvest by visual estimation such as size and skin color. Additionally, it needs laboratory analysis to obtain the internal quality of fruits. Water content of fruit is an important parameter to assess the quality of fruit, because it is related to the shelf life and also the texture of the fruit. The ripe sapodilla contains 72 to 78% of water content and total soluble solids ranged from 12 to 18 °Brix [1]. Composition of ripe sapodilla fruit is presented on Table 1.

Near infrared spectroscopy (NIR) is a non-destructive method to predict fruit quality. It has been used to predict a lot fruits quality since 1990s [2]. This technique is time and labor reducing alternatives. Blanco and Villarroya [3] studied that near infrared spectroscopy offers a rapid and accurate measurement for fresh commodities. Spectrometer records spectra data consist of physical and chemical content such as total soluble solids, firmness and color [4]. The non-destructive method using NIR has been previously performed to measure the fruit quality of mango [5], Jujube [6;7] with high accuracy. NIR also has been used for predicting soluble solid content, moisture content and hue color of on tree and after harvesting tomato [8] firmness, color and lycopene content of tomato [8] and chlorophyll of bitter gourd [9]. Ranganna [10] studied that once separated from the plant, fruits continue to respire until senescence phase because the harvested fruit can not access water from the plant anymore for the ripening process. Respiration determines the potential storage life of fruits [11]. Hence, this research objective aimed to focus on determinating of internal quality of intact sapodilla during storage using near-infrared spectrometer.

Table 1. Composition of ripe sapodilla fruit of 100 g of eatable part [12]

Constituents	Amount
Water content (%)	73.37
Protein	0.70 g
Fat	1.10 g
Mineral	0.05 g
Fiber	2.60 g
Carbohydrates	21.40 g
Energy	98.00 cal
Phosphorous	72.00 mg
Iron	1.25 mg
Calcium	28.00 mg
Vitamin B1	0.02 mg
Vitamin B2	0.03 mg
Carotene	97.00 mg
Vitamin C	0.06 mg

2. The Material And Method

2.1. *Sample Collective*

225 Sapodillas were all harvested from the orchad then stored for 0 day (H_0), 5 days (H_5), and 10 days (H_{10}). All samples were numbered before further analysis.

2.2. *Spectral Data Acquirement*

NirVana AG410 spectrometer with wavelength range of 312-1050 nm was used to measure the six separate measurements with three points on each side of fruits distributed along axial region (Figure 1.).

Figure 1. Irradiated points of fruits on each side

2.3. Reference Data Acquirement

Water content was calculated by the mass difference between the fresh fruit and the dried fruit. Sliced sapodilla were put into aluminum foil. Then The sample was dried using an oven until it reached a constant weight. The results were described as percentage of water content. The calculation of water content is described in the following equation :

$$\text{Water content} = (\%) \frac{W-Wa}{W} 100\%$$

where:
W : fresh fruit (g) Wa : dry fruit (g)

The method used in this research was multivariate data analysis. Then it continued by data processing using ISIS and The Unscrambler 9.7.

2.4. Multivariate Analysis

The calibration model was built using the partial least square (PLS) method. PLS is a chemometric technique that could analyze a large of hidden information in spectral data. It is used to evaluate a series of dependent variables (response) from immense number of independent variables (predictors).

The acceptable model is implied by R_{cal} of 0.71 or above as explained by Williams [13] that this value indicates rough screening ability, and is preferable for further prediction. Then, the acceptable calibration model was used to predict another sapodilla fruits (prediction set) in order to verify the prediction capability. The model was evaluated using coefficient of determination (R^2) which explains the capability of independent variables explaining variations of dependent variables.

Root mean square errors of calibration (RMSEC) is the difference in values between predicted and reference value of calibration set. Root mean square errors of prediction (RMSEP) is the difference in values between predicted and reference value of prediction set.

Nicolaï *et al.* [14] advised that an effective model should present a high R^2 value of prediction set and low value of errors in calibration set and prediction set, either. Nevertheless both error values should not have a high difference from each other. The calculations of RMSEC and RMSEP are described in the following equations [15]:

$$\text{RMSEC}=\sqrt{\frac{1}{n_c}\sum_{i=1}^{n^c}(\hat{y}i-yi)^2}$$

$$\text{RMSEP}=\sqrt{\frac{1}{n_p}\sum_{i=1}^{n^p}(\hat{y}i-yi)^2}$$

where:
$\hat{y}i$:predicted value,
yi :measured value,
n^c :amount of samples in calibration set,
n^p :amount of samples in prediction set.

3. Result And Discussion

3.1. Destructive Data Analysis of Sapodilla

Water content analysis by destructive measurement was used as reference data for the development of calibration model. Any kind of measurement surely produce errors, as well as the destructive measurements. Hence, to diminish the errors, it was performed special handling to the samples. Sapodilla samples that have been irradiated by near infrared spectrometer, then as quickly as possible should be measured of it's water content at the laboratory. This is necessary to avoid the changes of internal chemical content of sapodilla. Reference data measurement of sapodilla samples are described on Table 2.

Table 2. Statistical Data of Sapodilla Samples By Destructive Method

Water Content	H_0	H_5	H_{10}
Minimum	67%	64%	63%
Maximum	74%	77%	74%
Mean	70%	72%	70%
Standard Deviation	1.5%	2.8%	2.3%

The lowest and highest value of water content are found in H_{10} as 63% and H_5 as 77% respectively. Highest mean value was found in H_5 as 72% and the lowest mean value are found in H_0 and H_{10} as 70%.

Standard deviation indicates the diversity of amount of data being investigated. Largest variation of data was found in H_5 as 2.8% and the lowest data variation in H_0 value as 1.5%.

3.2. Near-Infrared Data Analysis of Sapodilla

Data analysis of water content was measured by NIR second derivative data using multivariate data analysis as partial least square (PLS). Total samples used was 225 fruits. Each fruit was irradiated by 6 points as illustrated in Figure 1.

3.3. Water Content Assessment

In calibration set the assessment of water content used 150 samples and 75 samples for prediction set with wavelength 312-1050 nm. Lowest to highest data range in calibration set starts from 63% to 77%, while in the prediction stage ranges from 64% to 76%. The mean value of calibration was 71% and 70% for prediction. The standard deviation values of calibration was 2% and prediction was 1%, respectively. Stastical data of sapodilla used in calibration and prediction are presented on Table 3. The calibration model of water content is showed on Figure 1. Prediction model of water content is showed on Figure 2.

Table 3. Statistical Data of Sapodilla Samples Used In Calibration and Prediction Set.

Set	n	Range	Mean	Standard Deviation
Calibration	150	63% – 77%	71%	2%
Prediction	75	64% – 76%	70%	1%

Coefficient of determination of calibration set (R^2_{cal}) for assessment of water content of sapodilla was 0,71 means high (closed to 1). This defined the NIR prediction for water content was close to the reference measurement by using destructive analysis. RMSEC of calibration set generated was 0.01. Considering the RMSEC of calibration set, it showed the calibration model was built well due to the standard error value close to zero (0). It is required to perform the prediction set to check the accuracy of calibration model.

The prediction set of water content of sapodilla used different samples from calibration set which 75 samples from second derivate absorbance data (d2a) NIR and destructive mesurement. The purpose of using a different samples was to test the capability of calibration model for predicting water content of new sapodilla fruit samples. Afterwards the calibration model would be reliable for predicting new samples. Coefficient of determination of prediction set (R^2_{pred}) of water content was 0.65. The accuracy level of prediction set was determined RMSEP value. It showed that the RMSEP generated was 0.01. The RMSEP was acceptable due to close to zero. Reliability of the calibration model could be examined by how close the value of coefficient of determination of the calibration set and the prediction set. The result explained the coefficient of determination and the error value in the calibration set and the prediction set were quite close. Hence calibration model that was built could be assumed to be reliable for predicting water content of sapodilla fruit.

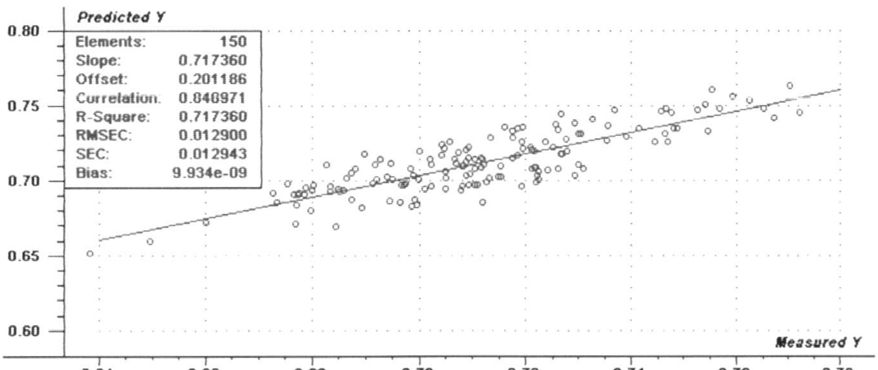

Figure 2. Calibration set of water content

IOP Conf. Series: Earth and Environmental Science **207** (2018) 012047 doi:10.1088/1755-1315/207/1/012047

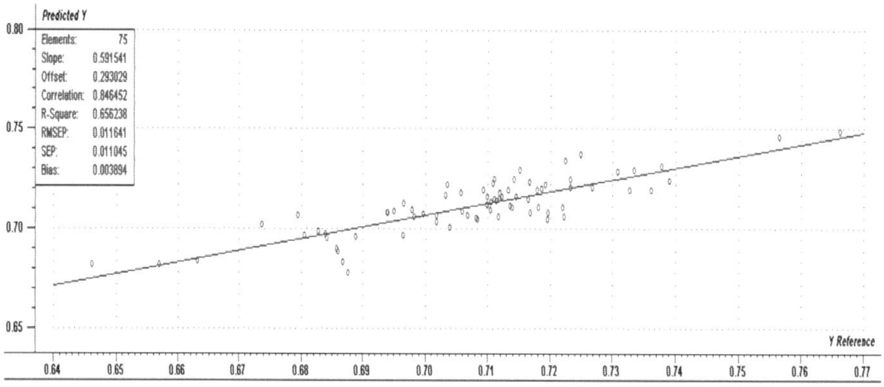

Figure 3. Prediction set of water content

4. Conclusion

Determination of intact sapodilla during storage using near-infrared spectrometer was able to measure water content value with good accuracy.

5. References

[1] Ganjyal G M, Hanna M A and Devadattam D S K 2005 Processing of Sapota (Sapodilla): Powdering *J. Food Technol.* **3** 326-330

[2] Kawano S 1998 New application of nondestructive methods for quality evaluation of fruit and vegetables in Japan *J. Jpn. Soc. Horticult. Sci.* **67** 1176–1179

[3] Blanco M and Villarroya I 2002 NIR spectroscopy: a rapid-response analytical tool *Trends Anal. Chem.* **21** 240–250

[4] Kusumiyati, Mubarok S, Hamdani J S, Farida, Sutari W, Hadiwijaya Y, Putri I E and Mutiarawati T 2018 Evaluation of sapodilla fruit quality using near infrared spectroscopy *J. Food Agric. Envir.* **16** 49-53

[5] Saranwong S, Sornsrivichai J and Kawano S 2004 Prediction of ripe-stage eating quality of mango fruit from its harvest quality measured nondestructively by near infrared spectroscopy *Postharvest Biol. Technol.* **31** 137–145

[6] Wang J, Nakano K and Ohashi S 2011a Nondestructive evaluation of jujube quality by visible and near-infrared spectroscopy *LWT-Food Sci. Technol.* **44** 1119–1125

[7] Wang J, Nakano K and Ohashi S 2011b Nondestructive detection of internal insect infestation in jujubes using visible and near-infrared spectroscopy. *Postharvest Biol. Technol.* **59** 272–279

[8] Kusumiyati, Akinaga T, Yonemori S, Kawasaki S and Tanabe T 2007 Evaluation of Tomato Quality on Tree and after Harvesting using portable NIR Spectroscopy *J. Soc. Agric Structures Japan (Nogyo Shisetsu)* **38** 117-126

[9] Kusumiyati, Akinaga T, Yonemori S, Shikanai T, Okamoto H, Tanabe T and Kawasaki S 2008b Internal quality sensor of fruit jagged surface: preliminary study of bitter gourd (*Momordica charantia* Linn) *Acta Hort.* **768** 391-398

[10] Ranganna S 2008 *Handbook of Analysis and Quality Control for Fruit and Vegetable Products* McGraw-Hill Publishing Company Limited, New Delhi p 444

[11] Varoquaux P and Sani O I 2005 *Packaging and produce degradation*, In: Lamikanra O, Imam S, Ukuku D (Eds.), Produce Degradation: Pathways and Prevention. CRC Press pp 117–153

[12] Swaminathan M 1979 Food Science and Experimental Foods. Madras, India: Ganesh and Co pp 45-63

[13] Williams P 2007 *Application to Agricultural and Marine Products*, In: Ozaki Y, McClure W F, Christy A A (Eds.), Near-Infrared Spectroscopy in Food Science and Technology, John Wiley & Sons, Inc. Publication, New Jersey pp 165–218

[14] Nicolaï B M, Beullens K, Bobelyn E, Peirs A, Saeys W, Theron K I and Lammertyna J 2007 Nondestructive measurement of fruit and vegetable quality by means of NIR spectroscopy: a review *Postharvest Biol. Technol.* **46** 99-118

[15] Yahaya O K M, MatJafri M Z, Aziz A A and Omar A F 2015 Visible spectroscopy calibration transfer model in determining pH of Sala mangoes *J. Instrum.* **10** T05002

IOP Conf. Series: Earth and Environmental Science **207** (2018) 012048 doi:10.1088/1755-1315/207/1/012048

Improvement of Sensory and Chemistry Quality of Fried Edamame by Freezing

B Hariono[1*], MF Kurnianto[1], A Bakri[1], M Ardiansyah[1], R Wijaya[1]

[1] Department of Agriculture Technology, Politeknik Negeri Jember, Indonesia

[*] budihariono1966@gmail.com

Abstract. Fried edamame is an edamame soybean product which is processed by vacuum frying with the peculiarities of crispy products which are crispy, savory, flavorful and green. Some research studies the freezing before vacuum frying fruit or vegetable can improve crispness and taste quality of the product. The duration of freezing before the vacuum frying process is expected to improve sensory qualities such as crispness, color, flavor and chemical characteristics of fried edamame products. The study was carried out using fresh edamame samples with 3 freezing time variations (temperature - 20 °C) which is 1 day; 2 days and 3 days, each with 4 replications, using a Completely Randomized Design. Observation on the quality of edamame samples in the form of Sensory Quality includes hedonic testing of Color, Texture, Aroma and Taste, and Chemical testing includes Water Content, Carbohydrates, Proteins, Fats, and Ash Content. The results showed that the longer the freezing time sequentially from 1, 2 and 3 days the hedonic color values were higher and the carbohydrate content was also higher, but for other quality attributes such as texture, aroma and taste and other chemical components din't show difference significantly.

1. Introduction

Edamame is green soybean or also known as (*Glycine max* (L.) Merr.) harvested as a vegetable when the pods are immature, Such as a local soybean, the varieties of edamame has rich in protein and highly nutritious. Edamame is consumed especially as snack, but also as vegetable, an additive in soups, or proceessed into Edamame milks, As snack, the pods are soaked in boiling water with sort time, or cooked directly with boiling water and then the seeds are separated with peel [1].

Edamame is produce a sweet flavor when cooked. At the peak of maturity, aromatic concentrates have much higher levels of (Z)-3-hexenyl acetate, linalool, acetophenone, and cis-jasmone [1]. Several components of beany odor were also identified: hexanal, 1- hexanol, (E)-2-hexenal, 1-octen-3-ol, and 2- pentylfuran. Cis-jasmone is found in edamame only at the peak of ripeness and may be the key component of aroma.

For edamame, the two most important components of flavor are sweet and savory. Its sweet taste is determined by sucrose content and its savory taste probably by amino acids like glutamic acid [1] Beany flavor increases with maturity and can be divided into two components--beany and bitter. The beany taste may come from linolenic acid oxidized by lipoxygenase and the bitter taste my be the lipoxygenase itself.

One of the methods to increase the shelf life of fruits and vegetable are processing technology. Processing technology is improved the food diversity and reduced of the losses in post-harvest. One of the product developments of fruits processing and a lot of markets are chips. Fruit chips are more resistant into storage than fresh condition due to the lower water content of fruit chips and the physiological processes are terminated, on the contrast, where are still occurring in a fresh fruit. Fruit Processing methods into chips need to design with a novel methodology to produce the chips be accepted for the consumers. Several methods to produce healthy food without change the native structure used by vacuum frying. Edamame Processing into Edamame Chips or fried Edamame are used vacuum frying which provides the products crispy and persistent the color and taste of a edamame [2]. Vacuum frying is a frying process to vapor or drying of Edamame also other ingredients use boiling oil at low pressure. Nevertheless, is also be required alternative strategy to improve the quality attributes of Fried Edamame.

The freezing is an alternative strategy of pretreatment to make the fruit matrix are crispy on vacuum frying process. Shyu and Hwang [3] are reported freezing process at temperature -30 C during overnight will be formed the fruit matrix as like sponge or porous matrix of vacuum-fried apple. These conditions due to rapidly of the heat transfer on a matrix. Water crystals in cells are sublimated under vacuum conditions when released from a pore of the food matrix, this situation is also a decrease of the humidity and water contents. Albertos et al [4] are reported that vacuum-fried carrot has lower water contents with under freezing conditions during overnight than without freezing. Furthermore, Rate of freezing has also affected the fruits. Slightly freezing is produce a large size of the crystals, where can be damage of cells. Then, it can increase the penetration of oil into the cell during frying process. Thus, rapid freezing is preferred to reduce oil absorption. Freezing is also preserving raw materials before fried. During the lightly freezing process, the large water crystals can be damage cell membrane, then it caused the water were melted. However, fruits have different sensitivity about a frozen injury. The difference was caused by the performance of cell membrane to adapt or counter for changed of phased during different freezing of each fruit. Apricots, bananas, and peaches are sensitive, Such as apples, grapes and pears are quite vulnerable and dates are most resistant with freezing damage.

The principle of a freezing is a heat of material was taken and decreased to reach under freezing point of the material. So all mechanisms of change in materials were inhibited and the shelf life was extended. Generally, the freezing mechanism was divided in 3 stages. The first stage of sensible heat of food was taken that decreasing temperature to freezing. The second stage in a freezing process was transferred a number heat energy, where food and water are contained it was a freeze. And the last stage is a freezing process, where heat energy remains released so that decreases temperature to certainly a temperature.

One of the biggest edamame producers in Indonesia is jember district. The objective study in this research is to improvement sensory and chemistry of fried edamame. Previously fresh edamame treated by freezing process to get a good condition for fried edamame.

2. Working Methodology

2.1. Research Area

The research area was carried out in the Jember State Polytechnic campus, starting from the treatment process in the processing of all parts that were done at UPT Food and Beverages, then the Sensory Analysis and Test study was conducted at the Food Analysis Laboratory.

2.2. Data Collection Method

The research data was obtained from primary data which was the result of direct observation from experiments at the Laboratory, including chemical analysis data and sensory test data on samples observed. Previously the edamame samples treated and observed were prepared as follows: Fresh Edamame were collected, after that were washed and then freezing at -20 oC with duration of Freezing variations as 1,2, and 3 days. Each treatment is used for three replications. The equipment used includes Airblast Freezers, Cold storage, Analysis Tools including Kjeldahl Analysis, Soxhlet Analysis, Furnaces, Ovens, and various glass tools for analysis.

2.3. Analysis Method

Analysis methods for research include:

2.3.1. The observation method for obtaining primary data includes chemical analysis and sensory testing, namely: water content, Ash, Fat, Protein, Carbohydrate

1. Detemination water content

Initialy, Cruscible was dried at 100℃ -105℃ until 30 mnt, than it was transfered into desikator, this step were duplicate. Then 2 g vacuum-fried edamame transfered into cruscible, furthermore sample was oven-dried at 105℃ until 2 hours or weight stabilized, water content was calculated as follows :

$$(\%) \text{ water content (wb)} = \frac{(\text{weight sample - dry weight})}{\text{Weight sample}} \; x \; 100\% \qquad (1)$$

2. Detemination of Ash content

Initially, a crucible dried at 85 ^{o}C during overnight, and then transfered into desikator until a dish temperature equal with ambient temperature or weight stabilized (G). 2000 g sample (W) transfered into placed into ashing crucibles were calcinated in a furnace at 600 ^{o}C until 3 hours. Sample was burned, than placed into desiktor until there weight is stable (C). As content are calculate with followed by :

$$\text{Ash content} = \frac{\text{C} - \text{G}}{\text{W}} \; x \; 100\% \qquad (2)$$

3. Fat content [5]

7-10 g sample placed into paper were mounted and covered with a cotton. Than, sample transfered into soxhlet instrument, a fat sample are extracted until 6 hours, then the flash was dried into oven at temperature 105°C. a fat content calculate as follows :

$$\text{Fat content} = \frac{\text{W2} - \text{W1}}{\text{W}} \; x \; 100\% \qquad (3)$$

Where : W1 = weight flash soxhlet before extraction, W2 = weight flash soxhlet after extraction

4. Detemination of Protein [6]

Determination of protein content was carried out using Kjeldahl steam distilation, where Kjeldahl steam distilation consist with 3 step : frist step is destruction, second step is distillation, and last step is titration.

Destruction step, 1 g placed into a 100 mL Kjehdahl flask, then added 10 mL of sulfuric acid. Add a catalyst (selenium). Then Kjehdahl flask is heated with a small flame, after that changed the level of flame with maximum level. Finally, collected the green solution. Distillation step, green solution diluted with distilled water up to 100 mL, pipetted as 5 mL, tranfered into a distillation flask. Add 10 mL of 30% sodium hydroxide solution through the wall in the distillation flask,. Distillation flask is installed and connected to the condenser, then the condenser front is soaked in a liquid reservoir. Steam from boiling liquid will flow through the condenser to the erlenmeyer reservoir. Erlenmeyer added 10 mL of 0.1 N hydrochloric acid solution which has been dripped with methyl red indicator. Check with lakmus paper, if the propeties of the mixture no alkaline, then the distilation step is stopped. After the distillation process, the next step is titration. The result of distillation which is colected into erlemeyer was contained 0.1 N hydrochloric acid with dripped 5 drops of methyl red indicator, then titrated with 0.1 N. sodium hydroxide solution. The end point of the titration is marked with pink to yellow. Each sample is carried out 3 times.

5. Determination of carbohydrate
Carbohydrate calculate with different methods as follow :

Carbohidrat = 100 % - (water content + ash content + Fat content + Protein content) (4)

6. Sensoric Evalution [7]
Sensory Evalution include a preference test (hedonic) of edamame samples. In this Sensoric Evalution, the criteria are using a hedonic test with a scale 1 to 7, 1 = Most dislike, 2 = very dislike, 3 = dislike, 4 = ordinary/soft, 5= likes, 6 = very like, and 7 = Most like. are Consumer preference was carried out 20 panelists and the test included color, taste, aroma, and texture.

2.3.2 Statistical Analysis Methods
The experiment was conducted as a completely randomized design, the tree experimental conditions were carried out with three replicates. Data were analyzed by analysis of variance (ANOVA) and means comparisons were performed according to the Tukey test. Where, a dependent variable or a response this experiments are Consumer preference and sensory evaluation, color, texture analysis, water content, carbohydrate content, Protein total, Fat conten, Ash content. Mathematic model for anlysis this experiment followed by this equation [8]:

$$\mathbf{Yij = \mu + \sigma_i \, \varepsilon ij} \quad (5)$$

Yij	=	response influence Freezing level at replication i to j
μ	=	Means value
σ_i	=	Long-term effect of freezing level
Eij	=	error value from i to j

3. Results and Discussion
The result of research as following a preference test (hedonic) components:

3.1. Colour
The results of analysis of variance in the freezing factor had a very significant effect on the colour of edamame, in Figure 1 presented the effect of freezing on the colour value of edamame with a range of 3, 65 to 4.8.

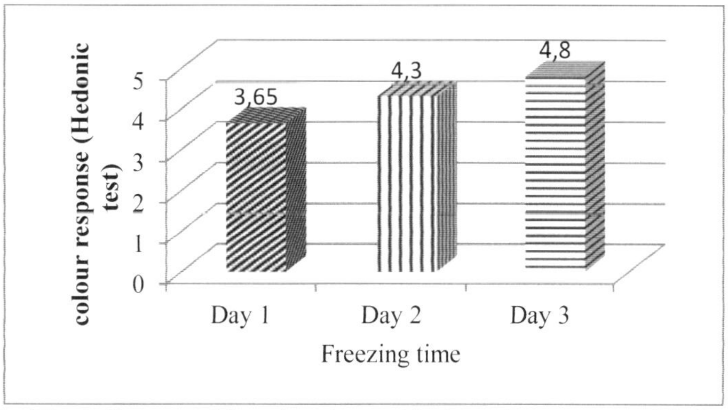

Figure 1. The influence of freezing time with the colour vacuum-fried edamame.

Based on Figure 1, the influence duration of Freezing on the hedonic level with a colour response is shown tend to be favoured. The result shows the longer duration of freezing produce the colour of fried edamame products have been brightly. Probably of freezing times variation are still influenced with an activity of enzymatic browning reaction, where the length duration of freezing up to 3 days showed that suppression process occurs for enzymatic browning reaction. Thus, fried edamame like as brighter.

3. 2 Aroma
Range of hedonic aroma values are 3.7 to 4.15 is like medium criteria. The results of the analysis of variance showed no significant difference, which are means that the duration of freezing does not affect an aroma of edamame.

3.3 Taste
The hedonic result of taste values have ranged from 3.8 to 4.6 is like medium criteria. The results of an analysis of variance showed no significant difference which means that the freezing treatment does not significantly affect the taste of edamame.

3.4 Texture
The preference test has shown the range of texture is 4.5 to 5.1 with criteria is Favored. The results of the analysis of variance showed no significant difference, which means that the duration of freezing did not significantly affect the edamame texture.

3.5 Water content
A moisture content of vacuum-fried edamame showed a range of 3.11% to 3.49%. The results of variance analysis showed that the duration of freezing did not significantly affect the water content.

3.6 Carbohydrate content
The result of Analysis of variance for the duration of freezing had a very significant effect to carbohydrates content are shown on Figure 2 with a range of 32.44% to 32.63.

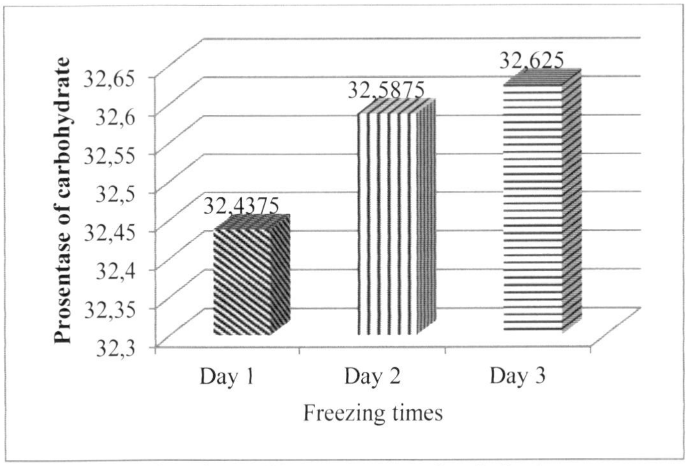

Figure 2. The influence of freezing time with persentase of carbohydrate on vacuum-fried Edamame.

In Figure 2 shows that the longest freezing is correlated with a highest of carbohydrate content, then hydrolysis process of starch is inhibited with a freezing process. Where is also known of a the hydrolysis process was initiated with enzymatic activity. Freezing process depressing effect of hydrolysed lactose and carbohydrates content increased.

3.7 Protein Content
The results of protein levels showed a range of 39.09% to 39.16%. The results of variance analysis show that the the duration of freezing did not significantly affect edamame protein levels.

3.8 Ash Content

The results of Ash content showed a range of 3,66 % to 3,75 %. The results of variance analysis show that the duration of freezing did not significantly affect edamame ash content.

3.9 Fat Content

The results of Fat content showed a range of 3,66 % to 3,75 %. The results of variance analysis show that the duration of freezing did not significantly affect edamame Fat content.

4. Conclusion

The results are shown that the influence of duration of freezing at day 1,2, and 3, to hedonic tests such as colour values and carbohydrate content was a highest, on contrast, freezing time showed no significant effect on other quality attributes such as texture, aroma and taste and chemical components are moisture, protein, fat and ash content of fried edamame.

Acknowledgments

Our thank goes to Department of Agricultural Technology, Politeknik Negeri Jember, who has helped support for this research.

5. References

[1] Konovsky J, Lumpkin T A and McClary D 1994 *Edamame: The Vegetable Soybean*, Washington State University

[2] Ayustaningwarno F, Dekker M V, Fogliano1 and Verkerk R 2018 Effect of Vacuum Frying on Quality Attributes of Fruits, *Food Engineering Reviews* ResearchGate https://doi.org/10.1007/s12393-018-9178-x.

[3] Shyu S L and Hwang L S 2001 Effects of processing conditions on the quality of vacuum fried apple chips. *Food Res Int* **34**:133–142

[4] Albertos I, Martin-Diana A B, Sanz M A, Barat J M, Diez A M, Jaime I, Rico D 2016 Effect of high pressure processing or freezing technologies as pretreatment in vacuum fried carrot snacks. *Innovative Food Sci Emerg Technol* **33**:115–122.

[5] Amriani H 2017 Verifikasi Metode Uji Lemak Pakan Buatan; Method Verification of Artificial Leed Lipid Analysis; Fishtech *Jurnal Teknologi Hasil Perikanan* ISSN: 2302-6936

[6] Zaim Z, Sayın TA, Sayın K, Karakaş D 2015 *Turkish Computational and Theoretical Chemistry* **2** 7-15.

[7] Ogles C 2016 Evaluation of Selected Edamame Cultivars for Isoflavone Content, Organoleptic Characteristics, and Production in Central Alabama. Crop Soils and Environmental Sciences. Auburn University. USA.

[8] Cheng X-F, Zhang M, Adhikari B 2015 The principles of ultrasound and its application in freezing related processes of food materials: A review. *Ultrasonics sonochemistry* 27 576-585.

Utilization of different nitrogen sources for the growth of microalgae isolated from mangrove leaves in Banda Aceh - Indonesia

S H Anwar[1*], S Harzaki[2], M I Sulaiman[1], T Rinanda[3]

[1]Department of Agricultural Product Technology, University of Syiah Kuala, Banda Aceh 23111, Indonesia
[2]Alumni of Agricultural Product Technology Department, University of Syiah Kuala, Banda Aceh 23111, Indonesia
[3]Faculty of Medicine, University of Syiah Kuala, Banda Aceh 23111, Indonesia

[*]E-mail: sri.haryani@unsyiah.ac.id

Abstract. The aim of this research was to investigate utilization of different nitrogen sources used for the growth of microalgae and determined the optimum period of the growth. The microalgae had been isolated previously from mangrove leaves in Banda Aceh, Indonesia and has been identified morphologically. Three different nitrogen sources i.e., yeast extract, monosodium glutamate or MSG and sodium nitrate were used. The biomass production was measured by optical density using spectrophotometer. The results showed that MSG stimulated microalgae to produce the highest amount of biomass (0.38 g/L/day) followed by yeast extract with 0.33 g/L/day, but both were statistically not significant. Sodium nitrate developed the lowest OD value with the amount of dried biomass only 0,16 g/L/day. Correlation analysis indicated the yield of biomass was positively related, r = 0.953, with the optical density of the cultures. The OD values of yeast extract, MSG and sodium nitrate cultures were 1.20, 1.18 and 0.82, respectively. The optimum time for microalgae growth was at day 9 in which 4.41 g/L of dried biomass was accumulated in the culture where MSG was used. The maximum yield of dried biomass for yeast extract was 4.22 g/L followed by 2.58 g/L for sodium nitrate.

1. Introduction

Previous researches indicated that polyunsaturated fatty acids (PUFAs) have positive benefits to human health. PUFAs are effective to prevent diabetes during pregnancy, helping development of infant, lowering cholesterol and blood pressure as well as reducing joints inflammation and paint in rheumatoid arthritis [1]. Moreover, both important polyunsaturated fatty acids such as Eicosapentaenoic Acid and Docosahexaenoic Acid are actually precursors of some important compounds such as hormones. In human body, these compounds are needed for many essential biological processes [2].

Omega-3 fatty acids have attracted intensive studies since early 1900s and the term "Vitamin F" has been created for these fatty acids [3]. The most important PUFAs are ALA (α-linolenic acid, C18:3n-3), EPA (eicosapentaenoic acid, C20:5n-3) and DHA (docosahexaenoic acid, C22:6n-3) [4]. Many investigations related to PUFAs enrichment into food products are challenged by the sensitivity of PUFAs toward oxidation and the fishy odour because most of PUFAs are extracted from seafood, particularly cold-water fish. In addition to these problems, overfishing, mercury contamination issue and fish sustainability have added the complexity of fish as the main source of omega-3 fatty acids [3].

Based on the facts that incorporation of fish PUFAs is difficult, alternative sources for omega-3 fatty acids must be found. Production of fatty acids in fish is actually originated from the microorganisms as the fish feed. Some microorganisms such as yeast, fungi and microalgae produce fatty acids in their cells. Some experts describe this alternative as microbial oils or Single Cell Oils (SCO) [5]. Microalgae that produce edible oil and the oil contains essential fatty acids as well as antioxidant (carotenoids and tocopherols) are *Thraustochytrium, Ulkenia and Schizochytrium* [6]. Moreover, thraustochytrids microalgae such as *Crypthecodinium cohnii, Botryococcus braunii,* and *Schizochytrium* sp. are informed to produce 20%, 25-75% and 50-77% oil, respectively [7].

Due to its flexibility for food application, investigation related to microbial oils particularly from microalgae has increases considerably. Studies to isolate microalgae and furthermore to identify them genetically are multiply every year worldwide. The outcomes were definite according to the area and environment (temperature and weather) where the microorganisms were isolated. In 2010, similar study had conducted in Indonesia to investigate production of PUFAs from microalgae by using glycerol. The authors claimed that seven strains of microalgae were able to produce docosahexaenoic acid in the range of 2.49-9.14 g/L biomass produced [8]. Previous research related to microalgae has also been done in Aceh Province. Two microalgae strains from mangrove area have been identified as potential PUFAs producers, morphologically. This investigation is still on going to identify these strains genetically [9].

Exploration to extract biodiesel from microorganisms are quite many but investigation to find new sources of omega-3 fatty acids particularly from microalgae are rare. Therefore this research was aimed to study the possibility of microalgae cultivation in laboratory scale from species that were isolated previously from mangrove leaves in Aceh Province, Indonesia. Two important factors investigated were nitrogen sources and the length of microalgae growth. In this research, three different nitrogen sources i.e., yeast extract, monosodium glutamate (MSG) and sodium nitrate were used for the growth of microalgae. The biomass produced were collected, dried and weight gravimetrically. The highest biomass production was an indicator for the optimum growth and measured by optical density using spectrophotometer.

2. Materials and methods

2.1. Materials

The materials to compose a so called YPGS media were yeast extract-peptone-glucose-seawater which then mixed in natural sea water (consist of yeast extract (Merck), mycological peptone (Himedia), glucose (Himedia), and bacto agar (Difco)). Other materials needed were ammonium nitrate (Merck), monosodium glutamate, antibiotics (penicillin and streptomycin), lugol staining and distilled water.

2.2. Methods

The experiments were divided into two stages: 1) the preliminary research, investigated the optimum growth of microalgae in liquid media to find the time (day) for microalgae to produce the highest biomass and 2) the main research, optimization of microalgae growth based on different nitrogen sources. Nitrogen sources tested were yeast extract, sodium nitrate and monosodium glutamate.

2.2.1. Inoculum preparation. Method for inoculum preparation is following the procedure developed previously [10] with modification. Briefly, one mL of the zoospore suspension of microalgae from previous investigation was transferred into 50 mL aliquot of liquid media in a 250 mL flask. The liquid media (YPGS media) was prepared in 1L natural seawater at a salinity of 15‰. The flask was covered with cotton wool and finally wrapped with aluminum foil. The mixture was then shaken at 150 rpm for

48 hours (at 25°C). The initial inoculum used was a 20 mL inoculum of the 48-hours old culture broth and was used for all subsequent growing time and nitrogen sources experiments.

2.2.2. Growth and nitrogen source experiments. The YPGS media was prepared like previously mentioned. The pH was adjusted to 6.0 using 2 N HCl. The culture flasks were shaken at 150 rpm at 25°C for 14-days period. Triplicate flasks harvested at 3 days intervals for biomass determination [10]. For finding the best nitrogen source, peptone was replaced by sodium nitrate ($NaNO_3$), yeast extract and MSG ($C_5H_8NNaO_4$).

2.2.3. Analysis of cell growth. Measurement of microalgal biomass was carried out by measuring the turbidity of microalgae suspension [11]. The culture medium (1.5 ml) was collected from each flask of growth from nitrogen source experiments every day during the incubation period (14 days) and measured the absorbance at 600 nm. The OD value accepted was $\leq$ 0.4 because if the OD value was too high, then it did not correlate well anymore with the cell densities. The values that were $\geq$ 0.4 (OD_{undil}) must be diluted at first and the absorbances were red as OD_{dil}. The value of OD_{dil} was then divided by its dilution factor and written as OD_{corr}. The final OD of microalgae cells' densities was recorded as the mean of the OD_{corr}.

2.3. Data analysis
All measurements were carried out in duplicate and from these results mean values and standard deviation are calculated. The quantification of microalgae oil was calculated gravimetrically.

3. Results and discussion

3.1. Preliminary research

3.1.1. Dried biomass. Determination of dried biomass was aimed to quantify the amount of biomass yielded from the growth of microalgae. The growth curve was built based on the biomass yield since the first day of growth until the 14th day of incubation time. The optimum growth of microalgae was the time when the biomass produced was at the highest amount. The result of the biomass achieved can be seen in Figure 1.

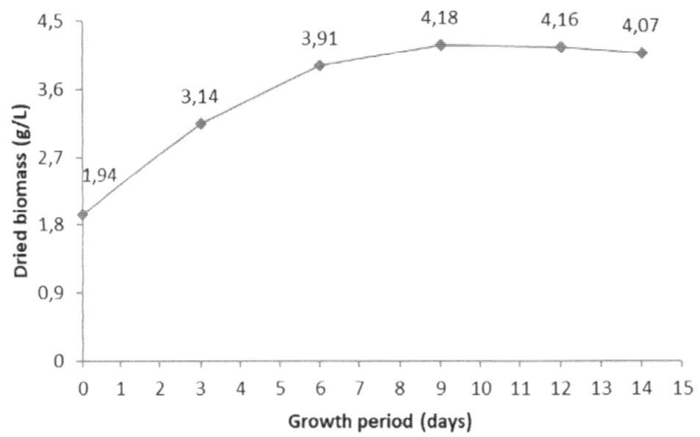

Figure 1. The amount of biomass produced during incubation time.

The highest amount of biomass produced by microalgae was at day 9 (4.18 g/L). At day 12, dried biomass started to decrease to become 4.16 g/L dan this declining continued until day 14. The exponential phase was begun on day 3 up until day 9, followed by the stationary phase and finally the falling growth period. The exponential phase is the phase where the biomass yield is increasing due to the rising of photosynthesis activities. The growing number of microalgae cells with maximum and constant cells' division are two important factors that determine the exponential phase [9].

3.1.2. Cultures' optical density. In order to confirm the microalgae growth rate based on the biomass dry weight, it is important to measure the culture optical density. This measurement is actually to quantify the cells density during cultivation. In this study, the sample was taken every three day from the culture broth and the cells density was measured using Shimadzu UV-1700 spectrophotometer at wave length of 600 nm (OD600). The results can be seen in Figure 2 which showed that the highest cell density was achieved at day 9 of cultivation which was 27.58. The densities were declined afterward at day 12 and 14 which only accumulated for 22.80 and 22.02 respectively. However, these results were in agreements with the outcomes of dried biomass quantification.

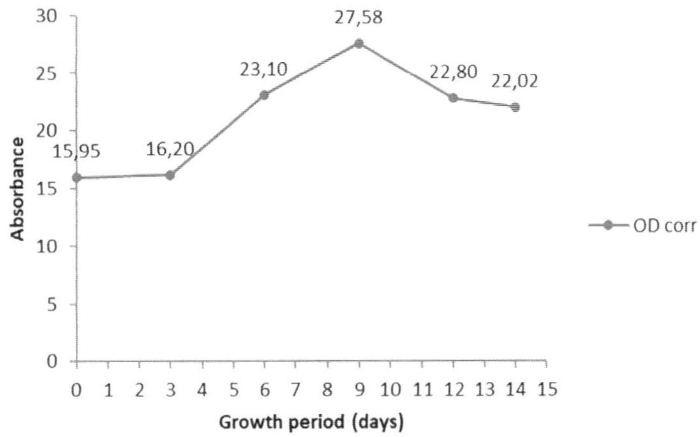

Figure 2. The microalgae cell density during cultivation

3.2. The main research

3.2.1. Microalgae morphology. Microalgae used in this research was isolated from mangrove forest in Banda Aceh. The colonies were homogeneous and had been identified morphologically and the isolate is belong to *Thraustochytriaceae* (unpublished result). The colonies are round in shape and yellowish white in color. The colonies were growth densely and the surface is smooth but is not slippery.

3.2.2. Dried biomass. Quantification of microalgae dried biomass was conducted every three days during 9 days of growing time. The usage of different nitrogen sources during cultivation resulted different amount of dried biomass, as can be seen from Figure 3. Yeast extract and MSG yielded higher amount of dried biomass compared to sodium nitrate. At day 9, dried biomass produced with yeast as nitrogen source was 4.22 g/L, while when MSG and sodium nitrate as nitrogen sources used resulted 4.41 g/L and 2.58 g/L dried biomass, respectively. These values were lower than the results published elsewhere [12] in which yeast extract produced 6.108 g/L, MSG yielded 6.585 g/L and 3.15 g/L of dried biomass if sodium nitrate was used [11].

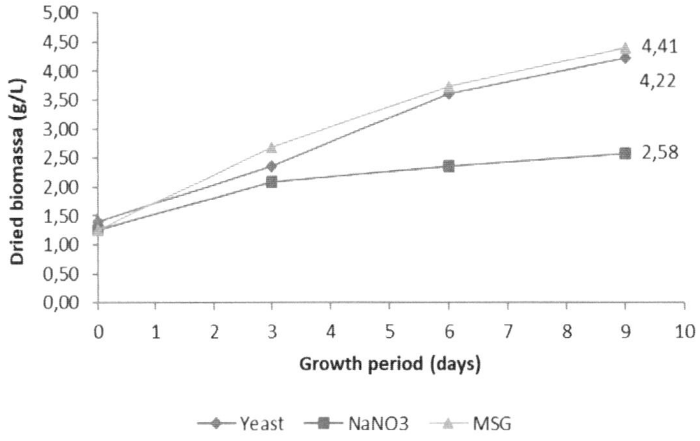

Figure 3. The yield of dried biomass produced by microalgae with three nitrogen sources.

Higher yield of dried biomass if MSG was used as nitrogen source can be caused by the ability of microalgae to absorb MSG was faster than to absorb yeast or sodium nitrate. Microalgae needs nitrogen to form chlorophyll [11]. When the nitrogen decreases to the lowest level, microalgae start to convert this rich nitrogen substance into protein, nucleic acids, the cells' wall and other important substances for the cells' growth. Therefore, the faster the nitrogen absorb, the better the growth of microalgae.

MSG and sodium nitrate are two sources of simple nitrogen which can be easily utilized by microalgae for its growth [12]. Complex nitrogen such as yeast extract also provides protein, amino acids, sugar, fatty acids and vitamins beside its main function as nitrogen source. The influence of nitrogen source to growth of microalgae is varied among the species. For *Thraustochytrid* in particular, MSG and yeast extract are the best nitrogen sources for microalgae to produce high yield of dried biomass [13]. Lipid production in microalgae cells is corelated well with nutrient availability particularly nitrogen and glucose which contained in medium growth. Figure 4 explains how do the cells stimulated to produce lipid [14].

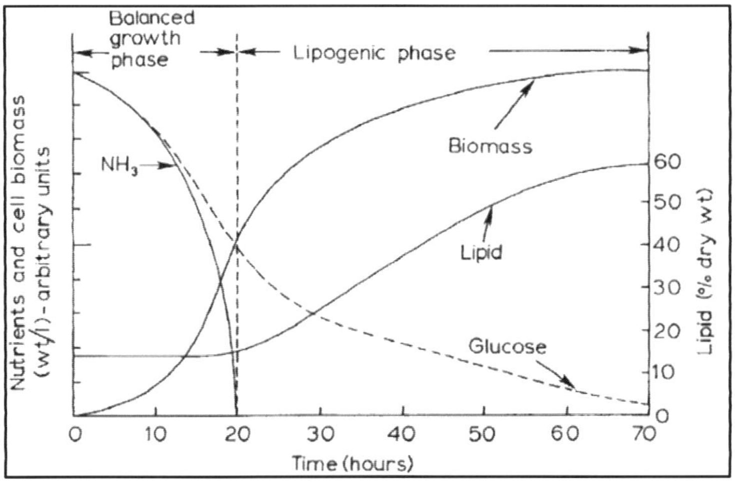

Figure 4.Nutrient availability during microalgae cultivation [14]

Figure 4 shows the relationship between nutrient availability in culture media with formation of biomass and lipid produced during cultivation time (hours). Nitrogen is absorbed fast and diminished in the first 20 hours followed by glucose in 70 hours during cultivation. Biomass and lipid production increase significantly after 20 hours and lipid accumulation in the cells start to begin when nitrogen (commonly in form of ammonium salt) in cultivation media is finished. At the same time, carbon supply (in the form of glucose) start to decrease. The nutrients depletion causes the cells growth and division to discontinue although the cells are able to absorb glucose. These conditions accelerate lipid biosynthesis in microalgae cells which is actually the conversion of glucose into lipid. When the external nitrogen in media was ceased then biomass concentration increases and nitrogen rich substances converted into other important substances such as protein, nucleic acids and cells wall which can help the growth process [11].

3.2.3. Cultures' optical density. The optical density (OD) of the culture is measured to predict the metabolic activity and the growth of cells microorganism. The principle of OD measurement by spectrophotometer is based on light absorption or radiation energy by a solution or liquid in a cuvette. The greater solution concentration in a sample, the higher the absorbance reading will be.

Measurement of optical density from microalgae cultures varying in nitrogen source resulted the density in the range of 0.82 – 1.20. Statistical analysis indicated that optical densities were influenced by different nitrogen source used, significantly ($P \leq 0{,}05$). As can be seen from Figure 5, yeast extract resulted the highest OD value (1.20) which was significantly different with OD value of sodium nitrate (0.82). The OD values were directly proportional to the biomass yielded. Yeast extract and monosodium glutamate developed the highest OD values as well as the amount of dried biomass. These values consistently higher than that produced by cultures which used sodium nitrate as nitrogen source.

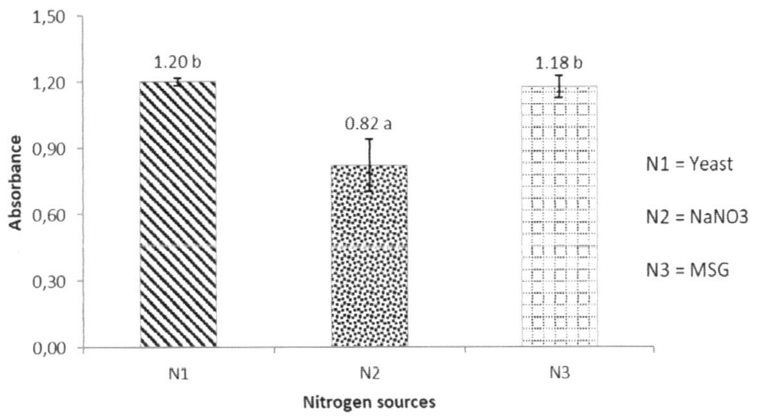

Figure 5. Influence of nitrogen sources to the cultures optical densities

3.2.4. Correlation analysis between dried biomass and cultures optical density. Analysis of correlation was performed to study the degree of linear relationship between two or more variables. In this research, the analysis was conducted for dried biomass and optical density of the culture from microalgae cultivation. The relationship value achieved was 0.953 which mean that the correlation is positive with high degree of relationship. This also means that increasing the amount of dried biomass give rise to higher optical density of the culture which was in accordance with Figure 6. Figure 6 demonstrates comparison of dried biomass with optical density of cultures from three nitrogen sources used in this research. The trend is similar where intensifying the density of cultures optically giving impact to accumulation of dried biomass.

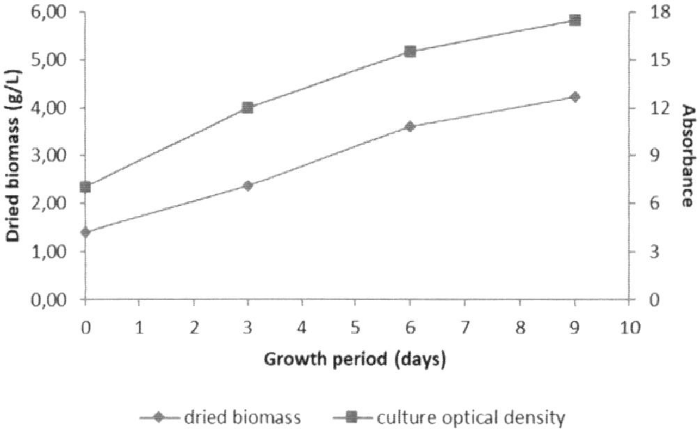

(a)

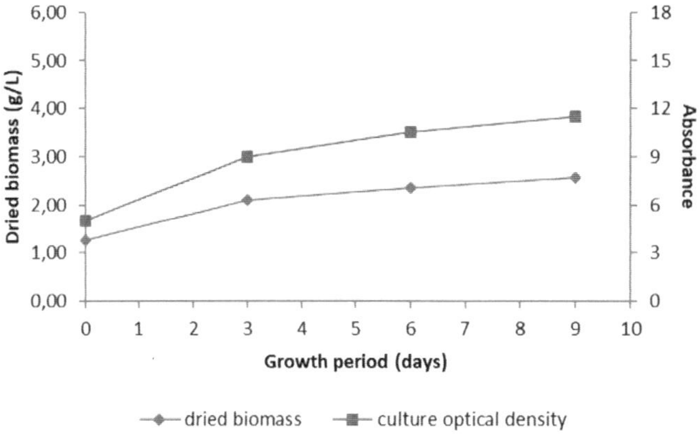

(b)

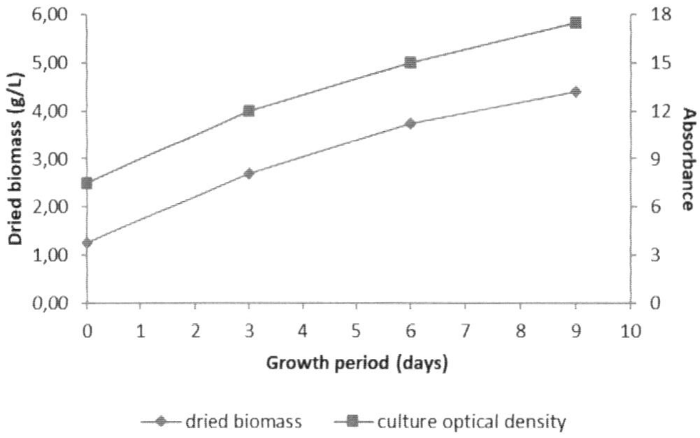

(c)

Figure 6. Comparison of dried biomass yield and cultures optical density from different nitrogen sources: (a) yeast extract, (b) sodium nitrate and (c) monosodium glutamate.

4. Summary
Nitrogen sources affected the growth of microalgae cultures. Monosodium glutamate stimulated microalgae to produce the highest amount of biomass during cultivation as proved by yield of dried biomass (0.38 g/L/day). Yeast extract gave results slightly differ with MSG but statistically not significant (0.33 g/L/day). Sodium nitrate on the other hand, developed the lowest OD value and the amount of dried biomass (0,16 g/L/day). Correlation analysis indicated that the yield of biomass was positively related with the optical density of the cultures ($r = 0.953$). The yeast extract culture had the highest OD value of 1.20 which was significantly not different with the OD value of MSG culture (1.18). Sodium nitrate had the lowest OD value which was only 0.82. The optimum time for microalgae growth was at day 9 in which 4.41 g/L of dried biomass was accumulated in the culture where MSG was used. The maximum yield of dried biomass when yeast extract was as nitrogen source was 4.22 g/L and followed by 2.58 g/L of dried biomass produced by culture added with sodium nitrate.

Acknowledgement
The authors thank University of Syiah Kuala for funding this research and also special appreciation to Muhammad Birri, Armanisah, Yusmaini and Rima Ristianti for their valuable laboratory assistances.

References
[1] McMurray I 2007 *Int. Food Ingred.* **2** 23–6
[2] Trautwein E 2001 *Eur. J. Lipid Sci. Technol.* **103** 45–55
[3] Cannon D 2009 *Explore J. Sci. Heal.* **5** 299–303
[4] Augustin M A and Sanguansri L 2003 *Food Aust.* **55** 294–6
[5] Ratledge C 2010 Single cell oils for the 21st century *Single cell oils, microbial and algal oils* ed Z Cohen and C Ratledge (Urbana: AOCS Press) pp 3–26
[6] Armenta R E and Valentine M C 2013 *J. Am. Oil Chem. Soc.* **90** 167–82
[7] Singh A, Nigam P S and Murphy J D 2011 *Bioresour. Technol.* **102** 26–34
[8] Basuki W 2011 *Pengembangan Teknologi Produksi Poly Unsaturated Fatty Acid (PUFA) dan Gliserol Hasil Samping Biodisel Crude Palm Oil (CPO) Secara Fermentasi Menggunakan Mikroalga* (Jakarta: Badan Penerapan dan Pengkajian Teknologi.)
[9] Haryani S, Sulaiman M I, Rohaya S and Safriani N 2013 *Isolasi dan Identifikasi Mikroalga Penghasil Asam Lemak Omega-3 di Perairan Aceh* (Banda Aceh: Laporan Penelitian Unggulan Daerah Universitas Syiah Kuala)
[10] Fan K W, Vrijmoed L L P and Jones E B G 2002 *Bot. Mar.* **45**
[11] Li Y, Horsman M, Wang B, Wu N and Lan C Q 2008 *Appl. Microbiol. Biotechnol.* **81** 629–36
[12] Wu S-T, Yu S-T and Lin L-P 2005 *Process Biochem.* **40** 3103–8
[13] Chen G, Fan K-W, Lu F-P, Li Q, Aki T, Chen F and Jiang Y 2010 *New Biotechnol.* **27** 382–9
[14] Wynn J P and Ratledge C 2005 Oils from Microorganisms *Bailey's Industrial Oil and Fat Products, Part 3. Edible Oil & Fat Products: Specialty Oils and Oil Products* (New York: Wiley-Interscience)

IOP Conf. Series: Earth and Environmental Science **207** (2018) 012050 doi:10.1088/1755-1315/207/1/012050

Preparation and characterization of biodegradable film based on skin and bone fish gelatin

N Arpi[1*], Fahrizal[1], Martunis[1] and E Hardianti[1]

[1]Agricultural Product Technology Department, Faculty of Agriculture, Syiah Kuala University, Jl. Krueng Hasan Kalee 3, Kopelma Darussalam, Banda Aceh, 23111 Indonesia

*Corresponding author's e-mail: normalina.arpi@unsyiah.ac.id

Abstract. Biodegradable films can act as excellent oxygen, water, and lipid barrier. Gelatin is utilized extensively as a gelling agent, emulsifier, stabilizer, adhesives, viscosity agent, binder agent and film-forming agent. Fish gelatin as a film-forming agent has unique characteristics but varies depending on fish species. The study aims to characterize biodegradable film made using skin and bone fish gelatin with the addition of plasticizer. Gelatin of ocean fish, spotted oceanic triggerfish *(Canthidermis maculata)*, and freshwater fish, tilapia *(Oreochromis niloticus)* were used. As a plasticizer, glycerol or sorbitol with concentrations of 0.25 and 0.5 % was added. Gelatin of spotted oceanic triggerfish resulted in film with higher tensile strength, whereas tilapia gelatin with glycerol produced biodegradable film with higher elongation and water vapor permeability. The best characteristics of biodegradable film made from gelatin of spotted oceanic triggerfish with 0.25% sorbitol.

1. Introduction

Biodegradable films have obtained attention because of their advantages over traditional synthetic films. Biodegradable films use renewable ingredients which are environmentally friendly. The films can act as barriers to gas, moisture, solute movement, aroma, and also can function as carriers of food antimicrobial and antioxidant agents [1]. Nevertheless, biodegradable films have some drawbacks, such as low physical and mechanical properties, lack of thermal resistance, and high cost compared to synthetic plastic films. Improving and developing excellent biodegradable films that can compete against plastic films has become research attention. Using waste product as film raw materials, not only benefits from less expensive by-products, but also reduces environmental pollution.

Gelatin as bio-macromolecule protein polymer has characteristics of colorless, translucent, brittle, edible, and flavorless. Gelatin is extracted from collagen tissue found in skin, bones, and connective tissue of various animal and fish by thermal hydrolysis [2, 3]. Commercial gelatin extracted mostly from porcine and bovine skin and bone, by-products of meat industries. Gelatin is one of functional biopolymers, owns good film-forming ability with good carbon dioxide and oxygen barrier properties, transparent, water-soluble, and highly extensible film [4, 5, 6]. Fish gelatin has gained increasing demand due to many socio-cultural, religious, and environmental reasons. Nevertheless, in general, fish gelatin films have lower mechanical properties than mammalian gelatin, lower oxygen permeability, and lower water vapor permeability [6, 7]. Moreover, fish gelatin properties vary depending on fish species, fish living environment, and gelatin extraction procedures.

The mechanical properties of gelatin films have been increased by using various plasticizers. The properties improved due to the decreasing of protein chains intermolecular forces. The reduced forces support the construction of strong covalent bonds in the protein network of the film or lead to the reduction of hydrophilic characteristic. Plasticizers intensify the flexibility of films and reduce the rigidity, and may assist in obtaining desirable mechanical properties [3, 8].

The aim of this research was to analyze the effects of different fish gelatin sources, and plasticizers on some properties of fish gelatin biodegradable film. Oceanic triggerfish *(Canthidermis maculata)* which is epipelagic fish inhabit tropical/subtropical oceans, and fresh water fish, tilapia *(Oreochromis niloticus)* were used in the study. Sorbitol and glycerol were used as plasticizers in an attempt to improve the characteristics of biodegradable film.

2. Method

2.1. Preparation of material

Fresh tilapia *(Oreochromis niloticus)* and triggerfish *(Canthidermis maculata)* skin and bone were obtained from small fish processing industry in Banda Aceh, Indonesia. The fish skin was cleaned from meat remnants with a knife and then washed and packed in plastic bags and stored at -24°C until further use. Frozen fish skin and bone were cut into small pieces (about 2-3 cm^2). Then, the skin was thawed overnight in a refrigerator at around 4°C and used for extraction. All chemicals used were analytical grade.

2.2. Fish gelatin extraction

Fish gelatin was extracted based on [9, 10] with some adaptations. The solution of 0.55N NaOH (1: 5 w/v) was used to soak fish skin and bone of triggerfish, whereas the tilapia was soaked in 0.77N NaOH for 1 hour. Then, for three times, using water (1: 5 w/v), the samples were washed, strained and then pressed. Subsequently, the triggerfish samples were submerged in 1.05 N HCl (1: 5 w / v) for 3 hours, and the tilapia sample in 0.59 N HCl for 1 hour. The samples were then washed three times with water (1: 5 w/v), strained, and pressed again. Next, distilled water (1: 4 w/v) were mixed with the samples in glass beakers and capped with aluminum foil, then extracted at 60°C for triggerfish and 66.8°C for tilapia in 3 hours by using a water bath. The samples were filtered using cheesecloth and the filtrate was dried in glass containers at 60°C for 72 hours.

2.3. Preparation of biodegradable film

Preparation of biodegradable film was carried out following [11, 12] procedures with some modifications. Gelatin of 3 g was dissolved in 100 ml of water and then stirred until homogeneous. Plasticizers (glycerol and sorbitol) were added with concentrations of 0.25% and 0.50% of the 100 ml water. The film solution then heated on a hot plate at a temperature of 45-50°C for 30 minutes, then added 0.1% corn oil and stirred for 5 minutes to obtain a good blend. Then 30 ml of the film solution was cast on rectangular teflon plates with a size of 13 cm x 18 cm and dried in an oven at 45°C for 15 hours. The film produced was released from teflon then stored for later analysis.

2.4. Film analysis and statistical analysis

The film thickness of 5 randomly selected spots on the film were measured by using a micrometer with 0.01 mm accuracy. Elongation and tensile strength at the breaking point were measured following the D882ASTM standard and by using Exceed Model E43 texture analyzer. A rectangular film of 4 cm x 1 cm was mounted between the two clamps of the machine at a specific initial distance, and the clamps moved at the rate of 10mm/min. Water vapor permeability (WVP) was measured following [13]. Statistical analysis was performed using Analysis of variance (ANOVA) to identify the effects of fish gelatin and plasticizer types. The least significance differences (LSD) was used to identify the differences.

3. Results and Discussion

3.1. Film thickness

The thickness of the fish gelatin biodegradable films was 0.10-0.15 mm with an average of 0.13 mm. The film was thicker compared to the film of [8]. As in this research, using the same amount of 3 grams of gelatin in 100 ml water, their film thickness of fish skin gelatin containing clay nanoparticles were 60-70 micrometer. The film thickness is different allegedly due to differences in the amount of film solution used to produce the same film area. The findings of [14] show that barley bran protein-gelatin composite film thickness increased from 0.05 to 0.09 mm as gelatin content was increased from 1 to 4%. Film thickness went up along with the increase in the amount of solid in the film solution.

3.2. Tensile strength

Biodegradable film from triggerfish gelatin has higher tensile strength with a value of 2.34 MPa compared to tilapia gelatin film which is of 1.72 MPa (Figure 3.1). Atmaka et al. (2018) [15] produced a film using tilapia bone gelatin also show a low tensile strength of 0.58 MPa. Depending on the waters where fish live, the imino acid content of the fish gelatin may vary. Cold water fish derive gelatin which has lower imino acid content compared to warm water fish. Gelatin with lower imino acids of proline and hydroxyproline content possess low gel modulus and lower gelling temperature [7]. The triggerfish (*Canthidermis maculata*) is epipelagic fish inhabit tropical and subtropical oceans, often occur at an offshore island. Whereas tilapia, considered as warm water fish, is freshwater fish inhabiting rivers, lakes, ponds, and shallow streams, and less commonly found living in brackish water. Sutono and Pranoto (2013) [11] using a higher amount of black tilapia gelatin (8 grams vs 3 grams in this study) with the addition of 0.8% glycerol shows a film with a higher tensile strength of 3.08 MPa which presumably due to higher gelatin concentration used. Much higher film tensile strength of 17.3 MPa reported by Jiang et al. (2010) [16] using catfish gelatin. Lee et al. (2016) [17] also reported that film made of starfish gelatin had a high tensile strength of 19.61 MPa which increased more with the addition of vanillin.

Biodegradable fish gelatin film containing glycerol had lower tensile strength compared to the film that had sorbitol as a plasticizer (Figure 2). The results are in accordance with the findings of [8] who reported the similar behavior of the plasticizer in the gelatin film containing clay nanoparticles. The low tensile strength of films using glycerol may be due to the rise of glycerol matrix lubrication and lateral branches fluidity. Figure 2 shows that increasing concentration of glycerol to the fish gelatin biodegradable film caused an increase in tensile strength. In contrast with sorbitol, as the concentration increased, the tensile strength of the film decreased. [8] found that increasing both plasticizers, glycerol and also sorbitol, reduced the tensile strength of fish gelatin containing clay nanoparticles film. The decrease in tensile strength of the fish gelatin film may due to the crippling of gelatin structure by weakening the bonds through coming in between the lateral branches.

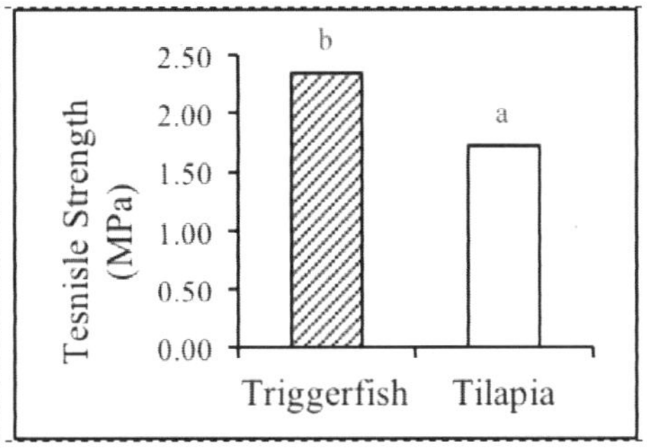

Figure 1. The tensile strength of fish gelatin biodegradable film based on the type of fish.

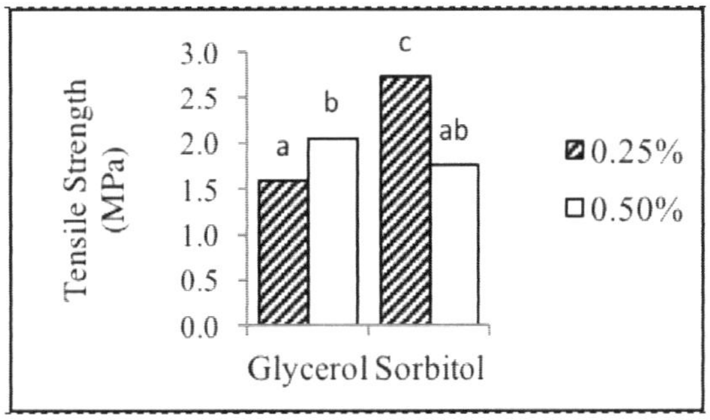

Figure 2. The tensile strength of fish gelatin biodegradable film as affected by the plasticizer.

3.3 Elongation

Figure 3 shows fish gelatin biodegradable film using glycerol had higher percent elongation than that using sorbitol. For both types of fish (triggerfish and tilapia) gelatin films, increased in glycerol concentration led to the increase of percent elongation. However, when sorbitol was used, mixed results were found, increasing sorbitol concentration induced a reduction in the elongation of tilapia gelatin film, in contrast, it caused a rise in the elongation of triggerfish gelatin film. The latter finding is in line with the results of Song et al. 2012 [14], using sorbitol in the porcine gelatin composite film. Plasticizers decrease intermolecular strengths and increase the movement of biopolymer chains which resulted in a decrease of brittleness and the increase of flexibility and strength of film [14]. Chemical and physical properties of gelatin vary with gelatin origin whereas, for fish species, the differences of derived gelatin are more significant [6]. The strength of the gel in gelatin affects the tensile strength of the edible film, the higher the strength of the gel, the higher the elongation of the edible film increases and the lower the tensile strength decreases.

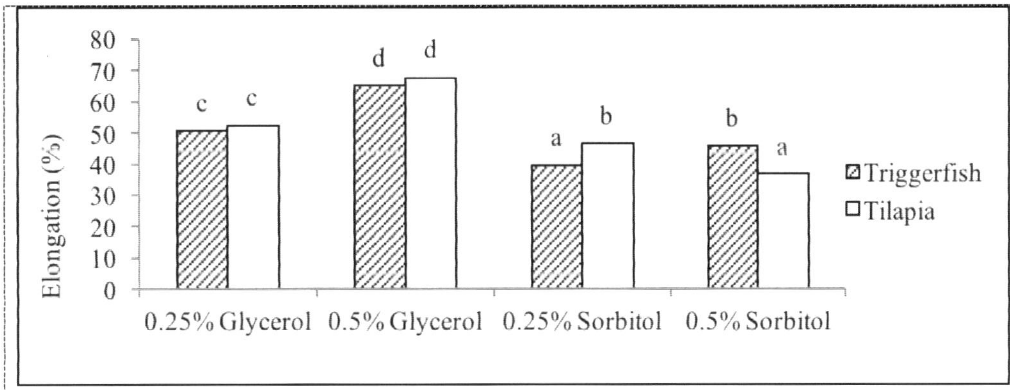

Figure 3. Percent elongation of fish gelatin biodegradable film as affected by the type of fish and concentration and type of plasticizer.

3.4. Water vapor permeability

Triggerfish gelatin film containing glycerol had lower water vapor permeability (WVP) compared to that with sorbitol. On the contrary, tilapia gelatin based film using glycerol had higher WVP than that containing sorbitol (Figure 4). The difference may due to differences in gelatin properties which depend on fish species, fish living environment, and gelatin extraction procedures which render gelatin with different composition of amino acid and distribution of molecular weight that will affect the interaction reaction [6].

Figure 5 shows that as the concentration of glycerol increased, the WVP of the film increased. Contrariwise, fish gelatin film WVP decreased as sorbitol concentration increased. These findings in accordance with that obtained by [8] who deduced that gelatin (bovine and fish-skin) films with sorbitol hold the minimum WVP and films with glycerol had the maximum, possibly because sorbitol is less hydrophilic than glycerol.

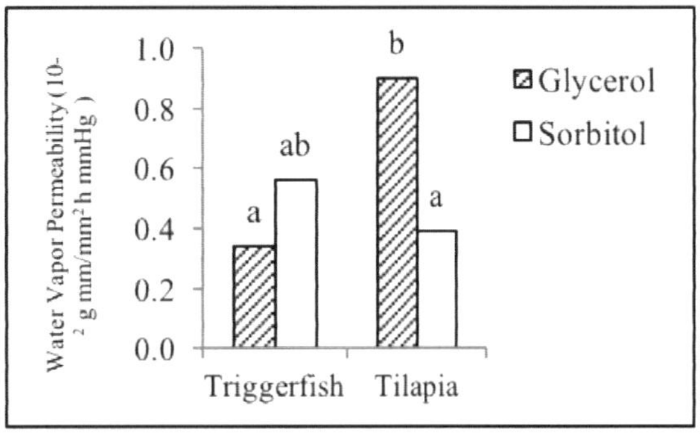

Figure 4. Water vapor permeability of fish gelatin biodegradable film affected by fish species and plasticizer.

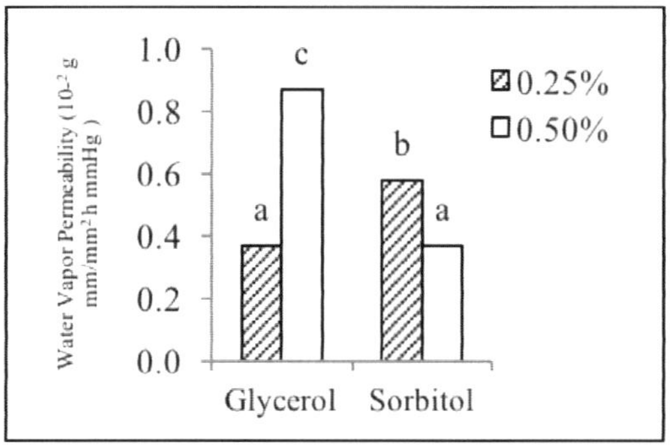

Figure 5. Water vapor of fish gelatin biodegradable film affected by plasticizer concentration.

4. Conclusion

For both types of fish (triggerfish and tilapia) gelatin films, increased in glycerol concentration led to the increase of tensile strength, percent elongation and also water vapor permeability (WVP). Contrarily, increasing sorbitol concentration decreased tensile strength of triggerfish and tilapia gelatin films, and increased elongation of triggerfish gelatin film, but decreased elongation of tilapia gelatin film. Triggerfish gelatin film has a high tensile strength and percent elongation, with moderate water vapor permeability (WVP), in contrast with tilapia gelatin film which had low tensile strength, and high percent elongation also WVP.

5. Acknowledgment

We would like to express appreciation to Syiah Kuala University, Indonesia and Ministry of Research, Technology, and Higher Education of Republic of Indonesia "Fundamental Research Project" which provide financial support to complete this work.

References

[1] Wang Q, Cao J, Yu H, Zhang J, Yuan Y, Shen X and Li C 2019 *Food Chem.* **271** 204-210
[2] Ahmad T, Ismail A, Ahmad SA, Khalil KA, Kumar Y, Adeyemi KD and Sazili AQ 2017 *Food Hydrocolloids* **63** 85-96
[3] Alfaro AT, Balbinot E, Weber CI, Tonial IB and Machado-Lunkes A 2015 *Food Eng. Rev.* 7 33-44
[4] Chen H, Shan ZH, Woo MW and Chen XD 2017 *Int J. of Biotechnol. Macromolecules* **94** 326-334
[5] Ahmada M, Benjakul S, Prodpranb T and Augustin WT 2012 *Food Hydrocolloids* **28** 189-192

[6] Gomez-Guillen MC, Turnai J, Fernandez-Diaz MD, Ulmo N, Lizarbe MA and Montero P 2002 *Food Hydrocolloids* **16** 1 25-34
[7] Hanani ZAN, Roos YH and Kerry JP 2014 *Int. J. of Biotechnol. Macromolecules* **71** 94-102
[8] Rezaei M and Motamedzadegan A 2015 *World J. of Nano Sci. and Eng.* **5** 178-193
[9] Arpi N, Fahrizal, Satriana and Edward S 2016 *Int. J. on Adv. Sci. Eng. and Inform. Technol. (IJASEIT)* **6** 561-569
[10] Apri N, Fahrizal and Novita M 2018 *IOP Conf. Series: Materials Sci. and Eng.* **334** 012061
[11] Sutono D and Pranoto Y 2013 *J. Agritech* **33** 2 169
[12] Gomez-Estaca J, Montero P, Fernandez-Martin F, Aleman A and Gomez-Guillen MC 2009 *Food Hydrocolloids* **23** 1334-1343
[13] Sukkunta S 2005 Thesis Department Technology of Environmental Management Faculty of Graduate Studies Mahidol University Thailand
[14] Song HY, Shin YJ and Song KB 2012 *J. of Food Eng.* **113** 541-547
[15] Atmaka W, Yudhistira B and Putro MIS 2018 *IOP Conf. Series: Earth and Env. Sci.* **142** 012028
[16] Jiang M, Liu S, Du X and Wang Y 2010 *Food Hydrocolloids* **24** 105-110
[17] Lee KY, Lee JY, Yang HJ and Song KB 2016 *Food Sci. Biotechnol.* **25** 1023-1028

IOP Conf. Series: Earth and Environmental Science **207** (2018) 012051 doi:10.1088/1755-1315/207/1/012051

The characteristics of Aloe vera gel as an edible coating

L Suriati[1*], I G P Mangku[1] and I N Rudianta
[1] Department of Food Science and Technology, Warmadewa University, Denpasar Bali Indonesia

*Corresponding author : suryatiluh1@gmail.com

Abstract. Preserving fruits using edible coating already long known. The edible coating serves to improve the appearance, as gas exchange barrier, retain moisture, antimicrobial and extend shelf life. The edible coating is widely used because it does not harm human health, can be eaten and biodegradable. One of the natural ingredients potentially as an edible coating is Aloe Vera leaf gel containing compounds such as polysaccharides, glucomannan, sugar reduction, tannins, organic acids, minerals, proteins. But the weakness is the gel that easily becomes diluted. The purpose of this research is to find out the best characteristics of Aloe vera gel as an edible coating that can be applied at the post-harvest handling of fruits. This study used a randomized complete design. From the results obtained that the gel is stored at cool temperatures are relatively stable, as seen from variable weights and water levels. The gel is stored at room temperature a little decline. The longer it is stored at room temperature, the moisture content of Aloe Vera gel. The gel is stored at cool temperatures increasing pH up to day 5 after it declined again. While the gel is stored at room temperature that is relatively stable at pH range 2.

1. Introduction

The development of the post-harvest handling of agricultural products in the world is increasingly developed and still developing. Some way has done among others with the use of low temperature, atmospheric modification, delivery of chemicals, coating wax, and edible coatings [1]. The edible coating is a thin layer of edible and easily definable natural (biodegradable), serves to prevent changes to the chemical, physical and biological. In addition, an edible coating can prevent mass displacement, as an additive, is a barrier against moisture and gas exchange of O_2 and CO_2 [2]. The edible coating gives the advantage that is some active ingredient can be incorporated into polymer matrices and can maintain the nutritional and sensory attributes [7].

The number of harmful preservatives causes concern in the community, because of the high risk of health. Some natural ingredients can be used as a preservative (edible coating) in particular on direct fruit is eaten without peeling his skin. Aloe Vera gel is a potential natural material that can be used as an edible coating because it consists of polysaccharides containing many functional components. The nature of antimicrobial and antioxidant ability of Aloe gel is able to inhibit the post-harvest damage [1][5]. Gel Aloe Vera can make such a layer of wax and easily applied. The disadvantage is the gel still unstable, easily become diluted due to the nature of the viscosity. If the gel is directly contacted with the air and light color, it will turn the color from pink to brownish [5], and arising unpleasant flavor. The Aloe Vera gel viscosity is decreasing the water viscosity drastically when stored at room temperature for 24 – 26 hours, this is due to hydrolysis of polysaccharides by an enzyme that still had high activity [6]. Based on this, the further research about treatment conditions and prolonged storage against the stability of Aloe gel as an *edible coating* is needed. This research aim is to study the influence of storage conditions and the duration in the storage against the characteristics Aloe Vera gel as *an edible coating*.

2. Research methods

The Implementation of the research began in January to May 2018 in the laboratory of Agriculture Faculty Warmadewa University. The materials that are used are citrate acid, chlorine, and the leaves of the Aloe Vera is obtained from farmers in Taro Village District of Gianyar Bali Province. This study used a randomized complete design with two factors and two repetitions: Factor 1 is the storage conditions of the aloe vera gel which consists of T_1 (cold temperature) and T_2 (room temperature). Factor 2 is the treatment prolonged storage consisting of P_1 (1 day), P_2 (2 days), P_3 (3 days), P_4 (4 days), P_5 (5 days), P_6 (6 days), P_7 (7 days), P_8 (8 days). The treatment of the combination is obtained 32 experimental unit.

The Implementation of research is starting from processing the Aloe Vera gel. Optimation of leaching techniques to eliminate mucus is yellow (yellow zap) and odor that can degrade the quality of the gel. To prevent Aloe Vera leaf gel from being contaminated, it must be handled with care and soaked in chlorine which can reduce the number of microbes outside the leaves [4]. After that, the homogenization is carried out for 10 minutes. If the homogenization is carried out for too long, it will cause the browning reaction of enzymatic in Aloe Vera gel [3]. The next step is to do the Stabilization of gel Aloe Vera with a heating treatment at a temperature of 85° C for 5 minutes and the addition of citric acid 4%. High-Temperature Short Time (HTST) is an effective method to reduce odor and biological activity of Aloe Vera gel [3]. The observed parameters include changes in weight, water content, pH, color, and thickness. Followed by shelf life, to know the characteristic of Aloe Vera gel during storage at cold temperature and room temperature for 8 days. The parameters observed were the moisture content, pH, changes in weight, color, and viscosity. Data from the measurement results are statistically tested using BNT media processing with SPSS. The edible coating made from Aloe Vera Gel with the best formulations will be applied on fruits (different studies).

3. Result and discussion

3.1. Shrink weights

Based on the results of the research, the weighting of the gel Aloe Vera without warming up stored on cold temperatures are relatively constant. Because storage at cold temperatures causes the activation of enzymes to decrease so that the process of oxidation and breakdown of gel tissue is smaller, therefore the losing weight because of old storage can be avoided [3]. Gel without or with heating stored at room temperature slightly decreases. A drastic decrease was not seen in the gel with the treatment of heating stored in cold temperatures. Storage at room temperature affects the loss of some gel components due to evaporation. The results showed that the maximum stability of Aloe Vera polysaccharide gel at 70 ° C will decrease at higher or lower temperatures [3]. While weights gel wrote first heated and added citric acid and then stored at cool temperatures on the 6th day of storage dropped dramatically. The weighting of gel at different storage conditions and times is shown in Table 1.

3.2. Color

The results of research on visual colours were taken as shown in Table 2 and Table 3. Aloe Vera Gel without heating stored at room temperature on day 4 was brown, this was due to contact with oxygen so that the reaction became brown. Leaves that are processed shortly after harvest start decomposition of the gel degradation matrix, because the enzyme reaction is natural in the presence of oxygen [3]. Aloe Vera gel if exposed to air or light after two hours of extraction, the colour is pink and then it will become darker [6]. Colour has a relationship with the effectiveness and stability of the gel. Some products psychologically users cannot accept colour changes. In the future, a simple but efficient processing process needs to be developed. Aloe Vera gel by heating in an open container stored at room temperature changes colour faster than the gel stored at cold temperatures. If Aloe Vera gel contact with oxygen at room temperature will stimulate the colour change process [6]. Supported by the opinion [11] that the colour changes to brown to sorrel, sometimes grey or yellow. Brown colour occurs due to carbohydrate oxidation by the peroxidase enzyme and catalase. Browning reactions depend on oxygen levels and pH can be prevented by heating.

Table 1. the shrink weights of aloe vera gel at different storage conditions

Day	Gel without heating (gr)		Gel with heating (gr)	
	Room	Cold	Room	Cold
1	4, 093	4, 420	4, 380	4, 780
2	4, 093	4, 434	4, 354	4, 778
3	3, 993	4, 488	4, 325	4, 763
4	3, 942	4, 433	4, 271	4, 747
5	3, 863	4, 408	4, 146	4, 746
6	3, 794	4, 369	4, 083	2, 725
7	3, 620	4, 371	3, 997	1, 596
8	3, 603	4, 367	3, 283	0, 799

Table 2. The color of Aloe Vera gel without heating at different conditions

Day	Gel without heating (gr)	
	Room temperature	Cold temperature
1	Transparent white	Transparent white
2	Transparent white	Transparent white
3	Pink	Pink
4	Brown	Pink
5	Brown	Pink
6	Brown	Pink
7	Brown	Pink Fade
8	Brown	Yellow

Table 3. Colors of Aloe Vera gel with heating at different conditions

Day	The gel in an open container		The gel in a closed container	
	Room	Cold	Room	Cold
1	Transparent white	Transparent white	Transparent white	Transparent white
2	Transparent white	Transparent white	Transparent white	Transparent white
3	White Fade	White Fade	Transparent white	Transparent white
4	Yellow	Yellow	Transparent white	Transparent white
5	Yellow Fade	Yellow Fade	Yellow Fade	Yellow Fade
6	Yellow Fade	Yellow Fade	Yellow Fade	Yellow Fade
7	Yellow Fade	Yellow Fade	Yellow Fade	Yellow Fade
8	Yellow Fade	Yellow Fade	Yellow Fade	Yellow Fade

3.3. The degree of acidity (pH)

The degree acidity of the gel which stored at room temperature is relatively stable, while stored at cold temperatures is increased (Table 4). An increased amount of acid during storage is caused by the activity of the microorganisms which can convert carbohydrates into acids [12]. Hydroxyl carboxylic polysaccharides undergo ionization reaction from -COOH became^{-} -COO and H^{+} [10]. A decrease in pH values causes a decrease in the level of viscosity. Viscosity is a measure of the relative to the movement of the fluid resistance of the parts.

Table 4. The Degree of Acidity Aloe Vera Gel on the conditions and Long Storage Different.

Day	Gel	
	Room temperature	Cold temperatures
1	2.04	2.00
2	2.03	2.11
3	2.04	2.16
4	2.18	4.05
5	2.09	3.86
6	2.00	5.13
7	2.00	3.93
8	2.04	6.79

3.4. Viscosity

The results of the research on the viscosity are taken as shown in Tables 5 and 6. The purpose of heating at 80°C is to reduce or suppress active phenolase which greatly affects the Aloe Vera gel physicochemical. From table 5 it can be seen that longer heating results in greater gel viscosity. This is because the heating time causes more water to evaporate. The water in the gel is mechanically immobilized. Gel has variations in the level of hardness, elasticity, and fragility. It depends on the gelling agents, salinity, pH and temperature. Aloe Vera gel is very unstable and easily influenced by air, light, heat, and microbes [13]. The longer the storage the greater the possibility of fluid coming out of the three-dimensional structure of the gel. The liquefaction gel involves crosslinking which involves the dismantling of the carbohydrate polymer chain [10]. Most polysaccharides have a reactive functional group forming carboxylates [9]. Gel stored at cold temperatures also decreases activity, even though the loss of activity is getting lower. Loss of activity caused by enzymatic activity after the leaves are separated from the tree. This means that even if stored at cold temperatures will also cause biological activity to decrease.

Table 5. Viscosity at the different time of heating

Heating (minutes)	Viscosity (cp)
0	275
5	397
10	448

Table 6. Viscosity at different storage

Day	Viscosity (cp)
1	394.8
2	376.7
3	290.1
4	221.3
5	105.5
6	69.0
7	30.4
8	28.8

3.5. Moisture

The results of the study of water content retrieved data as shown in Table 7. Aloe Vera gel is comprised of 99.5% water and solid materials (0.5 – 1%) which contain a variety of compounds including the compound that is soluble in water and fat, vitamins, minerals, enzymes, polysaccharides, phenolic compounds and organic acids [5]. The moisture of gel which is deposited at room temperature tends to increase until day 3 and dropped back on day 4. While the gel is stored at cool temperatures prevail instead.

Table 7. Moisture on the different conditions.

Day	Gel	
	Room temperature	Cold temperature
1	95.05	98.37
2	99.66	96.08
3	99.46	96.36
4	94.79	98.12
5	95.13	99.03
6	96.95	99.56
7	94.84	99.48
8	95.08	99.35

4. Conclusion

The conclusions of this research are Aloe vera gel with heating and stored at cold temperatures until day 5 relatively stable, as seen from the variables shrink weight, color, pH, viscosity and moisture content. While the gel is stored at room temperature decline started day 2. Aloe Vera gel should be stored at cool temperatures with a sealed container. The best characteristics of Aloe Vera gel as an edible coating that can be applied at the post-harvest handling of fruits.

Acknowledgments

I would like to thank the members of the team and all parties at the University of Warmadewa, for their participation in the research and help me get the best quality result.

Reference

[1] Raghav P K, Agarwal N and Saini M 2016 *International Journal of Modern Scientific Research and Education* **1** 2455–5630
[2] Bourtoom T 2008 *International Food Research Journal* **15** 237-48
[3] Ramachandra C T and Rao P S 2008 *American Journal of Agricultural and Biological Sciences* **3** 502-10
[4] Eshun K and He Q 2010 *Journal Food Science and Nutrition* 91-6.
[5] Hamman J H 2008. *Molecules* **13** 1599-1616
[6] Suriati L 2000 Studies the resistance against oxidation and phase change characteristics of the physicochemical Aloe Vera gel [translation] (Thesis) Bogor Agriculture University. Bogor
[7] Dhall R K 2013 *Journal: Critical Review Of Food Science Nutrition* **53** 435-50
[8] Serrano_M, Valverde J M, Guillén F, Castillo S, Romero D M and Valero D 2006 *J. Agric Food Chem.* **54** 3882–86
[9] Fennema O R 1976 *The principle of food science* Marcell's Dekker Inc. New York
[10] Glicksman M 1984 *Food hydrocolloid* CRC Press Inc. Boca Raton Florida.
[11] Stuckey B N 1981 *Antioxidants as Food Stabilizers in* TE Furia (ed). CRC. *Handbook of Food Additivies* CRC Press Inc. Boca Raton Florida
[12] Eskin N A M, Henderson H M and Townsend R J 1971 *Biochemistry of Foods* Academic Press New York.

[13] Suryowidodo C W 1992 *Wartà IHP* **9** 1–5

IOP Conf. Series: Earth and Environmental Science **207** (2018) 012052 doi:10.1088/1755-1315/207/1/012052

Thermal penetration study for the purpose of formulating sterilization procedures of yellowfin tuna canning

H Hasan[1*], S H Anwar[2], S Rohaya[2] and Martunis[2]

[1]Electrical Engineering Department, University of Syiah Kuala, Darussalam, Banda Aceh 23111 Indonesia
[2]Agricultural Product Technology Department, University of Syiah Kuala, Darussalam, Banda Aceh 23111 Indonesia

[*]E-mail: hafidh.hasan@unsyiah.ac.id

Abstract. This study aims to provide the community-based business with simple to follow procedures in the sterilization of canned yellowfin tuna. Two filling mediums, brine and palm oil were used. The thermal treatment process was performed using a small business scale retort of 24L. The process is divided into three stages: heating, holding temperature and cooling stage. The targeted temperature at the holding stage was set at 121.1°C for 50 minutes and measured using a thermocouple gland placed inside the retort. Two more thermocouple glands were placed inside the cans to record the product core temperature. During the heating stage, a time lag of 20 minutes was observed for the palm oil tuna to reach the same holding temperature as the brine tuna. The lethality rate F_h during heating stage was 2.52 and 8.69 minute for brine and palm oil respectively. Lethality rate at the holding temperature stage $F_{121.1}$ was 49.23 for brine and 38.96 for palm oil. The lethality rate F_{cl} of both products were closed at 2.3 and 1.93 minutes respectively for brine and palm oil. The total F_0 value were 54.05 and 49.58 minutes respectively for tuna in brine and tuna in palm oil.

1. Introduction

Food preservation practices have been widely adopted for human living since the beginning of nineteenth century, including preservation by canning. Canned food is produced by introducing heat into food or using thermal processing, either in a sealed container or by passing it through heat exchanger and followed by packaging. The heating process can be in a batch or continuous using heating for the most part of the process and cooling the product as quickly as possible to end the process. The heating medium can be saturated steam or hot water which at 100°C both are above atmospheric pressure [1].

The main reason for heating foods is to kill pathogenic and spoilage microorganisms as well as to inactivate enzymes that they produced. Heating also induces physical changes and facilitated chemical reactions which can affect color, flavor, texture and most importantly nutritional value of food. There are two major issues related with thermal processing of food, i.e., food safety and food quality, where the conflicts between these two issues cannot be avoided. The spore of *Clostridium botulinum* is one of dangerous heat-resistance pathogenic bacteria that most likely to be present in canned fish. However, the most heat-resistance one is actually *Bacillus stearothermophilus* [2]. These bacteria and their active spores are not allowed to be present in canned food. Therefore, before applying heat treatment and preserving foods, it is very critical to understand the reaction kinetics and its relation to microbial inactivation, chemical damage, and physical changes of food as well as on how to minimize sensory and nutritional reduction [1]. Significant reduction of essential fatty acids due to different heat treatment on tuna processing had been reported [3].

In order to produce a sterile product, the length of sterilization time at a recommendation temperature (121.1°C) must be calculated carefully [4]. This is related with the destruction of microorganisms that takes place logarithmically during canning process. Therefore, the term "commercial sterility" is crucial to be understood. This means "in practice that the heat processing inactivates substantially all microorganisms and spores which, if present, would be capable of growing in the food under defined storage conditions" [5]. The main concern during canning is to destroy bacteria and their spores completely and in case the spores are still exist in a canned product, they are not able to reproduce new bacteria but remain inactive throughout storage period in room temperature. The rate of survival spoilage microorganism is determined by the type of microorganisms that possibly contaminated the raw fish before canning.

Thermal processing of fish must consider the calculation of D value, z value, F_0 value and lethality rate. According to [4] and [5], the D value is known as the decimal reduction time of a specific microorganism. The z value is the temperature change which results in a tenfold change in the decimal reduction time. A reference F_0value is used to describe processes that operate at 121°C which are based on a microorganism with a z value of 10°C.

This study was designed to determine the sterilization process through temperature build-up and processing time for canning the yellowfin tuna that are suitable for community-based business in a small scale, particularly in Aceh Province. One of the study related with heat penetration in home canning had been done previously [6] as well as a study that investigated different filling mediums in canned tuna [7]. This investigation used specific retort volume, can dimensions, type of mediums, the volume of medium added and a certain weight of pre-cook tuna. The arrangement of cans inside the retort chamber might be different with other canning practices. Therefore, a heat penetration study was a fundamental step that must be carried out in the beginning of the whole process including determination of lethality rate in canned tuna, calculation of the F_0 value and validation of the sterilization process.

2. Materials and methods

2.1. *Sterilization equipment and thermal recording process*

A large diameter food steamer was used for exhausting prior to seaming. This step allowed oxygen to escape from tuna and filling materials avoiding the product from having air pocket. For the thermal treatment. it was conducted using a 24-liter All American (USA) pressure canner. Two temperatures were measured during sterilization, i.e. the retort interior temperature and the core of the canned tuna temperature. The measurements were taken using Ellab thermocouple gland with built-in logger (TSP Micro, Ellab, Denmark). To measure the retort temperature, the thermocouple was placed right above the water level in the lower part of the retort. While the core product temperature was measured by placing the thermocouple inside the cans before seaming. Thanks to the thermocouple miniature size. This thermocouple was positioning in the base of the can pointing upward to the geometric center and was held in the middle of the can by stuffing tuna meat. The recorded temperatures were converted into data using a docking system (Pro Multi Reader, Ellab, Denmark) and an application software (ValSuite Pro, Ellab).

2.2. *Tuna preparation, seaming and thermal processing*

Tuna was handling as quickly as possible before canning. All of the canning process was carried out in a laboratory, particularly for tuna preparation (washing, filleting, peeling, cutting), pre-cooking, filling, brining, can-seaming and sterilization process. The laboratory is located within 5 minutes from the nearby fishing port (Lampulo, Banda Aceh) where a 48 kg freshly landed yellowfin tuna (*Thunnus*

IOP Conf. Series: Earth and Environmental Science **207** (2018) 012052 doi:10.1088/1755-1315/207/1/012052

albacares) was purchased. After cleaning the fish which included beheaded, gutted, skinned, tuna was cut into blocks (weighing 0.5 - 1 kg) and frozen. At the day of canning and sterilization process, tuna was thawed first. When it reached the room temperature, the yellowfin tuna was pre-cooked for 15 minutes at 80°C. Pre-cooking process was actually a steaming process, particularly aimed at partially dehydrate the flesh and reduce the strong fishy odor. The pre-cooked tuna was covered and brought to room temperature for red meat removal stage. Only the white meat was selected and used for canning process. Portions of white meat (80 ± 0.01 g), which consist of tuna chunks and flakes, were positioned into a can (7.3 cm in diameter and 5.6 cm in height). Then 80 ml filling medium was added to each can and the necessary headspace (± 1.5 cm) was left to absorb the exerted pressure during thermal treatment. Two kinds of filling mediums were used. Each batch was fills with the same filling medium for consistency and uniformity. The first batch was filled with 5% brine and the second batch was using palm oil. The cans were loaded into the large steamer for exhausting. The steamer was operated to transfer the heat from generated steam into the core of the product to at least 80 °C for the duration of 10 minutes. Immediately after exhausting, the cans were covered with the lids and ready for double seaming process using *semitro* can seamer.

Prepared cans with the same filling medium were loaded into the retort. The sterilization process started when the core temperature of canned tuna reaching 82.2°C. This was called the heating stage which ended when the core reaching 121.1°C (holding temperature). The later stage was the holding temperature stage where the core temperature was kept constant at the holding temperature 121.1°C through maintaining the retort at the same temperature. This was the stage where the lethality was predictable and hence became controllable through regulating the time. The last stage was cooling stage, where the retort was removed from the heat source and cans were cooled under running tab water. The success of the sterilization process was determined by the duration of time where the product core temperature at the slowest heating point (SHP) reached the holding temperature at 121.1°C and it was called the F_0 value. F_0 value was determined by the heat penetration mechanism from the retort into the core of the product. While the retort temperature can be easily measure, the core temperature cannot be measured without penetrating and thus forfeiting the product. However, if similar equipment, materials (including can dimension and the amount of filling) as well as methods were consistently repeated in every process, the same thermal penetration could be reproduced. Thus ensuring the desired F_0 value was obtainable in every sterilization process.

The holding temperature was also introduced to minimize uncertainty. In this study, the holding temperature was set to last for 50 minutes. This would ensure adequate lethality rate was obtained that can be used in controlling the process to obtain the targeted F_0 value. Thermocouple glands were used to record the dynamic of temperature changes through the entire process with the resolution of 10 second.

2.3. *Bacteria thermal destruction and F_0 value*

The important factor to consider in this canning process is in obtaining the sufficient heat treatment. This is quantified as F_0 value. F_0 has a unit in minute to indicate lethality that has been accumulated during heat treatment with regard to bacterial destruction. The F_0 value was determined by the product core temperature plot versus time according to the following formula [2].

$$F_0 = \int_0^t 10^{\frac{(T-121.1)}{z}} dt \quad (1)$$

where t = time (min), T = core temperature (°C) and z = required temperature addition to accelerate decimal reduction time (D) by tenfold (°C). The constant 121.1 is the reference temperature where the controllable rate of destruction is expected. Notice that when T equals 121.1, F_0 is simply identical to time t in minute.

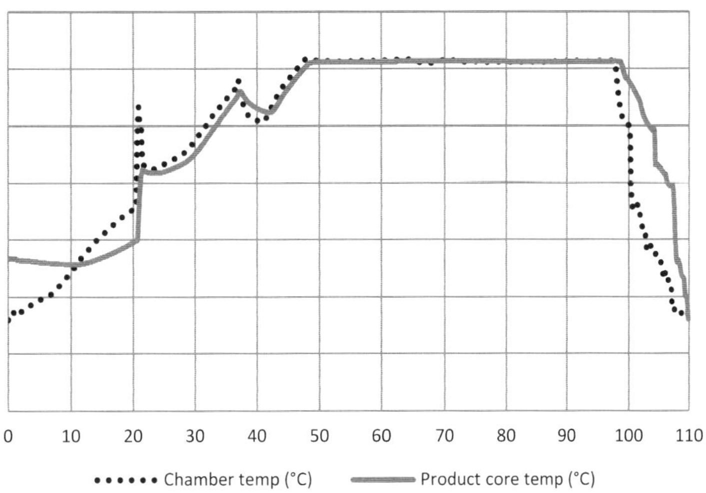

Figure 1. Thermal penetration characteristic of yellowfin tuna in brine with the corresponding retort temperature

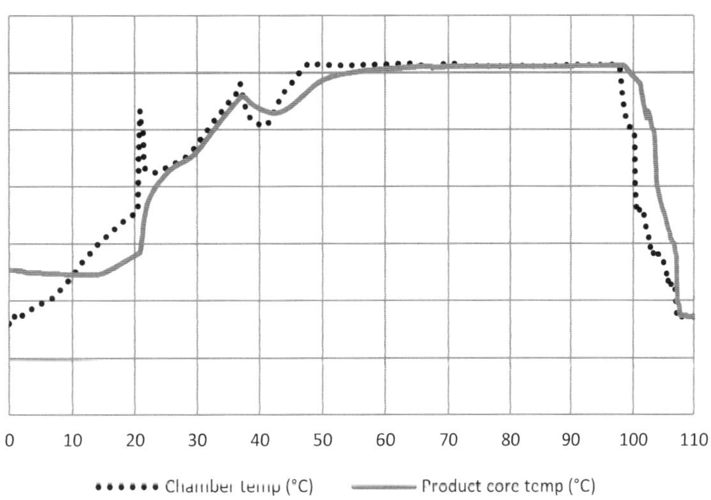

Figure 2. Thermal penetration characteristic of yellowfin tuna in palm oil with the corresponding retort temperature

3. Results and Discussion

Characteristic of thermal penetration for different filling mediums

The recorded temperature versus time in the retort (dashed line) and the core (solid line) of canned yellowfin tuna in brine and palm oil are depicted in Figure 1 and Figure 2 respectively. The recorded retort temperature in both experiments show consistency of the equipment used for sterilization. Which is important in formulating a standard procedure for canning. In addition, both graphs show the canned yellowfin tunas preserved the remaining heat (solid line) acquired from the previous exhausting process. Thus the products temperature were higher initially than the canner chamber temperature. It required more than 10 minutes for the canner to take over the product temperature. But the sterilization did not begin until the core temperature reached 82.2°C. This was the temperature where the destruction of bacteria, in particular *Clostridium botulinum* spores, starts to take place. The minimum F_0 value recommended for *Clostridium botulinum* is 2.52 min [7], which associated to 12 D (see Equation 1) as the minimum process to achieve adequate inactivation of this particular bacteria [8].

The sterilization process started with the heating stage when the products core temperature reached 82.2°C. The thermal penetration into the products continued as the retort temperature increased to reach 121.1°C where it was maintained for the next 50 minutes as the holding temperature. The heating stage still continued and ended when the core temperature matched the holding temperature and the holding temperature stage began. This was occurred around 47 and 59 minutes from the initial process time respectively for brine and palm oil.

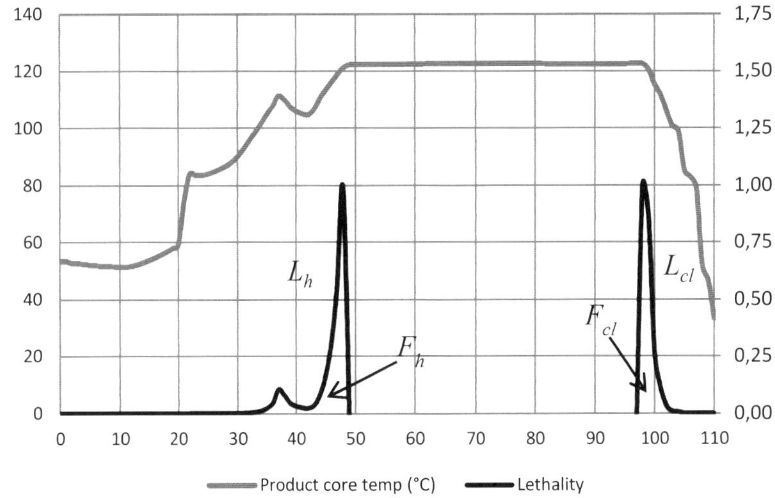

Figure 3. Lethality during heating and cooling of yellowfin tuna in brine in accordance with the product core temperature

The calculation of lethality at any particular time in the sterilization process can be done using the products core temperature. If T_h and T_{cl} are the core temperature during heating and cooling stage, the expression for lethality during heating stage L_h and cooling stage L_{cl} are:

$$L_h = log^{-1}\left(\frac{T_h - 121.1}{10}\right) \quad (2)$$

$$L_{cl} = log^{-1}\left(\frac{T_{cl} - 121.1}{10}\right) \quad (3)$$

It can be easily seen from Equation 2 and Equation 3 that when the core temperature equaled the chamber's (holding temperature stage) the lethality value becomes constant

$$L_{121.1} = log^{-1}\left(\frac{121.1 - 121.1}{10}\right)$$

$$L_{121.1} = 1 \quad (4)$$

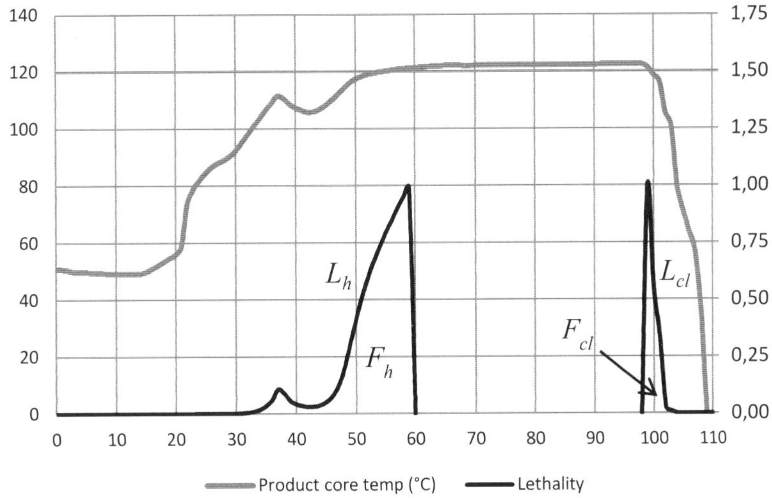

Figure 4. Lethality during heating and cooling for yellowfin tuna in palm oil in accordance with the product core temperature

The lethality during heating and cooling stages were calculated with Equation 2 and Equation 3. The results are charted in Figure 3 for tuna in brine and Figure 4 for tuna in palm oil. The area under the lethality curves can be calculated as F_0 value expressed in Equation 1. This value reflects the lethality rate imposed by thermal penetration during heating F_h and cooling stage F_{cl}. The results of the calculation of lethality rate are summarized in Table 1. The lethality rate during the holding temperature stage $F_{121.1}$ will be equaled with time in minute and are given in Table 2.

Table 1. Thermal penetration properties of yellowfin tuna in brine and in palm oil

Product	F_0 heating stage, F_h (minute)	F_0 cooling stage, F_{cl} (minute)	$F_h + F_{cl}$ (minute)	actual process time, t_h and t_{cl} (minute)
Yellowfin tuna in brine	2.52	2.30	4.82	15 | 6
Yellowfin tuna in palm oil	8.69	1.93	10.62	26 | 4

Table 2. Lethality rate of holding temperature stage for 50 minutes retort holding temperature

Product	$F_{121.1}$ (minute)	actual process time, $t_{121.1}$ (minute)
Yellowfin tuna in brine	49.23	49.23
Yellowfin tuna in palm oil	38.96	38.96

The yellowfin in palm oil exhibited slower heat accumulation and required longer time than tuna in brine to reach temperature uniformity in the retort. As the result, its F_h was significantly higher than tuna in brine in the heating stage. While the differences in fillings did not exhibit significant time lag in the cooling stage since the cooling was forced immediately under running tab water hence the nearby F_{cl}. The total F_0 value were 54.05 minute and 49.58 minute respectively for tuna in brine and palm oil. As the purposes of thermal process are to meet the safety requirement as well as sensory characteristic and preserving nutrition benefits, a certain F_0 value has to be met. Following the results in this section the total F_0 value can be easily adjusted at the holding temperature stage providing with knowledge of F_h and F_{cl}.

4. Conclusion

The thermal treatment of yellowfin tuna is intended to make the food safe for consumption to a finite period of time while continuously stored in room temperature. F_0 value is the standard parameter used for indication of success of a thermal treatment. F_0 values were greatly affected by filling material during heating and cooling stage. But when the core reached the holding temperature at 121.1 °C, F_0 value was identical with time in minute. In the laboratory experiment, yellowfin tuna in brine showed better heat penetration than yellowfin in palm oil hence it was faster to reach the holding temperature in 15 minutes compared to 26 minutes. But the former only registered about a third F_0 value of the latter (2.52 to 8.69 minutes). Despite the differences during heating stage, the rate of cooling was almost similar between the two. The results from this experiment can be reproduced to obtained similar F_0 value providing the same equipment and materials are used. In order to obtain a recommended F_0 value, which is higher than $F_h + F_{cl}$, can be added through regulating the holding temperature time. However, F_0 is not the only parameter to decide the safety of a fish canning product. Further studies in the quality of the material such as assesment in the fish quality is equally important.

References

[1] Lewis M J 2006 Thermal Processing *Food Processing Handbook* ed J G Brennan (Wiley-Blackwell) pp 33–70
[2] Warne D 1988 Manual on fish canning - 1. Principles of canning *FAO Fish. Tech. Pap. - 285 FOOD Agric. Organ. U. N.*
[3] Stephen N M, Jeya Shakila R, Jeyasekaran G and Sukumar D 2010 *J. Food Sci. Technol.* **47** 174–81
[4] Fellows P 2009 *Food Processing Technology: Principle and Practice* (London: Woodhead Publishing)
[5] Brennan J G 2011 *Food processing handbook* (Weinheim: Wiley-VCH)
[6] Etzel M R, Willmore P and Ingham B H *Food Sci. Nutr.* **3** 25–31
[7] Mohan C O, Remya S, Murthy L N, Ravishankar C N and Asok Kumar K 2015 *Food Control* **50** 320–7
[8] Sevenich R, Rauh C and Knorr D 2016 *Innov. Food Sci. Emerg. Technol.* **38** 65–75

The effect of native chicken legskin gelatin concentration on physical characteristics and molecular weight of edible film

M Sompie[1,2*], S E Surtijono[3] and Ch Junus[3]

[1]Laboratorium of Animal Production Technology, Faculty of Animal Husbandry, Sam Ratulangi University, Manado 95515, Indonesia
[2]Departement of Animal Production, Faculty of Animal Husbandry, Sam Ratulangi University, Manado 95515, Indonesia
[3] Faculty of Animal Husbandry, Sam Ratulangi University, Manado 95515, Indonesia

*Corresponding E-mail : meitysompie@yahoo.com

Abstract. Edible film is a thin layer made from polysaccharides, proteins, and lipids. These research was aimed to determine the effect of gelatin concentration on physical characteristics and molecular weight distribution of edible film produced from native chicken legskin gelatin. This research materials were used native chicken legskin gelatin and plasticizer gliserol. This study used Completely Randomized Design (CRD) with different concentration of gelatin (T1 = 5%, T2 = 10%, T3 = 15% and T4 = 20%) and five replications. The result of study described that the different gelatin concentrations had significant effect ($P<0.05$) on tensile strength, elongation and thickness of edible film. SDS-PAGE result showed that the band patterns of the molecules was dominated by the protein molecules of gelatin, SDS-PAGE pattern showed that do not changed of bands the type $\alpha 1$ and $\alpha 2$ of edible films. The molecular weight distribution of edible film from chicken legskin gelatin had ranges from 140 -148 kDa. The molecular weight distribution of edible film from chicken legskin gelatin had ranges from 140 -148 kDa.

1. Introduction

The development of edible films as coating and packaging materials is increasing, due to the higher human awareness of packaging materials that can be degraded to replace plastic materials that cannot be degraded [10] . Packaging is one way to maintain food quality, protect products from chemical and biological contamination, as a protector of food because it can prevent the migration of water vapor, gas, fat and aroma from ingredients to the environment and prevent damage caused by microbes [8].. One type of packaging that is environmentally friendly is edible packaging, it can protect food products, be edible and safe for the environmentIn the most recent years, food and packaging industries have been joining efforts to reduce the amount of food packaging materials. An edible film could be defined as primary packaging made from edible components. Edible film is a thin layer of edible coatings and are often used as food, and able to be a barrier of moisture, oxygen, mechanical properties, sensory, convenience, and prolong the shelf life of various products [3, 7, 11].

Edible polymers such as polysaccharide, protein, and lipid are the three main ingredients used to produce edible films [12]. Edible films have the same properties as packaging films such as plastics, which must have water retaining properties so they can prevent product moisture, control the transfer of dissolved solids to maintain color, natural pigments and nutrients and improves food quality. The applications of edible films and coatings include fresh produce coatings of sausage casings from collagen [10]. The main ingredient for making edible films is gelatin [1, 21, 22, 27]. Gelatin is a protein obtained by boiling skin, tendons, ligaments, and bones with water [16] , that can be obtained from collagen. Its functional properties depend on processing conditions as well as the raw material [17, 21, 24]. The gelatin quality based on its physical and chemical properties and rheological properties [20, 23]. Gelatin-based films are thin. The addition of plasticizer, make changes some of the functional and physical properties of these films. _Edible films from hydrocolloid have a good ability to protect products from oxygen, carbon dioxide and fat and have mechanical properties that can improve the structural integrity of products that are easily damaged. Edible films from gelatin or collagen have good potential [26]. Plasticizer from the group of polyols are widely used to improve the mechanical. This polyol gliserol as plasticizer can lower the modulus properties of tensile strength and increase elongation film [9].. Effects of different concentration of gelatin to produce edible film from native chicken legskin was limited information. If viewed from the chemical composition, native chicken legs skin has a protein content of 22.98%, 5.60% fat content and 3.49% ash content [9] This study was investigated the effect of native chicken legskin gelatin concentration on the physical characteristics and molecular weight of edible film.

2. Materials and methods

Three thousand grams of native chicken legskin gelatin were used as a raw material, plasticizer glycerol and distilled water.

2.1. Process of gelatin

Gelatine was prepared by the acid extraction method with 3% acetic acid (v/v) and distilled water [4].. 100 grams native chicken legskin as raw material were soaked for 24 hours. After soaked, samples were neutralized to pH 6 and extracted [5]. The extraction temperature were performed at 55^0C. Solubilized gelatin was separated from residual skin fragments by filtration through a filter. The extracted gelatin was concentrated at 70^0C for 5 hours then dried at 60^0C for 24-48 hours until the gelatin sheet was dried and solid. Gelatin sheets were milled and packaged in vacuum plastic, stored in a desiccator for subsequent process [9, 22] .

2.2. Preparation of edible film

The process of edible film made from native chicken leg skin gelatin according to the method of Sobral *et al.* (2001) as follows: the concentration of film-forming solution was prepared based on treatment by dissolving 5, 10, 15 and 20 grams gelatin in 100 ml distilled water in accordance with the treatment specified, then added plasticizer glycerol with concentrations 10% (w / w) then dissolved in a water bath at a temperature of 50 ° C while stirring for 30 minutes until the gelatin granules and homogeneously mixed plasticizer [2] . Homogenized for 5 minutes at room temperature. The solution was subsequently cast films formed in a Teflon and dried using an oven at a temperature of 55^oC for 18-20 hours. Sheet edible film formed was wrapped with clear plastic and stored in a container with silica gel, then analyzed according to the treatment [18]

2.3. Experimental Design.

This research were determined by analysis of Completely Randomized Design (CRD) 4x5 with different concentration of gelatin (T1 = 5%, T2 = 10%, T3 = 15% and T4 = 20%) and five replications. The significant differences were determined using Duncan's new multiple range test [25] .

2.4. Parameters.
The characteristics parameters of this study were film thickness, tensile strength, elongation at break and molecular weight distribution. The thickness film was measured at five different points for each sample to be tested. Tensile strenght is the maximum pull that can be achieved through film to survive before breaking, the elongation percentage is calculated based on the length of the film.

The molecular weight distribution of gelatin was determined according to Carvalho [4] using Sodium Dodecyl Sulfate Polyacrylamide Gel Electrophoresis (SDS-PAGE). Edible film 0.01 g was added with 1 ml of distilled water then homogenized. Then dilution of dissolved protein was carried out as much as 500 g at every 0.1 ml of the solution which was added with 0.4 ml of distilled water, then continued with a diluted sample of centrifuged at 3000 rpm at room temperature for 10 minutes. The supernatant 0.015 ml which was added with 0.015 ml of the buffer sample then heated at 100°C for 2 minutes and cooled at room temperature. The sample was taken 0.01ml, and running ends after the sample dye reaches the bottom of the gel.

Molecular weight was identified by means of gel soaked with commasie brilliant blue dye solution for 12 hours, slowly agitated, then the gel was dried and pressed with plastic until it became stiff and could be stored for a long time.

3. Results and discussion

3.1. Thickness
The film thickness will affect the physical properties and the rate of water vapor edible film (Bergo and Sobral, 2007). Statistical analysis showed that different gelatin concentration had significant effect (P<0.05) on edible film from native chicken legskin gelatin. This structure film is formed by the matrix protein interactions proteins catalyzed by heat that is hydrophobic bonds, hydrogen bonds and disulfide bonds [6] . Factors affecting edible film thickness is the concentration of dissolved solids in solution films. The higher concentration of dissolved solids, the film thickness increases.

Table 1. The Average values of physical characteristics edible film

Gelatin concentration (%)	thickness (%)	Tensile Strength (MPa)	Elongation at Break(%)
5	0.112 ± 0.02^{a}	5.713 ± 0.218^{a}	57.110 ± 0.230^{a}
10	0.123 ± 0.05^{b}	4.512 ± 0.018^{b}	59.210 ± 0.210^{b}
15	0.127 ± 0.01^{b}	4.271 ± 0.005^{b}	61.310 ± 0.061^{c}
20	0.128 ± 0.01^{b}	4.751 ± 0.203^{b}	61.712 ± 0.170^{c}

Different letters indicated the significant differences (P<0.05), Sd – standard deviation

3.2. Tensile Strength
The tensile strength is the ability of a material to resist breaking under tensile stress The average tensile strength of edible film from native chicken legskin gelatin was displayed in Table 1. Statistical analysis showed that the different concentration of native chicken legskin gelatin had significant effect (P<0.05) on edible film. An increase in the concentration of gelatin in a solution decreases the tensile strength of the film. This is due to the destability of the film matrix by an increase in the components of hydrophilic proteins in the edible film structure [2]

3.3. Elongation at break
The elongation at break of an engineering material is the length increase, expressed as a percentage, which occurs before it breaks under tension. Edible film with gelatin 20% (61.712±0.170) was higher EB than edible film with gelatin 15% (61.310±0.061), gelatin 10% (59.210±0.210) and gelatin 5% (57.110±0.230) The more powerful the film is formed, the more difficult for the film to elongate, that

it will reduce the value of percentage elongation, increasing TS is accompanied by a decreases elongation at break.

3.4. *Molecular Weight Distribution*

Based on Figures 1, it can be seen that in general the protein molecular pattern is clearly visible for the four samples (A,B,C,D and M as marker). It can be seen that some bands have not been able to provide real line information, but in general the band's appearance can already illustrate the pattern of molecular weight distribution produced by gelatin. The whole band showed that from the treatment applied, the molecular weight distribution of edible film from native chicken legskin gelatin had ranges 140-148 kDa. Color thickening in some bands can be caused by a less than optimal protein preparation process, but at least that information on the BM distribution pattern of the tested product can still be estimated based on the comparison band of marker (M). SDS-PAGE result showed the band pattern of the molecule was dominated by the protein molecules of gelatin, SDS-PAGE pattern showed no changed of bands the type α1 and α2 of edible films. The distribution of molecular weight was related to the length of the amino acid bond chain and the gel strength ([15, 19] . The dark colored ribbon shows the molecular weight of edible film was good. The magnitude of this molecular weight will affect all the physical characteristics of edible film. The bigger molecular weight had a longer amino acid bond chain and higher gel strength [19, 20].

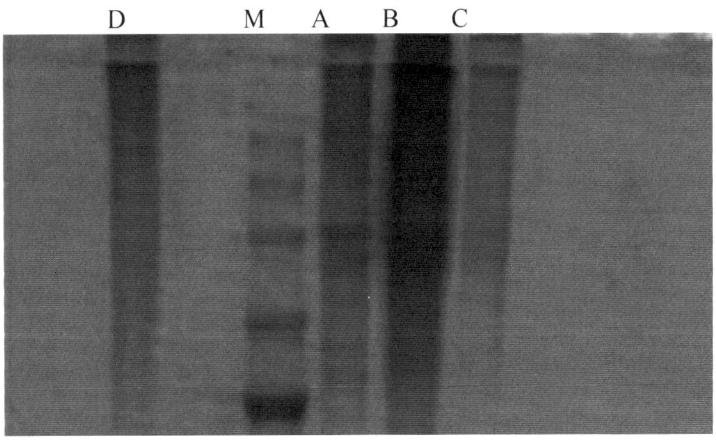

Figure 1. SDS-PAGE (Sodium Dodecyl Sulfate Polyacrylamide Gel Electrophoresis) edible film from native chicken legskin gelatin

M = Marker
A = gelatin concentration 5%
B = gelatin concentration 10%
C = gelatin concentration 15%
D = gelatin concentration 20%

4. Conclusion

It was concluded that the use of native chicken legskin gelatin 10% had the best physical characteristics of edible film (thickness mm 0.123±0.05, tensile strength 4.512±0.018, elongation at break 59.210±0.210^{b} % and molecular weight distribution 148 kDa).

5. Acknowledgements
This research was supported by the Ministry of Research Technology and Higher Education, Republic of Indonesia based on Research Grant Implementation, No. 087/SP2H/L /DPR/2018.

6. References
[1] Agustin A T and Sompie M 2014. *Proc. Int. Conf Challenges of Biotechnological Research in Food and Health Solo Indonesia* p. 103-104
[2] Bergo P and P.J.A. Sobral. 2007. *Food Hydrocolloid* **21** : 1285 – 1289
[3] Bourtoom T. 2008. *Int. Food Res. J.* **15**(3) : 1-9.
[4].Carvalho R.A., P.J.A. Sobral, M. Thomazine, A.M.Q.B. Habitante, B. Giménez,M.C. Gómez-Guillén and P. Montero. 2007. *Food Hydrocolloids*, **22** (6) : 1117-1123.
[5] Giménez B, Gómez - Guillén M C and Montero P 2005. *Food Hydrocolloids*. Vol. **19** (6) : 958-963.
[6] Gennadios A, Brandenburg A, Weller L C and Testin R F 1993. *J. Agr. Food Chem.* 1835–1839.
[7] Gontard N S. Guilbert and Quq J L 1992. Edible wheat films : *J. Food Sci.* **57** :190-195.
[8].Guilbert S and Biquet 1996. *Food Packaging Technology* Vol **I**. VCH Publisher, Inc. New York.
[9] Hasdar M 2012. Tesis. Fakultas Peternakan Universitas Gadjah Mada Yogyakarta.
[10]Kang H.J, Jo C, Kwon J H, Kim J H, Chung H J and Byun M W. *Food Control* 18 : 430-435
[11]Krochta J.M and Johnson M 1997. Edible and biodegradable polymer film : challenges and opportunities. *J. Food Tech.* **51** : 61-74
[12]Liu HY, Han J and Guo S D 2007. Extraction and properties of gelatin from channel catfish (*Ietalurus punetaus*) skin. *J Food Sci. and Tech.* Vol. **41**(3) : 414-419.
[13] Liu Z and Han J H 2005. Film forming characteristics of starckes. *J. Food Sci.* **70** (1) E.31-E36
[14] López-Carballo G, Hernández-Muñoz P, Gavara R and Ocio M J 2008. *Intern. Journal. of Food Microbiology* 65-70.
[15]Muyonga J H, Cole C G B and Duodu K G 2004. Extraction and physico- chemical characterization of Nile perch (*Lates niloticus*) skin and bone gelatin. *Food Hydrocolloids* **18** : 581-592.
[16]Ockerman H W and Hansen C.L. 2000. Animal by product processing and utilization. CRC Press, USA.
[17] Payung L 2001. Karakteristik edible *film* komposit gelatin daging buah pala (*Myrisitica ragran houtt*) dengan tapioka. Tesis. Program Studi Pascasarjana Universitas Gadjah Mada. Yogyakarta.
[18]Pranoto Y, Lee C M and Park H J. 2007.Characterizations of fish gelatin films added with gellan and k-carrageenan. *Swiss Society of Food Science and Technology*. Published by Elsevier Ltd. LWT **40** : 766–774.
[19]Sims T J, Bailey A J and Field D S 1997. *The Imaging Sci. Journal* **45** : 171-177
[20]Sobral P J A and Habitante A M Q B 2001. Phase transitions of pigskin gelatin. *Food Hydrocolloids* **15**: 377-382.
[21]Sompie M, Triatmojo S, Pertiwiningrum A, and Pranoto Y, 2012. The effect of animal age and acetic concertration on pigskin gelatin charateristic, *J. Indonesian Tropical Animal Agriculture* **37**(3) : 176-182
[22]Sompie M, Surtijono S E, Pontoh J W and Lontaan N 2015. *Procedia Food Science*, **3** (1): 383-388.
[23]Sompie M, Siswosubroto S E and Pontoh J H W 2015^{b}. *Proc The 6th ISTAP* (**2**): 714-718
[24]Steel R G D and Torrie J H 1991. Principles and Procedures of Statistics. McGraw-Hill Book Co. Inc. New York.
[25]Tapia-Blácido D, Sobral P, Menegalli F 2005. *J. Chem. Eng* **212** : 249-256.

[26]Ward A G and Courts A 1977.The Science and Technology of Gelatin, illustrated (ed). Academic Press. New York.
[27]Wang L, Auty M A E, Rau A, Kerry J F and Kerry J P 2008. *J .of Food Engineering*, 90 (1), 11-19

Physical properties and cellular structure of bread enriched with pumpkin flour

Agung Wahyono[1], A Z Tifania[1], E Kurniawati[1], Kasutjianingati[2], W W Kang[3,5] and S K Chung[4]

[1)] Food Industrial Technology Department, State Polytechnic of Jember, Jl. Mastrip PO. Box 164 Jember, Indonesia
[2)] Horticulture Plant Production Department, State Polytechnic of Jember, Jl. Mastrip PO. Box 164 Jember, Indonesia
[3)] Department of Food and Food Service Industry, Kyungpook National University, 2559, Gyeongsang-daero, Sangju-si, Gyeongsangbuk-do 37224, Korea
[4)] School of Food Science and Biotechnology, Kyungpook National University, Daegu 41566, Korea
[5)] Corresponding author, e-mail: wwkang@knu.ac.kr

Abstract. Production of a variety of bread is driven by the consumer interest in bringing a unique and healthier bread to the table. This study was aimed to evaluate the effect of partial substitution of pumpkin flour on physical properties and cellular structure of pumpkin bread. Pumpkin breads were made by partially substituting wheat flour with 5% to 20% of pumpkin flour. The physical properties and cellular structures of breads were evaluated. Breads specific volume decreased significantly at 10% and at higher level of pumpkin flour enrichment. Pumpkin flour did not affect the texture of enriched bread compared to that of control breads. The lightness of crumb decreased significantly with 10% of pumpkin flour enrichment. Inversely, the redness and yellowness increased significantly at the level of enrichment of 5% and 10%, respectively. Pumpkin flour enrichment affected significantly the cellular structured of enriched breads except of cell density. Mean cell area of enriched breads was smaller than that of control breads. At the level of 20%, pumpkin flour enrichment increased significantly the cell area of enriched bread. We suggest the use of pumpkin flour less than 10% in manufacturing pumpkin enriched bread without compromising its physical properties. At higher level, the pumpkin flour enrichment can be done to improve the cellular structures of bread.

1. Introduction

Pumpkin is a food that is rich in dietary fiber, especially pectin, functional compounds, vitamins (A, B6, K, C, E) and minerals (K, Mg, P, Se and Fe). High food fiber content in the form of pectin can control serum insulin levels, lower blood sugar levels, increase glucose tolerance and combating of various diseases such as heart attack, diabetes, high-blood pressure, constipation and colon cancer [1;2;3].

Public awareness to consume healthy bread continues to increase along with the high incidence of diseases caused by advanced civilization such as coronary heart disease, obesity, and diabetes. This encourages the development of the bread industry for certain purposes such as high nutritious bread and for diets [4]. Some studies were reported on the use of ingredients other than wheat flour for certain purposes include; the influence of fungal mycelial supplementation on the quality of bread [5], the effect of the use of inulin soluble fiber and beta-glucan oats on the quality of bread [6], the effect of adding Jerusalem artichoke flour to the quality of bread [7], and the effect of adding pumpkin flour to the volume and quality of organoleptic bread [8].

Previously Ptitchkina et al. [8] reported that adding pumpkin flour gave a very large increase in the volume and organoleptic characteristics of the bread produced. Based on this, the addition of pumpkin flour in bread making is very promising to improve its quality.

Bread making with the addition of pumpkin flour which has been optimized for the functional component content has never been done to date. Therefore, in this study, we made breads partially substituted with pumpkin flour at different concentrations. The objective of this study was to evaluate the effect of partial substitution of pumpkin flour on the physical properties and cellular structure of bread.

2. Materials and Methods

2.1 Materials

The yellow pumpkin used was *bokor* or *cerme* cultivar with flat rounded shape having orange yellowish color and does not rotten, obtained from Jember area. Other ingredients were wheat flour (high protein contents), sugar, salt, butter, yeast (Saf-Instant), water, and xanthan gum.

2.2 Methods

2.2.1 Bread making procedure

Bread formula is presented in Table 1. Bread making was performed by substituting wheat flour with pumpkin flour (PF). PF substitution was done at concentration of 0, 5, 10, 15 and 20%, based on the weight of wheat flour. Bread formulations are shown in Table 1. A bread maker (National SD-BT102, Panasonic, Osaka, Japan) was used to bake breads using 4-h bread making setting. Baking was performed three times (triplicate). Before analysis, the bread loaves were placed at room temperature (28–30 °C) for tempering.

Table 1. Bread formulations

Ingredients	**Sample**				
	C%	**PF5%**	**PF10%**	**PF15%**	**PF20%**
Wheat flour (g)	280	266	252	238	224
Pumpkin flour (g)	0	14	28	42	56
Xanthan gum (g)	1,4	1,4	1,4	1,4	1,4
Sugar (g)	16,8	16,8	16,8	16,8	16,8
Salt (g)	5,6	5,6	5,6	5,6	5,6
Water (mL)	200	200	200	200	200
Yeast (g)	2,8	2,8	2,8	2,8	2,8

2.2.2 Physical properties analysis

Specific volume of bread was measured using the seed displacement method [9]. The crust and crumb colors were determined by using color reader (CR-400, Konica-Minolta, Tokyo, Japan); Hunter's values were then analyzed (lightness, redness, yellowness) from six regions of bread crust and crumb (Wahyono et al., 2015). The bread texture was analyzed using texture analyzer. The flat probe of Warner-Bratzler with a 3.2 mm thickness aluminum blade was used to test the level of hardness of bread.

2.2.3 Bread crumb image analyses

The ImageJ software (1.47v, National Institutes of Health, Bethesda, MD, USA) was used to analyze the cellular structures of the bread crumb. Then, mean cell area, cell area relative distribution, and the fraction of cell area to total area were calculated. The crumb images were captured using a scanner (Epson Perfection V370 Photo, Epson, Suwa, Japan) at a resolution of 600 × 600 dpi. To reflect the actual size, crumb's image was calibrated using a known scale and cropped to 60 × 60 mm. Image processing was done using a bandpass filter. To produce binary images, convert to mask feature was used for differentiating the cell and non-cell area. The threshold was adjusted using a black and white and a dark background. To analyze particles (cells), the particle size was adjusted from 0.01 mm2 to infinity and the circularity was set from 0 to 1 [7]. The results were then tabulated and calculated using Microsoft Excel 2007v.

2.2.4 Statistical analysis

Statistical analysis was performed using SPSS for windows (ver. 19, SPSS, Inc., Chicago, IL, USA).

Statistical significance among the data were analyzed using one-way analysis of variance. Then, Duncan's multiple range test (DMRT) at $p < 0.05$ was performed to compare means between samples.

3. Results and Discussion

3.1 Physical Properties

Specific volume of bread indicates the development of bread dough after baking. The greater the specific volume value indicates the more inflated bread mixture after baking. The effect of Pumpkin Flour to the specific volume of enriched bread is presented in Table 2. Based on the statistical analysis showed that pumpkin flour enrichment had a significant effect ($P < 0.05$) on the specific volume of enriched bread.

Tabel 2. Effect of pumpkin flour (PF) enrichment on the specific volume and texture of bread

Sample	Specific Volume (cm^3/g)	Texture (N)
C	$6{,}98 \pm 2{,}54^{a}$	$16{,}34 \pm 3{,}14^{a}$
PF5	$4{,}92 \pm 1{,}04^{ab}$	$13{,}99 \pm 2{,}35^{a}$
PF10	$2{,}78 \pm 0{,}51^{b}$	$14{,}15 \pm 1{,}51^{a}$
PF15	$2{,}92 \pm 0{,}23^{b}$	$14{,}74 \pm 1{,}01^{a}$
PF20	$2{,}17 \pm 0{,}36^{b}$	$15{,}8 \pm 1{,}02^{a}$

Values presented in table are the mean of the three replications ± Standard Deviation. Values presented in column with different letters imply statistically differences ($p<0.05$), as determined by DMRT test.

Increased level of pumpkin flour causes a decreased in specific volume of enriched bread. The highest specific volume (6.98 ± 2.54 cm3/g) was obtained in bread without enrichment of pumpkin flour. Whereas, the lowest specific volume (2.17 ± 0.36 cm3/g) was obtained by adding 20% of pumpkin flour. Enrichment of 5% pumpkin flour resulted in a specific volume of 4.92 ± 1.04 cm3/g which was not significant compared to that of control bread (6.98 ± 2.54 cm3 / g). See et al. [10] reported that increased level of pumpkin flour in bread making decreased the specific volume. Adding pumpkin flour as much as 5% significantly increased the specific volume of bread compared to that of the control bread. It was suggested that water content was the main cause of an increase in the volume of bread (loaf volume).

The higher level of pumpkin flour reduced the specific volume of enriched bread. The addition of pumpkin flour weakened the gluten-starch matrix which can further reduced gas retention in the dough. The reduced ability of gas retention caused a decrease in the specific volume of bread [11]. In addition, the presence of pumpkin flour might increase solid-like behavior that can inhibits the expansion of dough during fermentation and in turn decreases the specific volume of bread. Peressini and Sensidoni [12] found a decrease in bread dough added with inulin. Similarly, Wahyono et al. [7] reported a decrease in the specific volume of bread with the addition of higher level of Jerusalem artichoke powder.

Based on the Table 2 shows that the enrichment of pumpkin flour had no significant effect ($P> 0.05$) on the texture of enriched bread. Thus, there is a chance to enrich the bread with a greater level of pumpkin flour without compromising its texture properties.

Enrichment of pumpkin flour had no significant effect ($P> 0.05$) on lightness (*L*), redness (*a*) and yellowness (*b*) of crust of enriched bread (Table 3). In addition, enrichment of pumpkin flour did not significantly affect the lightness (*L*) of crumb of enriched bread, but significantly increased the redness (*a*) and yellowness (*b*) as presented in Figure 1. Increased of redness compared to the controls bread occurred when higher level (>5%) of pumpkin flour was added. While the yellowness (*b*) increased significantly when 10% of pumpkin flour was added. See et al. [10] reported that adding pumpkin flour significantly increased ($P < 0.05$) the redness and yellowness of bread crumb. This is presumably due to the addition of pumpkin flour which has a high value of redness and yellowness. Color is an important attribute for bread products for consumers. Crust color is depending on the physical and chemical characteristics of the mixture such as water content, pH, amino acid content and reducing sugar. Besides that, it is also influenced by regulation during bread baking such as temperature, airspeed, relative humidity, and heat transfer type [13].

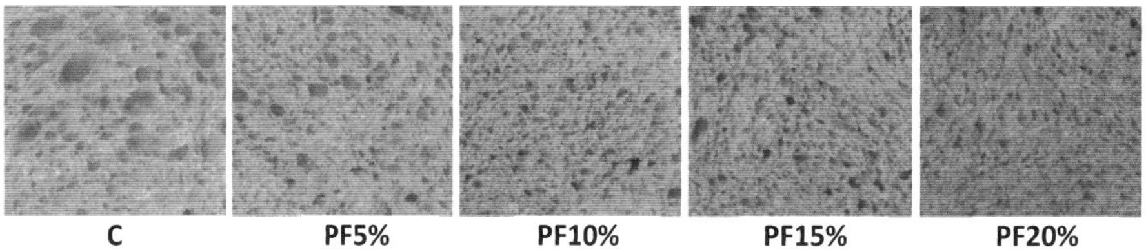

Figure 1. Color of bread's crumb made with different level of pumpkin flour (PF)

Tabel 3. Color of crust and crumb of bread made with pumpkin flour (PF)

Sample	Crust color			Crumb color		
	L	*a*	*b*	*L*	*a*	*b*
C	50,01 ± 4,72 ^{a}	27,87 ± 22,06^{a}	33,42 ± 2,86 ^{a}	73,35 ± 1,7^{a}	0,87 ± 0,64^{d}	21,66 ± 1,48^{b}
PF5	50,53 ± 6,68 ^{a}	17,27 ± 0,55 ^{a}	38,41 ± 5,76 ^{a}	69,02 ± 3,09^{a}	5,27 ± 0,32^{c}	41,81 ± 5,17^{ab}
PF10	55,35 ± 7,84 ^{a}	15,47 ± 0,97 ^{a}	47,33 ± 7,72^{a}	62,90 ± 1,64^{b}	9,68 ± 0,84^{b}	50,95 ± 9,36^{a}
PF15	43,31 ± 1,83 ^{a}	18,34 ± 0,64 ^{a}	33,82 ± 0,84^{a}	59,96 ± 1,55^{b}	11,94 ± 1,5^{ab}	53,65 ± 9,74^{a}
PF20	47,17 ± 2,83 ^{a}	18,44 ± 1,19 ^{a}	41,43 ± 6,89^{a}	57,52 ± 1,76^{b}	14,02 ± 1,06^{a}	55,45 ± 9,43^{a}

Values presented in table are the mean of the three replications ± Standard Deviation.
Values presented in column with different letters imply statistically differences ($p<0.05$), as determined by DMRT test.

3.2 Cellular Structures

Bread is manufactured in random physical processed resulting in a heterogeneous cellular structure. That heterogeneous structure is considered to affect the mechanical behavior of bread's crumbs. The contribution of cellular features to bread quality can be quantified by using image analysis, [14].

Based on the Table 4 it was observed that PF substitution significantly decreased ($P<0.05$) cell density and cell area fraction of resulting bread but no significant effect was observed in cell density. Cellular structure of bread's crumb enriched with pumpkin flour is shown in Figure 2. Incorporation of other ingredients into wheat bread tend to decrease mean cell area and cell area fraction. Previously, we found that incorporating Jerusalem artichoke powder into wheat bread significantly decreased mean cell area and cell area fraction of bread's crumb. It has been reported that physical texture and the cellular structure of bread crumb are strongly correlated. The two parameter are related quality factors that should be considered as a single entity [15]. Pyler [16] suggested that the crumb feels to the mouth is strongly associated with the cell structure of the crumb. The crumb having finer, thin-walled, uniform cells size produced a softer and more elastic texture than that of the coarse and thick-walled cell structures.

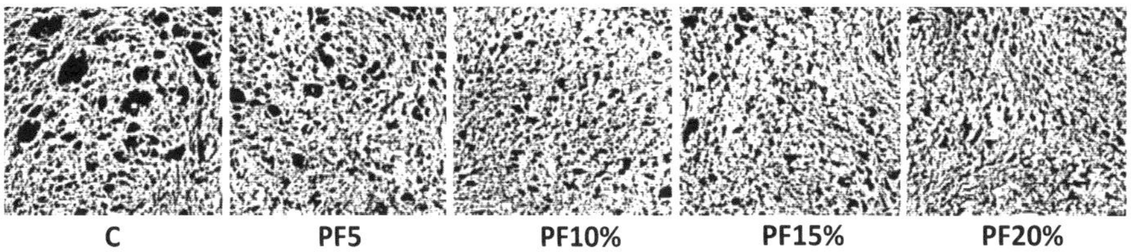

Figure 2. Cellular structure of bread made with various level of pumpkin flour (PF)

Table 4. Crumb cellular structure of bread made with pumpkin flour (PF)

Perlakuan	Cell Density(1/cm^2)	Mean cell area (mm^2)	Cell Area Fraction (%)
C	38,88±6,7^a	0.93±0,1^a	35.75±3,5^a
PF5	43,09±4,3^a	0.75±0,0^b	32.36±1,5^{ab}
PF10	42,76±3,8^a	0.74±0,1^b	31.09±1,1^{ab}
PF15	43,12±2,2^a	0.72±0,0^b	31.10±3,0^{ab}
PF20	48,89±12,0^a	0.63±0,1^b	30.12±3,1^b

Values presented in table are the mean of the three replications ± Standard Deviation.
Values presented in column with different letters imply statistically differences ($p<0.05$), as determined by DMRT test.

4. Conclusions

Pumpkin flour was significantly affect the physical properties and cellular structure of enriched bread. Increased level of pumpkin flour decreased the specific volume of enriched bread. The redness and yellowness of bread's crumb increased in line with the level of pumpkin flour enrichment. But, no effect was observed on the color of bread's crust. Mean cell area and cell area fraction decreased as the level of pumpkin flour increased. Thus, we suggest the use of pumpkin flour less than 10% in manufacturing pumpkin enriched bread without compromising its physical properties. At higher level, the pumpkin flour enrichment can be done to improve the cellular structures of bread.

Reference

[1] Aydin E and Gocmen D 2015 *LWT - Food Science and Technology* 60 (1) 385–392. doi:10.1016/j.lwt.2014.08.025
[2] Rakcejeva T, Galoburda R, Cude L and Strautniece E 2011 *Procedia Food Science 1* 441–447. https://doi.org/10.1016/j.profoo.-2011.09.068
[3] Adams G G, Imran S, Wang S, Mohammad A, Kok S, Gray D A and Harding S E 2011 *Food Research International* 44 (4) 862–867.
[4] Hager A S, Zannini E and Arendt E K 2012 In: Cauvain S P (Ed.) CRC Press, Boca Raton, Cambridge, England Pp. 711-735.
[5] Ulziijargal E, Yang J H, Lin L Y, Chen C P and Mau J L 2013 *Food Chemistry* 138: 70–76.
[6] Hager A S, Ryan LAM, Schwab C, Gänzle M G, O'Doherty J V and Arendt E K 2011 *Eur. Food Res. Technol* 232: 405-413.
[7] Wahyono A, Lee S B, Yeo S W, Kang W W and Park H D 2016 *Emirate Journal of Food and Agriculture* 28 (4): 242-250.
[8] Ptitchkina N M, Novokreschonova L V, Piskunova G V, and Morris E R 1998 *Food Hydrocolloids 12* (3): 333–337. https://doi.org/10.1016/S0268-005X(98)00024-1
[9] AACC 10-05. 2000 St. Paul :American Association of Cereal Chemists.
[10] See E F, Wan Nadiah W A and Noor Aziah A A 2007 *International Food Research Journal 14* (2): 123–130.
[11] Rubel I A, Pérez E E, Manrique G D and Genovese D B 2015 *Food Struct.* 3: 21-29.
[12] Peressini D and Sensidoni A 2009 *J. Cereal Sci.* 49: 190-201.
[13] Zanoni B, Peri C and Bruno D 1995 *Lebensm. Wiss. u.Technology* 28: 604- 609.
[14] Zghal M C, Scanlon M G and Sapirstein H D 2002 *J. Cereal Sci.* 36: 167-176.
[15] Kamman P W 1970 *Bakers Digest* 44 (2): 34–38.
[16] Pyler E J 1988 volume II (pp. 850– 910). Merriam, KS: Sosland.

Acknowledgements

Authors would like to thank to the Ministry of Research, Technology and Higher Education the Republic of Indonesia for financing this study. Authors also wishing to thank Kyungpook National University for supporting this study through research fund.

Formulation of Artificial Rice Cereal by using Fermentation of *L Plantarum* POLIJE 15420 for Diabetes Mellitus Patients

H Warsito[1*], A Santoso[2] and Y Wibisono[2]

[1]Health Department, Politeknik Negeri Jember, Mastrip PO Box 164 Jember 68101, Indonesia

[2]Agricultural Technology Department, Politeknik Negeri Jember, Mastrip PO Box 164 Jember 68101, Indonesia

[*] Corresponding author : hersit@gmail.com

Abstract. The rheological characteristics of POLIJE's artificial rice cereal have been examined by using a single screw extruder with a fermentation process using a new strain of *L. plantarum* bacteria POLIJE 15 420 (HI.05.01.03.050.0226/2010/US Patent) on the raw materials used. Nowadays, the development of artificial rice cereal variants has been investigated as a form of serving instant healthy food, using various combinations of the formulation and optimizing POLIJE extruder machine (Patent number: P00201100694 in 2011). The research objectives are optimizing the screw extruder cereal, examining the extruder performance on produced formula, and evaluating the characteristics of the products developed commercially. The results showed that POLIJE's single screw extruder machine reached the maximum speed at 225 $\pm$ 25 rpm with a fluid speed of 0.19 x 10-157 Pa. Based on FLUENT, it indicated that the biggest speed of fluid flow from the screw angle of 2.5 degrees was 0.780727403 m/s, while the screw angle of 5 degrees and 7.5 degrees were 0.184929482 m/s and 0.534818848 m/s.

1. Introduction

The development of artificial rice becomes one of the priorities and it plays a pivotal role in reducing the rate of rice consumption and functioning as food alternatives for controlling *diabetes mellitus*. Indonesia as a developing country has the largest rice consumption rate compared with other Asian countries. This position makes Indonesia becomes the largest net importers of rice. On the other hand, Indonesia has high potential of crops such as cassava and corn which is classified as indigenous natural wealth. Indonesian people have not maximized the consumption of these indigenous crops, even put them inferiorly. Some previous research result show that indigenous crops have a complex carbohydrate with a low GI value and contain a component for lowering blood sugar.

In 2014, a research team has developed encapsulation of POLIJE 15420 L. Plantarum. Further research has been done in 2015, and pursued the development of further generation artificial rice by using a single screw extruder machine with the formulation of utilizing bacteria. The results of the studies provided a recommendation to develop artificial rice in the form of instant cereal breakfast which meets the necessity of today's demand. However, only few studies had been conducted especially on the cereal production with the concept of single screw extruder and utilizing bacteria in it.

The steps of producing cereal also involves gelatinization process. It is the conversion of a raw material to a cooked and digestible material by the application of heat and less water of 12-22% [1]. The design of the screw has been modified to affect the cutting process of the starch [2]. She further gives the example of amylopectin on corns and soybeans which can be cut through friction forces. From this process, it also indicates the degradation of molecular weight both amylose and *amylectin*. [3] adds the information that the degradation of molecular weight on corn starch frequently happens in amylopectin

Cereals also undergo protein changing. Extrusion on soy shows changes in protein solubility and the composition of globulin [4]. Due to a non-covalent interaction and a formation of *disulfide*, the protein is not dissolved during extrusion. That is why cereals made from artificial rice technique is new and unexplored, mainly because it involves indigenous raw ingredients.

Nowadays, the study which focuses on the composition formula and optimization of single screw extruder of the rheological characteristic of artificial rice cereal has not been done, including studies to decrease blood sugar. It is expected that the research results can develop artificial rice cereal which provide pivotal benefits to the community.

2. Research method

Determination of screw and die designs that will produce rice-shaped grains with optimum properties. For this reason, computer simulations are used using CATIA and ANSYS computer programming. Several screw and die models were tested in the computer simulation. Finally four screw and die pairs were obtained which will be tested in the experimental stage.

At the experimental stage the parameters observed are Shanon enthrophy (Shanon, 1948) to measure the level of stirring from the screw extruder and the yield (% of rice produced): measures the ability of threads and dies in printing rice.

Artificial rice cereal variants can be made with the formulation of 40% corn and 50% mocaf (modified cassava flour) and other additional ingredients (in the process of drafting a patent). The first phase of research was examining the rheological aspects of the composition formula as shown in *Shanon entrophy* data, contribution, and feed density by optimizing screw exstuder. It also covered the pace of incoming ingredients, the speed of screw rotary, and cooking temperature. The second phase was investigating proximate analysis aspects, the digestibility of *granulas*' micro- structure and the consistency of gel granules of artificial rice cereal as well as components in decreasing blood sugar.

3. Result and discussion

The artificial rice cereal is easy to prepare. Pour the cereal into your bowl and add boiled water slowly, until the cereal begins to float a bit and wait for about 15 minutes. Finally, it is ready to be served.

This artificial rice cereal had rheological characteristics which has been identified by using *chromameter*. It had a specific color with the value of L 81.11, a + b + 3.21 and 26.03. It indicated that the color of the artificial rice cereal was brown. This product contained 11.02 protein, 27.2 % amylose, and 155% expansion value. The water content in this product was 11%. It was under the maximum water content required by SNI (Indonesian National Standardization) or BSN with the level of 13 %. The storage life of this product which relates to Aw and air RH was 5 weeks without showing increasing water content. It is assumed that this product can be consumed safely until the eighth week at room temperature.

Further results showed that the POLIJE's screw extruder machine reached the maximum speed at 225 + 25 rpm with a fluid speed of 0.19 x 10-157 Pa. Based on FLUENT result, it indicated that the biggest speed of fluid flow from the screw angle of 2.5 degrees was 0.780727403 m/s, while the screw angle of 5 degrees and 7.5 degrees were 0.184929482 m/s and 0.534818848 m/s. It was indeed well known that the smaller angle of the screw extruder, the bigger friction force produced. This happened because the increase of viscosity of the material had the greatest effect on a screw with a smaller angle.

The Artificial rice cereal variants could be made with the formulation of 40% corn and 50% mocaf (modified cassava flour) and other additional ingredients (in the process of drafting a patent). The preliminary characteristics which was essential for POLIJE's artificial rice cereal was development index of 155 % and amylose of 27.2 %. The pre-test which has done on mice using this product for 4 (four) weeks proved the decreasing level of sugar by 30 % and suspected to show improvement of insulin sensitivity. The value of IG for artificial rice cereal POLIJE variant is 70.

4. Conclusion

The artificial rice cereal can be produced by using the formulation of indigenous ingredients with a particular proportion and optimizing a screw extruder. When cutting the artificial rice cereal, it is urgent to consider about temperature accuracy, screw speed and water content of the dough. This artificial rice cereal can be developed commercially with the advantages of having a low GI, a longer shelf life of 4 (four) weeks, and ability to lower blood insulin levels by 30 percent

5. References

[1] Colonna P, Tayeb J and Mercier C 1989 *Extrusion cooking of starch and strachy products* In Extrusion cooking. C. Mercier, P.Linko and JM Harper (Ed.). Am. Assoc. Cereal Chem., Inc., St. Paul, MN.

[4] Mouquet C, Salvignol B, Van Hoan N, Monvois J and Treche S. 2003 Ability of a "very low cost extruder" to produce instant infant flour at a small scale in Vietnam. *Food Chem.* **82**: 249-255

[3] Juliano BO 1971 A simplified assay for milled rice amylase *Cereal Sci.* **16**: 334-338

[2] Fichtali J and van de Voort FR 1989 Fundamental and practical aspects of twin screw extrusion. *Cereal Food World* **34** (11): 921-928.

IOP Conf. Series: Earth and Environmental Science **207** (2018) 012056 doi:10.1088/1755-1315/207/1/012056

Improving Soil Fertilizer Through Application of Organic Fertilizer Humid Acid and Mikoriza in Supporting Growth and Production of Chilli Plants In Sand Land

M Z Sukri[1*], V K Sari[1] and R Firgiyanto[1]

[1] Departemen of Agricultural Production, State Polytechnic of Jember, Jl. Mastrip, Jember, Indonesia

*Corresponding author : mzayinsukri@gmail.com

Abstract. Availability of agricultural land has decreased, especially fertile rice fields and potential with technical irrigation caused by the use of land into non-agricultural land. Therefore, it is necessary to expand the area of agriculture by utilizing marginal land such as sand land. The purpose of this study is to get the best treatment combination of organic fertilizer, humic acid and mycorrhiza to support the growth and production of chili on sand land. This research activity was carried out in May to November 2018 in the greenhouse, Plant laboratory of the State Polytechnic of Jember, East Java at an altitude of 90 m above sea level. The research design used in this study was a non factorial complete randomized block design with twelve treatment combinations of doses from organic fertilizer, humic acid and mycorrhizae with three replications. Data from observations were analyzed by variance analysis with SAS version 9.0 software. If there is a real effect from treatment, the results are further tested using Duncan multiple range test at the level of 5%. The experimental results show that application of combination of organic fertilizer + humic acid + mycorrhizae can increase plant growth at 14 days to 35 days after planting. The best growth results of the best plants were found in organic fertilizer at a dose of 20 tons / ha + humic acid with a concentration of 200 ppm + mycorrhizae with a dose of 30 g/plant.

1. Introduction

Land is one of the basic needs in agricultural cultivation. The availability of agricultural land at this time has decreased, especially fertile rice fields with technical irrigation systems caused by changes in land use to non-agricultural land. According Istiyanti [1] in the period 1991 to 2020 is estimated at about 680,000 hectares of agricultural land in Java will be transformed into non-agricultural land. Therefore, it is necessary to expand the area of agriculture by utilizing marginal land. One land that has potential is sand land near the beach with almost 95% sand content. The main constraint of cultivation in sandy land is that it is very easy to drain water with loose structures, poor in nutrients and has strong wind conditions, with a high rate of evaporation that contains lots of salt. The advantage is that the amount of land is wide, flat, close to ecotourism, rarely floods with shallow water surface, and abundant sunlight [2]. According to Istiyanti *et al.* [1]; Sutardi and Wirasti [3] that the introduction of technology packages for cultivation in sand land is technically easy to implement, economically profitable and socially culturally easy to be accepted by farmers.

The use of sand land for cultivation has been carried out, including for onion cultivation [2], [4], corn [5], sesame [6], and watermelon [7]. Increasing soil fertility in sandy land is a major aspect that must be immediately improved to optimize agricultural production. These improvements include through application of organic fertilizer, humic acid and mycorrhizae. Application of organic matter will improve soil fertility, both physical and chemical. Application of organic matter also provides a source of energy for soil microorganisms so that the population of microorganisms increases.

The results of research by Purwaningsih *et al.* [7] and Nurhayati *et al.* [8] reported that the application of cow manure at a dose of 10 tons / ha can increase the yield of sesame crops on sand land in Purworejo. Application of organic fertilizer also needs to be supplemented with the addition of humic acid. Humic acid is the active ingredient from the extraction of organic matter that has passed the humification process and its function as a growth stimulating agent for plants [9]. The application of humic acid has been able to increase the growth and yield of some plants, including spinach [10]; corn, rice [9], and tomatoes [11]. Arbuscular Mycorrhizal Fungi (AMF) is one of the fungi groups that play a role in improving soil quality [12]. Mycorrhizal fungi live in the rhizosphere by forming a symbiotic mutualism relationship through intensive hyphae production [13] so as to increase host plant resistance to drought conditions [14], increasing the ability of plants to absorb nutrients including elements P [15], [16], [17], and increased resistance to plant diseases [18]. One of the horticultural plants that has great prospects to be developed and cultivated in the sandy land of Puger Sub-district is the Chili Plant. This plant is one of the plants that has broad adaptability that can be planted in the lowlands to the highlands and can grow and produce in the rainy and dry season [19]. Chilli plants also have a short life and have high economic value with good market opportunities so that attention to chili commodities becomes very important [19]. The consumption of chili every year is always increasing, therefore high chili production is needed. Through the application of amelioration technology on sandy land in the form of organic matter, humic acid and mycorrhizae will be able to increase land productivity so that it will be able to improve the economy of the community in Puger sub-district based on recommendations from the results of the research.

2. Methods

This research activity was carried out from May to November 2018 in a greenhouse from the plant laboratory, Jember State Polytechnic, East Java. The tools used in this study were rulers, shovel runners, scissors, colling boxes, calipers, stationery, sprayers, cameras, hoes, buckets, analytic scales and knives. The research materials used were chili seeds, organic fertilizer, urea, SP36 and KCL, mycorrhizal, humic acid, pesticides, raffia, bamboo and black silver plastic. The design used in this study was a complete randomized block design with twelve dose treatment combinations of organic fertilizer, humic acid and mycorrhizae with three replications. The treatment combination i.e.:

The experimental observation variable from chili growth refers to research conducted by Hapsoh *et al.* [20], Ernawati *et al.* [21], and Zakia *et al.* [22]. Observation variables include plant height measured starting from the base of the stem to the point of growing plants from the soil surface. The plant height variable is measured when the plant has flowered. Observation of the stem diameter is done by measuring the diameter of the plant stem at a position of 5 cm from the surface of the soil by using calipers. Data were analyzed by variance analysis (Anova) with SAS version 9.0 software. If there is a significant effect, then proceed with the Duncan multiple range test (DMRT) at the level of 5%.

Tabel 1. Combination of treatment

No.	Code	Treatment
1	P0 A0 M0	Organic Fertilizer 0 ton / ha + Humic Acid 0 ppm + Mycorrhizae 0 g / plant
2	P0 A0 M1	Organic Fertilizer 0 ton / ha + Humic Acid 0 ppm + Mycorrhizae 30 g / plant
3	P0 A1 M0	Organic Fertilizer 0 ton / ha + Humic Acid 200 ppm + Mycorrhizae 0 g / plant
4	P0 A1 M1	Organic Fertilizer 0 ton / ha + Humic Acid 200 ppm + Mycorrhizae 30 g / plant
5	P1 A0 M0	Organic Fertilizer 10 ton / ha + Humic Acid 0 ppm + Mycorrhizae 0 g / plant
6	P1 A0 M1	Organic Fertilizer 10 ton / ha + Humic Acid 0 ppm + Mycorrhizae3 0 g / plant
7	P1 A1 M0	Organic Fertilizer 10 ton / ha + Humic Acid 200 ppm + Mycorrhizae 0 g / plant
8	P1 A1 M1	Organic Fertilizer 10 ton / ha + Humic Acid 200 ppm + Mycorrhizae 30 g / plant
9	P2 A0 M0	Organic Fertilizer 20 ton / ha + Humic Acid 0 ppm + Mycorrhizae 0 g / plant
10	P2 A0 M1	Organic Fertilizer 20 ton / ha + Humic Acid 0 ppm + Mycorrhizae3 0 g / plant
11	P2 A1 M0	Organic Fertilizer 20 ton / ha + Humic Acid 200 ppm + Mycorrhizae 0 g / plant
12	P2 A1 M1	Organic Fertilizer 20 ton / ha + Humic Acid 200 ppm + Mycorrhizae 30 g / plant

3. Result and Discussion

Analysis data from the experiment showed that the treatment of the combination of organic fertilizer, humic acid and mycorrhizae significantly affected the variables of plant height at 14, 21, 28 and 35 days after transplanting with the best combination treatment found in organic fertilizer application with a dose of 20 tons / ha + humic acid with a concentration of 200 ppm + mycorrhizae with a dose of 30 g / plant (**Table 2**).

Table 2. Results of variance analysis of experimental data

No.	Variables observed	Treatment
1	Height of plants 7 Dat (cm)	Not significant
2	Height of plants 14 Dat (cm)	significant
3	Height of plants 21 Dat (cm)	Very significant
4	Height of plants 28 Dat (cm)	Very significant
5	Height of plant 35 Day (cm)	significant
6	Height of plant 42 Day (cm)	Not significant
7	Height of plants 49 Day (cm)	Not significant
8	Diameter of stem (cm)	Not significant
9	The time to start flowering (Day)	Not significant

Note:
Day = The Day after transplanting

The use of organic fertilizer on critical land such as coastal sand land is used to improve soil structure which affects the increase of water infiltration into the soil, the ability to bind water increases, and the use of chemical fertilizers is more efficient [5]. application of organic fertilizer, humic acid and mycorrhizae can increase plant height, allegedly due to an increase in soil fertility so that the soil is able to supply nutrients for the needs of plants that were not previously available. Increased nutrients will affect the production of chlorophyll in plants so that they can produce proteins and enzymes. The chlorophyll content will stimulate metabolism in plants so that there is also a process of increasing photosynthesis rate which affects vegetative growth such as plant height. According to Heil [23] and Ferrara and Brunetti [24] and Victolica *et al.* [11] that the administration of humic acid will be able to improve metabolic processes in plants, such as improving the process of photosynthetic rate of plants due to increased chlorophyll content in leaves. In the experiments conducted by Sarno et al 2015 the application of humic acid at a dose of 150-200 mg / liter was able to increase the growth of tomato plants at a concentration of 150-200 mg L-1. The symbiosis between FMA with high-level plants that occur in the root area and outside the plant roots also causes an improvement in soil structure [25] which is beneficial for plants. In addition to the height of plants aged 14 to 35 after transplanting is a condition where plants occupy the peak of growth in the sigmoid curve, so that the additional intake of nutrients is thought to also affect the high increase in that age.

Variable of stem diameter and flowering start time did not show a significant influence after various treatments were applied. This is presumably because the influence of the combination treatment is still not optimal for plant growth because the initial purpose of the combination is to improve the soil fertility level of the Puger District which is classified as very low. Soil in Puger District based on the results of soil analysis that has been carried out has a soil texture in the form of clay sand, very fast soil drainage, effective depth is very shallow to moderate, salinity is very low, has a low cation exchange capacity (CEC), base saturation (KB) high to very high, pH slightly sour and neutral, C-Organic is very low, total N is very low, P content is moderate. These characteristics are thought to cause the need to add a combination of organic fertilizer, humic acid and mycorrhizae with a higher dose than before so that the effects of the treatment will show better results not only to only add to the high growth of the chili plant.

Table 3. Results of DMRT analysis influence the combination of organic fertilizer, humic acid and mycorrhiza on the growth of chili

No.	Treatment	Variables of Growth								
		HoP 7 Day (cm)	HoP 14 Day (cm)	HoP 21 Day (cm)	HoP 28 Day (cm)	HoP 35 Day (cm)	HoP 42 Day (cm)	HoP 49 Day (cm)	DoS (cm)	TF (Hst)
1	P0 A0 M0	15,38	16,42 c	18,03 cd	22,00 bcd	26,64 ab	30,43	47,28	2,76	46,78
2	P0 A0 M1	13,26	14,54 abc	17,91 cd	21,90 bcd	26,41ab	28,83	43,89	2,51	44,17
3	P0 A1 M0	11,80	13,71 abc	16,73 abc	20,86 abc	26,26 ab	29,70	47,11	2,83	44,83
4	P0 A1 M1	11,18	13,80 abc	17,89 cd	22,76 cd	28,72 abc	30,66	49,72	2,81	44,11
5	P1 A0 M0	9,99	11,98 a	14,28 a	18,77 a	24,54 a	27,70	46,06	2,70	44,22
6	P1 A0 M1	12,31	14,21 abc	16,87 abc	22,13 bcd	28,19 abc	29,07	50,11	2,69	43,33
7	P1 A1 M0	12,08	15,09 bc	18,36 cd	23,02 cde	28,92 abc	29,74	47,67	2,86	41,50
8	P1 A1 M1	11,00	14,57 abc	17,50 bcd	23,34 cde	30,07 bc	30,53	55,17	2,98	42,56
9	P2 A0 M0	10,86	12,89 ab	14,72 ab	19,44 abc	24,44 a	27,81	46,39	2,81	45,33
10	P2 A0 M1	13,46	15,00 abc	17,34 bcd	22,36 bd	27,47 abc	29,73	49,56	2,80	44,00
11	P2 A1 M0	13,63	16,70 c	19,21 cd	24,36 de	30,66 bc	31,56	55,44	2,99	42,78
12	P2 A1 M1	13,81	16,66 c	19,98 d	25,89 e	30,48 c	31,88	57,44	3,03	41,06
F Hitung P		1,87	2,64*	3,50**	4,20**	2,47*	0,71	2,32	1,95	1,27

Note:
HoP= Height of plant (cm); DoS= Diameter of stem (cm); TF= The time to start flowering; Day= The day after transplanting. The numbers followed by the same letter in the same column are not significantly different in the 5% DMRT and (**) shows very real differences and (*) shows significantly different

4. Conclusion

Application of combination of organic fertilizer + humic acid + mycorrhizae can increase plant growth at 14 days to 35 days after planting.The best growth results of the best plants were found in organic fertilizer at a dose of 20 tons / ha + humic acid with a concentration of 200 ppm + mycorrhizae with a dose of 30 g / plant.

Acknowledgment

The author would like to thank Ristekdikti for the BOPTN research fund in 2018 through the Research and Community Service Center of Jember State Polytechnic.

References

[1] Istiyanti E, Khasanah U and Anjarawati A 2015 *J. Agraris* **1** 6-10
[2] Rajiman, Yudono P, Sulistyaningsih E and Hanudin E 2008 *J. Agrin* **12** 67-77
[3] Surtadi C A and Wirasti 2017 *J. Pengkajian dan Pengembangan Teknologi Pertanian* **20** 125-139
[4] Nugroho W A and Firmansyah M A 2016 *Prosiding Seminar Nasional Inovasi Teknologi Pertanian Banjarbaru, 20 Juli 2016.*
[5] Suwignyo B, Pratomo B W, Umami N, and Suhartanto B 2015 *Buana Sain* **15** 13-18.
[6] Irawanata M H Y 2017 *Evaluasi kesesuaian lahan pasir pantai untuk Pengembangan tanaman wijen (Sesamum Indicum L.) di Kecamatan Puger Kabupaten Jember* (Yogyakarta: Fakultas Pertanian Universitas Muhammadiyah Yogyakarta)
[7] Purwaningsih H S, Muhartini S and Kurniasih B 2014 *Vegetalika* **3** 25–34.
[8] Nurhayati D R, Sarwono A E and Haryono B 2013 *Bul. Tanaman Tembakau, Serat & Minyak Industri* **5** 31−39.
[9] Suwardi H and Wijaya 2013 *J. Ilmu Pertanian Indonesia* **18** 79-84.
[10] Sarno and Eliza F 2011 *Prosiding SNSMAIP III:* 289-293.
[11] Victolica H, Sarno and Ginting Y C 2014 *J. Agrotek Tropika* **2** 297-301.
[12] Agustin W, Ilyas S, Budi S W, Anas I and Suwarno F S 2010 *J. Agron. Indonesia* **38** 218 – 224.
[13] Hildebrandt U, Janetta K and Bothe H 2002 *Appl. Environ. Microbiol.* **68** 1919-1924.
[14] Echave M, Conti M, Clua A, Ruscitti M and Beltrano J 2005 *Lotus Newsletter* **35** 182-186.
[15] Nurhayati N 2012 *J. Agrista* **16** 80-86.
[16] Musafa K M, Aini L Q and Prase B 2015 *J. Tanah dan Sumberdaya Lahan* **2** 191-197.
[17] Musi R, Ghulamahdi M., Melati M, Purwono and Mansur I. 2016 *J. Penelitian Pertanian Tanaman Pangan.* **35** 229-238.
[18] Khairani H S, Sinaga M S and Mutaqin K H 2017 *J. Fitopatol. Indonesia* **13** 17-25.
[19] Surtadi C A, and Wirasti. 2017 *J. Pengkajian dan Pengembangan Teknologi Pertanian* **20** 125-139.
[20] Hapsoh, Gusmawartati, Amril A I and Diansyah A 2017 *J. Hort. Indonesia* **8** 203-208.
[21] Ernawati P, Rahardjo, and Suroso B 2017 *Agritrop* **15** 71-83.
[22] Zakia A, Ilyas S, Budiman C, Syamsuddin and Manohara D 2017 *J. Hort. Indonesia* **8** 171-182.
[23] Heil C A 2005 *Harmful Algae* **4** 603–618.
[24] Ferrara G, and Brunetti G 2010 *J. Agric. Res.* **8** 817-822.
[25] Nichols K A 2008 *Springer* 177-194.

IOP Conf. Series: Earth and Environmental Science **207** (2018) 012057 doi:10.1088/1755-1315/207/1/012057

Going Digital: A New Concept of Indonesian Food Composition Databases

M Iqbal [1,*], ***M R Permadi*** [1]

[1] Department of Clinical Nutrition, Politeknik Negeri Jember, Indonesia

*E-mail: iqbalbasagili@polije.ac.id

Abstract. Current Indonesian nutritional value information is only in the shape of books (TKPI) and not very easy to be carried in mobile. No media or reference that provides information on nutritional value unless we buy processed food products in the market. The purpose of this research is to develop digital database of food and packaged food using waterfall method (requirement engineering, design and implementation, testing, release and maintenance). Evaluation uses descriptive statistics in the form of table and diagrams. In judging satisfaction, the digital TKPI was preferred by users who previously used TKPI manually or used book nutrition content food. Users were given a questionnaire to give aspects judgment on TKPI digital including usefulness, satisfaction, and ease of use by providing a score with value of 1-7. TKPI could be designed and made in digital form and seemed providing good experience for users.

1. Introduction

Nutrition label reference or known as nutritional value information is a reference for stating information about nutrient contents on the labels of food products [1]. Nutritional value information is one of labels to be included on food and drink packs. Nutritional value information labels assist consumers to decide foods to be consumed by considering their nutrient contents and health choices beforehand [2].

Currently, foods' nutritional value information is only stated on an Indonesian Food Composition Table (TKPI) in form of a book which is not easy to carry around - previously DKBM [3]. Moreover, processed foods' nutritional value information can only be read on their packs. There is no media or reference providing information about their nutritional values except if we directly purchase the processed food products in markets.

The impracticability to obtain nutritional value information in TKPI urges commoners and nutritionists to utilize internet's search functions, for example through Google [4]. However, the quality of obtainable information cannot be accounted for [5]. Referring to the issue, media/instruments to ease in obtaining reliable and valid nutritional value information need to be created.

Currently, the provision of nutritional value information can no longer relies on conventional methods like books and other printed media. Media/instruments to be created in the future have to be digital-based and online since Indonesia has numerous internet users. According to the survey of

Indonesian Internet Service Implementation Association (APJII) in 2017, the most internet users for a decade in Java Island has surged by 6.58% from 51.5% to 58.08%. In national level, Indonesia has seen drastic internet use hike for 3 consecutive years from 110.2 million users in 2015 to 143.26 million users in 2017 or 54.68% from Indonesia's total population of 262 million [6]. The data shows that to address the challenge of the age, it is essential to make an information system to display foods' nutrient contents online.

The purpose of this research is to develop digital Indonesian food and processed food composition tables. Moreover, it also aims to know the usefulness, ease, and satisfaction of Indonesian food and processed food composition table websites.

2. Method and material

This research is conducted in 2 stages namely application development and application evaluation. Application development uses waterfall method comprising the stages of requirement engineering, design and implementation, testing, release and maintenance [7].

After application development was completed, a survey will be conducted. The survey will implement usability testing using USE Questionnaires. The test will USE Questionnaires by Arnold M. Lund [8] since his instrument has been used in various researches so its validity has been tested. Website test using USE method is a must to perfect the website itself. The research by Chung [9] states that USE method has the highest measurement value than other usability measurement methods. Therefore, USE method is better to be used in measuring the feasibility of nilaigizi.com website. USE can cover 3 aspects namely usefulness, satisfaction, and ease of use. All items are measured using seven points of Likert scale from 'totally disagree (= 1)' to 'totally agree (= 7)' adapted from TAM model [10]. After that, the total values of all items in each USE aspect are divided by the maximum value multiplied by one hundred to obtain the percentage of every aspect. To know website's feasibility from three USE aspects, the following feasibility category reference is used [11].

Table 1. Website Feasibility Category

No	Category	Interval Class
1	Highly Feasible	81 – 100 %
2	Feasible	61 – 80 %
3	Fairly Feasible	41 – 60%
4	Not Feasible	21 – 40 %
5	Highly Not Feasible	0 – 20 %

3. Result

3.1 Website development

When developing a website for the first time, the main things needed are concept, domain, hosting, and theme. The concept under development is a website focusing on nutritional value information of a food. The website has a simple concept without too many features to meet the focused aspect, but is enriched with information. The provided information is not only limited to providing how much nutrients to be stated on Indonesian Food Composition Table (TKPI) or packed foods' nutritional value information labels, but also providing nutritional value data compared to its fulfillment against Recommended Dietary Allowance (RDA). Moreover, the percentage of RDA fulfillment and featured nutrients in a food product are also displayed.

IOP Conf. Series: Earth and Environmental Science **207** (2018) 012057 doi:10.1088/1755-1315/207/1/012057

Figure 1. Display of Main Page of Foods' Nutritional Value Information

The website domain selected to ease user visits is nilaigizi.com. The domain selection is based on the ease of remembering and '.com' is the most common domain for users. Meanwhile, the website uses Indonesia's best website hosting as a place to develop it [12]. The selected theme is dominated with white and soft colors since according to research, contrast colors often disturb users.

3.2 Website evaluation

The respondents of this research are college students and nutritionists who are taking/have took nutritional education either D3, D4, S1, S2, or S3 of nutrition field in various polytechnics and universities across Indonesia. Before filling research questionnaires, respondents are required to visit and use nilaigizi.com website. The characteristics of subject are shown in table 2.

Table 2. Respondent Frequency Distribution

Characteristic	Frequency	Percentage (%)
Sex		
Man	19	18.1
Woman	86	81.9
Educational Level		
Associate Degree (D3)	26	24.8
Bachelor's Degree (D4)	13	12.4
Bachelor's Degree (S1)	48	45.7
Master's Degree (S2)	16	15.2
Doctoral Degree (S3)	2	1.9

The characteristics of respondent show that most respondents are women. Respondent's educational level shows that most of respondents' educational level is S1 and the least of respondents' educational level is S3.

Table 3. Questionnaire Scores of Usefulness

Item	Score (%)
1. Nilaigizi.com assists respondents to know foods' nutritional value information more effectively.	87.9
2. Nilaigizi.com is useful for me.	85.4
3. Nilaigizi.com can provide good controls in selecting foods' nutrients for daily life.	82.7
4. Nilaigizi.com can save my time.	83.6
5. Nilaigizi.com meets my needs.	78
6. Nilaigizi.com is suitable with my expectation.	77.4

Table 3 shows respondents' assessment in terms of usefulness of nilaigizi.com website. It can be seen that all statement items' scores are more than 61% which means the website is useful for the subjects. The statement item that nilaigizi.com assists respondents to know foods' nutritional value information more effectively has the highest score of 87.9%. Meanwhile, the statement of nilaigizi.com is suitable with respondents' expectation has the lowest score of 77.4%.

Table 4. Questionnaire Scores of Satisfaction

Satisfaction	Score (%)
1. I am satisfied with nilaigizi.com.	81
2. I will recommend nilaigizi.com to my friends.	82.9
3. Nilaigizi.com is interesting to be used	84.7
4. Nilaigizi.com works based on my expectation.	80
5. Nilaigizi.com is amazing.	79.8
6. Nilaigizi.com is the one I need.	81.6

Table 5. Questionnaire Scores of Ease of Use and Learning

Ease of Use and Learning	Score (%)
1. Nilaigizi.com is easy to use.	86.5
2. Nilaigizi.com is simple to use.	85.9
3. Nilaigizi.com needs simple steps to use.	85.7
4. I can use nilaigizi.com without written instructions.	85.1
5. Nilaigizi.com is consistent in providing information.	80.8
6. First and regular users will favor nilaigizi.com.	81
7. I can correct nutritional value information mistakes quickly and easily (specifically for nutritionists registered in ahligizi.id).	81.6
8. I can use nilaigizi.com any time seamlessly.	81.7
9. I learn to use nilaigizi.com quickly.	87.4
10. I easily remember how to use nilaigizi.com.	87.4
11. I quickly master the use of nilaigizi.com.	85.7

Table 4 shows respondents' assessment in terms of satisfaction of nilaigizi.com website. It can be seen that all statement items' score are more than 61% which means the website provides satisfaction to respondents. The statement item that nilaigizi.com is interesting to be used has the highest score of 84.7%. Meanwhile, the statement of nilaigizi.com is amazing has the lowest score of 79.8%.

Table 5 shows respondents' assessment on 11 statement items about Ease of Use and Learning of nilaigizi.com website. It can be seen that all statement items' scores are more than 61% which means the website provides ease of use to respondents. The statement items about quickness and ease of learning to use nilaigizi.com and ease of remembering how to use nilaigizi.com have the highest scores of 87.4% each. Meanwhile, the statement of nilaigizi.com is consistent in providing information has the lowest score of 80.8%.

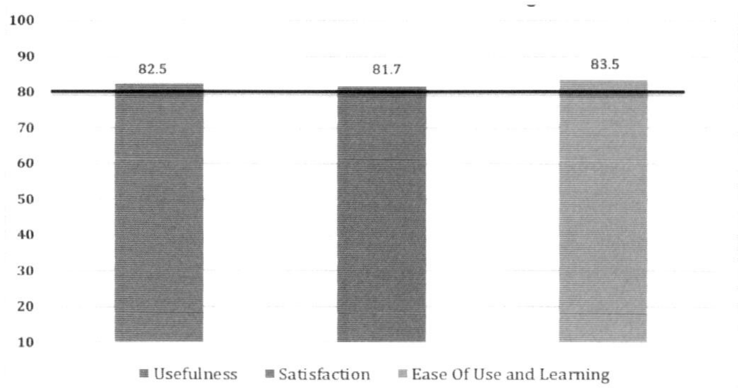

Figure 2. Nilaigizi.com's Assessment Distribution Based on the Criteria of Usefulness, Satisfaction, and Ease of Use

The average scores of three assessment criteria implemented to test the usability of nilaigizi.com website through USE Questionnaire method namely usefulness, satisfaction, and ease of use show that the three have over 80% scores. The highest score is on ease of use and learning criteria of 83.5% and the lowest score is on satisfaction criteria of 81.7%.

4. Discussion

Nilaigizi.com website is a website to know the nutritional values of ingredients and food products. After going through website development and design, testing stage is conducted. The website is tested to users who, in this case, are college students and nutritionists. This website is often used to seek nutritional value information of a product used in nutritional consultation activities. The testing stage shows that from 105 nutritionists participated in website assessment using usability testing with USE (usefulness, satisfaction, and ease of use and learning) method, the score of 80% is obtained which means nilaigizi.com website provides benefit, satisfaction, and ease of use to subjects [13].

In usefulness assessment, the statement of nilaigizi.com assists in providing effective nutritional value information has the highest score. Nutritional value information on ingredients and food products is one of compulsory information on food labels if they state specific information. Nutritional value information provides information about nutrient content whose format is determined by Food and Drug Control Agency (BPOM) [1]. Nutrition label is a standard displaying nutrients of an ingredient or food product based on specific nutrient contents for food. Consumers will decide whether they will purchase the product or not after reading the information stated on the label [14].

In satisfaction aspect, the statement of nilaigizi.com is interesting has the highest score. The website's attractive display in terms of color, writing, and menu layout contributes to provide interesting experience to subjects. In creating a website, there are two things to be considered namely website layout and design. Layout is the process of arranging or organizing objects located on a website page while design is the process of someone or website developer's imagination and creation in expressing their works and creating beauty by considering high aesthetics and artistry. Layout also includes arrangement and division of place on a page. Good and understandable layout and arrangement will make a page more interesting, organized and balanced so it is attractive to see and easy to read [15].

In Ease of Use and Learning aspect, the statements of I quickly learn using nilaigizi.com and I easily remember how to use nilaigizi.com have similar score. Nilaigizi.com website is designed in such a way so it can be easily learnt by people. Its ease of use also contributes to its feasibility. Perceived ease of use is "The degree to which an individual believes that using a particular system would be free of physical and mental effort" [16]. The statement can be interpreted as a level in which an individual believes that using a particular system will free them from physical and mental efforts. This opinion can be measured through indicators or items on the statements about ease of learning, controllability, clarity & understandability, flexibility, and ease of access [10].

5. Conclusion

TKPI can be designed and created digitally. TKPI digitalization in an online format through the development of nilaigizi.com website is acceptable for college students in nutrition field and nutritionists and provides good experience. It is shown by the average score of usefulness, satisfaction, and ease of use aspects of over 80 percent. Nilaigizi.com can be made as reference to obtain foods and processed foods' nutritional value information based on people's needs.

Acknowledgements

This thank-you note is delivered to Institute of Research and Community Service of State Polytechnic of Jember for the funding granted to researcher through the institute's competitive granted fund (PNBP).

References

[1] BPOM RI, 2016. *Peraturan Kepala Badan Pengawas Obat dan Makanan Republik Indonesia Nomor 9 Tahun 2016 tentang Acuan Label Gizi*, (Jakarta: BPOM RI)

[2] Lewis S 2017. *How food labels help us make better food choice*s. KSL.com. Available at: https://www.ksl.com/?sid=44653326&nid=1010&title=how-food-labels-help-us-make-better-food-choices [Accessed May 10, 2018].

[3] Persatuan Ahli Gizi Indonesia 2013 *Tabel komposisi pangan Indonesia* (Jakarta: Elex Media Komputindo)

[4] Cahyono L A, Winarno W, Nugroho H A 2015 *Semnas Teknomedia Online* **3**(1) 1–2

[5] Fahy E 2014. *The Australian medical journal* **7**(1) 24–8

[6] Asosiasi Penyelenggara Jasa Internet Indonesia 2017 Penetrasi & Perilaku Pengguna Internet Indonesia 2017 *Asosiasi Penyelenggara Jasa Internet Indonesia*, 1–39.

[7] Petersen K, Wohlin C, Baca D 2009 *In Springer* Berlin Heidelberg 386–400.

[8] Lund AM 2001 *Measuring usability withthe USE Questionnaire*. Usability anduser experience SIG.

[9] Chung TK and Sahari N 2015 *International Journal of Computer Theory and Engineering* **7** (2)167-171

[10] Davis F D 1985. *A Technology acceptance model for empirically testing new end user information systems: theory and results*. [*Final Project*] (United States Of America Massachusetts Institute of Technology)

[11] Amanda RF. *Evaluasi kinerja sistem informasi inventori studi kasus: PT. Titian Sinar Mustakim* [*Final Project*] (Riau: Universitas Islam Negeri Sultan Syarif Kasim Riau)

[12] Willya R. 2018 .*Web Hosting terbaik di 2018* : review &perbandingan. https://penasihathosting.com/hosting-terbaik.

[13] Iswiara H 2016 *Pengembangan aplikasi mobile sistem pendukung keputusan dengan metode simple additive weighting berbasis web untuk mencari SMK yang sesuai dengan kebutuhan siswa di Kota Yogyakarta* [*Final Project*] (Yogyakarta: Technology Information Department Univeritas Negeri Yogyakarta)

[14] Wahyu E, Widyastuti I, Setiyobroto 2018 *Pengaruh pemberian informasi nilai gizi pada buku menu terhadap persepsi konsumen dalam pemilihan menu di Rumah Makan* [*Final Project*] (Yogyakarta: Departement Of Nutrition, Poltekkes Kemenkes Yogyakarta)

[15] Hakim Ld and Uus Musalini 2004 *Cara CerdasMenguasai Layout, Desain Dan Aplikasi Web* (Jakarta: PT Elex Media Komputindo)

[16] Chuttur M Y 2009 *Sprouts: Working Papers on Information Systems*, **9** (37)

Development of Low Cost Toxic Gas Explosive Modeling System using Wireless Array Sensor Netwok

D P S Setyohadi[1*], H Y Riskiawan[1], S Kautsar[1], P Destarianto[1]

[1]Department of Information Technology, Politeknik Negeri Jember, Indonesia

*Email : dwi.putro@polije.ac.id

Abstract. The Natural disasters such as floods, earthquakes, erupting volcanoes and toxic gases affect life can even result in permanent ecosystem and environmental changes. Monitoring of toxic gases is one of the important issues, it is based on the nature of the gas that is difficult to detect movement, especially in large areas. Ijen crater reportedly began spewing toxic gas on March 21, 2018, and poisoned 30 people who lived in four villages on the slopes of Mount Ijen. This study aims to develop a tool with an array sensor application that is able to adapt to environmental conditions. This detection device providing 24hour surveillance of toxic gases. In the application, the sensors will be placed in areas affected by toxic gas disasters, it aims to provide early warning if there is a toxic gas wave from Mount Ijen. It is also able to give warning to visitors and prospective visitors of Mount Ijen. So that ultimately no longer found victims of toxic gas, or at least reduce the impact of this toxic gas disaster.

1. Introduction

Natural disasters such as floods, earthquakes, volcanic eruptions and toxic gases affect life, property, livelihoods, industry, and even better. [1]. Monitoring of toxic gases is one of the most important issues in disaster management, this is done from the nature of the gas which is difficult to detect its movement, especially in large areas. Difficulties that arise cannot be seen directly, movements are difficult to read and change at any time [2].

Mount Ijen / Gunung Kawah Ijen the administrative locations of Banyuwangi and Bondowoso Regencies, East Java 8 ° 03 '30 South Latitude and 114 ° 14 '30 "East Longitude [16] Ijen Crater began to emit toxic gases on March 21, 2018, and poisoned 30 residents who live in four villages on the slopes of Mount Ijen. The vomit of poison gas from Ijen crater is unpredictable [3].

Research on toxic gases generally uses ideal conditions / conditions limited as a reference, or in other words does not reflect a changeable / adaptive environment [4]. Research conducted by [5] carried out Boundary Diffusion analysis of toxic gases in petrochemical plants, Sun Research from four distance graphs that predicted its accuracy to detect Diffusion Boundary, after the limit detection algorithm. Research [6] modeling of a growing milling environment. Research [7] developed a heat sensor designed as an array sensor, each sensor isolated and specifically designed for MEMS. Research by [8] developed a device that uses an array sensor with a variety of gas sensors (methane, carbon dioxide, carbon monoxide, hydrogen sulfide, ammonia) which serves to analyze the gas in underground channels. Research conducted by [9] [10] [11] [12] [13] [14] [15] still uses limited environmental conditions. In reality, reading about toxic gases with changing environmental conditions [4] or in other words the developed tools should have capabilities with unstable environmental conditions.

This study aims to develop an array sensor device that is able to adapt to environmental conditions. The array sensor is a self-organizing sensor, each sensor itself, then to model the gas condition that will be used for all sensor readings, using an array sensor to provide a more complete description of the condition of a room containing smoke gas or not. In its application the sensors will be placed in areas affected by toxic gas disasters, this aims to provide early information in the event of a wave of toxic gas from Mount Ijen. In addition, it is also able to provide information to visitors and prospective visitors of Mount Ijen. Nothing else you find from poison gas.

2. Research Methods

This research method consists of several stages, such as literature review, data collection, parameter identification and data processing, application development, results and discussion, conclusions and recommendations.

1. Literature Review

Literature review is carried out to collect information from several references related to the issues to be discussed. Theories related to research problems are used as a basis for processing data. At this stage, identification and problems formulation will be conducted which will be the objectives of the research. Problem formulation to be examined based on the background of the problem.

2. Hardware Design

Based on literature studies, the deadly toxic gases found in the Gunung Ijen region are carbon monoxide and hydrogen sulfide gas. Therefore, in the hardware design, sensors MQ-7 and MQ-136. The MQ-7 sensor is used to measure carbon monoxide gas levels in the air. While the MQ-136 sensor is used to measure carbon hydrogen sulfide gas levels in the air. DHT-11 sensor is also used which serves to measure the temperature and humidity values in the ijen region.

All sensors are connected to the ATMEGA-2560 controller. The output of the system is the LED bar display, status LED, graphic LCD, 7 segment and buzzer. LED bar serves to display pollution levels in a range: good, medium, unhealthy, very unhealthy, and dangerous. Status LED functions to display the status of Mount Ijen in Normal, alert, standby, alert condition. 7 segment display is used to display the degree of temperature in the Gunung ijen area. Figure 1 is a system block diagram of the tool.

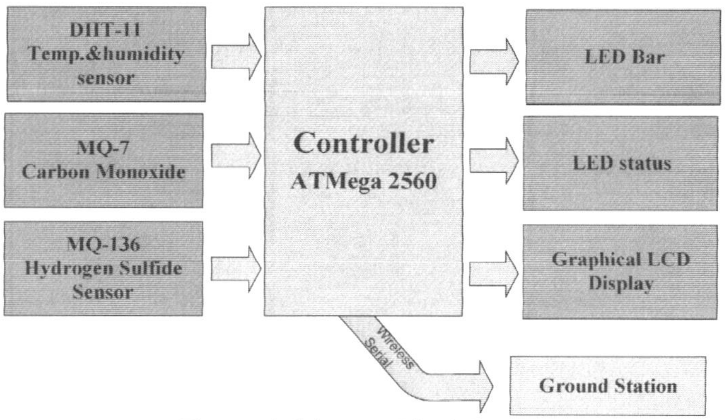

Figure 1. Diagram Block System

3. Application Development

The controller programming algorithm is made according to the flowchart in figure 2. In the initial stage, the controller will wait for 2 minutes to pause the sensor heating. Gas-based sensors, require heating for several minutes before they can be used. After the heating process is complete, the controller will read the data from the MQ-7, MQ-136 and DHT-11 sensors. Furthermore, based on the data read, the controller will give the output value in the form of air pollution level status, temperature degree (in centigrade), and the status of the Mount Ijen area. After the entire status is successfully displayed, the controller saves the data from the sensor reading on the memory card. Before, the controller will also read the time data from the RTC module. This functions as a data logger system on the device. All processing data is then sent to Ground Station through serial wireless communication.

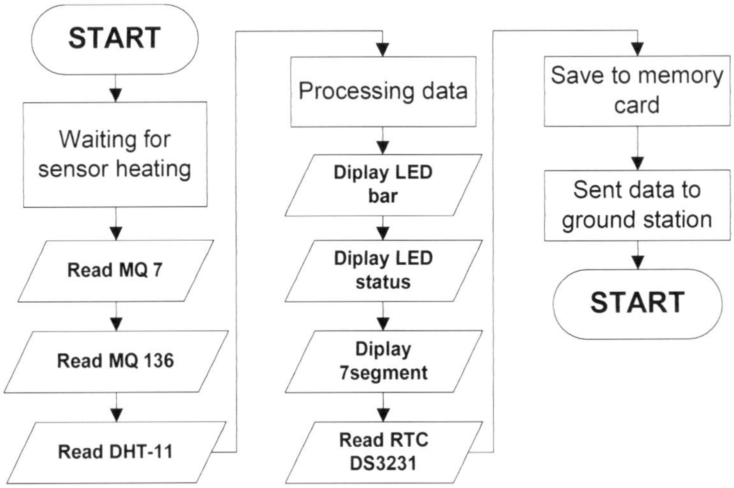

Figure 2. Flow Chart System

4. Wireless Network & Data logger

Ground station is a device that is placed at the bottom of the climbing area. In this case, visitors can monitor the condition of the Mount Ijen area before climbing. Wireless data communication module uses xbee with antenna modification. Data transmission distance can reach 2 KM. The data logger system is added using an SD memory card type device. The controller will store data on environmental conditions every 1 minute. Furthermore, data from the memory card can be read through the Excel application.

5. Results and Discussion

The results produced by the tool are then compared with the results of expert assessors. The equation that comes out shows the value of accuracy possessed by the tool. The method used by the research is Ground Turth.

6. Conclusions and Recommendations

This stage is the final stage of the research that draws conclusions from the results of the analysis of the discussion and provides suggestions for further research. This stage reviews the implementation of related technology / research, this is carried out to ensure that the research has a novelty contribution to the research fields.

3. Result and Discussion

The results of the tool design realization are shown in Figure 3. The device is designed on an aluminum box measuring 50 cm x 30 cm. The device uses a 12volt battery power supply. The device is designed to work within 24 hours non-stop. Based on testing on location, the equipment can work properly. However, the condition of the gas at a dangerous level cannot yet be monitored, because the air quality is at a safe level during the testing process.

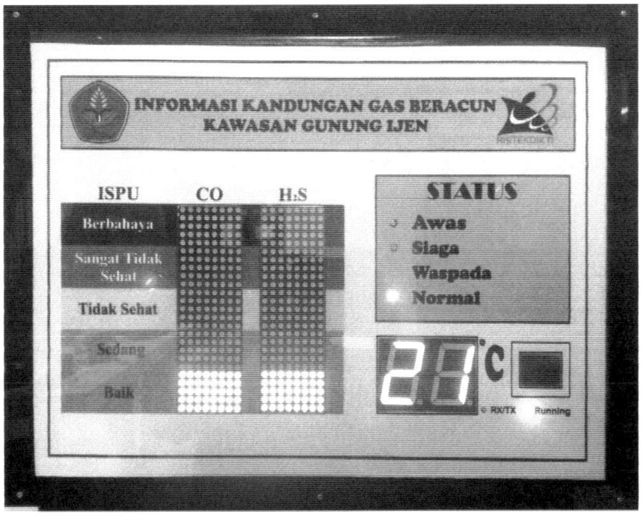

Figure 3. System Testing

Table 1. Data Logger

Tanggal	Jam	Suhu	Hidrogen Sulfida (ppm)	Status	Carbon Monoksida (ppm)	Status
23/09/2018	10:12	18	8,0	baik	2,0	baik
23/09/2018	10:13	18	8,0	baik	2,0	baik
23/09/2018	10:14	18	8,0	baik	2,0	baik
23/09/2018	10:15	18	8,0	baik	2,0	baik
23/09/2018	10:16	18	8,0	baik	2,0	baik
23/09/2018	10:17	18	9,0	baik	2,0	baik
23/09/2018	10:18	18	9,0	baik	2,0	baik
23/09/2018	10:19	18	9,0	baik	2,0	baik
23/09/2018	10:20	18	8,0	baik	2,0	baik
23/09/2018	10:21	19	8,0	baik	2,0	baik
23/09/2018	10:22	19	8,0	baik	2,0	baik
23/09/2018	10:23	19	8,0	baik	2,0	baik
23/09/2018	10:24	18	7,0	baik	3,0	baik
23/09/2018	10:25	18	7,0	baik	3,0	baik
23/09/2018	10:26	18	7,0	baik	3,0	baik
23/09/2018	10:27	18	7,0	baik	3,0	baik
23/09/2018	10:28	19	8,0	baik	3,0	baik
23/09/2018	10:29	18	8,0	baik	3,0	baik
23/09/2018	10:30	19	9,0	baik	3,0	baik
23/09/2018	10:31	19	9,0	baik	3,0	baik
23/09/2018	10:32	18	9,0	baik	3,0	baik
23/09/2018	10:33	18	9,0	baik	3,0	baik

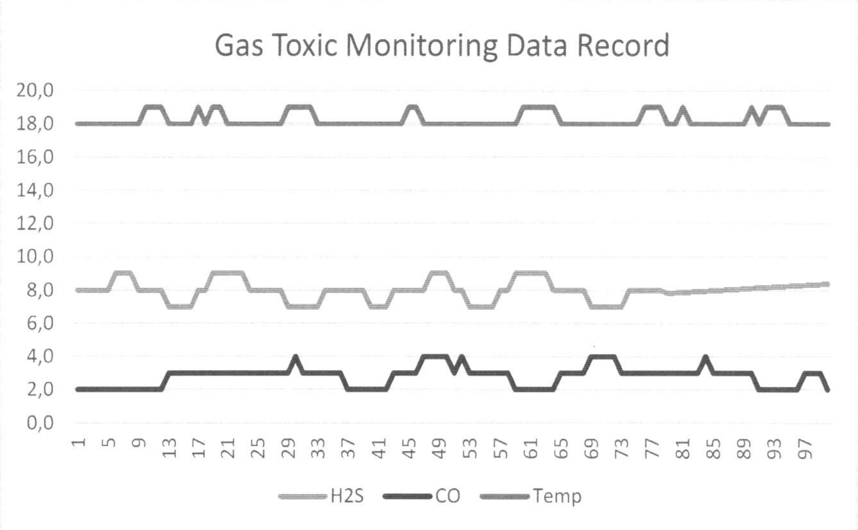

Figure 4. Data record

Besides sensors, the device is also equipped with a timing and data storage system. This consists of accurate timing by backing up the battery so that the microchip can still work even though the main power supply is off. Data storage space uses an 8GB SD card which is sufficient for data storage for more than a year. Sensor data is stored every minute on the memory card. Users can access the results of data recording through the excel application.

The tool is designed to operate in cold temperatures with a range of 2-32 degrees Celsius. The sensor mounting is designed so that it is not disturbed by splashes of rain water and dew. Table 1 is an example of the results of recording data from a tool. Based on these data, it can be made into a graph like Figure 4. The data obtained can also be used for forecasting toxic gas disasters in the Ijen area.

4. Conclusion

Based on the research that has been conducted, it can be concluded several things. The toxic gas monitoring system in the Gunung Ijen area can work 24hours a day. The data communication system at the ground station can work normally in real-time.

Acknowledgment
The authors would like to acknowledge the financial support of this work by grants from PNBP, State Polytechnic of Jember. The author also thanked the P3M and Information Technology Department, State Polytechnic of Jember, which has provided support and assistance in completing this research.

References

[1] M. S. K. Awan, A. Nadeem, and S. Amer, "DMSim: A Virtual Environment for Managing Natural Disasters," 2017 14th Int. Conf. Smart Cities Improv. Qual. Life Using ICT IoT, pp. 26–30, 2017.

[2] M. Mukherejee, L. Shu, L. Hu, G. P. Hancke, and C. Zhu, "Sleep Scheduling in Industrial Wireless Sensor Networks for Toxic Gas Monitoring," IEEE Wireless Communications, 2016.
[3] R. Aman, "Ijen Crater Tourist Site Closed Due to Toxic Gases," The Jakarta Post, 2018. [Online]. Available: http://www.thejakartapost.com/travel/2018/03/22/ijen-crater-tourist-site-closed-due-to-toxic-gases.html. [Accessed: 25-Apr-2018].
[4] E. Science, "Dynamic Simulation Method of Toxic Gas Diffusion in Complex Geographic Environment."
[5] Z. Sun, H. Wang, Y. Chen, L. Shu, and M. Mukherjee, "Understanding the Impact of Planarized Proximity Graphs on Toxic Gas Boundary Area Detection," in Proceedings - 2017 International Conference on Recent Advances in Signal Processing, Telecommunications and Computing, SigTelCom 2016, 2017, pp. 109–114.
[6] L. I. N. Qiang, "Study on Wellsite Toxic Gas Leakage and Dispersion of High Temperature High Pressure Gas Wells with High Sulfur Content," no. 1, 2010.
[7] S. Ghosh, S. Chatterjee, A. Kundu, S. Maity, and H. Saha, "Thermal Analysis of Cantilever MEMS Based Low Power Microheater Array for The Selective Detection of Explosive and Toxic Gases," Proc. - ISPTS-1, 1st Int. Symp. Phys. Technol. Sensors, vol. 1, pp. 290–293, 2012.
[8] S. Ghosh, A. Roy, S. Singh, H. Saha, V. K. Ojha, and P. Dutta, "Sensor Array for Manhole Gas Analysis," Proc. - ISPTS-1, 1st Int. Symp. Phys. Technol. Sensors, pp. 9–12, 2012.
[9] L. C. L. Chen, S. Y. S. Yang, and Y. X. Y. Xi, "Based on ZigBee Wireless Sensor Network The Monitoring System Design for Chemical Production Process Toxic and Harmful Gas," Comput. Mechatronics, Control Electron. Eng. (CMCE), 2010 Int. Conf., vol. 4, pp. 425–428, 2010.
[10] G. H. Reddy, P. Chakrapani, A. K. Goswami, and N. B. D. Choudhury, "Fuzzy Based Approach for Restoration of Distribution System during Post Natural Disasters," IEEE Access, vol. 6, no. c, pp. 3448–3458, 2017.
[11] Prihandoko, Bertalya, and M. I. Ramadhan, "An Analysis of Natural Disaster Data by Using K-Means and K-Medoids Algorithm of Data Mining Techniques," 2017 15th Int. Conf. Qual. Res. Int. Symp. Electr. Comput. Eng., pp. 221–225, 2017.
[12] J. Sweeney, "Integration of Toxic Gas Monitoring Systems Into Building Fire Alarm Systems at Harvard University," Bienn. Univ. Microelectron. Symp. - Proc., 2010.
[13] B. Geng and Y. Bao, "Monitoring System about Temperature, Humidity and Toxic Gas," no. Iccasm, pp. 632–634, 2010.
[14] A. Kroll, "Quantification of Methane Gas Leakages using Remote Sensing and Sensor Data Fusion," 2017.
[15] G. Wei, J. W. Gardner, M. Cole, and Y. Xing, "Multi-Sensor Module for a Mobile Robot Operating in Harsh Environments," Proc. IEEE Sensors, pp. 1–3, 2017.
[16] Badan Geologi, "Pusat Vulkanologi dan Mitigasi Bencana Geologi - Badan Geologi," Kementerian Energi dan Sumber Daya Mineral, 2014. [Online]. Available: http://www.vsi.esdm.go.id/index.php/gunungapi/data-dasar-gunungapi/522-g-ijen. [Accessed: 25-Apr-2018].
[17] D. Kim, S. Kim, J. An, and S. Kim, "A Portable Colorimetric Array Reader for Toxic Gas Detection," in ISOEN 2017 - ISOCS/IEEE International Symposium on Olfaction and Electronic Nose, Proceedings, 2017, pp. 0–2.
[18] I. Concina, M. Falasconi, and V. Sberveglieri, "Electronic Noses as Flexible Tools to Assess Food Quality and Safety: Should We Trust Them?," IEEE Sens. J., vol. 12, no. 11, pp. 3232–3237, 2012.
[19] H. R. Estakhroueiyeh and E. Rashedi, "Detecting Moldy Bread Using an E-nose and The KNN Classifier," in 2015 5th International Conference on Computer and Knowledge Engineering, ICCKE 2015, 2015, pp. 251–255

IOP Conf. Series: Earth and Environmental Science **207** (2018) 012059 doi:10.1088/1755-1315/207/1/012059

The Precision Agriculture Based on Wireless Sensor Network with MQTT Protocol

Y Syafarinda[1], F Akhadin[1], Z E Fitri[1], Yogiswara[1], B Widiawan[1], E Rosdiana[2]

[1]Department of Information Technology, Politeknik Negeri Jember, Mastrip PO.BOX 164 68101 Jember, East Java, Indonesia
[2]Department of Agriculture Production, Politeknik Negeri Jember, Mastrip PO.BOX 164 68101 Jember, East Java, Indonesia

*Email: yunisyafarinda@gmail.com

Abstract. Precision agricultural techniques are widely developed to monitor and control growth of plant in greenhouse. The use of greenhouses is suitable for the countries that have unpredictable weather e.g. Indonesia. The main monitors and controls in greenhouse are sensor networks and actuators. Some examples of the precision agriculture technique are monitoring of fertilizer needs and smart irrigation. This paper presents a develpment of precision agriculture based on Wireless Sensor Network (WSN) with Message Queuing Telemetry Transport (MQTT protocol. The WSN system involves Internet of Things (IoT) technology to connect devices, collect and distribute information. IoT integration with cloud computing is aimed to overcome limitations of the IoT devices in remote monitoring and control. WSN development uses MQTT protocols that have advantages that are lightweight, simple, can run on a small bandwidth and more advanced compared to Hypertext Transfer Protocol (HTTP). In this paper the SHT3x and TSL2561 sensors are used to observe temperature, humidity and light intensity of the greenhouse. Test results of three sensors in data transmission speeds indicate that the MQTT protocol run with very small bandwidth. It can transmit data on the 4G network of AXIS provider with download capability of 0.02 kbps and upload 1.65 kbps. The average data transmission speed is approximately 1 second.

1. Introduction

A greenhouse is a construction building as a place to grow and protect plants from extreme weather (wind, rain, ultraviolet radiation) and insect disturbances and allows for urban farming [1][2]. Usually, the farming technique applied to the greenhouse is hydroponic techniques. Hydroponic techniques are techniques for processing plants without using soil as a medium of growth. This technique supplies the nutrients needed by plants through aqueous solutions. Some environmental factors that must be considered include oxygenation, salinity, temperature, light intensity, photoperiod, air humidity, pH and conductivity of nutrient solutions [3].

During the plant growth period, hydroponic farmers must monitor these factors.. During this time, plant monitoring is carried out manually with a measuring instrument and comes to the location directly and the provision of plant nutrition based on the age of the plant, but this does not fully suit the needs of the plant. Because the plant cannot convey its condition directly at any time to the farmer. Efforts to adjust the needs in agriculture are known as precision agriculture. Precision agriculture is

IOP Conf. Series: Earth and Environmental Science **207** (2018) 012059 doi:10.1088/1755-1315/207/1/012059

information and technology in agricultural processing systems by providing precision treatment to obtain optimum and sustainable benefits [4]. The realization of precise environmental conditions with human power is considered ineffective. Humans often forget and are inconsistent when doing repetitive work. Therefore, the automation control system is needed and can be consistently working to meet the needs of plants properly.

Several precision agricultural researches have been carried out including automation of fertilizer, intelligent irrigation, and smart greenhouse. The research utilizes technology including the application of the Wireless Sensor Network (WSN) in agriculture. WSN is a solution to facilitate the realization of precision agriculture in the field of information technology. WSN consists of 3 main components, namely sensor nodes, gateways, and software. Sensor nodes serve to provide data on environmental conditions. The data is delivered wirelessly through the gateway to be collected, processed and displayed using the software. [5]. The basic principle of using internet technology and WSN will be the basis for the application of the Internet of Things (IoT).

Internet of Things (IoT) connects objects to share data via the internet. IoT is often integrated with cloud computing to overcome the limitations of IoT devices and requires communication protocols in the process of sharing data. [6]. Communication protocols that support the internet of things are the MQTT Protocol (Message Queuing Telemetry Transport). MQTT protocols have advantages over other protocols such as HTTP (Hypertext Transfer Protocol). MQTT is lighter, able to run in low bandwidth, simpler, and more modern than HTTP which has low overhead protocol and runs on high latency [7].

Based on these descriptions, this research combines the advantages of similar research in the use of technology for agriculture precision. The purpose of the research is to build precision farming based on wireless sensor networks with the MQTT protocol. The system used to monitor and control environmental conditions consisting of temperature, humidity, lighting and plant nutritional needs in the greenhouse in real time. Integration using cloud computing provides broad access for users to monitor and control anywhere and anytime via the website.

2. Material and Method

This research utilizes the MQTT protocol for communication between wireless sensor network devices for precision agriculture. Communication protocols play an important role in realizing the system runs according to planning. The following is a description of system architecture and theories that support research.

2.1 The Proposed System Architecture

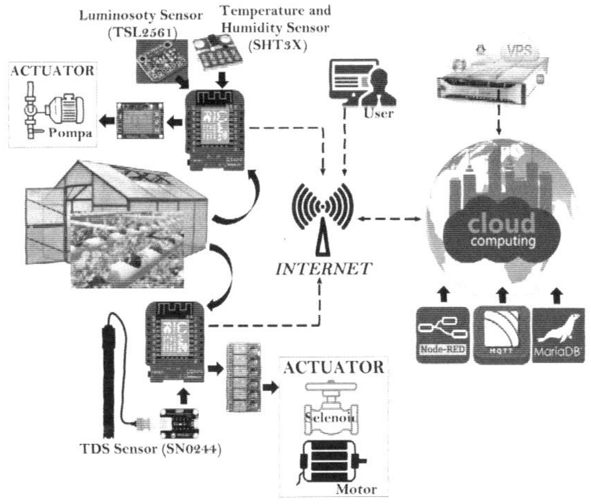

Figure 1. Architecture of System

The communication systems in precision agriculture design based on Wireless Sensor Network are very important to provide precision treatment, collect and receive monitoring data. In this research, we use the MQTT (Message Queuing Telemetry Transport) protocol. The MQTT protocol bridges the sending of sensor data to the cloud server for later data is stored in a database and displayed on a web page so that users can access it. In controlling greenhouse environmental conditions, the actuator will work with the help of sensor nodes. All activities in this precision agriculture system can be monitor by users through the web.

2.2 Wireless Sensor Network with Internet of Things
Wireless Sensor Network is an advancement in the field of micro-electromechanical system technology, wireless communication, and digital electronics. WSN consists of sensor nodes that are spread across a particular area and collaborate with each other to collect, process, and communicate through wireless channels about some physical conditions or environmental conditions[8][9]. Wireless transmission at WSN makes wiring simple and the cost of using wires decreases. Wireless technology will eliminate 20% - 80% of the physical costs of industrial installations[4]. The wireless network with internet technology in WSN node communication is a form of the Internet of Things application which is a form of M2M (machine to machine) communication development through internet communication [8][10]. Figure 2 describes the wireless sensor network structure in general. It consists of sensing units, processing units, transmission units, and power units[8][11].

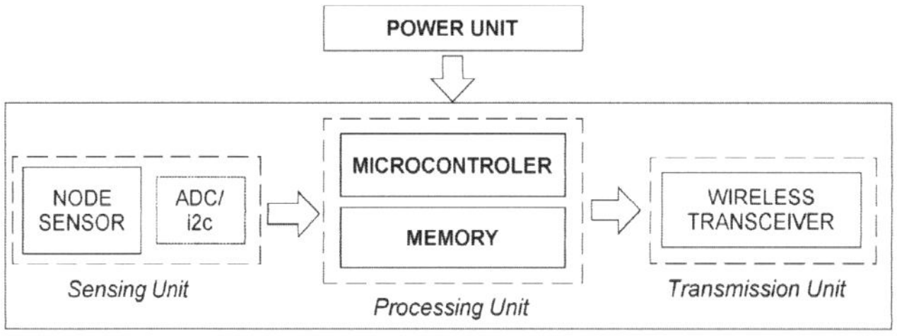

Figure 2. General Structure of Wireless Sensor Network

2.2.1 Sensing Unit. This section consists of sensor nodes to monitor environmental conditions. This research uses three sensor nodes namely the SHT3x to monitor the temperature and humidity levels, then the TSL2561 sensor to monitor the level of light intensity, and the TDS sensor to monitor the level of plant nutrients.

2.2.2 Processing Unit. The processing unit consists of a microcontroller that has memory to process the program in it and perform commands to the device connected to it. The device is in the form of sensors and actuators as control objects. All sensor nodes are controlled by a microcontroller (Wemos D1 Mini) located at a certain position. This integrated Wi-Fi microcontroller will easily communicate with the internet of things. Other than that, it can also control the environment by controlling the actuator.

2.2.3 Transmissing Unit. Transmitting unit is a part consisting of communication media for sending and receiving data by WSN (Wireless Sensor Network) device. Devices used on WSN wireless networks such as television satellites, GPS, mobile units, wireless computers, Zigbee technology and other wireless communication devices. In this study using MiFi (WiFi Hotspot Router) as an internet client source (sensor node) to connect with cloud servers.

2.2.4 Power Unit. The power unit of precision agriculture based wireless sensor network (WSN) uses an AC power installation energy source. WSN actuator devices require the AC power while the processing device uses DC power. Therefore, this research uses an adapter 220V AC to 5 V DC. The use of AC power installation is right for a sustainable system such as the WSN design.

2.3 Wireless Sensor Network to Precision Agriculture
Precision agriculture is information and technology in agricultural processing systems by providing benefits to obtain optimal and sustainable benefits. Wireless sensor network on the development of precision agriculture can be used to regular monitoring of changes in environmental conditions such as climate, hydrology, plant physiology, pests and others. Other than that, it can also act as a control in the provision of inputs for seeds, fertilizers, pesticides, and others. The WSN sensor node will help the data collection process to produce information for farmers. WSN can take the form of a data collection system and control system on agricultural machinery. This system is widely used in research projects and commercial products to provide solutions for monitoring plant status, regulating irrigation, fertilizer management, pest control, and automatic harvesting[8].

2.4 Internet of Things with Cloud Computing
Internet of Things (IoT) means connecting devices via the internet. By using IoT, users can control the tool anytime, anywhere. The role of cloud on IoT aims to provide storage and computing resources to overcome the limitations of IoT devices[12]. IoT applications have limitations because operators do not have standard specifications to follow and each operator installation can vary significantly from each other. So that's important to have a standard architecture to encourage uniform IoT rectification on the cloud platform. Besides that, some other limitations that IoT can overcome by the cloud are interoperability, scalability, security, availability and big data[6].

2.5 Communication System on Cloud MQTT Broker
The MQTT protocol is designed to minimize bandwidth and energy usage and be as simple and easy to integrate as much as possible. The MQTT component consists of clients, servers or brokers, publishers, subscribers, and topics. MQTT Client is any device that acts as a publisher or subscriber. The MQTT broker is responsible for receiving all messages, filtering, determining who requests the message and sending messages to the client. The Publisher is a client that issues messages to clients who subscribe through brokers while subscribers are clients who receive messages through brokers according to the requested topic. The topic here is message identity[13][14].

MQTT provides three transfer modes based on reliability, namely QoS level 0 (Non assured transmission), QoS level 1 (Assured transmission) and QoS level 2 (Assured service on applications). Level 0 QoS is the fastest transfer mode. At level 0 QoS there is only the publishing process and the MQTT protocol does not require a server to forward publication or no acknowledgment status. There is a possibility that the publication will be discarded, depending on the server. QoS level 1 adalah mode transfer default. Level 1 QoS is the default transfer mode. At level 2 QoS messages are always delivered at least once. If the sender does not receive an acknowledgment, the message is sent again until the publisher gets the message recognition status from the client. Level 2 QoS is the safest but slowest transfer mode. At level 2 QoS there are two message transmissions. The first transmission, the sender transmits and receives recognition from the recipient. The second transmission of the sender notifies the recipient that he can complete the processing of the received message[7].

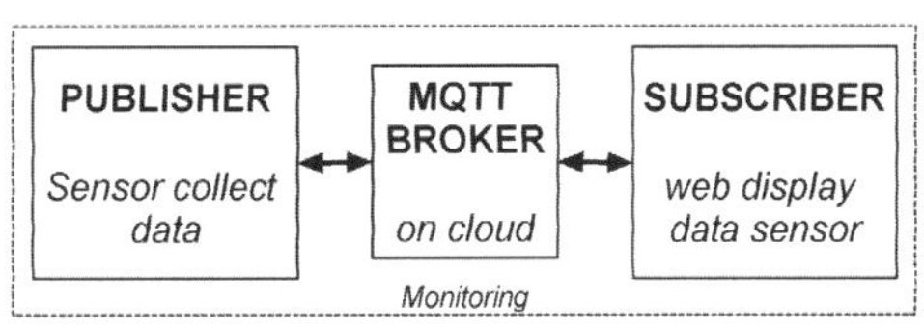

Figure 3. Communication with MQTT Protocol

In this research, the MQTT broker is installed on the cloud so that all messages are stored and processed in cloud storage. The monitoring and control process in this study uses the QoS level 2 mode. In the process of monitoring environmental conditions, the sensor (publisher) in the greenhouse sends reading data to the MQTT broker. Then the MQTT broker sends the sensor data readings to the web (subscriber) according to the message request topic so that the user can find out through the website.

3. Result

Figure 4 shows the monitoring data of precision agriculture environment conditions communicated by the MQTT protocol from the Wireless Sensor Network device. This website page is created using nodered software that has been configured on the cloud. The web displays monitoring data in graphical form in real time from sensor readings. Standard environmental conditions are also available there to facilitate users in monitoring precision treatment so users can compare the state of the plant at that time with the needs of plants.

Figure 4. Precision Agriculture Monitoring Website

Communication through the cloud with the MQTT protocol in publishing data can be monitor from the MQTT broker with the help of PuTTY software using the python command. Data transmission testing is done by tracking three sensor data simultaneously and displaying the timestamp and topic data for each track data that occurs. All data transmission activities use the MQTT protocol can be seen in figure 5.

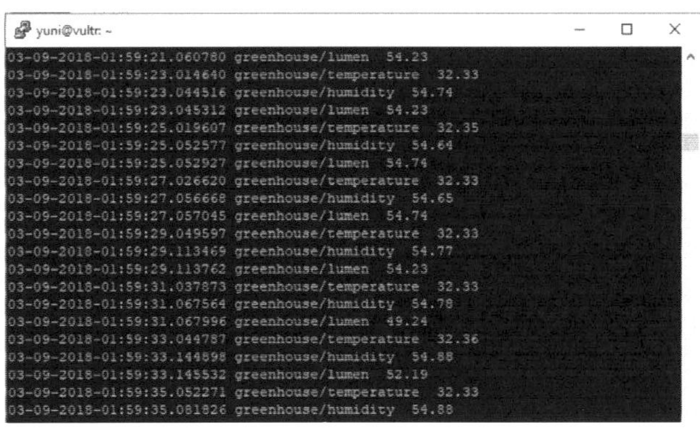

Figure 5. Screenshot of The Data Transmission On Cloud MQTT Broker

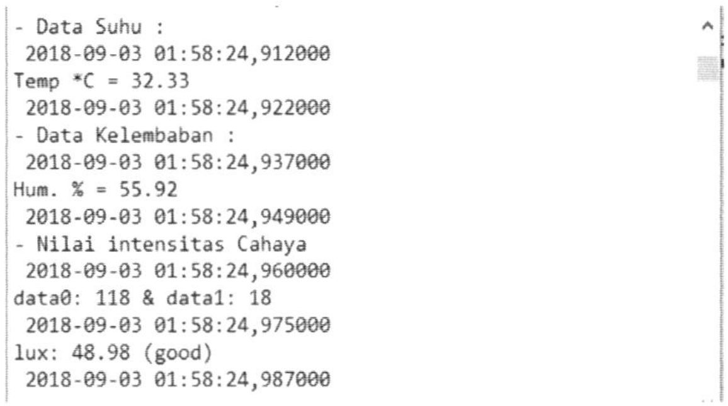

Figure 6. Screenshot of The Data Transmission From Serial Monitor Arduino

Testing the speed of data transmission with the MQTT protocol for monitoring sensor data to the website is done by comparing the serial monitor timestamp with the timestamp on the website database. The comparison object for the testing of this research uses three providers with different network speed conditions as described in table 1.

Table 1. List of each provider wireless speed

PROVIDER	TIME OF TRIAL	SPEEDTEST		DEF.
		DOWNLOAD (Kbps)	UPLOAD (Kbps)	
Telkomsel 3G	30 s	2413	228	Work
AXIS 3G 1	30 s	558	142	Work
AXIS 3G 2	30 s	0,65	0,02	Work
Ooredoo 3G 1	30 s	304	215	Work
Ooredoo 3G 2	-	16	90,1	Fail
Telkomsel 4G	30 s	7687	5001	Work
AXIS 4G 1	30 s	1688	1154	Work
AXIS 4G 2	30 s	0,02	1,95	Work
Ooredoo 4G 1	30 s	256	354	Work
Ooredoo 4G 2	30 s	2,29	2,35	Work

The use of a GSM provider is chosen to get the same network conditions that are used by many users. The results of the speed test indicate that on 3G and 4G networks from Telkomsel, AXIS and Ooredoo providers can run Wireless Sensor Network communications except when using Ooredoo providers with 16 Kbps and 90.1 Kbps upload capacity. This test shows AXIS 3G and AXIS 4G with the smallest upload and download capacity but can run WSN communication with this MQTT protocol.

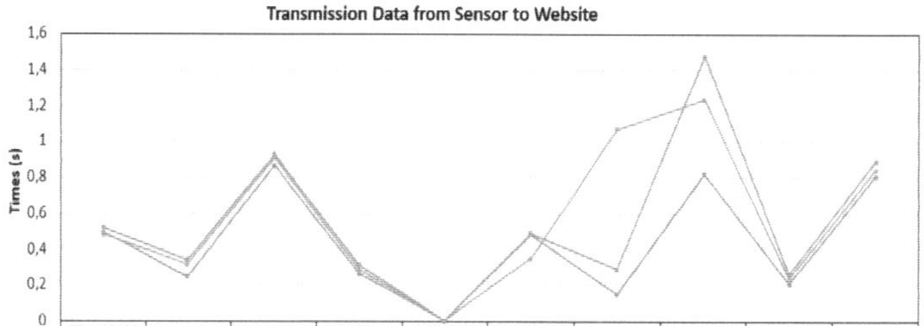

	TELKOMSEL 3G	AXIS 3G 1	AXIS 3G 2	OOREDOO 3G 1	OOREDOO 3G 2	TELKOMSEL 4G	AXIS 4G 1	AXIS 4G 2	OOREDOO 4G 1	OOREDOO 4G 2
Temperature	0,4926	0,2468	0,8664	0,2612	0	0,4824	0,1494	0,8216	0,2084	0,808
Humidity	0,5178	0,3418	0,9286	0,311	0	0,4898	0,2922	1,4772	0,2648	0,8886
Ilumination	0,4812	0,3136	0,9118	0,2838	0	0,3508	1,0718	1,2346	0,2386	0,8436

Figure 7. Screenshot of the average data transmission

Each data provider takes 30 seconds of data track with the current bandwidth as described in table 1. Then the time of data transmission from the data track for each sensor is taken on average. The graph on figure 7, it can be seen that the speed of transmission of data tracking sensors with the 3G / 4G network of each provider is almost the same or stable except when using Telkomsel and Axis providers on 4G networks that look unstable. From all provider, the speed of data transmission on Ooredoo 4G with a download speed of 256 kbps and upload 354 kbps is tested faster and more stable, while testing the most unstable and slow in the process of transmitting data using the Axis 4G provider with download speeds of 0.02 kbps and uploading 1.95 kbps.

4. Conclusions

The MQTT protocol deserves recommended as a communication protocol for the application of wireless sensor networks for precision agriculture. This matter is related to data communication in real time. In this research, the MQTT protocol can transmit sensor data approximately one second. This time depends on the ability to download and upload from each network provider.

The MQTT protocol can transmit sensor data to the smallest download and upload capabilities, namely on AXIS 3G networks (download = 0.65 kbps; upload = 0.2 kbps) and AXIS 4G network (download = 0.02 kbps; upload = 1.65 kbps) However, the MQTT protocol cannot run the data transmission process on Ooredoo 3G's smallest bandwidth (download = 16 kbps; upload = 90.1 kbps).

Other than that, the ability to download and upload from each provider also affects the speed and stability of sensor data transmission by the MQTT protocol. In 3G network providers, the third sensor data transmission graph tends to be stable while 4G network providers tend to be less stable. The MQTT protocol is capable of running the fastest and most stable transmission on the use of the Ooredoo 4G network with 256 kbps download capability and 354 kbps upload. But it looks unstable and slow to use AXIS 4G providers with a download capability of 0.02 kbps and uploading 1.95 kbps.

Acknowledgements

We would like to thank the Rowosari Village Tani Jaya II Group, Sumberjambe District, Jember Regency for providing a place for WSN-based precision agricultural development.

References

[1] R K Kodali, V Jain, and S Karagwal 2016 *IEEE Reg 10 Humanit Technol Conf* - 2016 *R10-HTC* - 2017 *Proc*

[2] H. Abbas, R. Syam, and B. Jaelani 2015 no. Snttm Xiv pp 7–8

[3] M F Saaid, A Sanuddin, M Ali, and M S A I M Yassin 2015 *ISCAIE* - 2015 *IEEE Symp Comput. Appl Ind. Electron* pp 186–190

[4] M R Mohd Kassim, I Mat, and A N Harun 2014 *Int Conf Comput Inf Telecommun Syst* pp 1–5

[5] R Susana, A R Darlis, and S Aqli 2016 *J Elektro dan Telekomun Terap* vol 2 no 1 pp 53–60
[6] L L Dhirani, T Newe, E Lewis, and S Nizamani 2017 *Proc Int Conf Sens Technol ICST* – 2018 Decem pp 1–6
[7] T Yokotani and Y Sasaki 2016 *ICCEREC - 2016 Int Conf Control Electron Renew Energy Commun* - 2017 *Conf. Proc* pp 1–6
[8] J P P and D K S S 2017 *IOSR J. Electron Commun Eng* vol 12 no 01 pp 23–28
[9] M S Dzulkifli and M Rivai 2016 *Tek. ITS* vol 5 no 2 pp 261–266
[10] V P Kafle, Y Fukushima and H Harai 2015 *Int. Conf. Recent Adv Internet Things RIoT* pp 7–9
[11] M A Matin and M M Islam 2012 *Wirel Sens Networks Technol Protoc* pp 3–24
[12] T T S Vatari and A Bakshi 2016 *IEEE 3rd World Forum Internet Things WF-IoT* pp 246–250
[13] S A Shinde, P A Nimkar, S P Singh, V D Salpe, and Y R Jadhav 2016 *Int J Res* vol 3 no 3 pp 240–244
[14] L Cruz-Piris, D Rivera, Marsa-Maestre, E De La Hoz, and J R Velasco 2018 *Sensors (Switzerland)* vol 18 no 3

An investigation of dummy load energy in gunung sawur 1 microhydro power plant - Lumajang East Java

DE Rahmanto[1*] and V Femintasari[1]

[1]Study Program of Renewable Energy Engineering, Politeknik Negeri Jember, Indonesia

E-mail: dedy_eko@polije.ac.id

Abstract. Dummy load are used for microhydro power plant which its work controlled by electronic load controller. Electrical energy supply to dummy load will increased or decreased automatically depends on consumer load. All this time, dummy load energy of microhydro are rejected to the environment without any use. This research was carried out by investigating on Gunung Sawur 1 micro hydroelectric power plant in Sumberwuluh Village, Candipuro Districts, Lumajang Regency, East Java, Indonesia. Purposes of this research are to find out the potency of dummy load electric energy on Gunung Sawur 1 microhydro and possibility of its utilization. The research was carried out by recording data of electrical energy of dummy load on microhydro using Fluke 1735 power logger. Research results show that voltage of dummy load was fluctuated, from 0 volt until higher than 100 volts. In the night, the average of dummy load power is 2502.5 watts. Then during the day, the average of dummy load power is 3688.5 watts. The average total electric energy of dummy load is 66.1 kWh per day. Utilization unstable voltage and energy of dummy load can only be used for heating process, such as water heating or water purification on distillation system.

1. Introduction

The dummy loads are needed for balancing of consumer loads and the generator electric power at the microhydro power plant (MHPP). Working of dummy load is controlled using electronic load controller (ELC). ELC work based on the amount of electrical energy in consumer loads. The supply of electric energy to the dummy load will increased automatically if the consumer loads decreases and will decrease if the consumer loads increases. This condition will keep the loads of MHPP generator still stable[1, 2, 3].

Dummy loads are generally in the form of water heating elements or air heating elements. The electric energy that is supplied to the dummy load will become heat energy which is generally just rejected into the environment without any use. Gunung Sawur 1 is a MHPP that uses an dummy load of air heater elements. The electric energy of the dummy in Gunung Sawur 1 MHPP also rejected into the environment and sometimes even disturbing when the operator of turbine is in the power house because the air in the power house gets hot. This study aims to determine the electrical energy potential of dummy load in Gunung Sawur 1 MHPP and its potential of utilization.

2. Electronics Controller of MHPP

Electronic Controller at MHPP needed to stabilize turbine rotation. A stable turbine rotation will produce a fixed voltage and frequency in accordance with the control settings. Stable electricity voltage and frequency will ensure the safety of electrical and electronics equipment used by MHPP consumers. The main components of electronics controller in the MHPP are ELC modules and dummy loads [4, 5, 6, 7].

2.1. ELC

ELC on MHPP is an electronics module that is used to regulated electrical energy that is supplied to a dummy load whose value is affected by consumer load. ELC will increase electricity supply to the dummy load automatically when the consumer load decreases. Conversely, if the consumer load increases, the electricity supply to the dummy load will be reduced by the ELC. This condition makes the load received by the generator of MHPP will always be stable so that the voltage and electrical frequency produced will be stable [4, 5, 6, 7].

Gunung Sawur 1 MHPP uses ELC for 3 phase systems. The ELC on Gunung Sawur 1 MHPP has two 3 phase outputs supplied to the dummy loads (Figure 1).

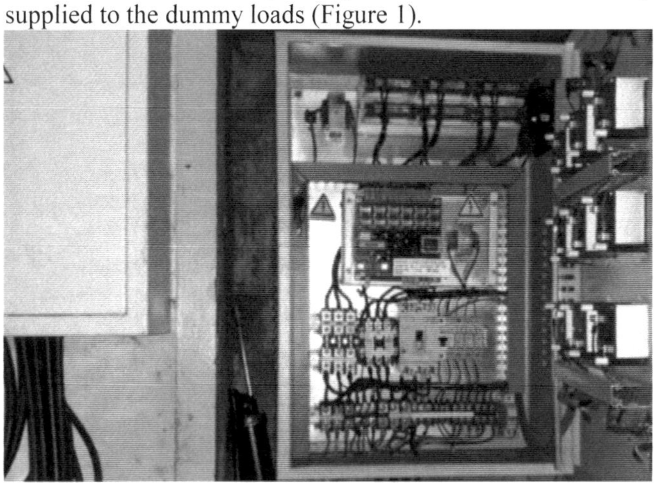

Figure 1. ELC module at Gunung Sawur 1 MHPP

2.2. Dummy load

Dummy load is the balancing load used on MHPP. Dummy load is usually a heating element. The heating element that is used as a dummy load can be a water heater element or an air heating element. Dummy load must have a power capacity that is greater than the maximum power generated by the micro hydroelectric generator, so that if the consumer loads is zero (0 watts) then all the power generated by the MHPP will be supplied to the dummy loads [6,7].

Gunung Sawur 1 MHPP uses a dummy load of 6 heating elements with each capacity around 3000 w at 220V voltage. Dummy load on the Gunung Sawur 1 MHPP is assembled in star circuit into 2 series. The first star circuit is connected to the output 1 of ELC and the second star circuit is connected to output 2 of ELC (Figure 2).

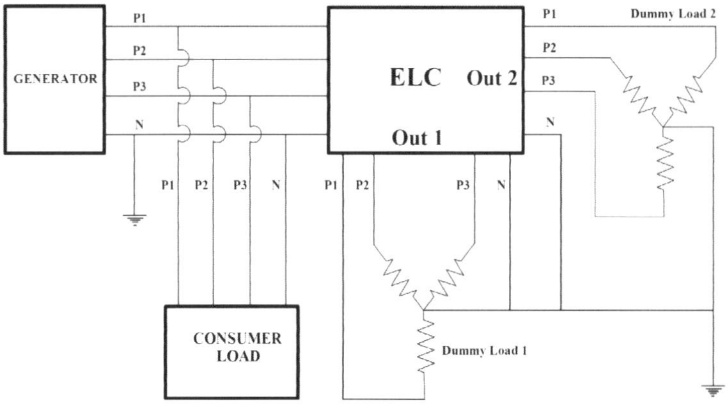

Figure 2. Dummy loads wiring diagram in Gunung Sawur 1 MHPP

3. Method

This research was carried out in Gunung Sawur 1 MHPP were located in Sumberwuluh Village, Candipuro District, Lumajang Regency, East Java, Indonesia at the elevation 650 m above sea level. The research started February until May 2018. The equipment used is the Fluke 1735 three phase power logger which is used to recording the electrical data on dummy loads in the Gunung Sawur 1 MHPP. The investigation of dummy loads energy in Gunung Sawur 1 MHPP use these following steps:

- Preparation of the power logger.
- Installation of probes and clamps sensors (Figure 3).
- The recording forevery 1 minute.
- Let the data logger record for several days.
- Recording stoped.
- Probes and clamps uninstalled.
- Downloading recorded data.
- Data processed using MS Excel [8].

Figure 3. Instalating of probes and clamps sensors on dummy loads

Observation data is displayed in graph and table. The standard deviation is calculated in order to determine the deviation value that accured [9].

4. Result and discussion

The electrical data resulted from investigation of dummy load in Gunung Sawur 1 MHPP . The electrical voltages on the dummy loads were fluctuate from 0 volt to more than 100 volts (Figure 4). The average voltage of dummy load is 78.9 volts with standard deviation of 38.3 volts. This voltage variation occurs because the consumer load is always changing from time to time. When the consumer load is high, the electrical voltage that is supplied to the dummy load will be low even when the peak voltage of the electricity supplying the dummy load is 0 volt because all electrical energy are supplied to the consumer loads. Based on the Figure 4, the peak load is in the morning and early evening. At that time the voltage supplying the dummy load is very low even 0 volt. The peak load that occurs in the morning is caused by high electricity consumption because some consumers cook rice using a rice cooker that requires a lot of electric energy. The peak load that occurs at the beginning of the night because electricity is widely used for lighting and electronic equipment such as television.

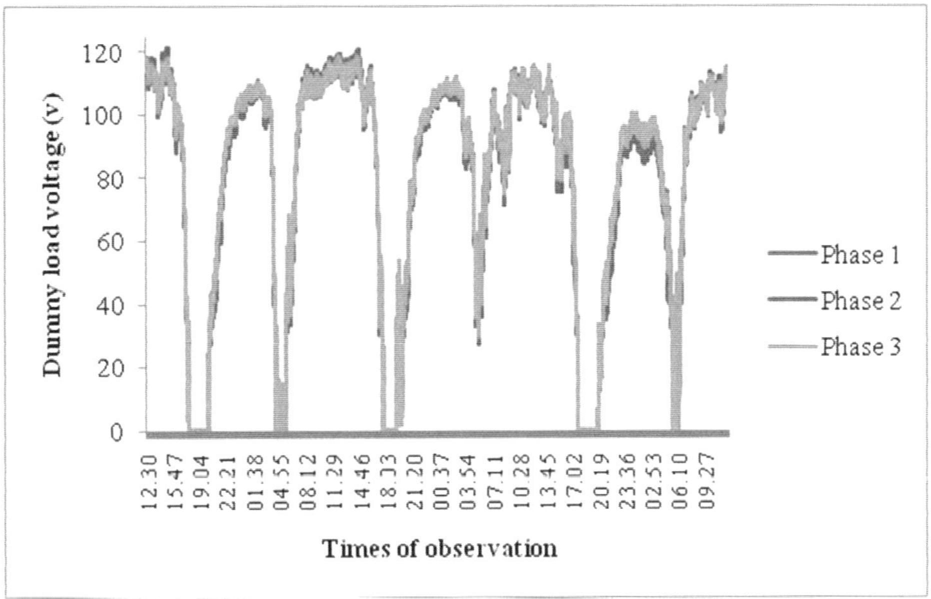

Figure 4. Voltages of dummy loads observed in Gunung Sawur 1 MHPP

Power and electrical energy dummy load during the day is greater than at night. Consumer loads during the day is smaller than at night. During the day there is almost no need for lighting. The Rural consumers generally go to the fields during the day so that the use of electricity is also low. The occurrence of fluctuations in electricity consumption during the day is more due to electricity consumption from the shop machine in the Hydro Cipta Mandiri workshop to produce some micro hydro turbines.

Unstable dummy load electrical condition only allows the opportunity to use for heating processes such as water heaters or water purification in distillation process, and also for drying process. Heating elements can accept the fluctuates voltage. The utilization of dummy load electric energy could be applied by changing the dummy load heating elements using the others for that utilization.

The average of electric power dummy load at night is 2502.5 with standard deviation of 1865.1 watts. Then the average of electric power dummy load during the day is 3688.5 with standard deviation of 1477.8 watts. A smaller standard deviation value during the day shows a smaller power variation. Power and dummy loads energy are tabulated in Table 1.

Table 1. Power and energy of dummy loads

Observation	Daily dummy power(w)	Dummy potency energy (wh)
1st day	2872.9	68949.6
2nd day	3238.7	66438.5
3th day	3184.1	64488.7
4th day	3052.8	59098.9
5th day	3219.2	70697.4
6th day	3185.9	69664.9
7th day	2914.6	63325.9
Average	3095.5	66094.9

The average of electric energy dummy load is 66094.6 wh or 66.1 kwh per day. This energy is equivalent to 237940.6 kJ[10]. This energy can be used to raise the temperature of the water from 20°C to boiled at 97°C and it will be able to evaporate about 92 kg of water. If only to increase the temperature of the water without evaporation, it would be able to heat 738.1 kg of water from a temperature of 20°C to 97°C. The energy to increased 1 gram of water temperature every °C is 4,19 J. The boiling point of water at 97°C occur at the about elevation 650 m above sea level [11]. Heat energy to evaporated of water at 97°C about 2264.4 kJ/kg [12].

5. Conclusion

The average electric energy of dummy load in Gunung Sawur 1 MHPP is 66.1 kWh per day. Utilization unstable energy of dummy load can only be used for heating process, such as water heating or water purification on distillation system.

6. Acknowledgement

This research was funded by the Ministry of Research, Technology and Higher Education of the Republic of Indonesia through PDP program 2018. Thanks to Mr. Sucipto as Owner of Hydro Cipta Mandiri and Gunung Sawur 1 MHPP manager for supporting to this research.

References

[1] Adhikari R, R Bhattarai. 2013. Fifth International Conference on Power and Energy Systems, Kathmandu, Nepal 28 - 30 October, 2013
[2] Aung NW, Aung Ze Ya. 2015. International Journal of Electrical, Electronics and Data Comunication. Volume-3. Issue-6. June-2015
[3] Bisht VS, YR Sood, N Kushwaha, Suryakant. 2012. International Journal of Scientific Engineering and Technology. Volume 1 Issue 2
[4] Kamble SV, DP Kadam. 2014. International Journal of Scientific and Research Publications. Volume 4 Issue 1
[5] Kathirvel. C, K Porkumaran, S Jaganathan. 2015. The Scientific World Journal. Volume 2015
[6] Rahayuningtyas A, Teguh Santoso, Maulana Furqon. 2012. Prosiding SnaPP 2012. Vol 3 No 1 Tahun 2012 (Indonesian)
[7] Roodsari BN, EP Nowicki, P Freere. 2014. Energy Procedia 57 (2014) 1465 – 1474
[8] Fluke 1735. 2006. Fluke Corporation. USA

[9] EMC Education Services. 2015. John Wiley & Sons. Canada
[10] Halliday D, R Resnick, J Walker. 2011. John Wiley & Sons. New Jersey
[11] Anonymous. 2018. https://www.engineeringtoolbox.com/boiling-points-water-altitude-d_1344.html. [accesed Sep 5- 2018].
[12] Parry WT, JC Bellows, JS Gallagher, AH Harvey, RD Harwoods. 2014. ASME. New York

Bioethanol Production from Sugarcane Molasses with Simultaneous Saccharification and Fermentation (SSF) Method using *Saccaromyces cerevisiae-Pichia stipitis* Consortium

Audiananti Meganandi Kartini[1], Yeny Dhokhikah[2]

Environmental Engineering Study Program, Engineering Faculty, University of Jember[1,2]
Tegalboto Campus, Jl. Kalimantan No. 37, Sumbersari, Krajan Timur, Sumbersari, Jember Regency, East Java 68121, Indonesia[1,2]

audiamega@unej.ac.id[1]

Abstract
The sugar-making industry from sugar cane has a byproduct of sugar cane as molasses. One alternative to reduce the amount of molasses waste is by reusing molasses as biomass for bioethanol production. Molasses can be reused as bioethanol that have economic value because of their high sugar content. The process of making bioethanol from molasses is done by Simultaneous Saccharification and Fermentation (SSF) method using 10% substrate (100 gr / L molasses) and 20% substrat (200 gr / L molasses). The process of hydrolysis of sugarcane molasses is carried out biologically using *Aspergillus niger* and *Trichoderma viride*, while the fermentation process uses the consortium *Saccaromyces cerevisiae-Pichia stipitis*. Ethanol production using the SSF methods had the highest bioethanol concentration of 14.38% v/v for 20% substrat and 10% consortium usage at 72 hours fermentation.

Keywords: Molasses, Bioethanol, Simultaneous Saccharification and Fermentation (SSF)

1. Introduction

Indonesia has a large population with sugar consumption reaching 4.2-4.7 million tons / year (PTPN X, 2015). The sugar-making industry from sugar cane in Indonesia has a byproduct of sugar cane or molasses. The high amount of molasses produced by sugar companies can be an environmental problem. Molasses waste decreases the dissolved oxygen content in water and disrupts the lives of aquatic organisms. The principle of Reuse, Reduce, and Recycle is still the best way to manage and handle waste with various problems including reducing the amount of molasses waste (Sarris *et al*, 2014).

Molasses waste is one of the wastes that can be reused as raw material which if processed will have high economic value. Most people use molasses as a raw material for making syrup, but the content of non-sugar impurities is very dangerous for health. The use of molasses in Indonesia is more widely used as raw material for making monosodium glutamate (MSG), but molasses has other features because the sucrose content ranges from 48 - 55% which has the potential to be used as raw material for ethanol production (Prescott and Dunn, 1959).

Ethanol is an alternative energy source that has several advantages, including renewable and environmentally friendly properties due to its low carbon dioxide emissions (Jeon, 2007). Ethanol can be used as a mixture of gasoline and can be used as fuel (McKetta, 1983). Ethanol production can be done by fermentation techniques (Hemamalini, 2012).

This large amount of molasses and high sugar content allows molasses to be converted into environmentally friendly fuel products. The role of the Indonesian government is expected to emerge as a policy maker for the use of ethanol as a national fuel mixture, this is needed to save the budget for crude oil purchases and support the national economy from ethanol production

2. Methods

2.1 Tools and Materials

The materials that will be used in this research are molasses, aquades, acetic acid buffer, sodium acetate, Potato Dextrose Broth (GDP), Potato Dextrose Agar (PDA), *A. niger* culture, *T. viride, S. cerevisiae* and *P stipitis*. The tools used are 2 liter reactor, erlenmeyer, petridish, shaker, analytic balance sheet, test tube, thermometer, indicator pH paper, spectrophotometer, gas chromatography-mass spectrometry (GC-MS).

2.2 Preparation

Sugarcane molasses are obtained from the Glenmore Sugar Factory in Banyuwangi, East Java. *A. niger* and *T. Viride* were bred at the Microbiology Laboratory of FMIPA, University of Jember. S. cerevisiae CC 3012 has been bred in the Food And Nutrition Culture Collection (FNCC), PAU Food and Nutrition, Gajah Mada University. *P. stipitis* has been bred in the Laboratory of Microbiology, Department of Biology, Airlangga University.

Breeding in Potato Dextrose Broth (PDB) was done by agitation of 180 rpm shakers and incubated based on the growth velocity of ½ of Vmax (log phase) *A. niger* and *T. viride* which were 3 days, while ½ of Vmax *S. cerevisiae* CC 3012 and *P. stipitis* is 24 hours.

2.3 Hydrolysis and Fermentation (SSF)

Inoculum *A. niger* and *T. viride* 3 days (w/v) incubation and the consortium *S. cerevisiae CC 3012-P. stipitis* (2:1) (v/v) 24 hour incubation was added to the sample in the reactor then incubated for 6 days. Every 24 hours measurement of temperature, pH and reducing sugar is carried out. During the process of hydrolysis and fermentation in the reactor, microbial growth is observed every 24 hours. Periodic temperature and pH measurements were taken during the hydrolysis and fermentation processes to maintain stable pH conditions at 5. Reduction sugar test was carried out by the Nelson Samogyi Method. The samples were measured for bioethanol content after commencement after addition of *A. niger* and *T. viride* isolates and *S. cerevisiae CC 3012-P consortium stipitis* counts from 24, 48, 72, 96 and 120 hours with the Gas Chromatography Method.

3. Results And Discussion

3.1 Characteristics of Molasses Substrate

Characteristics of C, N and P substrate molasses in this study need to be known to determine the characteristics and potential of nutrients contained in molasses (Table 3.1).

Table 3.1. C: N: P Mollases

Measurement Time	C:N:P Mollases (%)		
	C	N	P
Before hydrolysis	22.27	1.4	0.19
After hydrolysis	25.70	1.2	0.31

C: N: P molasses at the stage before hydrolysis in this study is 22.27: 1.4: 0.19 (%), whereas after going through the hydrolysis stage, the C: N: P ratio is known to be 25.70: 1.2: 0.31. The ideal comparison of C: N : P for biomass for microorganism growth is 100: 5: 1 (%). In this study the ideal comparison is quite good. The initial C / N ratio in this study was 15.90%, while the C / N ratio after hydrolysis was 21.41%. These results are close to the ideal ratio of 20-30% (Napon *et al.*, 2013).

3.2 Reduction Sugar Levels in SSF

The SSF process that combines the process of hydrolysis and fermentation has an effect on the reducing sugar produced. Hydrolysis aims to break down polysaccharides into monosaccharides so that they can be fermented directly by a yeast consortium.

3.2.1 Reduction Sugar Levels in SSF with 8% and 10% Consortium

The advantage of the SSF process is that hydrolysis and fermentation are carried out in one container or reactor so that it can run efficiently. The results of SSF reduction sugar using 10% substrate (100 gr

/ L molasses) and 20% substrat (200 gr / L molasses) with the addition of 8% yeast consortium found in Figure 3.1.

In SSF treatment with a consortium concentration of 8% there was an increase in reducing sugar from 0 hours to 48 hours in all treatments, but thereafter decreased. The increase in reducing sugar at 0 to 48 hours was caused by hydrolysis of sucrose molasses to glucose. During the hydrolysis process, an average sugar level increase of 1.12 mg / g for the substrate 10% and 1.33 mg / g for the substrate 20% on average.

In SSF treatment with the addition of 10% yeast *S. cerevisiae-P consortium. stipitis* also has a different amount of reducing sugar. The process of increasing and decreasing the reduction sugar in the treatment of 10% yeast has the same trend as the treatment of 8% of the consortium. These results can be seen in Figure 3.2.

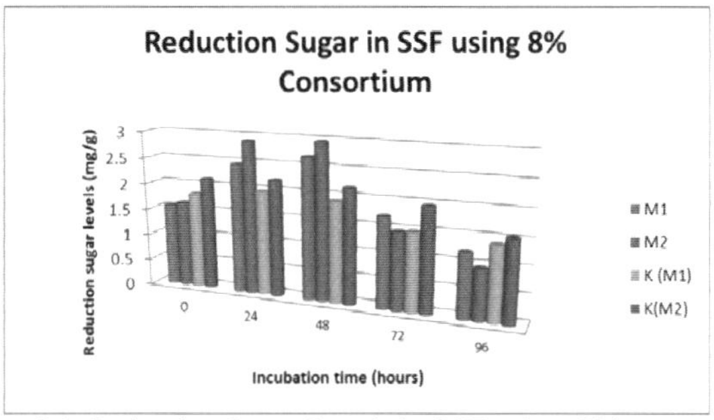

Figure 3.1. Reduction Sugar Levels in SSF using 8% consortium.
Description: M1 = 10% substrate concentration (100 g / L)
M2 = 20% substrate concentration (200 gr / L)
K = Control

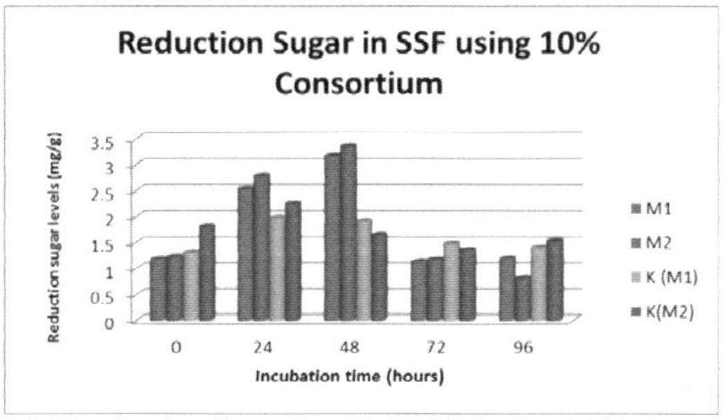

Figure 3.2 Reduction Sugar Levels in SSF using 10% consortium.
Description: M1 = 10% substrate concentration (100 g / L)
M2 = 20% substrate concentration (200 gr / L)
K = Control

During SSF hydrolysis process using 10% consortium, there was an increase in the average sugar level of 1.99 mg / g for the substrate 10% and 2.13 mg / g for the substrate 20%. Reduction in reducing sugar occurs in all treatments at 48 to 96 hours. The reduction in reducing sugar is proportional to the amount of ethanol produced, the longer the fermentation, the greater the ethanol produced (Hemamalini et al., 2012). The decrease in the amount of reducing sugar during the fermentation process is caused by sugar being used by microbes as a carbon source in fermentation. In the fermentation method simultaneously gives an effect on the reduction in the amount of reducing sugar in all treatments of SSF samples with 8% and 10% of the consortium.

In the process of hydrolysis of this study, *T. viride* plays a role in destroying high levels of cellulose and has the ability to synthesize several essential factors to dissolve parts of cellulose which are strongly bound to hydrogen bonds. In addition, *T. viride* produces cellulase complete with all the

IOP Conf. Series: Earth and Environmental Science **207** (2018) 012061 doi:10.1088/1755-1315/207/1/012061

components needed for total hydrolysis of crystalline cellulose with high sugar content produced (Esterbauer *et al.*, 1991).

A. niger plays a role in producing the enzyme hemicellulase which can hydrolyze hemicellulose complex into glucose, and also produce high cellulase. The advantages of *A. niger* from other fungi that is besides being able to produce cellulolytic enzymes also produce xyloglucanase (Pham *et al.*, 2010). Based on the advantages of these two fungi, the combination at the same time is able to provide better performance in reducing sugar. This is shown from the reduced sugar production which continues to increase until 48 hours when hydrolysis.

3.3 Microbial Growth

3.3.1 Microbial Growth During Hydrolysis

The stages of hydrolysis and fermentation in this study were carried out biologically (enzymatic hydrolysis using a combination of mold *T. viride* and *A. niger*). The results of growth observations of *T. viride-A. niger* using 10% substrate in this study found in Figure 3.3 and growth observations of *T. viride-A. niger* using 20% substrate found in Figure 3.4.

The results of growth observations of *T. viride* and *A. niger,* both on the usage of 10% substrate and 20% substrate, in this study showed that the combination of the two fungi experienced a significant increase at 48 hours. The significant increase in cell growth at 48 hours was a log (exponential) phase in this study.In this log phase the growth of microorganisms is the fastest.

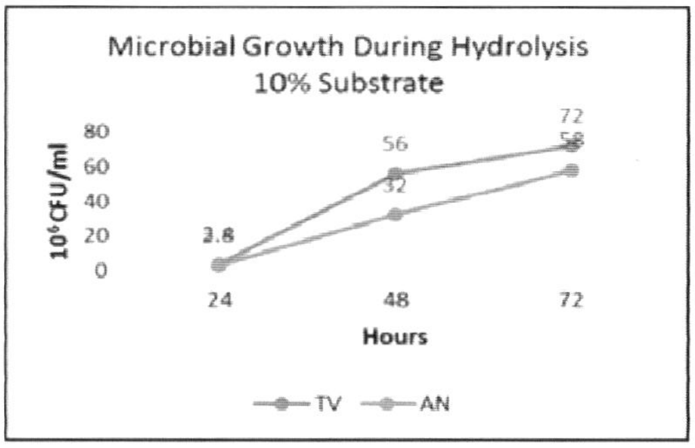

Figure 3.3. Microbial Growth During Hydrolysis with 10% Substrate..
Description: TV = *Trichoderma viride*
AN = *Aspergillus niger*

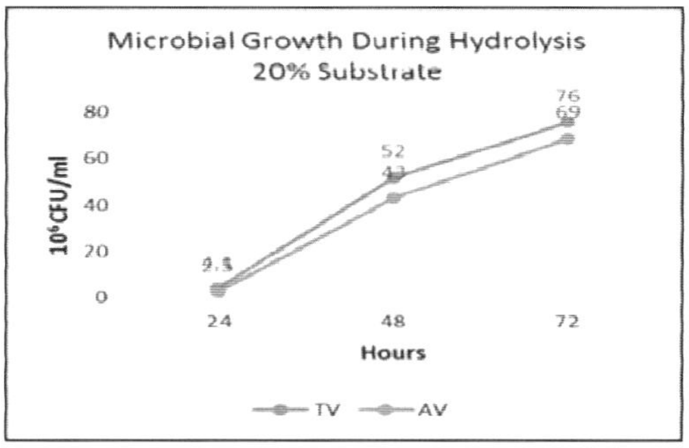

Figure 3.4. Microbial Growth During Hydrolysis with 20% Substrate..
Description: TV = *Trichoderma viride*
AN = *Aspergillus niger*

This is due to the large amount of substrate catabolism used for growth, enzyme synthesis and synthesis other compounds (Ernes et al., 2014), along with the high sugar content produced at the hydrolysis stage, but at 72 hours *T. viride* and *A. niger* began to enter the stationary period because the substrate that can be hydrolyzed decreases so that the substrate catabolism process also decreases, this is inversely proportional to reducing sugar produced. During the exponential phase the sugar levels continue to increase until the 48th hour. This is because the process of cellulose catabolism becomes simple sugars due to the cellulose enzyme activity of *T. viride* and *A. niger*.

3.3.2 Microbial Growth During Fermentation

The stages of hydrolysis and fermentation in this study are carried out biologically (enzymatic hydrolysis). The microorganisms used in the fermentation process are *S. cerevisiae-P consortiums. stipitis*. The results of growth observations of *S. cerevisiae-P. stipitis* consortium in this study found in Figure 3.5

The results of observations of *S. cerevisiae-P. stipitis* consortium growth in this study shows that *S. cerevisiae-P. stipitis* as a consortium experienced a significant increase in cell growth at 24 hours incubation since the consortium was inoculated into the reactor. Significant increase in cell growth at 48 hours was indicated as a log (exponential) phase in *S. cerevisiae* and *P. stipitis*. But on the 72 hour of *S. cerevisiae-P. stipitis* consortium no longer has a significant increase in cell numbers. This is because microorganisms enter the stationary phase where the number of cells is relatively fixed due to the limited number of substrates so that living cells are the same as dead cells (Wardani and Pertiwi, 2013).

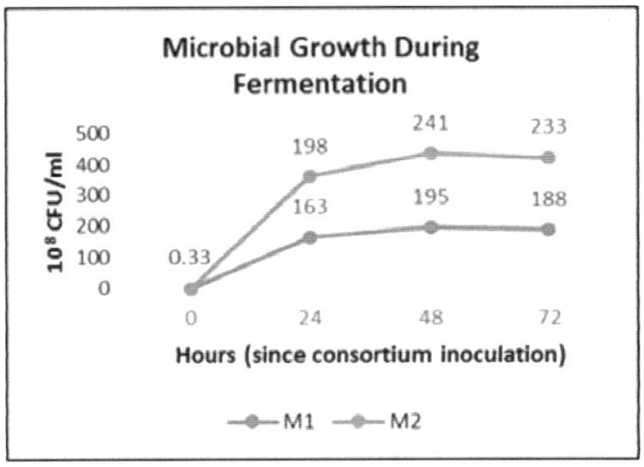

Figure 3.5. Microbial Growth During Fermentation
Description: M1 = 10% substrate concentration (100 g / L)
M2 = 20% substrate concentration (200 gr / L)

The growth of fermentation microbial cells is inversely proportional to the reduction in sugar content. The higher number of *S. cerevisiae* and *P. stipites* cells, the lower the reducing sugar level. This is because the simple sugars produced during hydrolysis are utilized by *S. cerevisiae-P. stipitis* consortium as a carbon source and converted to ethanol. In this fermentation stage, *S. cerevisiae* produces ethanol from hexose sugar contained in the substrate (Bayrakci and Kocar, 2014), while *P. stipitis* ferments hexose and pentose to ethanol (Ganguly *et al.*, 2012).

3.4 pH and Temperature Analysis of Reduced Sugar and Bioethanol

In this study, the acetate buffer pH 5 in the inoculum was adjusted to the optimum conditions of microorganisms. The average pH range in this study was pH 5. The initial temperature of the substrate before treatment was 30°C, while the temperature range after pretreatment in this study increased to 31°C and continued to increase up to 32.5°C at the end of the fermentation stage.

There was no significant increase in pH during the hydrolysis process and in the fermentation process, but the temperature continued to increase during the process of hydrolysis and fermentation. The decrease in pH that occurs during the fermentation process because the microorganisms used in addition to producing bioethanol also produce CO_2 and organic acids (Wignyanto *et al.*, 2001). While the increase in temperature that occurs in the hydrolysis and fermentation stages proves the occurrence of microbial activity in the reactor.

pH stabilization is used acetate buffer pH 5. pH 5 is the optimum pH of fungal growth, especially yeast *S. cerevisiae* and *P. stipitis*. According to Apriwinda (2013), yeast can grow at a pH range of 2.5-8.5 and can grow optimally at pH 4.0-5.0. In addition, growing yeast at low pH can prevent bacterial contamination. In fermentation, pH control is very important because optimum pH must be maintained during fermentation. Conditioning pH to be stable at 5 is to maintain yeast activity in fermentation more optimally. pH also affects enzyme activity and lignin degradation process (Patel *et al.*, 2009).

3.5 Ethanol Levels in the SSF Method

Microbes as fermenting agents are certainly very influential on the length of fermentation. In alcoholic fermentation generally yeast is used because yeast can convert sugar to alcohol in the presence of enzymes. In this study, the microbes used were *S. cerevisiae - P. stipitis* consortium. According to O'leary *et al.* (2004), *S. cerevisiae* is a yeast commonly used in alcoholic fermentation. *S. cerevisiae* has several advantages that can produce high ethanol yields compared to other microbes that can also form alcohol.

3.5.1 Ethanol Yield in SSF using 8% S. cerevisiae-P. stipitis Consortium.

The results of ethanol yield through the SSF stage are much greater than the control treatment because in the control treatment the microbial fermentation is not added. The results of the bioethanol yield of the SSF method using 8% consortium are shown in Figure 3.6.

The highest Ethanol yield of SSF using 8% consortium and 10% subatrate treatment in the experienced was done at 72 hours with result 8.2% (v / v). The highest ethanol yield of SSF using 8% consortium and 20% substrate was done at 72 hours with result 9.83% (v / v). The level of ethanol produced with a substrate amount of 20% is higher than the level of ethanol produced with a substrate amount of 10%. The results of this study indicate that the highest ethanol content on average is formed at 72 hours incubation with pH 5 and temperature 32°C.

According to Wahlbom and Hahn-Hägerdal (2002), *S. cerevisiae* can produce ethanol optimally in anaerobic facultative state at temperature 30-35 °C and pH range 4-5. Ethanol produced by *S. cerevisiae-P. stipitis* consortium tends to be higher. This is because the ability of *S. cerevisiae* which is able to ferment hexose sugar and *P. stipitis* is capable of fermenting xylose and other pentose sugars both aerobically and anaerobically (Ho *et al.*, 1998)

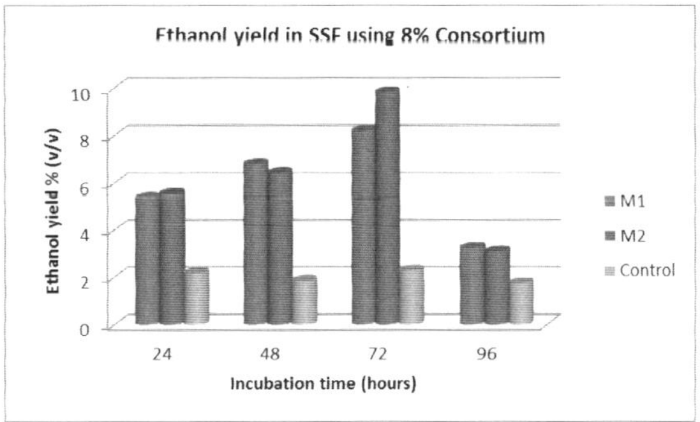

Figure 3.6. Ethanol Yield in SSF using 8% Consortium
Description: M1 = 10% substrate concentration (100 g / L)
M2 = 20% substrate concentration (200 gr / L)
K = Control

P. stipitis is able to ferment under aerobic and anaerobic conditions, and has the highest natural ability to ferment xylose, turning it into ethanol without producing xylitol (Nigam, 2002). The ability to ferment xylitol is not owned by *S. cerevisiae*. *P. stipitis* completes the performance of *S. cerevisiae* so that the consortium can produce ethanol better than using a single microorganism.

3.5.2 Ethanol Yield in SSF using 10% S. cerevisiae-P. stipitis Consortium.

The ethanol concentration obtained in this study is quite high. This is because in the fermentation process occurs the process of solving sucrose into monosaccharides or reducing sugars which can be utilized by the *S. cerevisiae-P. stipitis* for biological activities. During fermentation there is a decrease in the concentration of reducing sugar because it is used by the yeast consortium as a carbon source in its metabolic process. The results of bioethanol content using 10% consortium using the SSF method are shown in Figure 3.7.

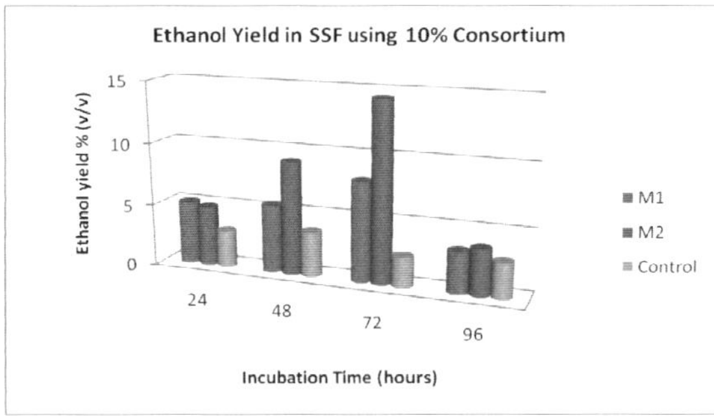

Figure 3.7. Ethanol Yield in SSF using 10% Consortium
Description: M1 = 10% substrate concentration (100 g / L)
M2 = 20% substrate concentration (200 gr / L)
K = Control

The 10% yeast consortium usage treatment produced ethanol reached 8.04% for the substrate 10% and 14.38% for the substrate 20% at 72 hours incubation. The concentration of ethanol produced was much greater compared to the control treatment that produced the highest ethanol with a level of 3.67%. The ethanol content obtained with the treatment of 10% yeast consortium yeast decreased at 96 hours. A decrease in ethanol in excess glucose concentration occurred as an inhibitory effect of substrates and products.

There was a decrease in ethanol levels at 96 hours up to 3.36% (v / v) ethanol yield for SSF using 10% substrate and 3.72% (v / v) ethanol yield for SSF using 20% substrate. This shows a relationship between fermentation stage and bioethanol produced. This fermentation time is also related to yeast cell growth. From the observations of microbial growth, the log (exponential) phase of the consortium is up to 48 hour incubation. Whereas at 72 hours incubation, the stationary phase begins to enter. In the stationary phase the microbes are in the stag phase, where nutrients have begun to run out and the number of cells tends to be stable until it enters the phase of death. During the phase there will be accumulation of toxins so that it can paralyze other microbes (Griffin, 1996).

4. Conclusion

Based on the research that has been done, the conclusions obtained are ethanol yield from sugarcane molasses using the SSF methods has the highest bioethanol concentration of 14.38% (v / v) for substrate 20% usage at 72 hours fermentation.

5. Acknowledgments

This research was supported by University of Jember DIPA Fund (SP Number DIPA-042.01.2.400922 / 2016). We thank to Dr. Yeny Dokhihah for assistance with particular technique and methodology for comments that greatly improved the manuscript and for the comments on an earlier version of the manuscript.

We also thank our colleagues from Environmental Engineering Study Program and Microbiology Laboratories Science Faculty University of Jember who provided insight and expertise that greatly assisted the research, although any errors are our own and should not tarnish the reputations of these esteemed persons.

References

[1] Apriwinda 2013 Studi Fermentasi Batang Sorgum Manis (*Sorghum bicolor* (L.) Moench) untuk Produksi Etanol, *Skripsi,* Fakultas Pertanian, Universitas Hasanuddin.
[2] Bayrakci A G and Koçar G 2014 Second-Generation Bioethanol Producti From Water Hyacinth and Duckweed In Izmir: A Case Study *Renewable and Sustainable Energy Reviews* 30 306–316.
[3] Esterbauer H, Steiner W and Labudova I 1991 Production of Trichoderma cellulase in laboratory and pilot scale *Biores Technol* 36 51-65.
[4] Ernes A, Ratnawati L, Wardani A K and Kusnadi J 2014 Optimasi Fermentasi Bagas Tebu Oleh *Zymomonas mobilis* CP4 (NRRL B-14023) Untuk Produksi Bioetanol *Agritech* 34: 3.
[5] Ganguly A, Chatterjeea P K and Dey A 2013 Studies On Ethanol Production From Water Hyacinth-A Review *Renewable and Sustainable Energy Reviews* 16 966–972.
[6] Griffin D H 1996 *Fungal Physiology* Wiley Science Paper Back Series.
[7] Hemamalini V, Saraswati S G E, Hema C and Geetha S 2012 Comparative Study of Continuous of Ethanol Fermentation from Molasses by Using *Saccharomyces cerevisiae* and *Schizosaccharomyces pombe*, *Internationl Jurnal Science* 219-228.
[8] Jeon B Y 2007 Development of a Serial Bioreactor System for Direct Ethanol Production from Starch Using *Aspergillus niger* and *Saccharomyces cerevisiae*, *Biotechnology and Bioprocess Engineering* 12 566-573.
[9] McKetta J J and William A C 1983 *Encyclopedia of Chemical Processing and Design* Marcel Dekker Inc. New York and Bessel Gadjah Mada University Press Yogyakarta.
[10] Nigam J N 2002 Bioconversion of Water Hyacinth (*Eichhornia crassipes*) Hemicellulose Acid Hydrolysate to Motor Fuel Ethanol by Xylose-Fermenting Yeast. *Journal Biotechnology* 97 107–16.
[11] O'Leary V S, Green B C, Sullivan V and H Holsinger 2004 Alcohol production by selected yeast strains in lactase Dhydrolyzed acid whey *Biotechnol Bioeng* 19:10 19–35.
[12] Pham T H, Berrin J G, Record E K A, Sigoillot J 2010 Hydrolysis of softwood by *Aspergillus mannanase*: Role of a carbohydrate-binding module. *Journal of Biotechnology* 148 163-170.
[13] Prescott *S C* and Dunn *C G* 1959 *Industrial Microbiology* Third edition revised by Cecil Gordon *Dunn* McGraw-Hill, New York.
[14] PTPN X 2015 Angka Produksi Unit usaha Gula [access date August 16th 2016].
[15] Sarris D, Matsakas L, Aggelis G, Koutinas A A and Papanikolaou S 2014 Aerated vs non-aerated conversions of molasses and olive mill wastewaters blends into bioethanol by *Saccharomyces cerevisiae* under non-aseptic conditions *Journal of industrial Crops and Products* 56 83-93.
[16] Wardani A K, Pertiwi, F and N Eka 2013 Produksi Etanol Dari Tetes Tebu Oleh *Saccharomyces cerevisiae* Pembentuk Flok (NRRL–Y265) *Agritech* 33:22 Mei 2013.
[17] Wignyanto, Suharjono and Novita 2001 Pengaruh Konsentrasi Gula Reduksi Sari Hati Nanas dan Inokulum Saccharomyces cerevisiae Pada fermentasi Etanol *Jurnal Teknologi Pertanian* 2 68-77
[18] Wahlbom C F and Hahn-Hägerda B 2002 Furfural, 5-hydroxymethyl furfural, and acetoin act as external electron acceptors during anaerobic fermentation of xylose in recombinant *Saccharomyces cerevisiae. Biotechnol. Bioeng* 78 172-178.

IOP Conf. Series: Earth and Environmental Science **207** (2018) 012062 doi:10.1088/1755-1315/207/1/012062

A novelty design of GHG Emission Reduction Cost for the Province of Nusa Tenggara Timur, Indonesia: A quantitative-based scientific view

A Amheka[1] **and Y Higano**[2]

[1] Head of Center for Research and Community Services, State Polytechnic of Kupang, Indonesia

[2] Professor at University of Tsukuba and past president of JSRSAI, Japan.

E-mail: adrianus.amheka@gmail.com

Abstract. At present GHG emissions is a strategic issue in the effort to achieve SDGs. The Nusa Tenggara Timur (NTT) Provincial Government through its new leader in 2018 emphasizes the development of infrastructure in the field of renewable energy must be a top priority. This study aims to introduce a structured model of GHG emission reduction costs in NTT Province by referring to carbon tax considerations as exogenous factors. The results show that the average cost a reduction in the constraints between 14% and 16% are the most optimal would increase GDP to 20% compared to BAU conditions. The implications of this study will support the government's efforts in embracing a policy relating to the General Regional Energy Plan of NTT Province called in English Regional Energy General Plan which is being promoted.

1. Introduction

Currently, the livelihoods of community life at the NTT Province are still largely dependent on natural resources as one of the economic supporting commodities which situation is affected by geography and geology condition, so it is extremely prone to various natural disasters in addition lack of public awareness to maintain a sustainable environment which indirectly contribute to the risk level of the threat of climate change impacts. The province has been a victim the impacts of climate change. As new leader of this Province Mr. Victor Bungtilu Laiskodat has ambition as one of the prone provinces to be at the forefront in addressing the global problem of climate change by introduction the renewable energy system to improve electrification ratio for the last few years ahead. While the central government in Jakarta so far was conducted many efforts to tackle the climate change impacts by intervention the policy mitigation actions to become a national driving force to optimize its influential position internationally through bilateral and multilateral cooperation. so that, the climate change adaptation and mitigation action strategic issues will be integrated into the national and provincial development planning [1, 2, 3]. The NTT Province has potential to reduce greenhouse gases (GHG) emission by 2030 significantly such as various policies and technology applications and strategic studies integrated to the regional economic development planning in line to achieve sustainable development goals (SDGs) while keep economy activities running well under GHG emission controlled.

2. Electricity and GHG emission profile of NTT Province.

Indonesia's national energy policy (INEP) targets as much as 23% and 31% of the primary energy mix will supply from renewable energy by 2025 and 2050 [4]. Strengthening the national energy buffer through regional primary energy supply, instead of costly fossil fuel imports. Every year electricity subsidies increase 5% and data available show in 2010, fossil fuel subsidies in NTT Province was achieved 447.5 Billion IDR which make the present subsidies for electricity supply affordable for the population, cost the province over 661.1 Billion IDR per year which does not consider increases cost of diesel oil, while dependence on fossil fuel imports is steadily growing [2, 5, 6]. Although Indonesia made great advances in the electrification across the country, some provinces including NTT Province is still hard to reach great electrification until today [7]. NTT Province is covered into the Indonesian energy blueprint shown in Fig.1 to achieve objectives the development of an integrated network management power plant by 2020 to achieve GHG emission reduction

Figure 1. Map of existing and planned integrated network management power plant [8]

Current electricity in NTT Province is generated from Diesel Power Plants (DPP) that is operated by a state power company called "PLN NTT Branch." PLN also serves as the administrator for distribution of electricity in the Province. Increased sustainable energy generation would benefit the NTT population, 5.12 million people and would help to support the population growth rate of 1.65%/year. The GDRP in 2015 was USD 3.8 Billion and GDRP per capita USD 742.2 and its growth rate 3.31% annually. However, the poverty rate was 22.19% which indirectly affects the good practice of sustainable use of renewable energy in achieving energy intensity of 1% or less [9, 10].

In order to pursue Indonesia GHG emission target 29% from BAU and 41% with foreign assistance [4], there was a series effort has been done by NTT province government through Regional Planning Agency (or BAPPEDA) as the institution is appointed for. As for economy of NTT province is lower compared among other therefore, a novelty model is necessary to overcome the challenge to support a General Regional Energy Plan which is being promoted. The certain amount of NTT Province's GHG emission for the base year 2010 is 1.381Mt or around 0.047% for a global warming potential (GWP) 100 years and 1.454Mt or around 0.049% for a GWP 20 years both compared to total GHG emission in Indonesia in 2010. This result is an adjustment from the previous results. Then verified use a linear program software and the result shown same percentage is 0.047% in comparison with total GHG emission in Indonesia by 2010 and 0.034% in comparison with the total GHG emission in Indonesia by 2020. These amounts are used as exogenous data to conduct for the model [4, 9, 11, 12].

3. Methodology

This research is done to integrate the design and statistical analysis to forecast economic growth trends and marginal GHG reduction cost. The methodology is used to estimate the costs and effectiveness of electricity produced by renewable energy technology (RET) sources and associated with carbon tax and reductions rate in GHG emissions in NTT Province for the period of 2014 as indicated in the hypotheses and explored the proof truth of the hypothesis. The economic I/O model has been adopted for this study which is investigated by Leontief showing relationship between economic sectors and environmental burden specifically it assesses consumption of energy final and environmental impacts of goods and services from a State by taking advantage of a country's economic I/O matrix [11, 12, 13, 14, 15]. The I/O model also feasible to sum up the energy consumption and environmental impacts of each life-cycle phase [16]. In this study the NTT Province I/O [17] model is used coupled with an environmental perspective can support local government to monitor and controlled their emission, then will be able to evaluate the effectiveness of a proposed energy policy.

3.1 Construction of the simulation

The simulation is constructed of some main indicators among others comprehensive social costs as indicated for GHG reduction cost (average and marginal) and carbon tax model with an objective function in maximize the gross regional product (GRP). The variables are determined based on endogenous (*en*) and exogenous (*ex*). The exogenous variables are taken from existing data, while the endogenous variables are obtained from the results raised by the simulation model. The intended indicator model for maximize the GRP as follow:

$$GRP = \sum vv(i) \cdot X(i) + \sum vv^{RT}(i) \cdot X^{RT}(i) \tag{1}$$

where, GRP (t) is the gross regional product in term t (time) (*en*); $X(i)$ is production vector of usual sectors (*en*); $X^{RT}(i)$ is production vector of RET sectors (*en*); $vv(i)$ is (gross) added value ratio of usual sectors (*en*); and $vv^{RT}(i)$ is (gross) added value ratio of RET sectors (*en*). Subscripts (i) are the industries in NTT Province. However, the GRP itself is determined endogenously and will not the main focus of this study due to the controlling is only focused on the social costs that need to be spent due to GHG emissions resulting from economic activities. With this type of economy structure, the GHG reduction cost has an important index. Generally, it is defined as follow:

$$\varphi = (Cost\ neccesary\ for\ reducing\ GHG\ by\ \Delta Z)/\Delta Z \tag{2}$$

where, φ is average GHG reduction cost which is obtained through inverse comparisons between the main costs needed to reduce GHG emissions and difference amount of GHG emission as labeled (ΔZ). The ΔZ is comparison between total amount of GHG emission and amount of GHG emission contributed by energy sector. With this simulation, the cost is measured in terms of reduction in GRP due to the GHG reduction constraint as indicated in formula below:

$$\varphi^i(r) = \{(GRP^i(0) - GRP^i(r)\}/\{\overline{ZZ}(1\text{-}\frac{r}{100})\} \tag{3}$$

where, $\varphi^i(r)$ (en) is average GHG reduction cost in scenario of i and when the reduction rate of GHG emission (n) is $r\%$ $r = 0\%$ up to 20% (ex); $GRP^i(r)$ is the GRP in scenario of i and when $r\%$ (en); $\overline{ZZ}$ is amount of GHG emission in NTT Province was emitted into the nature when the n is 0% and the amount is 14.51977Mton CO_2e (ex).

note: scenario (i) means, which is no carbon tax and no-RET available when $i = 0$; scenario in which carbon tax is charged and no-RET available when i=1; and scenario where carbon tax is charged and RET system available when i=2. The carbon tax is exogenously determined.

While estimation the marginal GHG reduction cost is an important thing for decision makers to know how much society would want to spend on GHG emission reduction [jstor *The Energy Journal*]. In this study, it can be calculated as follow:

$$\lambda^i(\mathrm{r}) = \{(GRP^i(r) - GRP^i(r+1)\}/14.51977 \quad (r = 0,2, \ldots, 19) \tag{4}$$

where, λ^i (r) is a comprehensive social for marginal GHG reduction cost in scenario of i when the n is r %.

3.2 Hypotheses

The hypothesis is based on verification the effect between the trend amount of GHG reduction cost in a setting of reduction rate of the emissions, which is referred to the statement formula below:

- $\varphi^i(r)$ decrease drastically as scenario increases from $i = 0$ (no carbon tax, no RET) to $i = 2$ (with both carbon tax and RET available) with same r of any n; and
- $\lambda^2(\mathrm{r})$ increases as r increases with fixed $i = 2$.

4. Results and discussion

4.1 Results

The Marginal costs every region is different each other including the global marginal costs. For the case study in NTT Province the simulation results indicated in Figure 2.

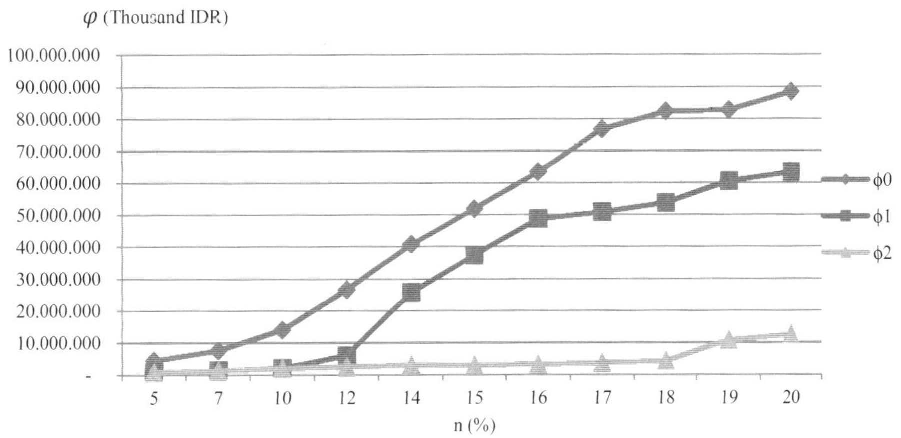

Figure 2. Average comprehensive social of GHG reduction cost

The simulation shows the social costs having great influence on GHG emission reduction in the province of NTT with BAU scenario (φ^0)((blue-line), for instance on the GHG constraints between

5% and 10%, the condition cost of averange GHG emission reduction quite spent the Province cost. The social costs continue to grow moving forward in line with efforts to reduce emissions between 10% and 18% and are able to stabilize in the constraint between 18% and 19% but these costs cannot exceed 20% reduction which will spent a lot of cost to burden the economy of NTT Province. This level shown the hypothesis is true. Further, on the conditions where there is no tax and RET is available (φ^1) (red-line), averange GHG reduction cost quite better than φ^0 especially for constraints between 5% and 12%, this results shows that the abatement cost of the NTT Province economic will be able to reduce GHG emissions between 5% and 10% if applying carbon tax without intervention of RET system application (this scenario, the carbon tax has been included as exogenous data in the model). However, after 10% of the contraints are apllied, the reduction cost increase drastically exactly on the constraints between 14% and 16% and flat up to 17%. This is reversed with the BAU conditions in the same constraint, and after 18% and above, the GHG will be uncontrollably large and not optimal, so that this level also shown the hypothesis is true.

While for scenario (φ^2) (green-line), the conditions of the carbon tax and RET are equally available. The simulation shown that the averange GHG reduction cost are very small and do not burden the current economic conditions in the Province. A very drastic comparison between the scenario φ^0 and φ^1. This optimal scenario shows that the emissions constraint after 18% raises the trend of social costs slightly compared to the previous cost trend and flatly after constraint 19%. So the government is strongly encouraged to choose the condition of the scenario, where both rhe carbon tax and RET system are availabe. This level of simulaton is meet the hypothesis mentioned that the average GHG reduction cost decrease drastically as scenario increases. The average cost of GHG reduction from the sample n=14% and 20% are 40.8 Billion IDR to 88.4 Billion IDR; 25.8 Billion IDR to 63.3 Billion IDR; and 3.01 Billion IDR to 12.5 Billion IDR compared among scenarios φ^0, φ^1 and φ^2 respectively.

The Marginal GHG reduction cost in the NTT Province is meet the hypothesis as indicated in Figure 3.

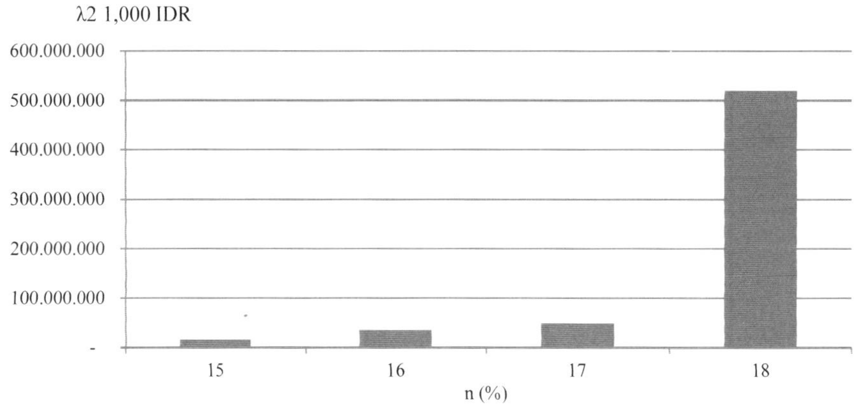

Figure 3. Marginal GHG reduction cost of carbon tax and RET are available

An example of marginal GHG reduction cost to answer hypothesis 2, for constrains between 15% and 17% are not so high between 16 Billion IDR and 48.7 Billion IDR, however raised drastically on the GHG constraint 18% and so on is 519.7 Billion IDR and above.

4.2 Discussion

The marginal costs are effectively rise by such amount referred to the sum of impacts into countries accounts associated with inequalities in income distribution or for risk aversion [18] at the condition of average GHG emission cost is introduced where no carbon tax and no RET system in available, with continued efforts in reducing emission in a sustainable way does not guarantee rapid acceleration of NTT emissions reduction without sacrifice the current economy. At this BAU level, conservation efforts through adaptation actions are consolidated and coordinated can still continue to be carried out. However, the GHG emission mitigation actions that necessary indicate financing cannot be applied after 20% constraints without conflict among stakeholders in term of economy perspective and emission control. Strictly, the emission reduction policy carried out at this level remains optimal, which can be done only under the constraints of 14% to 16%. Further, if we compared the average cost among scenarios by introduction a carbon tax mechanism without RET system and/ or if at the same time introduce both the carbon tax mechanism and RET system available will become more optimal especially the last scenario has mentioned as labelled φ^2. If this policy is implemented, definitely will attract special attention or it can become a dynamic for modern society and industry, but government authority is a good opportunity at the level of this scenario. Nevertheless, many studies have done mentioned that the total costs of climate change are small in the future and this definitely meet our two three scenario case studies present the sanguine view that seemingly expensive investments today result in lower costs in the future [19] in this sense, we could recommended the carbon taxes system must be introduced and implemented by consider any exogenous values that are not considered in the current model to make the results more strong and reliable.

The uncertainty about the marginal costs is large and exceed the costs that have been suggested by the policies and many assumption-based studies have been carried out with uncertainty [20] Similarly, studies that use equity weighing, have higher estimates and larger uncertainties [21, 22]. In our study found the marginal of GHG emission reduction cost is strongly correlated through introduction of RET and carbon tax mechanism, however this must be done for further research as mentioned by [23] said reliable of policy simulations in term of Marginal costs (abatement and welfare) for GHG emission reduction are differ from the conditions under which the marginal abatement costs were simulated.

5. Conclusions

Half important structure of economy in terms of the impact caused by introduction of a carbon tax and environmentally friendly subsidy to promote RET under GHG emission constraints, then analyze value of proposed scenarios are considered in this study. A mechanism and/or policy about carbon tax as special tax and subsidy distribution should become a concern for the government. This model can be optimized the NTT Province economy and flexible may applied to any other developing region which has the same economic characteristics to achieve the SDGs especially goals 7 said ensure access to affordable, reliable, sustainable and modern energy for all and the same time meet the provincial government mission to emphasize the development of infrastructure in the field of renewable energy as a top priority. However, deeply research regarding the carbon tax itself and contribute to tackle GHG emission and GRP improvement need to be further research in addition to count the efficiencies indexes to get more comprehensive social cost-benefit analysis of RET while GRP is strictly considered.

The implication of this study will equip the Nusa Tenggara Timur government trust in adopt a proper method in deciding a mechanism for environmental policy and to set technical steps for Indonesia government to overcome targets to reduce GHG emission in the level of national and regional by 2030 and become a strong reference to prepare General Regional Energy Plan and achieve SDGs in regional level match to the current government leader vision.

6. References

[1] BAPPENAS 2011 Guideline for developing local action plan for GHG's emission reduction (RAD-GRK), Official document of Ministry of National Development Planning (BAPPENAS), Republic of Indonesia.

[2] Amheka A, and NTT Province Government team 2018 "General Plan of Regional Energy: A draft" Vision and Mission of Mr. Victor Bungtilu Laiskodat: a Modul presentation for the Nusa Tenggara Timur (NTT) Province.
[4] General Plan for National Energy 2017 in Bahasa: Rencana Umum Energi Nasional), ESDM Ministry
[5] Amheka A, Higano Y 2015 *Regional Science Inquiry* **1**: 11-19.
[6] Statistik Ketenagalistrikan 2014, EBTKE, ESDM Ministry.
[7] IEA 2018, International Energy Agency, 2017 Southeast Asia Energy Outlook
[8] Ministry of Energy and Mineral Resources of the Republic of Indonesia (ESDM) 2008 Road-map of renewable energy development, ESDM press.
[9] Amheka A. Ph.D. Dissertation 2014. University of Tsukuba, https://core.ac.uk/download/pdf/56661548.pdf
[10] Amheka, Adrianus. 2018 ERSA Conference, <https://ersa.eventsair.com/QuickEventWebsitePortal/58th-ersa-congress-cork-28-31-august-2018/ersa2018/Agenda/AgendaItemDetail?id=80da5777-75f0-442b-9f7c-fd36214377ab> access 05 September 2018
[11] Amheka A, Higano Y, Mizunoya T and Yabar H 2015 *Studies in Regional Science* **45**(1):41-60.
[12] Amheka A, Higano Y, Mizunoya T, and Yabar H 2014 *Procedia Environmental Science* 79-88
[13] Leontief WW 1936 *Review of Economics and Statistics* **18** (3), 105–125.
[14] Leontief WW 1941. Structure of American Economy 1919-1929. Oxford University Press, New York.
[15] Leontief W 1986 Input-output Economics. Oxford University Press, New York.
[16] Hendrickson C, Lester B.L, Matthews H.S 2006. Resources for the Future, Washington, DC.
[17] Input- Output Table of NTT Province 2005 BPS Province of NTT.
[18] Richard SJ 1999 Tol, *The Energy Journal* **20** (1), 61-81.
[19] Kenneth G and James H. Stock 2018. https://scholar.harvard.edu/files/stock/files/gillingham_stock_cost_080218_posted.pdf access 05 November 2018
[20] Michael H and Ivar I 1995 *Control and Game-Theoretic Models of the Environment* **199**: 89-105.
[21] Onno K, Luke B, Richard S.J.T 2009 *Energy Policy* **37** (4), 1395-1403.
[22] Richard SJT 2005 *Energy Policy* **33** (16), 2064-2074.
[23] Jennifer M, Sergey P and John R 2012 *Environmental Modeling & Assessment* **17** (4), 325–336.

IOP Conf. Series: Earth and Environmental Science **207** (2018) 012063 doi:10.1088/1755-1315/207/1/012063

Implementation microcontroller-based bioethanol levels measurement tool on bioethanol purification equipment

E Antika[1*] and Y Susmiati[2],
[1]Department of Information Technology, Politeknik Negeri Jember, PO BOX 164 Jember East Java, Indonesia.
[2]Department of Engineering, Politeknik Negeri Jember, PO BOX 164 Jember East Java, Indonesia.

*elly_antika@polije.ac.id

Abstract. In bioethanol purification device, result of process has to measure the level of alcohol usually by alcohol meter or by picnometer but need more then 100ml sample and need much time to measure. The measuremet alcohol level by sensor detecting the vaporized alcohol in the air that will caught gas sensor TGS 2620. Therefore, development of simple and quick detection of alcohol level in purification result is strongly required. The alcohol equipment detection consists of a microcontroller module, a sensor module, output module and a bioethanol tube. There are two tubes that are used as a container for the bioethanol purification. The first container is automatic alcohol sensor tube, and the second container is storage tube for result bioethanol purification. Bioethanol data detection is displayed through the LCD and a computer. The microcontroller modul regulate the delay time, process sensing and display results. After alcohol levels are detected, then the alcohol sensor will be cleaned by blowing the fan so that the alcohol vapor attached will disappear. The delay time is set so that the sensor reading has an accurate value with the smallest error detection

1. Introduction

In bioethanol purification device, result of process has to measure the level of alcohol by alcohol meter or by picnometer. The measurement by that equipment need sample more then 100ml. the bioethanol purification device use rectification distillation sieve tray type and can produce 90% bioethanol[4].

The measurement alcohol level by sensor for detect the vaporized alcohol in the air that will be caught gas sensor TGS 2620 type. The sensing process, coverting analog to digital data and displaying data to LCD [3][5]. Alcohol level detection consist of a microcontroller module, a sensor module and output module. The microcontroller module regulates the detection time, detection delay, process sensing and display results. The sensor module consists of a sensor circuit and fan circuit. And the output module consists of LCDs, and ports for connection devices to the computer [1]. The detection alcohol content is affected by humidity and room temperature [2]. Therefore, development of simple and quick detection of alcohol level in purification result is strongly required.

In this paper, we have studied on making automatic detectiong alcohol on storage tube for bioethanol purification system using a TGS 2620 sensor.

2. Methodology

The implementation of bioethanol level detection devices to improve the performance of bioethanol purification equipment as follows:

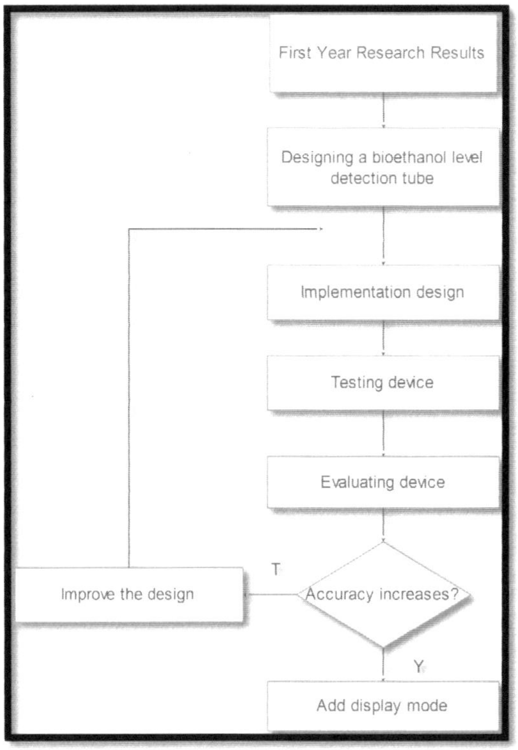

Figure 1. Research method for implementing bioethanol level detection devices.

Caption 1 as follows:

- The first research is determine alcohol sensor that has a faster response and good accuracy. Measurement bioethanol levels requires closed room and at normal room temperature condition.
- The design stage make a tube design with an alcohol sensor and an automatic faucet.
- The implementation of the design by making a continer tube from bioethanol purification result and placing a sensor to detecting alcohol levels automatically. The device is driven the sensor by motor forward and backward to the hole in the tube. The sensing process requires setting the timer detection.
- Testing has purpose to see the results of the implementation.
- Evaluation is carried out for repairing bioethanol level detection devices.
- If the sensor tube has an inaccurate test it will change the design and implementation of the sensor tube. If it has an accurate, the other output data display mode is added.

3. Experimental

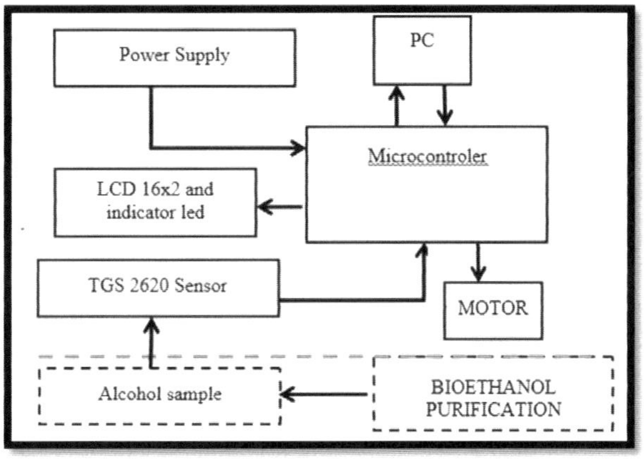

Figure 2. Bioethanol detection system

3.1 Results of the implementation the actuator instructions

The actuator is the output of the process, and it is a motor. It is useful for pushing and pulling the slider in scanning process. The implementation device by conducting instructions through the interface that has been connected to Arduino. The initial setting are time, date and scanning delay. The slider will move forward when starting scanning and re-enter the box when scanning delay or scanning alcohol ends. Time synchronization can be seen in table 1.

Table 1. Synchronize delay timer

Start/Stop scanning	*Scanning delay*	time(*stopwacth*)	Time difference
Start	00:15:00	00:14:59	1 second
Start	00:05.00	00:05.01	1 second
Start	00:10:00	00:10:01	1 second

3.2 Determination sensor height from the surface of bioethanol liquid

The position of sensor in the alcohol detection devide is determined by measuring the voltage value of sample bioethanol with a difference height on 1 cm, 2 cm, 3 cm and 4 cm from bioethanol liquid surface. Measurements were made 3 times and results that were not different from 1, 2, and 3 testing, see table 2 more clearly.

Table 2. Determination sensor height

Height (cm)	testing 1 (V)	testing 2 (V)	testing 3 (V)
1	3.02	3.06	3.07
2	3.04	3.06	3.06
3	3.03	3.06	3.06
4	2.85	2.90	2.90

From the results testing, the sensor placement is good between 1 cm to 2 cm from surface of bioethanol liquid.

3.3 The results of device

The bioethanol detection system can be seen in Figure 2. The flowchart process can be seen in Figure 3.

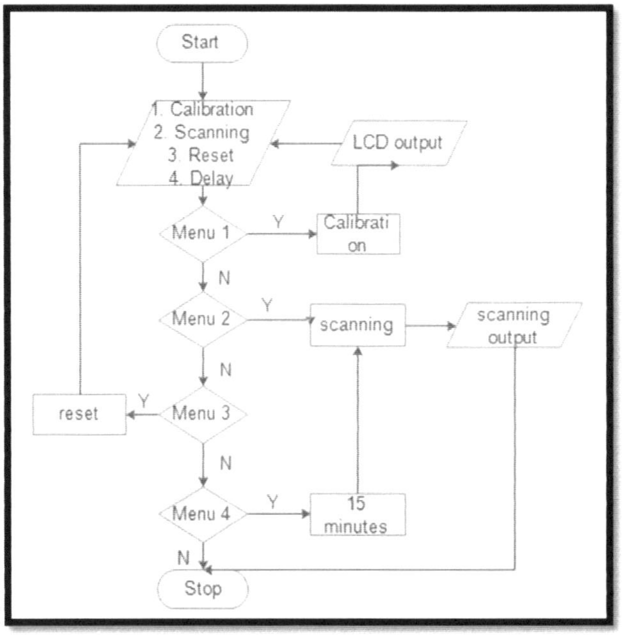

Figure 3. Flowchart process alcohol detection device.

Figure 4. Detection bioethanol level

3.4 Testing detection with and without delay time

The bioethanol sample is used in testing 5 times in an hour and the length of each measurement is 2000 times the arduino machine cycle.

a. Measurement 44% alcohol level

The data will be taken on average as a result of alcohol measurement, more clearly can be seen in table 3.

Table 3. Results alcohol measurement 44%

alcohol sample	testing	without delay	with delay 15"
	1	46.38%	47.82%
	2	49.31%	48.13%
	3	47.53%	50.20%

44%	4	46.80%	45.92%
	5	50.01%	47.66%

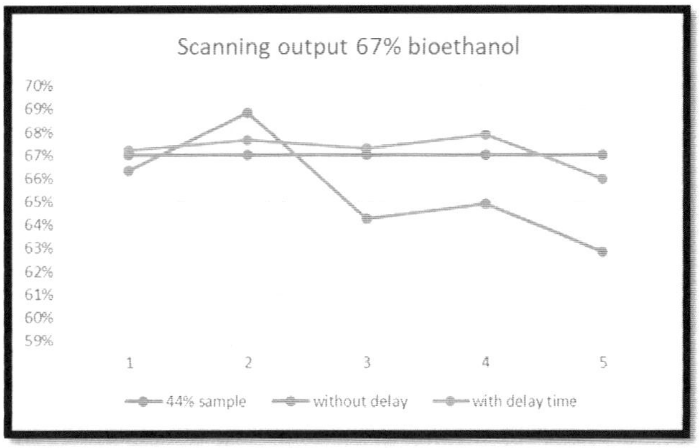

Figure 5. Scanning output 44% bioethanol

b. Measurement 67% alcohol level
60% alcohol level is measured with an alcohol meter and will be detected through a closed tube, the measurement results can be seen in table 4.

Table 4. Results of alcohol measurement 67%

alcohol sample	testing	without delay	with delay 15"
	1	66.33%	67.22%
	2	68.82%	67.63%
67%	3	64.28%	67.29%
	4	64.91%	67.89%
	5	62.83%	65.96%

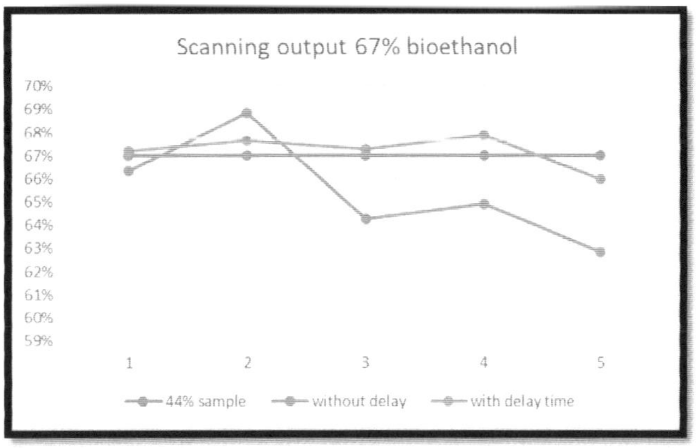

Figure 6. Scanning output 67% bioethanol

One of measurements result through the interface can be seen in Figure 7.

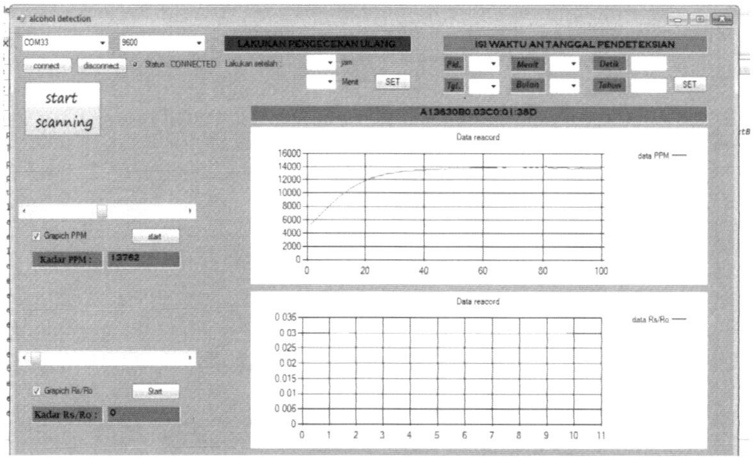

Figure 7 Measurement by interface

4. Conclusion

The height of sensor placement above the surface bioethanol liquid is 1cm to 2cm. Scanning alcohol level in bioethanol need delay time, it is 15 minutes. Device can display the scanning output by LCD and by computer interface.

Acknowledgements

The authors gratefully acknowledge the financial support given by RISTEK DIKTI 2017-2018.

References

[1] Antika, E., Susmiati, Y., 2017, *Design and Implementation of MQ3 Sensors and TGS2620 Sensors Based on Microcontrollers Supporting of Bioethanol Purification* , SEMNAS Hasil Penelitian dan Pengabdian Politeknik Negeri Jember 2017. (in Bahasa)

[2] Antika, E., Susmiati, Y., dan Nuha, F.U., 2018, *Comparison of Bioethanol Level Reading Devices Using MQ3 Sensors and TGS2620 Sensors*, KNSI 2018, Pangkal Pinang (in Bahasa)

[3] Simatupang, G.H.N., Sompie, S.R.U.A dan Tulung, N.M. 2015. *The design detector through ekhalasi alcohol levels using a sensor TGS 2620-based microcontroller arduino UNO.* E-Jurnal Teknik Elektro dan Komputer, Vol.4 No.7 (2015) ISSN : 2301-8402. (in Bahasa)

[4] Susmiati, Y. dan M. Nuruddin. 2015. *Bioethanol purification by destilation rectification sieve tray type for fuel grade ethanol (FGE).* Laporan penelitian Hibah Bersaing Tahun Anggaran 2015. Politeknik Negeri Jember. (in Bahasa)

[5] http://www.figarosensor.com/products/docs/TGS%202620C%280814%29%20pdf.pdf

[6] https://www.sparkfun.com/datasheets/Sensors/MQ-3.pdf

[7] https://www.arduino.cc/en/uploads/Main/ArduinoNanoManual23.pdf

Studies on inhibition of *Morinda citrifolia* leaf extract (*Morinda citrifolia* L) against the growth of *Aeromonas hydrophilla* in vitro

I A A Pongoh[1*]**, A A Gemaputri**[1]

[1]Departement of Agribusiness Management, Politeknik Negeri Jember, 68101, Indonesia

* Corresponding author : idapongoh@gmail.com

Abstract. *Morinda citrifolia* leaf contains active compounds that function as antibacterial substances, namely saponin, flavonoid, polyphenol, tannin and triterpen. The purpose of this research to discover the effectiveness of the *Morinda citrifolia* leaf extract in order to inhibit the growth of the bacteria *Aeromonas hydrophilla* as a cause of Motile Aeromonas Septicemia disease (MAS) on freshwater fish. This research is experimental research laboratory in vitro. The method used as an antibacterial test is a soft diffused method paper disc. The proportion of *Morinda citrifolia* leaf extract used in this research are 7.5%, 15%, 30%, 60% and negative control (ethanol 96%). The results of the study showed that *Morinda citrifolia* leaf extract is significantly ($p<0.05$) effective in inhibiting the growth of the bacteria *Aeromonas hydrophilla*. It shows that higher proportions of *Morinda citrifolia* leaf extract then inhibit zone formed when increasingly wide.

1. Introduction

Indonesia has the potential of fishery and aquaculture very potential, one of them is the aquaculture freshwater fish. This time intensive aquaculture system in freshwater fish cultivation activities implemented by fish farmer. This system is done to get maximum production results with widespread minimum land in a short time. This aquaculture system applies a high forbidding sowing solid that caused the fish more susceptible to illnesses.

The disease is vulnerable to attack freshwater fish namely infectious diseases such as Motile Aeromonas Septicemia disease (MAS) or often called with red spots disease on the body of fish. This infectious disease is a disease caused by bacteria *Aeromonas hydrophila*. It is often the problems in the aquaculture activities freshwater fish because bacterial pathogens can cause primary and secondary infections later will cause the death of fish and eventually cause loss for fish farmer

The Disease MAS began to be known in Indonesia around 1980, where these bacteria cause of disease in various freshwater fish that begins from the West Java Area and finally to spread throughout the region of Indonesia. Motile Aeromonas Sydrome primarily cause disease incultured warm water fishes, like the example Carp (*Cyprinus carpio*), Channel cat fish (*Ictalurus punctatus*), striped bass (*Morone saxalitis*), an large mouth bass (*Micropterussalmoides*). *Aeromonas hydrophila* may also affect a variety of cool water species, but is not only restricted to fresh water environment. *Aeromonas hydrophila* is the cause of frightening diseases in fish farming around the world and is considered to be a major problem, in being the causative agent of *Haemorrahagic septicemia* and *Epizootic ulcerative* syndrome (EUS) [1], [9].

Generally tackle disease on aquaculture activities using chemicals or antibiotics. But this time its use has been banned because it has a high risk that, namely can cause the resistance against bacteria, leave oily residues on the body of the fish that can be harmful to human health if consume the fish and also pollute the environment. To overcome this need to search for a replacement alternative more secure, effective and environmentally in controlling disease MAS In freshwater fish

One of disease control MAS with secure and friendly environment is the use of plants that contains natural active ingredients that are antibacterial. *Morinda citrifolia* one of the plant is as antibacterial. The substance contained in the plant pure Noni juice acts as a antibacterial one namely antrakuinon.

This substance is proven to suppress the growth of bacteria *Pseudomonas aeruginosa, Proteus morganii, Staphylococus aureus, Basillus subtilis and Escherchia coli* [2], [5].

Morinda citrifolia has many used by fresh water fish farmer to cope with infectious diseases in fish. The crops are often used to overcome the disease MAS is *Morinda citrifolia* juice. The *Morinda citrifolia* juice proved to be able to cope with the disease MAS on the seed of the nilem [3]

Morinda citrifolia leaf is also part of plants which are potential developed to suppress the growth of bacterial pathogens. *Morinda citrifolia* leaf proved to have active substances that can be against the bacteria infection, like *Straphylococcus aureus* [4] , *Salmonella sp. And Escherchia coli* [5]. Based on the research [4], *Morinda citrifolia* leaves proved to contain oil industry, fenol triterpenoid saponin, tannin and glycosides that function as antibacterial. With the potential of *Morinda citrifolia* leaf, does not close the possibility if the leaf pure Noni juice can also be used to inhibit the growth of bacteria in the *Aeromonas hydrophila* causes disease MAS In freshwater fish. Therefore, this this research was conducted to see the effectiveness of noni leaves in inhibiting bacterial growth causes disease MAS In freshwater fish in vitro

2. Material and Method

The ingredients in research among others are bacteria *Aeromonas hydrophila*, *Morinda citrifolia* leaf, Natrium Agar media, Aquadest, Alcohol, ethanol, sanitary napkins, aluminum foil. The method used in this research is an experimental method that is done in a laboratory, compiled using five treatment and three times in Deuteronomy. The treatment is done with soaking the discs paper in *Morinda citrifolia* leaf extract with different proportions. The main treatment used as the negative control treatment is (ethanol 96%) and *Morinda citrifolia* leaf extract in various proportion (7.5%, 15%, 30% and 60%). The results of these observations is data tested using descriptive data analysis and quantitative analysis of simple linear regression data. Quantitative descriptive analysis is the analysis of the function that gives the impression of research data described by the descriptive quantitative and presented in the form of tables and graphs.

2.1 Morinda citrifolia L. Leaves extraction procedure

706 gram of fresh *Morinda citrifolia L* washed leaves which were neither too old nor young, then dried in 50°C temperature 18 hours. Afterwards the leaves were mashed by using a blender. Then, the *Morinda citrifolia L* leaves which had become powder form, was extracted using maceration technique. *Morinda citrifolia L* leaves powder was put into a tightly closed erlenmeyer flask with addition of ethanol solvent 96% until the powder submerged for 3 days [6]

After the maceration completed, the powder was filtered with a *buchner* funnel that had been previously coated with filter paper and accommodated into the eruvier flask to obtain filtrate and residue. Then the filtrate was evaporated using a *rotary evaporator* at a temperature of 40°C to obtain a viscous extract. The results of liquid extraction were stored at temperature of 0-4 °C. According to [6] flavonoids were unstable against the influence of light,therefore a dark storage container is required to avoid the effect of light on the extraction results. Phytochemical test was performed for *Morinda citrifolia L* leaves extract results to determine the content of it's active compounds.

2.2 Inhibitory Test of Growth the Aeromonas hydrophilla

The method used in the bacterial growth inhibitory test is the paper disc diffusion method. The steps to test the growth inhibition of *Aeromonas hydrophila* bacteria in vitro starting from pour Natrium Agar media as much as 12.5-15 ml into a sterile petri dish then leave it at room temperature until it solidifies, then pour 0.1 ml of the bacterial suspension using a micropipette on the solid NA medium and flatten it using a bent pipe. Previously soaking strerilized disc paper (diameter 1.5 cm) into *Morinda citrifolia* leaf extract various concentrations according to negative treatment and control for 24 hours. Laying disc paper on the surface of the media Solid NA which has been spread by bacteria using sterile tweezers then incubated in an incubator at 37°C for 24-48 hours, then measuring the diameter of the clear zone formed around the disc paper using a caliper run [7]

3. Results and Discussion

The results of the measurement of the diameter of the clear zone or zone of inhibit the bacteria *Aeromonas hydrophila* in proportion (7.5%, 15%, 30%, 60%) and ethanol 96% as negative control can be seen in the table 1:

Table 1. Measurement Results Diameter inhibit the *Aeromonas hydrophila* Zone

Treatment (Concentration of Morinda citrofilia leaf extract)	Inhibit Zone (mm)
Control	0
7.5 %	3.3
15 %	18.3
30%	27
60%	36.1

The diameter of inhibit zone formed in each of the treatments shows on the control using the ethanol 96% could not have formed the inhibit zone, while in proportions 7.5% , 15%, 30% and 60%. Sequentially the average inhibit zone was formed by the increasing of 12.7 mm, 18.3 mm, 27 mm and up 36.1 mm.

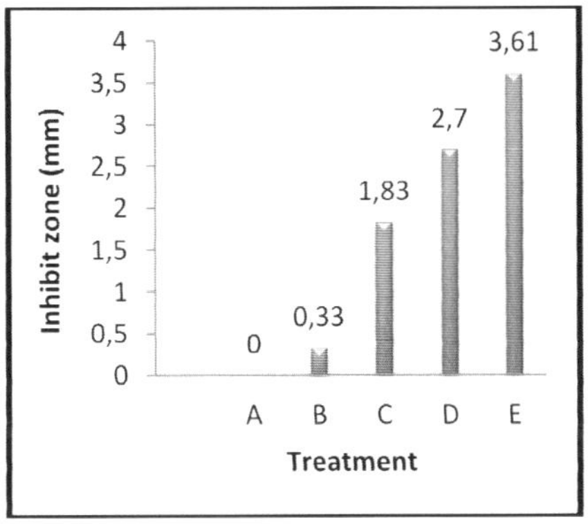

Figure 1. Graph of the relationship between the proportion of *Morinda citrifolia* Leaf Extract (Treatment) with the diameter of the Clear Zone Bacteria *Aeromonas hydrophila*

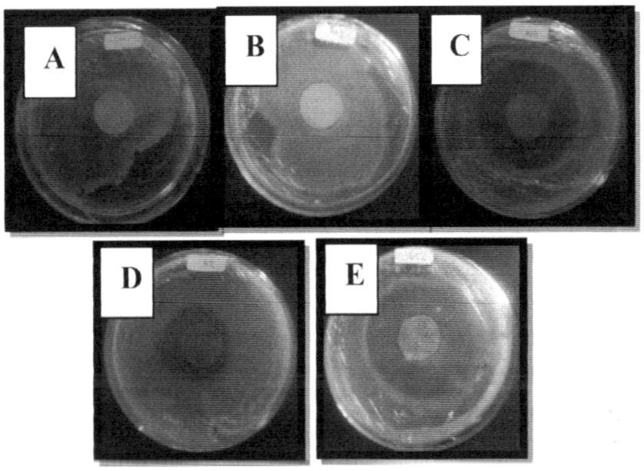

Figure 2. The Clear Zone Formed in Individual Treatment

A. Negative control (ethanol 96%)
B. The proportion of 7.5% *Morinda citrifolia* leaf extract
C. The proportion 15% *Morinda citrifolia* leaf extract
D. The proportion 30% *Morinda citrifolia* leaf extract
E. The proportion 60% *Morinda citrifolia* leaf extract

The diameter of the inhibit zone formed are indicators that the *Morinda citrifolia* leaf extract may inhibit the growth of *Aeromonas hydrophila*. Proportions of *Morinda citrifolia* leaf extract 30% and 60% (including in the category of the very strong antibacterial power), proportion *Morinda citrifolia* leaf extract 15% including in the category of strong antibacterial power and proportion *Morinda citrifolia* leaf extract 7.5 % (including in the category of antibacterial power weak)

The results of this research showed that the *Morinda citrifolia* leaf extract proved effective in inhibiting the growth of *Aeromonas hydrophila*. The results of these observations also show that the zone nodes start clearly visible on day two's incubation period. The clear zone is a guide to the sensitivity of the bacteria to other antibacterial materials used as the test material. It is stated with the width of the diameter of the inhibit zone formed around the paper discs. The higher the proportion of *Morinda citrifolia* leaf extract the more womb antibacterial substance plays an important role in the inhibition of the *Aeromonas hydrophila*, graph is of the relationship between the proportion of *Morinda citrifolia* leaf extract with the diameter of the inhibit zone *Aeromonas hydrophila* can be seen in figure 1

Figure 2. A on the negative control treatment shows no clear zone formation around the paper discs. This illustrates that ethanol 96% as a solvent in the manufacture of *Morinda citrifolia* leaf extract does not affect the power activity in inhibiting the growth of bacteria. While the picture B looks as if it is in the clear zone with a size that is not wide, in picture C visible zone of nodes that is quite clear. While the image of D and E shows the zone nodes more clear and the area around the paper discs

The formation of the inhibit zone around the paper (picture 2) shows the existence of the activity of antibacterial compounds against bacteria in the bacteria *Aeromonas hydrophila* test. The more the width of the diameter of the inhibits zone that formed more bacteria that died, is seen from the clear zone around the paper discs. This proves that the *Morinda citrifolia* leaves have a good capability in inhibitings the growth of the bacteria *Aeromonas hydrophila*. *Morinda citrifolia* leaf also has been proven to inhibit the growth of bacterias Escherichia coli, Salmonella sp [5], and Staphylococcus aureus [4].

According to Daviz and Stout in [4], the diameter of the inhibit zone categorized has the power to inhibit the very strong if the diameter of the clear zone ≤ 20 mm, strong if the zone nodes 10-20 mm, is if the zone nodes 5-10 mm and weak if the zone nodes 2-5 mm. From the results obtained the treatment proportion for *Morinda citrifolia* leaf extract 7.5% have inhibit zone with the category of weak, whereas proportion *Morinda citrifolia* leaf extract 15% has inhibit zone with the category of strong while the proportion *Morinda citrifolia* leaf extract 30% and 60% has inhibit zone with the category is very strong

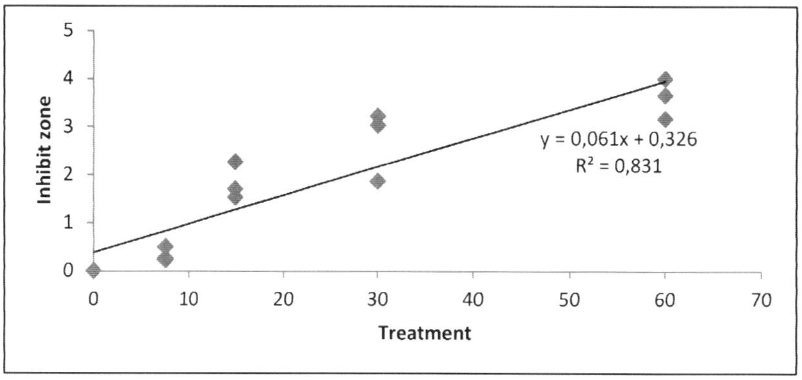

Figure 3. Graph of Linear Regression the diameter of the Inhibit Zone formed in each proportion of the *Morinda citrifolia* Leaf Extract

Based on the results of the statistical analysis provided on the linear regression the value, of correlation coefficient (R) of the *Morinda citrifolia* leaf extract of 0,912, was obtained which means that there is a very strong correlation between the proportion of *Morinda citrifolia* leaf extract and the diameter of the zone of inhibit where the *Aeromonas hydrophila* formed. This means that the diameter of the zone of inhibit where the bacteria *Aeromonas hydrophila* formed influenced 83,1% by the treatment (proportion *Morinda citrifolia* leaf extract), while the remaining 16.9 % were influenced by other factors outside of the treatment. This included factors such as the age of the leaf or plant species or noni used.

Based on figure 3, it can be informed that the influence of the proportion extract to inhibit zone looks very real by linear ($P<0.05$). The higher the proportion of extracting the higher is its ability to inhibit the growth of bacteria in the *Aeromonas hydrophila*. From the results it is known that effective *Morinda citrifolia* leaf extract inhibits the growth of the bacteria Aromonas hydrophila in vitro and level of proportion also effects the effectiveness of inhibitor.

The existence of antibacterial substances in the *Morinda citrifolia* leaf extract plays an important role in inhibitings the growth of or killings the bacteria *Aeromonas hydrophila* [10]. According to in [4], *Morinda citrifolia* leaf contains a wide spectrum of antakuinon as, flavonol iridoid glycosides and triterpen. The antrakuinon substance that was found in the *Morinda citrifolia* leaf extract is impregnation fenolik which can also inhibit the bacteria with protein denaturation [2]. [6] also added that the *Morinda citrifolia* leaf extract has the highest content of total flavonoid compared 90 to other plants amounting 254mg/100gram fw, flavonoid have microbial active. [2] explain flavonoid is the largest fenol compound in nature that have the nature of microbial on the plants.

The activity of antibacterial compounds to microorganisms can be caused by several factors among other disorders of the building blocks of cell walls and can increase cell membrane permeability that can cause loss of components of the building blocks of the cell, make it inactive enzymes and aim or damage the function of the genetic material [5]. The cell wall is the main target that is attacked by an antibacterial substance contained in the *Morinda citrifolia* leaf extract to ease antibakter compounds into the cell membrane. This is explained by [8] when they explain the antibacterial mechanism on flavonoids and its ability to cause damage to the bacterial cell wall permeability, mikrosom and lysosomes as the interaction between flavonoid with DNA bacteria

The wall of the cell that has the nature of not selectively permeable so easily in the penetration of antibacterial compounds that penetrate the cells that will cause disturbance intergritas cell walls. [2] also strengthened the analysis with his statement that the flavonoid can change the chemical and physical nature of the cytoplasm that contains protein and the denaturation wall of bacterial cells with how to bind to the protein through the hydrogen bond. This activity will interfere with the function of selective permeability, active transport function and control the order of proteins

Based on screening tests fitokimia, *Morinda citrifolia* leaf contains essential oil industry, saponnin, triterpenoid, fenol, tannin and glycosides. Saponnin and tannin include in antibacteria groups which also can interfere with cell membrane permeability microbes. Saponnin can cause damage to the cell membrane and cause escaping in various important components in the cell microbes which is a protein, nucleic acids, nucleotides and others [8]. Tannin is capable making inactive adhesin microbes, enzymes and transport proteins in the cell membrane. Oil industry is antibacterial which contains volatile compounds such as the monoterpen and sesquiterpen which is also antibacterial [4]

The diameter of the clear zone formed in every proportion of the *Morinda citrifolia* leaf extract has the size of the wide and is included in the category of inhibit zone weak, strong and very strong. This is according to [2] and can also be influenced by the acceptance of our extraction method. In this research the method used is the method macerasion with solvent ethanol 96%. Ethanol 96 % has high ability in the extract of the content of the plants which has dissolved compounds in it. In addition compounds in the *Morinda citrifolia* leaf is polar to dissolve easily ethanol also is polar [4] so that suspected compound can be bound perfectly in solvent ethanol

It has not yet been local communities who have the role of the materials most important in inhibitings the growth of the bacteria *Aeromonas hydrophila*, the active substances in the *Morinda citrifolia* leaf extract is suspected to work on its own or together in inhibitings the growth of *Aeromonas hydrophila*. Flavonoid allegedly has the ability in high antibacterial substances compared to other antibacterial substances. Flavonoid capability in effecting antibacteria among others by inhibiting the function of the membrane of the cytoplasm, inhibits the synthesis of nucleic acids and inhibits the antibacterial activity with how to inhibit the metabolism of energy and also inhibits the consumption of oxygen with how to interfere with electron transport chain respiration. While according to [4], noni leaves have the womb saponin, flavonoid, polyphenol, tannin and triterpen. The active substance is bactericidal and has its own method in inhibitings the growth of bacteria

The results of our research and the discussion of the *Morinda citrifolia* leaf extract proved effective in the inhibition of the growth of *Aeromonas hydrophila* in vitro. However the activity of the compound in the *Morinda citrifolia* leaves extract and the type of compounds had the most active role in inhibiting the growth of bacteria *Aeromonas hydrophila*. Therefore they, still need to be examined and identified with more information

The results of this research showed that the use of the *Morinda citrifolia* leaf extract is effective to inhibiting the growth of *Aeromonas hydrophila* in vitro, and the higher the proportion of *Morinda citrifolia* leaf extract, the more effective in its inhibition of the growth of *Aeromonas hydrophila* in vitro.

4. Conclusion

The results showed that in vitro, noni leaf extract proved capable of inhibiting the growth of Aeromonas hydrophila. However,the activity of the compound in the *Morinda citrifolia* leaf extract and the type of compounds had the most active role in inhibiting the growth of bacteria *Aeromonas hydrophila*. Therefore they, still need to be examined and identified with more information.

The research also shows that the higher the proportion of *Morinda citrifolia* leaf extract then it will be more effective at inhibiting the growth of Aeromonas hydrophila

Acknowledgments

Author would like to thank the Director and Research and Community Service Center Polytechnic of Jember who have facilitated international seminar activities, thanks also to the Director General of Research Services at the Ministry of Research, Technology and Higher Education that has funded the research and implementation of international food and agriculture conferences

References

[1] P Thiyagarajan, A L Bhavani, M G Ebbie, and G Chandra 2014 A study on the control of Aeromonas hydrophila infection in the cat fish by medicinal plants *Sch. Acad. J. Biosci* vol 2 no 2 www.saspublisher.com [16 September 2018]

[2] M S Kameswari, N K Besung and H Mahatmi 2013 Perasan Daun Mengkudu (*Morinda citrifolia*) Menghambat Pertumbuhan Bakteri *Escherichia coli* Secara In Vitro *Indones. Med. Veterinus* vol 2, no 2

[3] Nurrafita Humaira 2013 Efektivitas Ekstrak Buah Mengkudu Untuk Pengobatan Infeksi Bakteri *Aeromonas hydrophila* Pada Benih Ikan Nilem (*Osteochilus vittatus*) *Universitas Padjajaran*

[4] P Aryadi 2014 Pengaruh Ekstrak Daun Mengkudu (*Morinda citrifolia* L.) Terhadap Pertumbuhan *Staphylococcus aureus* Sebagai Penyebab Abses Periodontal Secara In Vitro

[5] Karmila 2016 Daya Hambat Ekstrak Daun Mengkudu (*Morinda citrifolia* L.) Terhadap Pertumbuhan Bakteri Penyebab Diare UIN Alaudin, Makasar www.repositori.uin-alaudin.ac.id [16 September 2018]

[6] S Rezeki, Sunnati, and N Vidirachmilla 2017 The effect of noni leaves extract (*Morinda citrifolia* L) on wound healing percentage of traumatic ulser in oral mucosa of wistar rats (*Rattus norvegicus*) by in vivo *Biomed. Pharmacol. J.* vol 10, no 4

[7] W N U Barus, H Sitorus, and I Lesmana 2010 Uji Efektivitas Antibakteri Ekstrak Daun Kamboja (*Plumiera rubra*) pada Konsentrasi yang Berbeda Terhadap Pertumbuhan Aeromonas hydrophila secara In Vitro *Univ. Sumatera Utara*, pp. 1–7

[8] G A A Permatasari, I N Besung, and H Mahatmi 2013 Daya Hambat Perasan Daun Sirsak Terhadap Pertumbuhan Bakteri Escherichia coli *Indones. Med. Veterinus Univ. Udayana* Vol. 2, No. 2

[9] D Zhang, G S A Moreira, C Shoemaker, J C Newton and D H Xu 2016 Detection and Quantification of Virulent *Aeromonas hydrophila* in Channel Catfish Tissue Following Waterborne Challenge *FEMS Microbiology Letters* Vol 263 No.9 https://www.ncbi.nlm.nih.gov/pubmed/27044300 [30 September 2018]

[10] Sundrarajan. M, Bama K, Bhavani M, Jegatheeswaran S, Ambika S, Sangili A, Nithya P and Sumathi R 2017 Obtaining Titanium Dioxide Nanoparticles With Spherical Shape And Antimicrobial Properties Using M. citrifolia Leaves Extract by Hydrothermal Method *Journal of Photochemistry and Photobiology B: Biology* Vol 171

IOP Conf. Series: Earth and Environmental Science **207** (2018) 012065 doi:10.1088/1755-1315/207/1/012065

Ethyl Acetate Fraction Of Secang As Anti Cervical Cancer By Inducing p53 and Caspase 9

N P L Laksmiani[1*], N M W Astuti[1], C I S Arisanti[1] and N L P V Paramita[1]

[1] Departement of Pharmacy, Faculty of Mathematics and Science, Udayana University

*E-mail: lindalaksmiani@gmail.com

Abstract. Secang (*Caesalpinia sappan L.*) contains potentially anti-cancer active compounds, namely brazilin and brazilein which are soluble in ethyl acetate. This study is conducted to evaluate the anticancer activity of secang wood in cervical cancer cells (HeLa) in vitro and molecular mechanisms that selectively mediate their cytotoxic effects on several cancer cells through molecular docking or in silico. The anticancer potential of the ethyl acetate fraction of secang wood through its cytotoxic activity on HeLa cells was carried out with MTT assay and calculated the IC_{50} value. The molecular mechanism is carried out in silico with the target protein p53 and caspase 9 using the autodock 4.2 program. The docking process includes protein and active compounds preparation, validation of molecular docking methods and docking brazilein as the active substance with the target protein. The docking score is measured by binding energy value between brazilein and the target protein. The fraction of secang wood was cytotoxic in HeLa cells with IC_{50} values of 48.71 μg/mL. The docking results showed that brazilein had an affinity for the target proteins p53 and caspase 9 with binding energies of -8.24 and -6.71 kcal / mol, respectively.

1. Introduction

In Indonesia, approximately 15.000 cases of cervical cancer occur every year. That makes cervical cancer referred to the most cause of death disease in women in Indonesia. WHO sets Indonesia as the country with the highest number of cervical cancer patients in the world due to its cases number[1]. Cervical cancer is classified as carcinoma, which is a cancer that occurs in squamous cells in the uterine lining of the woman's uterine epithelial tissue. This cancer is caused by infection with HPV (Human Papilloma Virus) [2]. Oncogenous viruses often exploit infected cell working systems, to carry out the life and replication of the virus itself. As part of the process, the virus will affect cell control mechanisms that will cause abnormal cell growth, genetic changes, even malignant transformation. Human papilloma virus 18 encode the formation of 2 oncogens, namely E6 and E7 which interact with the regulator of cell cycle. E6 protein bind to p53 and accelerate p53 degradation [3]. Thus, when p53 is degraded, nothing inhibition process mechanism of cell proliferation and cervical cancer develop and become malignant [4].

Cancer treatment using chemotherapy has many side effects, especially in normal cells. So that research is developed on compounds derived from nature as anticancer agents to increase the sensitivity of cancer cells and reduce side effects due to chemotherapy agents. Many plants contain potential compounds as co-chemotherapy agents [5] [6].One of which is secang wood (Caesalpinia sappan L.). Secang wood contains brazilian, brazilien, 4-O-methylsappanol, protosappanin A, and caesalpin J homoisoflavonoids [7]. Which have cytotoxic activity in some cells such as Hep G2 (liver) and A549 (lung) [8] [9]. Brazilein increases the apoptosis mechanism in liver cancer cells [10]. Brazilin and brazilein are known to be soluble in ethyl acetate solvents, so that in this study the development of ethyl acetate fraction of secang wood (Caesalpinia sappan L.) as an active anticancer through cytotoxic assay on cervical cancer cells (HeLa) and molecular evaluation to determine the molecular mechanism of the anticancer potential possessed by ethyl acetate fraction of secang wood through in silico on the target proteins p53 and caspase 9. These three target proteins play a role in the apoptosis process.

2. Materials and Methods

2.1 Fractination of secang wood

Secang (Caesalpinia sappan L.) heartwood was derived from Center for Research and Development of Medicinal Plants and Traditional Medicine. The dried heartwood of secang was macerated with methanol (Merck) at a volume ratio of 1:10 (for 48 hours). The methanolic extract was then evaporated with reduced pressure by rotary evaporator. The extract was partitioned with n-hexan and ethyl acetate (Sigma Aldrich co. USA.) by liquid liquid extraction (LLE). The ethyl acetate fraction was evaporated using rotary evaporator to get the yield of ethyl acetate fraction [11]

2.2 Cell culture

HeLa cells belong to Cancer Chemoprevention Research Center (CCRC), Pharmacy Universitas Gadjah Mada, obtained from Prof. Masashi Kawaichi, Nara Institute of Science and Technology (NAIST), Japan. HeLa cell lines were cultured in DMEM-low glucose on incubator at 37°C and 5% CO_2, supplemented with 10% (v/v) FBS, 1.5% (v/v) penisilin-streptomisin, and 0.5 % (v/v) fungizone.

2.3 Cytotoxicity and Cell viability assay: 3-(4,5-dimethylthiazol-2-yl)-2,5-diphenyltetrazolium bromide (MTT) assayp

The potency of ethyl acetate fraction of secang on HeLa cells was evaluated by using MTT cytotoxicity assay. HeLa cells with a concentration of 1×10^4 cells / wells were planted in a 96 well plate and incubated for 24 hours. Test solution (ethyl acetate fraction of secang) and media of 100 µl each were added to the well then incubated for 24 hours. Cells were washed with 100 µl PBS, then added 100 µl of culture medium containing MTT 5 mg / ml into each well and then incubated again for 4 hours at 37 ° C. Living cells will convert MTT to purple formazan. After 4 hours of media containing MTT removed, sodium dodesylsulfate stopper solution is added, then the plate is left overnight. The plate is shaken on the shaker for 10 minutes then absorbance will calculate using an ELISA reader at λ 595 nm. Absorption absorbance data is converted into percent viability and used to calculate IC_{50}

The data result in the form of absorbance of each well is then converted into percent cell viability. Equation (1)[13] was used to calculate the percentage of HeLa cells viability

$$\%\ \text{Viable cells} = \frac{\text{Treated cells abs} - \text{medium control abs}}{\text{Cells control abs} - \text{medium control abs}} \times 100\% \quad (1)$$

The chart of the concentration of the cytotoxic assay and cell viability compound is presented as an average ± SE of 3 experiments. Data in the form of cell viability were then analyzed with Microsoft Excel 2007 program to obtain linearity regression Y=BX+A and correlation coefficient (r) between concentration and percentage of viability cells and to calculate IC_{50} values. The r value of the calculation results is compared with r table ($p < 0.01$)[14]. IC_{50} is a concentration which causes 50% inhibition of cell proliferation so that its cytotoxicity potential is known.

2.4 Molecular docking
Molecular docking techniques can be used to predict the affinity of an active compound to the target protein that its activity can be predicted. Molecular docking was applied to determine the potency and mechanism of brazilein as an active compound in ethyl acetate fraction of secang as an anticancer agent in cervical cancer by in silico through p53 and caspase 9 target proteins.

2.4.1 Preparation of p53 and caspase 9 target proteins. The process of preparation of p53 and caspase 9 by using Chimera 1.10.1 software. The protein will separate from its ligand (native ligand).

2.4.2 Validation methode of molecular docking. Validation of the molecular docking method is done by redocking the native ligand back on the target proteins that native ligand have been removed from the proteins using the Autodock 4.2 program. The results of the analysis described which the compound with conformation had the lowest binding energy to bind to the target protein. If the root mean square distances (RMSD) $\leq$ 3 Å, the protocol is received and docking of the active compound on the target proteins can be done.

2.4.3 Optimization of 3D brazilein. The 3-dimensional structure of brazilein compounds is optimized using the HyperChem 8 program. Optimization is performed on the 3-dimensional structure of brazilein compound complete with hydrogen atoms, semi-empirical AM1 computation methods, and calculations with single points and optimization geometry.

2.4.4 Molecular docking of brazilein and target proteins. Docking was conducted using the Autodock 4.2 program between brazilein from the optimization results and the target proteins that its native ligand has been removed. The results of the analysis will show the conformation of the lowest binding energy to bind to the target protein as score docking.

2.4.5 Data anaylisis. The results of molecular docking are score docking. Score docking in the form of binding energy score. The value of the binding energy shows the strength of the bond between the compound and the target protein. Lower of the binding energy value means stronger and more stable of the bond. Lower of binding energy score reflect stronger interaction that occurs between brazilein and the target proteins.

3. **Result**

3.1 Cytotoxicity assay of ethyl acetate fraction of secang
The activity assay of ethyl acetate fraction of secang on heLa cells was carried out by in vitro using MTT cytotoxicity assay through the measuring parameter was IC_{50}. The ethyl acetate fraction of secang has an IC_{50} value of 48.71 µg / mL. The cytotoxicity assay result is described according to figure 1. In figure 2 seen morphologically, higher concentration of ethyl acetate fraction of secang cause more cells undergoing changes in shape. Condensation occurs, fragmented, membrane blebbing and shrinkage of cells. This indicates that ethyl acetate fraction has a potency to be developed into a natural anticancer agent or later it can be developed into a co-chemotherapy agent as a companion to chemotherapy agents.

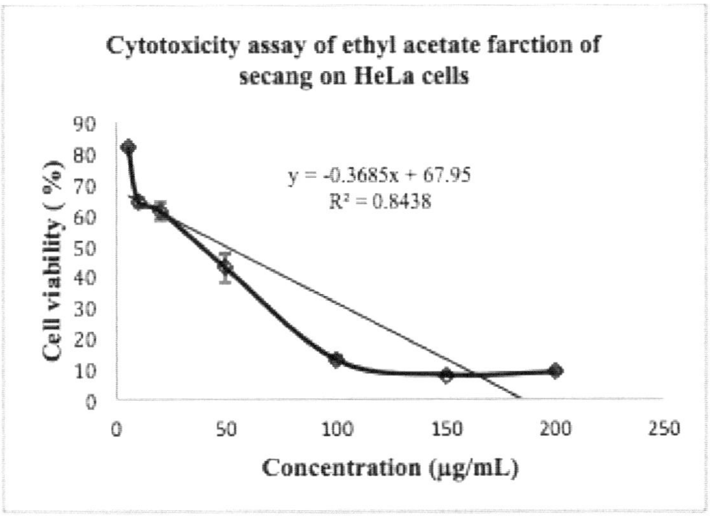

Figure 1. Percentage of HeLa cells viability due to ethyl acetate fraction of secang

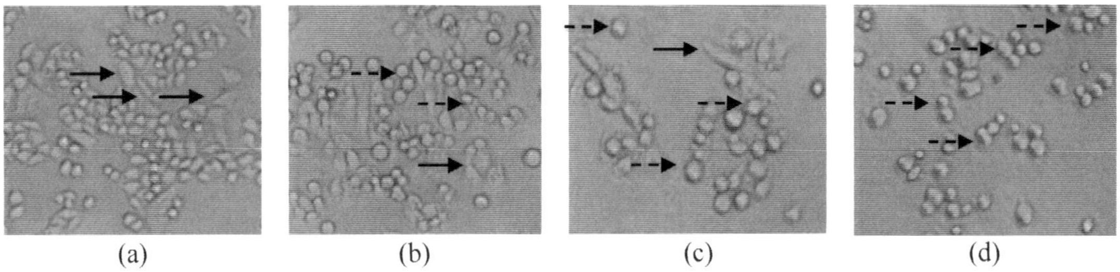

Figure 2. HeLa cells morphological changes after ethyl acetate fraction exposure in 24 hours. (a) control cell; (b) 25 µg/mL ethyl acetate fraction; (c) 50 µg/mL ethyl acetate fraction and 100 µg/mL ethyl acetate fraction. → normal cell - - ▸ morphology cell changing

3.2 Brazilein from ethyl acetate fraction of secang increase apoptosis process by inducing p53 and caspase 9.

In silico evaluation with molecular docking using the autodock method was carried out to determine the molecular mechanism that mediates cytotoxic activity in HeLa cells. The target proteins evaluated were p53, and caspase 9 which the target proteins responsible for apoptosis. In HeLa cells had characteristics, namely the presence of mutations in p53 and interference with the apoptosis process [15]. Before molecular docking was carried out, validation was necessary to ensure that the molecular docking stages using autodock provided appropriate and valid results. Validation could be seen from the RMSD value which is <3 Å (table 1). Molecular docking result between brazilein and protein target were showed on table II dan III.

Table I. RMSD value of molecular docking validation method

Protein	RSMD (Å)	Binding Energy (kcal/moL)	Hydrogen Bond
p53	2.26	-8,24	SER152; GLU124
Cappase 9	2.44	-3.62	THR181; ARG 355

Table II. Binding energy of brazilein with p53 and caspase 3 using molecular docking

No.	Target Proteins	Ligand	Binding Energy (kcal/moL)	Hydrogen Bond	(Ligand-Protein) Group
1	p53 (2J8Z)	Native ligand	-6,29	ARG 42	HH11
				ALA 149	HN
				LYS 178	HZ2
				ASN 321	HD21
				ASN 323	HD21
		Brazilein	-8,24	SER 152	HG
				GLU 124	OE2
2	Caspase 9 (1JXQ)	Native ligand	-3.62	ARG355	HN-O2
				ARG355	HE-O4
		Brazilein	-6.71	THR181	HN-O
				ARG355	HN-O

4. Disscussion

IC_{50} value is the concentration needed by a compound to inhibit growth and proliferation of cancer cell by 50% of the cell population. Lower IC_{50} value means more potent of compound has anticancer activity. In this study the ethyl acetate fraction of secang had an IC_{50} value of 48.71 μg/mL (figure 1) and the cytotoxic activity of the fraction showed a positive correlation. Higher concentration of ethyl acetate fraction of secang cause greater the percentage of cell death with r = 0.918 (r table = 0.874) [14]. Correlation (r test) value is higher than r table showed concentration of secang treatment on heLa cells related to percentage of cell death and many cells to be morphological changes (figure 2). According to figure 1 and 2, ethyl acetat fraction of secang had a potent activity to cervical cancer that HeLa as cell line model. This fraction has IC_{50} values below 100 μg / mL. A compound to be potent as an anti-proliferative if it has IC_{50} below 100 μg / mL [16].

Molecular mechanism of brazilein in secang had been further observed in silico using autodock 4.2 program to evaluate molecular capability of it that mediates anticancer activity. RMSD (Root Mean Square Deviation) is a measurement of two poses by comparing the atomic position between an experimental structure with a predicted structure or a deviation between a ligand conformation that is tethered to a protein compared to a site that has a native ligand bind to a protein. Getting greater RMSD value means greater deviation which indicates the position of the ligand predicton father away to native confirmation caused greater prediction error of ligand interaction with protein [17]. The results of the validation of the molecular docking method show that the RMSD value is ≤ 3 Å which demonstrated that the method used has met the requirements of its use [17]. The affinity of the compound can be known by comparing the energy value of the compound bond with native ligand on the target protein [18]. The value of binding energy obtained from the target protein docking with native ligand and brazilein was then compared to determine the potency of brazilein as an anticancer agent through the mechanism of inducing p53 and caspase 9. Table 1 and table 2 showed that brazilein compounds in secang have lower binding energy values than native ligand on p53 and caspase 9. Brazilein had lower binding energy than native ligand with the target protein exhibited stronger bind formed between compound and the target proteins [18]. The affinity of compound for the target protein that is directly proportional to the strength and stability of the bond. Comparison of binding energy values in Table 2 shows that brazilein has greater potency as an anticancer agent compared to its native ligand with the mechanism of inducing p53 and caspase 9 in silico by molecular docking method so that the apoptosis process is accelerated and kills cancer cells.

Table III. Interaction between brazilein and target proteins (p53 and caspase 9) with amino acid residue in target proteins

No	Target proteins	Interaction between brazilein and amino acid residu of target proteins
1	p53	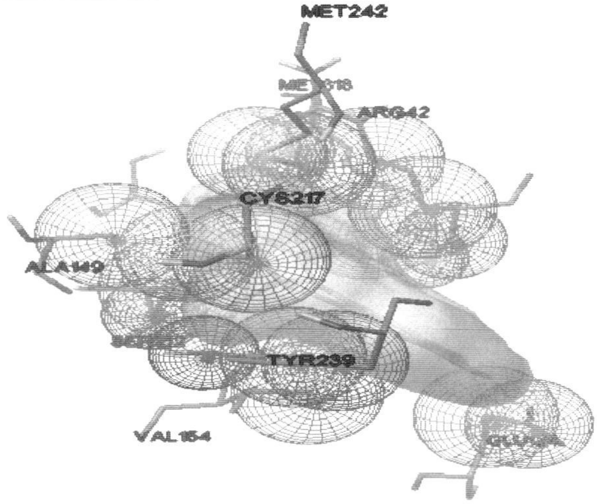
2	Caspase 9	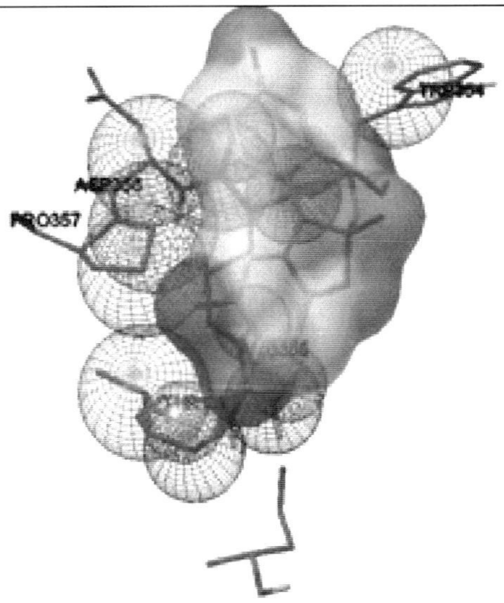

5. Conclusion

Ethyl acetate fraction of secang had a potent cytotoxic activity as cervical anticancer agent with HeLa cell lines as a model through inducing p53 dan caspase 9. This fraction promising to be developed into an anticancer agent or co-chemotherapeutic agent.

6. Acknowledgement

Ethyl acetate fraction of secang had a potent cytotoxic activity as cervical anticancer agent with HeLa cell lines as a model through inducing p53 dan caspase 9. This fraction promising to be developed into an anticancer agent or co-chemotherapeutic agent.

References

[1] Ministry of Health Republic of Indonesia 2013 Riset Kesehatan Dasar (RISKESDAS). Jakarta : Badan Litbang Kemenkes RI.

[2] Castellsagué X 2008 *Gynecologic Oncology*. **110**: S4–S7.

[3] Georgieva, S., V. Iordanov, and S. Sergieva 2009 *J Buon*. **14**(3): p. 391-8.

[4] Kines, R.C. and Cole A 2009 *Proc Natl Acad Sci USA*. **106**(48): p. 20458-63

[5] Greenwell, M. and Rahman, P.K.S.M 2015 *Int J Pharm Sci Res*. **6**(10): 4103-4112.

[6] Ho JW, LYK, Pong C 2002 *Anticancer Agents* **2**: 209-214

[7] Chun Fu, Xin-an Huang, Zhen, yuan Lai, Ying-jie Hu , Hong, jiao Liu, *and* Xiao,ling Cai 2008 *Molecules* **13**: 1923-1930.

[8] Yen,C., Kyoko Nakagawa-Goto, Tsong-Long Hwang, Pei-Chi Wu, Susan L.Morris-Natschke, Wan-Chun Lai, Kenneth F. Bastow, Fang-Rong Changb, Yang-ChangWu, and Kuo-Hsiung Lee 2010 *Bioorg Med Chem Lett*. **20**(3): 1037–1039.

[9] Ren L, Yang X, Wang G, Zhang H, Zhao L, Mi Z 2011 *World Academy of Science, Engineering and Technology* **60**: 215-219

[10] Zhong , B. Wu, Y. J. Pan, S.Zheng 2009 *Neoplasma* **56** (5).

[11] Laksmiani, N.P.L., Meiyanto, E. and Susidarti, R.A 2017 *Int J Pharm Pharm Sci*. **12**, 124-130.

[12] Mosmann T 1983 *Journal of immunological methods* **65**(1): 55-63.

[13] Meiyanto E, Fitriasari A, Hermawan A, Junedi S, Susidarti RA 2011 *Orient Pharm Exp Med* **11**:183-190.

[14] Alhusin, S 2002 SPSS 10 for Windows, Edisi 1, J & J Learning, Yogyakarta.

[15] Buitrago-pérez Á, Garaulet G, Vázquez-carballo A, Paramio JM 2009 *Current genomics* **10**: 26-34.

[16] Teng WY, Yu LH, Chien CS, Ray LH 2005 J Chin Chem Soc **52**:1253-5.

[17] Jain, A. J., dan A. Nicholls 2008 *J. Comput. Aided Mol.* Vol.**22**: 133-139.

[18] Laksmiani, N. P. L., N. L. P. V. Paramita, and I. M. A. G. Wirasuta 2016 *International Journal of Pharmacy and Pharmaceutical Sciences* **8**(8): 177-181.

The Influence of Sediment Loads on the Irrigation Discharge in The Upstream and Downstream of the Major River in Banyuwangi Regency

Z Erwanto[1], D Dwi Pranowo[1], D S W P J Widakdo[2] and N S R Wilujeng[3]

[1] Department of Civil Engineering, Politeknik Negeri Banyuwangi, Banyuwangi, Indonesia
[2] Study Program of Agribusiness, Politeknik Negeri Banyuwangi, Banyuwangi, Indonesia
[3] Study Program of Animal Product Processing Technology, Politeknik Negeri Banyuwangi, Banyuwangi, Indonesia

E-mail: zulis.erwanto@poliwangi.ac.id

Abstract. In Banyuwangi Regency there are four major rivers include the Tambong, Bomo, Setail and Baru river. Those rivers often overflow due to the high sediment deposition in the weir. The purpose of the study is to determine the influence of sediment loads upstream and downstream of the four rivers on irrigation impact. Suspended load analysis used method of Engelund and Hansen; Einstein; and Chang Simons & Richardson. Bed load analysis using methods of the Meyer-peter; Duboys; and Shields. Loads of sediment, both suspended load and bed load on the irrigation discharge in both the upstream and downstream of the major rivers in Banyuwangi greatly influences the sedimentation of irrigation channels which affect on fluctuations of irrigation discharge to irrigate land of Agriculture. The Baru River carried the largest sediment load in irrigation intake of Karangdoro Weir that carried Qs of 385.96 tons/year and Qb of 6.37 tons/year. It is necessary to have policies and efforts to control sediment and study about the maintenance of weirs periodically to maintain the volume of irrigation water so that it can increase agricultural productivity in Banyuwangi Regency.

1. Introduction

1.1. Background
One of the decreasing capacity of a river is the sediment buildup called sedimentation in the river either in the middle flow of the river or in the estuary of the river. Sedimentation can lead the rise and fall of the riverbed elevation (river bed alteration). One of them also affects irrigation watering because the reservoir storage capacity is full of sediment deposits. Based on the priorities of regional development programmes in the field of food security in 2014 – 2019, that one of them is increasing the development and rehabilitation of reservoirs, dams and reservoirs with geo-membrane technology, maintaining the volume of water sources to maintain water availability throughout the year, and increasing the acceleration of development, maintenance and repair of infrastructure that supports the development of the agricultural sector and rural areas. From one of the problems in the agricultural sector, namely the fluctuations in climate change that disrupt agricultural production, where one of the

factors is the influence of the existence of floods and sediment loads that affect sedimentation in irrigation building infrastructure. The phenomenon of changes in sediment load has an impact on the danger of flooding and at the same time on the achievement of production and agricultural productivity [2]. Therefore, in maintaining the volume of water sources, it is necessary to study the impact of river sediment load on irrigation water in order to increase agricultural productivity.

Based on the Central Statistics Agency [1], there are 35 watersheds in Banyuwangi Regency itself. The largest watersheds include Tambong, Baru, Setail, and Bomo watershed. The majority of water springs and surface runoff come from Raung and Ijen Mountain with steep topographic relief and go directly to the Sea. This natural condition could potentially occur deliverable flood from upstream to downstream and if the watershed conditions are not well maintained it will result in easily eroded land due to high rainfall which causes high sediment loads in the four major rivers of the Banyuwangi Regency. Based on research [3], the evaluation of the flood hydrograph in the Major Rivers in Banyuwangi District has a 5-10 year flood return period. From these conditions, it shows the influence of discharge fluctuations in these rivers on sediment loads and impacts on irrigation. Therefore, it is necessary to study the influence of sediment load on both suspended load and bed load in upstream and downstream of the large rivers of Banyuwangi Regency on irrigation discharge in the technical weir.

2. Theoretical Basis

2.1. Sediment Loads

2.1.1. Suspended Load. To find the size of the suspended load, it is used the following 3 methods [4]:

- Engelund and Hansen. *Engelund and Hansen* assume that suspended load can be calculated using the equation:

$$q_{sw} = 0{,}05\,\gamma_s\,V^2 \left[\frac{d_{50}}{g\left(\frac{\gamma_s}{\gamma}-1\right)}\right]^{1/2} \left[\frac{\tau_o}{(\gamma_s-\gamma)d_{50}}\right]^{3/2} \quad (1)$$

$$\tau_o = \gamma\,D\,S \quad (2)$$

$$Qsw = W.qsw \quad (3)$$

$$G_w = \gamma.W.D.V \quad (4)$$

$$C_t = \frac{Q_{sw}}{G_w}\ (ppm) \quad (5)$$

$$Q_s = 0.0864.C_t.Q \quad (6)$$

With, q_{sw} is suspended load charge (ton/day), γ_s is specific gravity of sediment (g/cm^3), γ is a specific gravity of water (1 ton/m^3 = 1 gr/cm^3 = 1000 kg/m^3), V is the average flow velocity (m/s), g is gravity force (9,81 m/s^2), d_{50} is the result of sieve test of 50 mm in diameter (mm), D is the depth of the river (m), S is the slope of the river bed, τ_o is friction tension (kg/m^2), W is the width of a river (m), Ct is sediment concentration (ppm), Q_{sw} is charged suspended load per unit of width (ton/day), Gw is water pressure (ton/s), Qs is sediment loads (ton/day).

- Einstein. *Einstein* assumed that suspended load can be calculated using the equation:

$$q_{sw} = 11{,}6\,U'_*C_a a\left[\left(2{,}303\log\frac{30{,}2D}{\Delta}\right)I_1 + I_2\right] \quad (7)$$

With, q_{sw} is suspended Load charge (ton/day), D is the depth of the river (m), Ca is the concentration of suspended load at a distance y and x above bed load each or laboratory test results of sediment concentration (mg/l).

$U^{'}$ = $U_* = (gRS)^{0.5}$ or $U_* = (gDS)^{0.5}$
a = $2.d_{65}$, meter
Δ $= k_s/x = d_{65}/x$

Calculate the value of $\delta' = \frac{11.6 \times v}{U'_*}$ with v, is a viscosity kinematic of water based on water temperature °C (cm^2/s). Then calculate the value of $U'_* = \sqrt{gR_bS}$, with R_b is a radius of hydraulic or depth of the river in meter, S is the slope of the river bed, and g is the gravity force = 9,81 m/s^2. Calculate value of $\frac{k_s}{\delta} = \frac{d_{50}.(R)^{1/2}}{\delta\prime}$ in **Figure 1** to get the value of x.

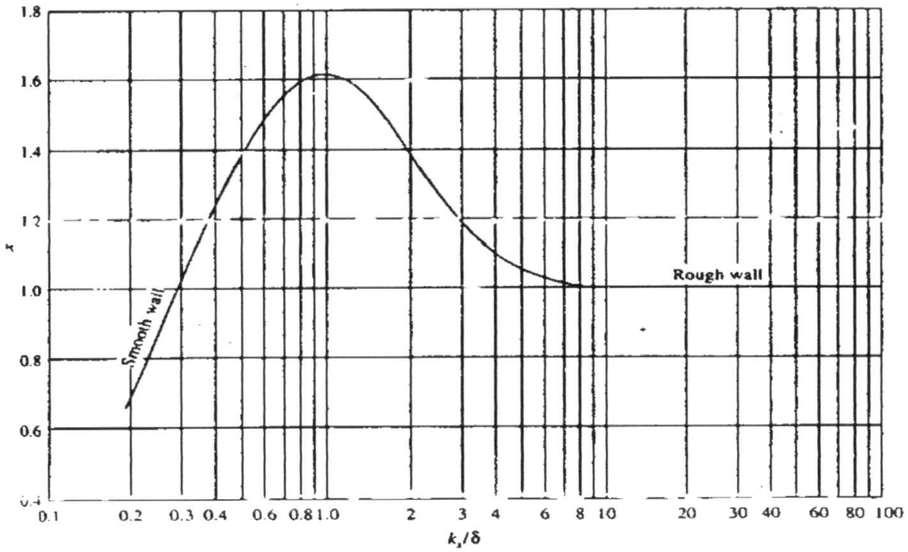

Figure 1. Correlation between $\frac{ks}{\delta}$ and x [5]

I = Integrity numeric can find with **Figure 2** for a value of I_1 and **Figure 3** for a value of I_2

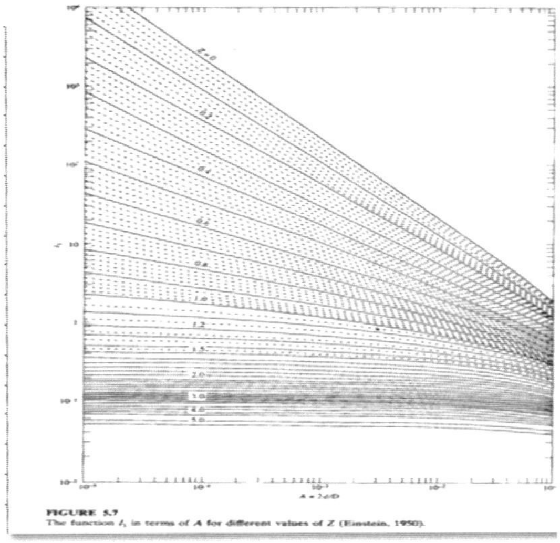

Figure 2. By using the value of I_1 it is found the value of A and Z [5]

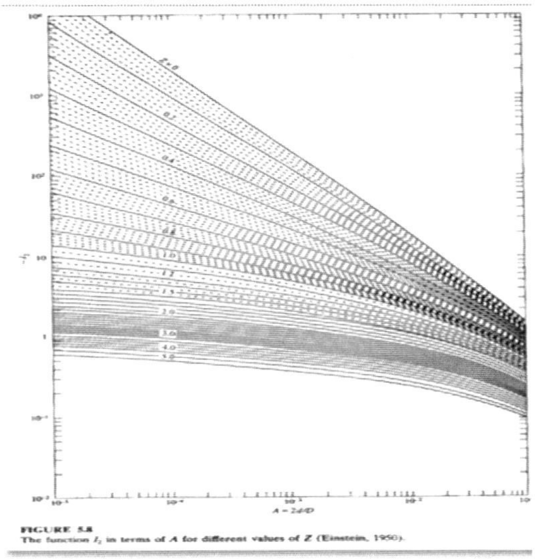

Figure 3. By using the value of I_2 it is found the value of A and Z [5]

Calculate value of Z $= \frac{\omega}{0.4 \times U'_*}$ and A $= {}^{2d_{65}}/_{D}$ at equation in **Figure 2** and **Figure 3.** To find the value of the fall velocity can be searched by using the equation $\omega = \frac{1}{8}\frac{\gamma_s - \gamma}{\gamma} g \frac{{d_{35}}^2}{V}$.

With, ω is fall velocity of sediment (m/s), γ_s is specific gravity of sediment, γ is a specific gravity of water (1 ton/m^3), d_{35} is diameter of granules 35 mm (mm), g is gravity force and V is the average flow velocity (m/s) with equation $V = 5.75. U'_*. \log(12.27. \frac{D'}{ks}. x)$ and then continue calculation with **equation (3), (4), (5),** and **(6).**

- Chang Simons and Richardson. *Chang and Simons Richardson* assume that suspended load can be calculated using the equation:

$$q_{sw} = \gamma D C_a \left(V. I_1 - \frac{2U_*}{k} I_2 \right) \quad (8)$$

With, q_{sw} is *suspended Load Charge* (ton/day), γ is specific gravity of water, k is Prandtl – Von Karman Constanta = 0,4, V is the average flow velocity (m/s), D is the depth of the river (m). And calculate $U'_* = U_* = (gRS)^{0,5}$ or $U_* = (gDS)^{0.5}$.

Ca is the concentration of suspended load at a distance y and x above bed load each or laboratory test results of sediment concentration (mg/l), and I is integrity numeric can be searched using **Figure 5** for I_1 value and **Figure 6** for I_2 value.

$$q_{bw} = K_t. V. (\tau_o - \tau_c) \quad (9)$$

$$\tau_o = \gamma. D. S \quad (10)$$

$$\tau_c = \tau_o.(\gamma_s - \gamma).d_{50} \tag{11}$$

To get the value of Kt, calculate the value of $\frac{V}{U*}\frac{\tau_o.S}{(\gamma_s-\gamma).d_{50}}$ and use **Figure 4**.

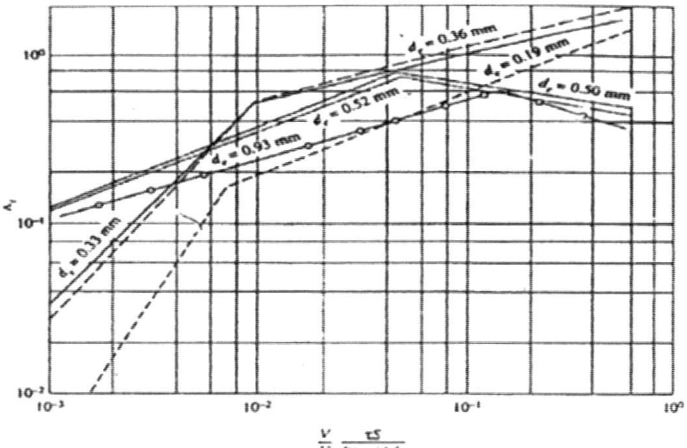

Figure 4. Correlation between $\frac{V}{U*}\frac{\tau_o.S}{(\gamma_s-\gamma).d_{50}}$ and Kt [5]

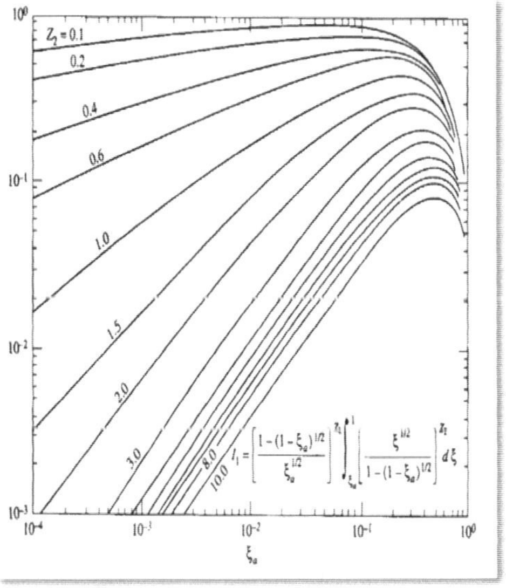

Figure 5. Correlation between ξ_a with Z_2 it is found a value of I_1 [5]

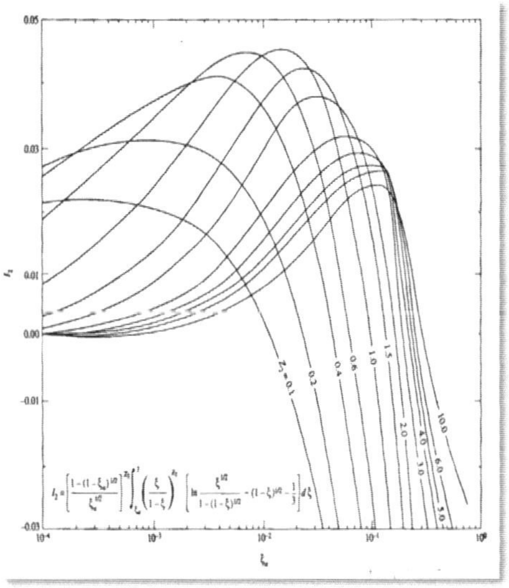

Figure 6. Correlation between ξ_a with Z_2 it is found a value of I_2 [5]

With, γ_s is specific gravity of sediment, γ is specific gravity of water, q_{bw} is bed load charge (ton/day), d_{65} is diameter of granules 65 mm (mm), d_{50} is diameter of granules 50 mm (mm), D is the depth of the river (m), V is the average flow velocity (m/s), τ_o is basic tension (kg/m^2), τ_c is critical base friction tension (kg/m^2), S is the slope of the river bed, B is the length of the river (m), g is gravity force. Calculate $q_{sw} = q_{bw}.R_s$, with $R_s = \frac{D}{0.8.a.V}\left(V.I_1 - \frac{2U_*}{k}I_2\right)$, than calculate value of $\xi_a = \frac{a}{D} = \frac{2.d_{65}}{D}$ and $z_2 = \frac{\omega}{U_*k}$ at equation **Figure 5** and **Figure 6.**

2.1.2. Bed Load. To find the bed load using the following 3 methods [4] :

- Meyer-Peter (1948) [5]. The basic equation of bed load using the Meyer-Peter can be written as follows:

$$q_b{}^{2/3} = 39{,}5\, q^{2/3}\, S\, 9{,}95\, d_{90} \quad (12)$$

$$Q_b = q_b.W \quad (13)$$

With, q_b is *Bed Load charge* (ton/day), q is the discharge of river (m^3/s), S is the slope of the river bed, d_{90} is diameter of granules of 90 mm (m), W is width of river (m), Q_b is charge *Bed Load* per unit of width (ton/day).

- Duboys [5].

$$q_b = \psi\left(\frac{\tau_o - \tau_c}{\gamma}\right) \quad (14)$$

$$\tau_o = \gamma\, D\, S \quad (15)$$

$$Q_b = q_b.W \quad (16)$$

With, q_b is *Bed Load charge* (ton/day), S is the slope of the river bed, τ_o is friction tension (kg/m^3), W is width of river (m), Q_b is charge *Bed Load* per unit of width (ton/day), ψ is the coefficient that depends on the size of a grain of the average sediment. Value of ψ can be found using the equation $\psi = \frac{\gamma_s - \gamma}{\gamma}\frac{d_{90}}{R_b.S}$, with τ_c is a critical base friction tension (kg/m^3), can be searched using the equation $\tau_c = \frac{\tau_0}{\gamma(\gamma_s - 1)D}$.

- Shields [5]. *Shields* determine the magnitude of the bed load with empirical formula, can be seen in the equation:

$$q_b = 10\, q\, S\left(\frac{\tau_o - \tau_c}{(\gamma_s - \gamma)^2}\right) \quad (17)$$

$$\tau_o = \gamma\, D\, S \quad (18)$$

$$Q_b = q_b.W \quad (19)$$

With, q_b is *Bed Load charge* (ton/day), q is the discharge of river (m^3/s), τ_c is critical base friction tension as shown in **Figure 7**, D is the depth of the river (m), γ_s is specific gravity of sediment, V is the average flow velocity (m/s), γ is specific gravity of water, S is the slope of the river bed, v is kinematic viscosity (cm^2/s), W is width of river (m), $Q_b$is charge *Bed Load* per unit of width (ton/day). Calculate the value of $\frac{V\,D}{v}$ at the equation in **Figure 7.**

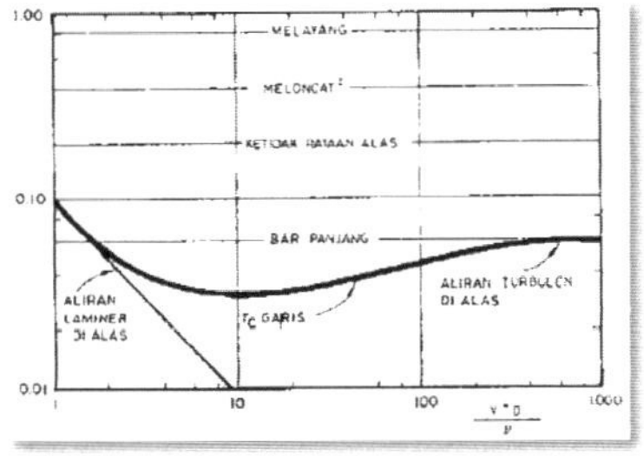

Figure 7. Critical Base Friction Tension [5]

3. Methodology

3.1. Location Of Study

In the upstream of the river among others were Tambong River (Poncowati Weir), Bomo River (Gembleng Weir), Setail River (Jambewangi Weir), and Baru River (Karangdoro Weir). In the downstream of the river among others were Tambong River (Pondok Nongko Bridge, Sukojati), Downstream Bomo River (Mangir Village, Wonosobo Bridge), Setail River (Dam 1, Kradenan), and Sungai Baru River (Watu Kemeloso Weir, Barurejo).

3.2. Data Collection

Primary data collection was obtained directly to study location, in the form of river condition survey, suspended load sampling with a 3 bottle sediment sampler each, 15 kg of bad load sample taken with a Van Veen Grab Sampler tool, measurement coordinates and elevation with GPS, and measurement of flow discharge with the Current Meter tool. Sediment sampling was carried out 6 times in a row. As well as testing sediment samples in the laboratory in the form of sieve analysis, hydrometer, specific gravity, and sediment concentration with filter paper.

Secondary data collection was obtained from other sources such as irrigation discharge data in the Banyuwangi Regency Water Service, and daily river discharge data in the Technical Management Unit of the Water Resources Management of the Sampean Baru River Region in Bondowoso Regency.

4. Results And Discussion

4.1. The Results Of The Sediment Load In The Major Rivers

4.1.1. Tambong River. The results of measurements of the average river discharge per segment in the Upstream of Tambong river is 1.57 m^3/s and in the Downstream of Tambong river is 2.11 m^3/s. The correlation between sediment load and discharge at upstream and downstream Tambong River can see in **Figure 8 – 11**.

Based on **Figure 12,** it shows that in the irrigation intake in the Poncowati weir with a total annual average irrigation discharge of 6.54 m^3/s carries an annual suspended load of 21.89 tons/year while the annual bed load is 6.92 tons/year. Fluctuations in sediment load follow the irrigation discharge and are still stable and the irrigation conditions are rather turbid. Based on **Figure 13,** it shows that in the irrigation intake in the Downstream of Tambong River with a total annual average irrigation discharge of 63.35 m^3/s carries an annual suspended load of 166.06 tons/year while the annual bed load is 35.98 tons/year. It is known that in January-April during the rainy season the suspended load condition is higher than the intake discharge of the irrigation causing high turbidity and sedimentation, while during the dry season the river conditions seem clear because the content of suspended load is very low. The condition of bed load follows the fluctuation of irrigation discharge and is still stable.

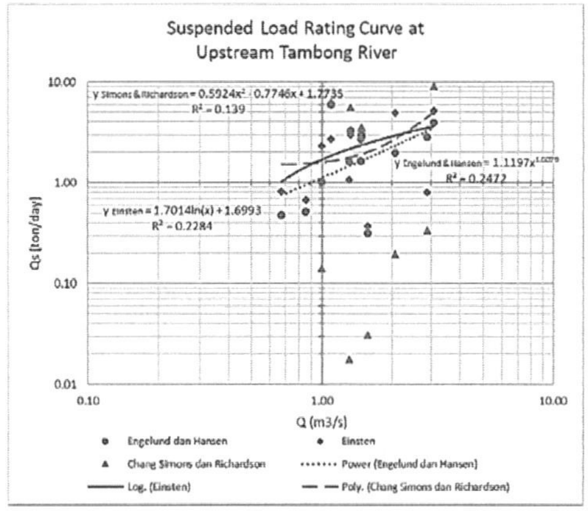

Figure 8. Suspended Load Rating Curve at Upstream Tambong River

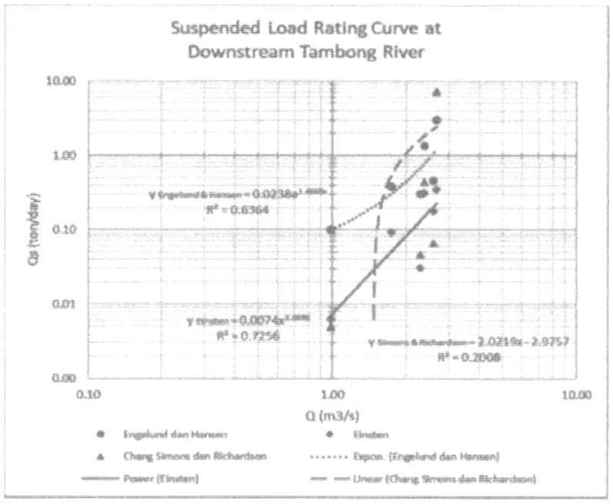

Figure 9. Suspended Load Rating Curve at Downstream Tambong River

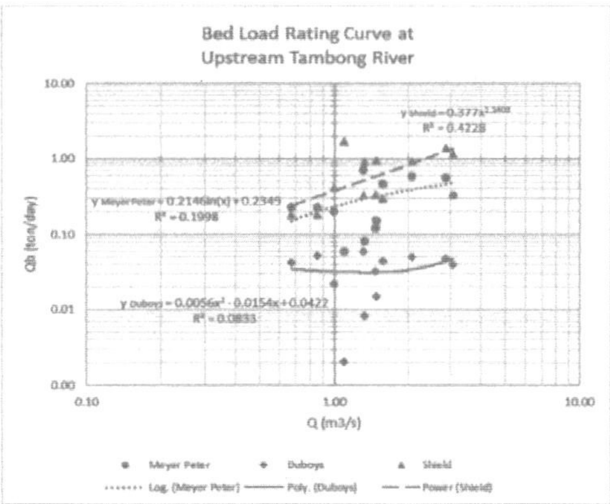

Figure 10. Bed Load Rating Curve at Upstream of Tambong River

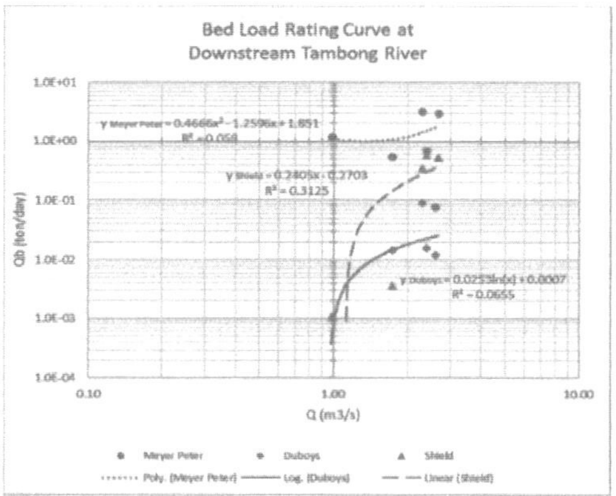

Figure 11. Bed Load Rating Curve at Downstream of Tambong River

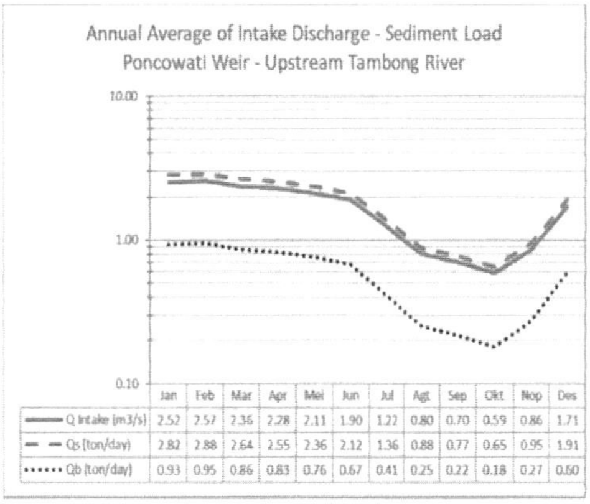

	Jan	Feb	Mar	Apr	Mei	Jun	Jul	Agt	Sep	Okt	Nop	Des
Q Intake (m3/s)	2.52	2.57	2.36	2.28	2.11	1.90	1.22	0.80	0.70	0.59	0.86	1.71
Qs (ton/day)	2.82	2.88	2.64	2.55	2.36	2.12	1.36	0.88	0.77	0.65	0.95	1.91
Qb (ton/day)	0.93	0.95	0.86	0.83	0.76	0.67	0.41	0.25	0.22	0.18	0.27	0.60

Figure 12. Annual Average of Intake Discharge and Sediment Load at Poncowati Weir Upstream of Tambong River

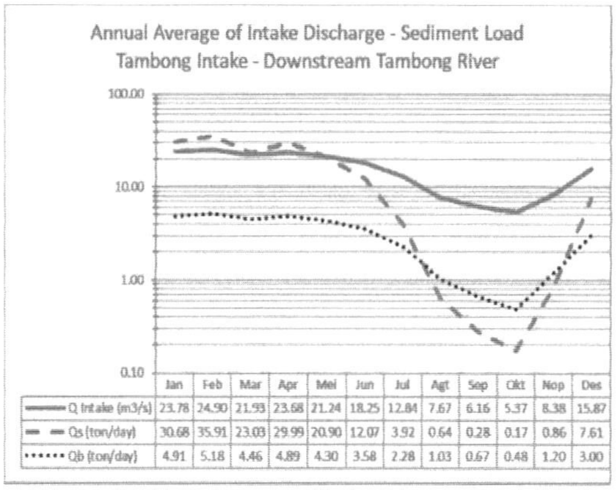

	Jan	Feb	Mar	Apr	Mei	Jun	Jul	Agt	Sep	Okt	Nop	Des
Q Intake (m3/s)	23.78	24.90	21.93	23.68	21.24	18.25	12.84	7.67	6.16	5.37	8.38	15.87
Qs (ton/day)	30.68	35.91	23.03	29.99	20.90	12.07	3.92	0.64	0.28	0.17	0.86	7.61
Qb (ton/day)	4.91	5.18	4.46	4.89	4.30	3.58	2.28	1.03	0.67	0.48	1.20	3.00

Figure 13. Annual Average of Intake Discharge and Sediment Load at Tambong Intake Downstream of Tambong River

4.1.2. Bomo River. The result of measurement of the average river discharge per segment of the Bomo river Upstream is 2.56 m^3/s and in the Bomo river Downstream is 0.37 m^3/s. The correlation between sediment load and discharge at upstream and downstream Bomo River can see in **Figure 14 – 17**.

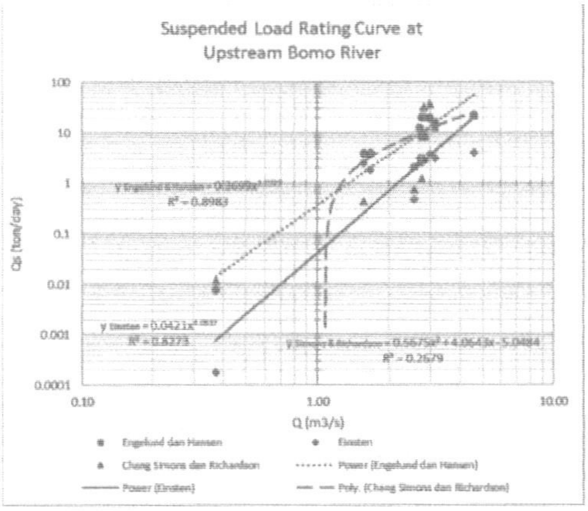

Figure 14. Suspended Load Rating Curve at Upstream Bomo River

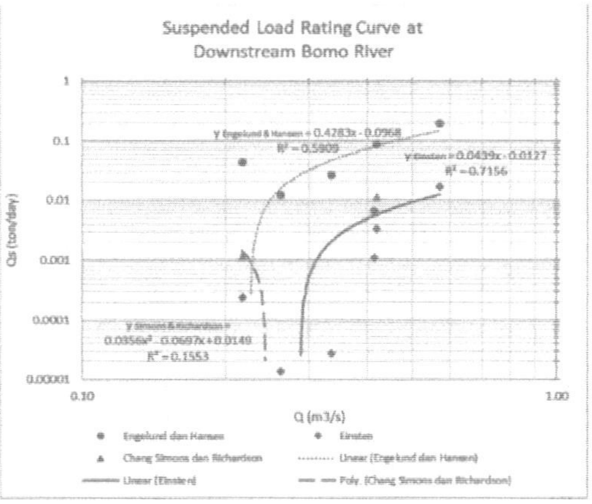

Figure 15. Suspended Load Rating Curve at Downstream Bomo River

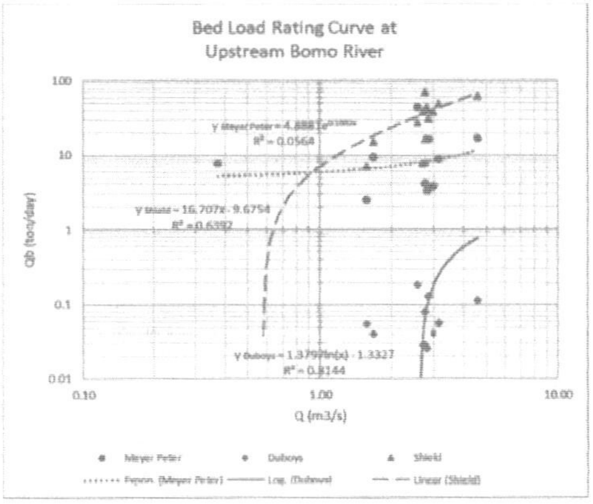

Figure 16. Bed Load Rating Curve at Upstream Bomo River

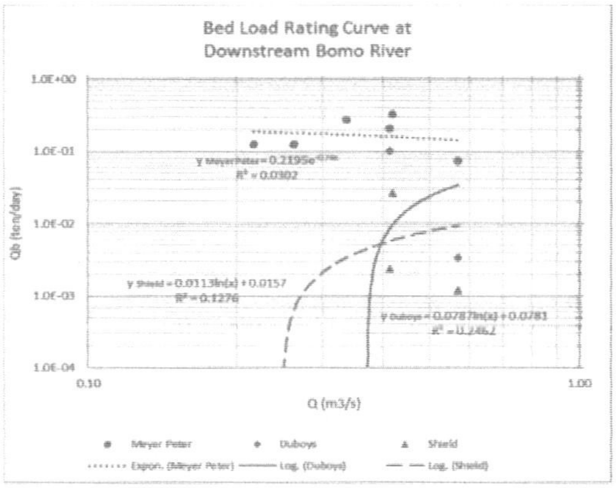

Figure 17. Bed Load Rating Curve at Downstream Bomo River

Based on **Figure 18,** it shows that in the irrigation intake in the Gembleng wier with a total annual average irrigation discharge of 20.76 m^3/s carries an annual suspended load of 196.24 tons/year while the annual bed load is 35.81 tons/year. It is known that in July and December the suspended load conditions are very high and bed load sedimentation occurs in October. River and irrigation intake categories in the Gembleng irrigation area are very high in sediment loads. There are also many sand mining in the upstream Bomo river. Based on **Figure 19,** it shows that in irrigation intakes in Maron of Wonosobo in the Downstream of Bomo with total annual average irrigation discharge of 2.43 m^3/s carries the annual suspended load of 0.18 tons/year while the annual bed load is 2.45 tons/year. It is known that the conditions of the suspended load is lower than intake irrigation discharge so that it looks clear and has low sedimentation, while for bed load condition in October or in dry season, there is large deposition and the discharge of the river recedes, so that it is widely used for mining of sand in the vicinity of the downstream of Bomo river.

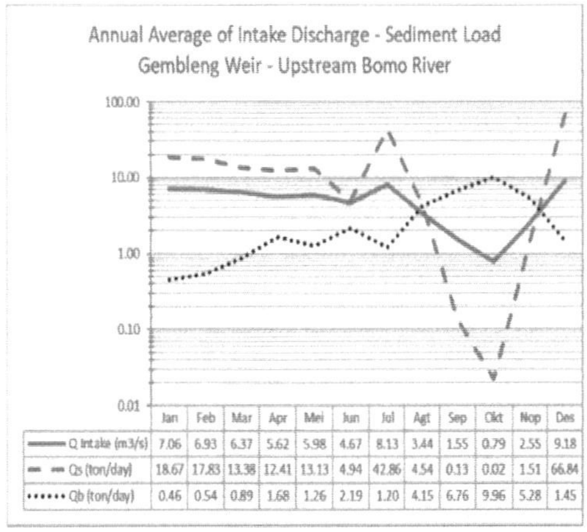

Figure 18. Annual Average of Intake Discharge and Sediment Load at Gembleng Weir Upstream Bomo River

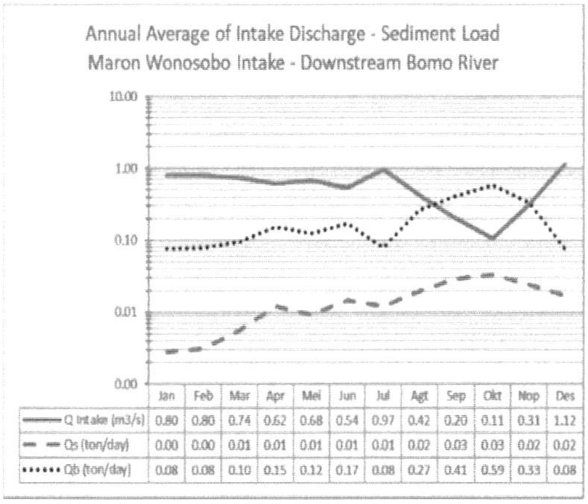

Figure 19. Annual Average of Intake Discharge and Sediment Load at Maron Wonosobo Intake Downstream Bomo River

4.1.3. Setail River. The results of measurements of the average river discharge per segment at Setail River Upstream is 4.29 m^3/s and at Setail River Downstream is 0.69 m^3/s. The correlation between sediment load and discharge at upstream and downstream Setail River can see in **Figure 20 – 23**.

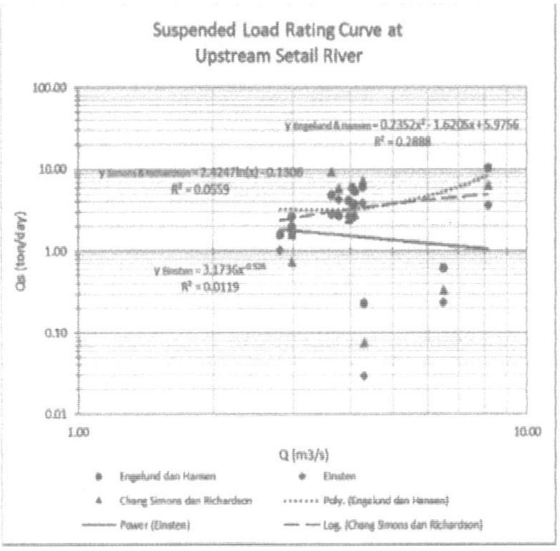

Figure 20. Suspended Load Rating Curve at Upstream Setail River

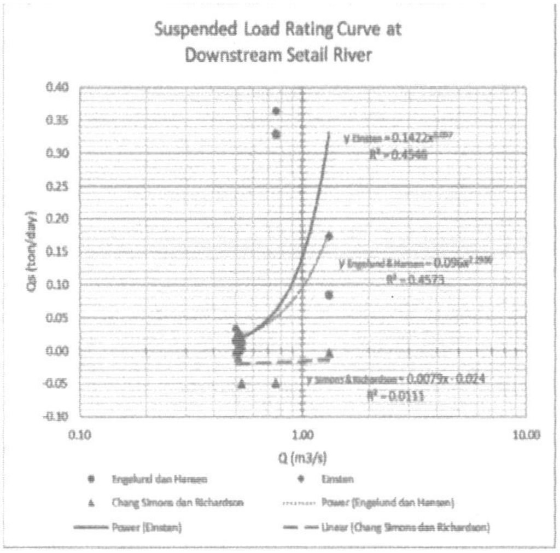

Figure 21. Suspended Load Rating Curve at Downstream Setail River

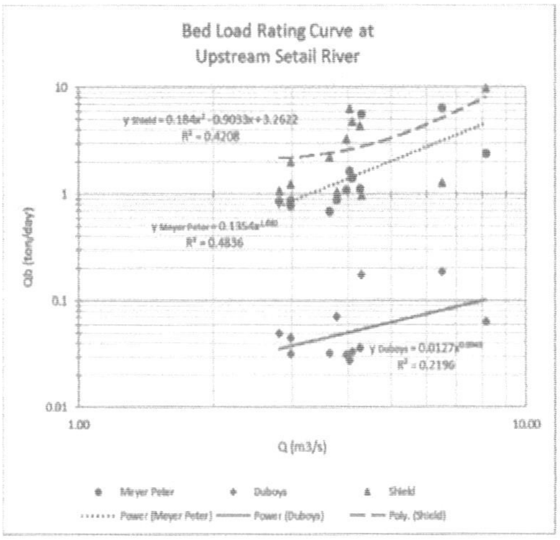

Figure 22. Bed Load Rating Curve at Upstream Setail River

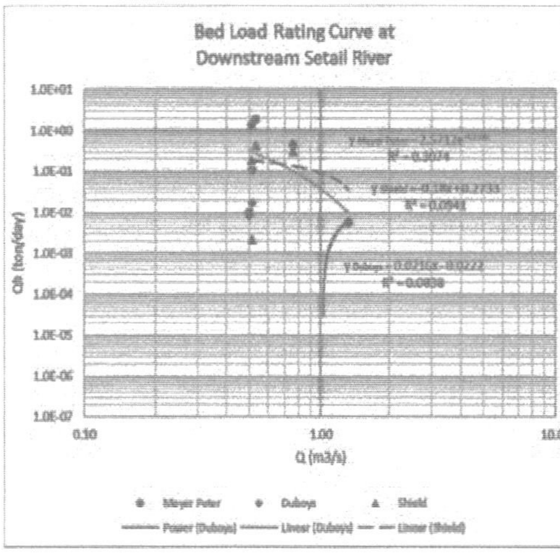

Figure 23. Bed Load Rating Curve at Downstream Setail River

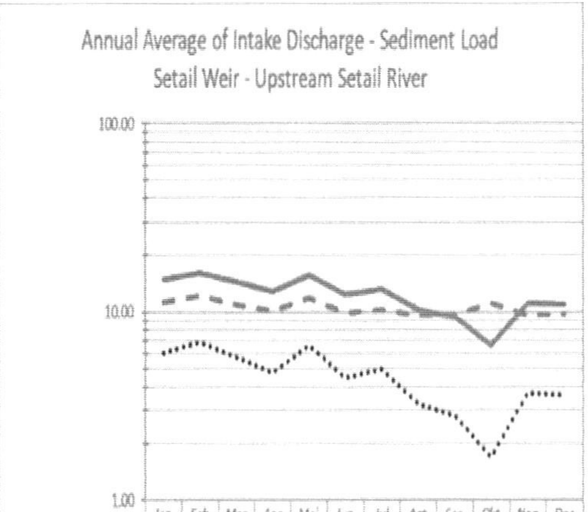

	Jan	Feb	Mar	Apr	Mei	Jun	Jul	Agt	Sep	Okt	Nop	Des
Q Intake (m3/s)	14.90	16.22	14.51	12.95	15.78	12.45	13.27	10.32	9.42	6.66	11.17	10.93
Qs (ton/day)	11.25	12.27	10.92	10.10	11.92	9.91	10.25	9.56	9.65	11.17	9.66	9.78
Qb (ton/day)	6.03	6.94	5.76	4.76	6.64	4.45	4.96	3.25	2.79	1.70	3.72	3.62

Figure 24. Annual Average of Intake Discharge and Sediment Load at Setail Weir Upstream Setail River

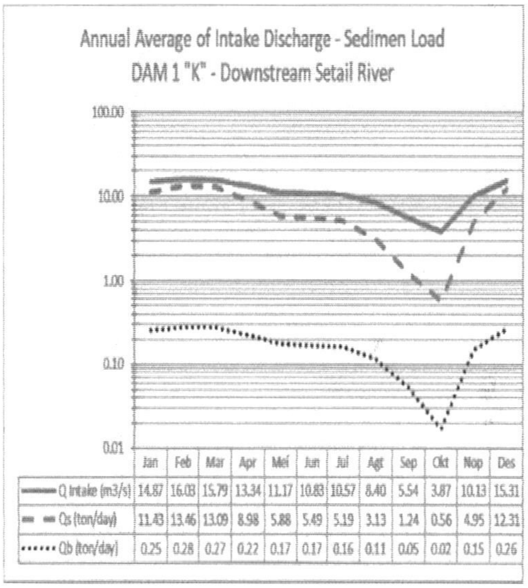

	Jan	Feb	Mar	Apr	Mei	Jun	Jul	Agt	Sep	Okt	Nop	Des
Q Intake (m3/s)	14.87	16.03	15.79	13.34	11.17	10.83	10.57	8.40	5.54	3.87	10.13	15.31
Qs (ton/day)	11.43	13.46	13.09	8.98	5.88	5.49	5.19	3.13	1.24	0.56	4.95	12.31
Qb (ton/day)	0.25	0.28	0.27	0.22	0.17	0.17	0.16	0.11	0.05	0.02	0.15	0.26

Figure 25. Annual Average of Intake Discharge and Sediment Load at DAM 1 "K" Downstream Setail River

Based on **Figure 24,** it shows that in the irrigation intake in the Setail weir with total annual average irrigation discharge of 49.53 m^3/s carries an annual suspended load of 126.44 tons/year while the annual bed load is 54.61 tons/year. It is known that in October the suspended load condition is greater than the irrigation intake discharge, causing high turbidity and sedimentation and bed load conditions follow the fluctuation of the irrigation discharge and are still stable. Based on **Figure 25,** it shows that in irrigation intakes in DAM 1 "K" in the Downstream of Setail River with total annual average irrigation discharge of 45.29 m^3/s carries an annual suspended load of 85.70 tons/year while the annual bed load is 2.14 tons/year. It is known that the condition of the sediments load is still stable and follows the fluctuations of discharge of irrigation.

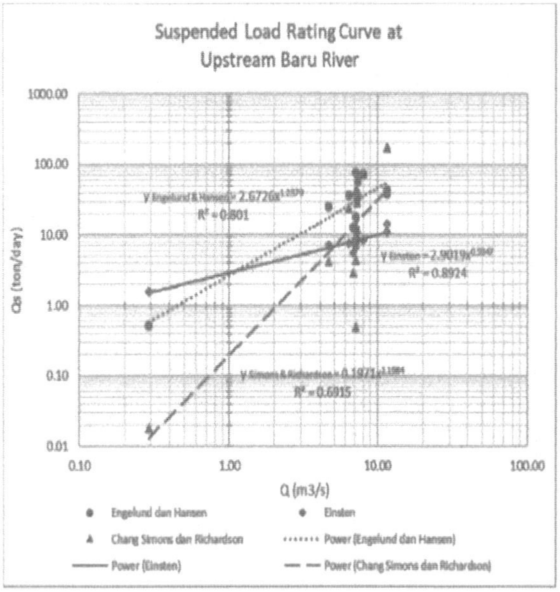

Figure 26. Suspended Load Rating Curve at Upstream Baru River

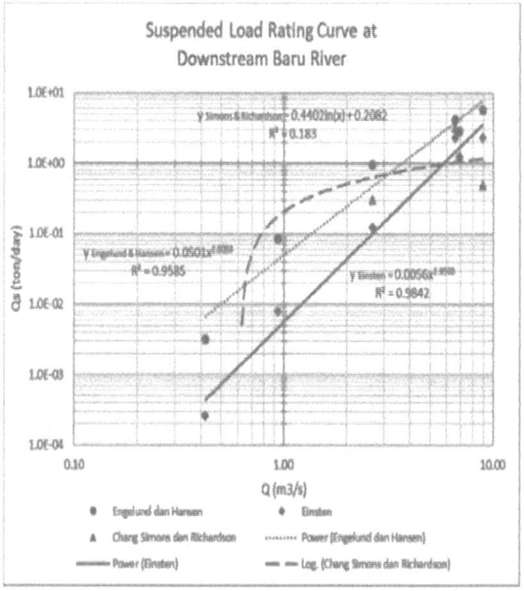

Figure 27. Suspended Load Rating Curve at Downstream Baru River

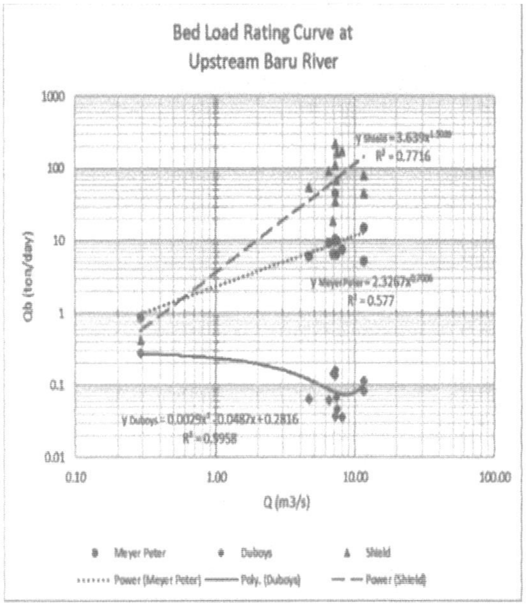

Figure 28. Bed Load Rating Curve at Upstream Baru River

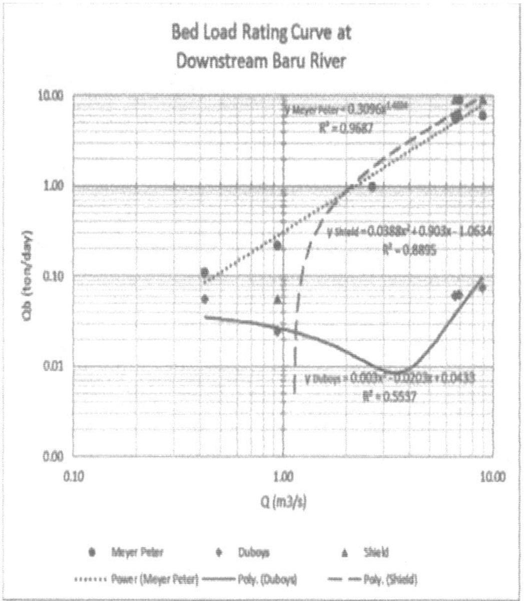

Figure 29. Bed Load Rating Curve at Downstream Baru River

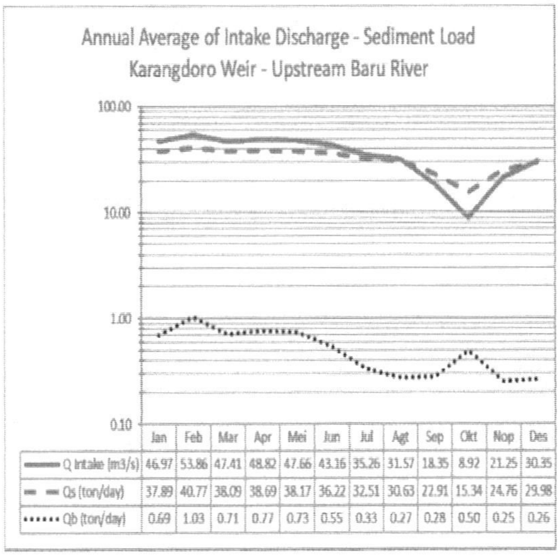

Figure 30. Annual Average of Intake Discharge and Sediment Load at Karangdoro Weir Upstream Baru River

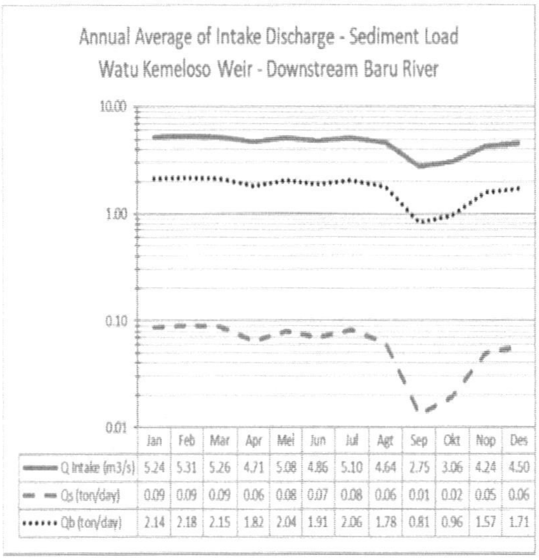

Figure 31. Annual Average of Intake Discharge and Sediment Load at Watu Kemeloso Weir Downstream Baru River

4.1.4. Baru River. The results of measurements of the average river discharge per segment of the Baru River Upstream is 7.10 m^3/s and in Baru River Downstream is 4.40 m^3/s. The correlation between sediment load and discharge at upstream and downstream Baru River can see in **Figure 26 – 29**.

Based on **Figure 30,** it shows that in the irrigation intake in the Karangdoro weir with a total annual average irrigation discharge of 144.53 m^3/s carries an annual suspended load of 385.96 tons/year while the annual bed load is 6.37 tons/year. It is known that in October the suspended load condition is greater than the irrigation intake discharge, causing high turbidity and sedimentation and bed load conditions follow the fluctuation of the irrigation discharge and are still stable. Based on **Figure 31,** it shows that in the irrigation intake in the Watu Kemeloso Weirs in Downstream of Baru River with a total annual average irrigation discharge of 18.25 m^3/s carries an annual suspended load of 0.76 tons/year while the annual bed load is 21.13 tons/year. It is known that the condition of the suspended load is smaller than the bed load. It can be concluded that the process of erosion is greater so the sedimentation is very high. Many sand minings are also found from the impacts of sedimentation.

5. Conclusions

Loads of sediment, both suspended load and bed load on the irrigation discharge in both the upstream and downstream of the major rivers in Banyuwangi greatly influences the sedimentation of irrigation channels which affect on fluctuations of irrigation discharge to irrigate land of Agriculture. The irrigation intakes in Poncowati weir carried Qs of 21.89 tons/year; Qb of 6.92 tons/year. Irrigation intakes in the Downstream Tambong river have Qs of 166.06 tons/year; Qb of 35.98 tons/year. Irrigation intakes in Gembleng weir have Qs of 196.24 tons/year; Qb of 35.81 tons/year. Irrigation intakes of Maron Wonosobo in the Downstream Bomo river have Qs of 0.18 tons/year; Qb of 2.45 tons/year. Irrigation intakes in Setail weir have Qs of 126.44 tons/year; Qb of 54.61 tons/year. The irrigation intakes of DAM 1 "K" in the Downstream Setail river have Qs of 85.70 tons/year; Qb of 2.14 tons/year. Irrigation intakes in Karangdoro weir have Qs of 385.96 tons/year; Qb of 6.37 tons/year. The irrigation intakes of Watu Kemeloso Weirs in Downstream Baru river have Qs of 0.76 tons/year; Qb of 21.13 tons/year. It is necessary to have policies and efforts to control sediment and study about the maintenance of weirs periodically to maintain the volume of irrigation water so that it can increase agricultural productivity in Banyuwangi Regency.

Acknowledgments

Thanks to Regional Development Planning Agency of Banyuwangi Regency for the funding that has been given in the scheme research on regional government cooperation with universities from Banyuwangi Regency Budget funding sources in 2018. Our respect is given to Politeknik Negeri Banyuwangi for their support for our research. The appreciation is also given to the research team for their cooperation so that we can complete our research.

References

[1] Badan Pusat Statistik 2017 *Banyuwangi Dalam Angka Tahun 2017* (Banyuwangi)

[2] Dewan Riset Daerah Provinsi Jawa Timur 2015 *Jakstrada Iptek, Kebijakan Strategis Pembangunan Daerah Iptek, Agenda Riset Daerah Provinsi Jawa Timur 2014-2019* (Surabaya: Jalin Matra)

[3] Erwanto Z and Ulfiyati Y 2017 *Evaluasi Hidrograf Aliran Dengan Agiel NN Pada Sungai-Sungai Besar Kabupaten Banyuwangi* (Bali: Logic Jurnal Rancang Bangun dan Teknologi) [S.l.], v. 16, n. 1, http://ojs.pnb.ac.id/index.php/LOGIC/article/view/122

[4] Wardani F Y, Erwanto Z and Ulfiyati Y 2018 *Studi Muatan Suspended Load Dan Bed Load Pada Upstream Bendung Di Hulu Sungai-Sungai Besar Kabupaten Banyuwangi* (Bali: Logic Jurnal Rancang Bangun dan Teknologi) [S.l.], v. 18, n. 1, p. 12-19, mar. 2018 ISSN 2580-5630, http://ojs.pnb.ac.id/index.php/LOGIC/article/view/789

[5] Yang C T 2003 *Sediment Transport Theory and Practice* (Malabar, FL: Krieger Publishing Company)